Technical English

Technical English

WRITING, READING, AND SPEAKING

FIFTH EDITION

Nell Ann Pickett
Hinds Community College
Raymond Campus

Ann A. Laster
Hinds Community College
Raymond Campus

HarperCollins*Publishers*

For Harry, Bob, and Ryn

Sponsoring Editor: Lucy Rosendahl
Project Editor: Carla Samodulski
Text Design: Joan Greenfield
Cover Design: Hudson River Studio
Text Art: Hudson River Studio/Vantage Art, Inc.
Production Manager: Jeanie Berke
Production Assistant: Beth Maglione
Compositor: Waldman Graphics, Inc.
Printer and Binder: R. R. Donnelley & Sons Company
Cover Printer: NEBC

TECHNICAL ENGLISH: Writing, Reading, and Speaking, Fifth Edition

Library of Congress Cataloging in Publication Data

Pickett, Nell Ann.
   Technical English.

   Includes index.
   1. Readers—Technology.   2. English language—Technical English.   3. Technical Writing.   I. Laster, Ann A.
II. Title.
PE1127.T37P5   1988   808'.0666   87-26976
ISBN 0-06-045204-8

      92 9 8 7 6 5

Contents

Inside front cover: General Format Directions for Writing

A detailed table of contents precedes each chapter.

v

Preface

Since 1970, when *Writing and Reading in Technical English* was published, some 25 faculty members and hundreds of students at our college have used the book in technical writing classes. During this time, these teachers and students have discussed with us additions and revisions that would improve the book for classroom use. Many faculty members and students at other institutions using the text have also shared with us their ideas on additions and revisions. At the same time, we have continued to talk with instructors in technical fields and with industrial personnel, listening to their ideas and suggestions. We have studied all these ideas carefully, and the result is *Technical English*, Fifth Edition.

What's New in the Fifth Edition

The fifth edition includes:

☐ An introductory section on making a message easier to read and understand
☐ Stronger emphasis on layout and design
☐ More coverage on memorandums, current practices in letter writing, additional specific types of letters
☐ Expanded discussion of microforms and library research tools
☐ Documentation style from the MLA *Handbook* (1984)
☐ Additional attention to practices in business communication
☐ Revision of the chapter on report writing, including the addition of the feasibility report and the proposal
☐ New readings

Rationale

This fifth edition continues to fill two definite needs: the need of teachers in two-year and four-year colleges and technical schools to have texts written on a level appropriate to their students, and the need of technical students who, out of necessity and interest, require guidance in technical writing. It was written with two underlying principles in mind, both of which have been formulated from our formal training in English and in technical education, our teaching experiences, and our visits to industries, businesses, and service facilities.

1. Emphasis in an English course for technical students should be on *practical* application.
2. The preparation and selection of class materials for such an English course require more discrimination than do those for the regular academic English course because the technical student frequently has only two years to prepare formally to become a productive member of society.

We reaffirm, with even greater conviction, our original philosophy about the technical student and the English language: The future technician must be made aware of industry's demand that workers be able to communicate effectively.

Organization

Technical English retains the three-part division of the previous editions. Part I presents the basic principles and forms of communication that any student needs to know, but there is an emphasis on industrial communication demands. Included are chapters on instructions and process, description of a mechanism, definition, analysis through classification and partition, analysis through effect-cause and comparison-contrast, summary, memorandums and letters, reports, the research report, oral communication, and visuals. The material on layout and design is expanded; and the chapters on instructions and process description, the summary, memorandums and letters, reports, and the research report have been expanded and updated. Part II, Selected Readings, retains pieces that instructors have found especially useful and includes several new pieces. Introductory material suggests a method for selective reading, and questions at the end of each reading will stimulate thought and initiate responses. The readings, chosen for their inherent interest and pedagogical utility, can have real value for the technical student—the person whose background, experience, ability, interests, and ambition have led to a specialized program of higher education. Applications at the end of each chapter in Part I suggest ways to relate chapter content to selected readings. Part III, a handbook of usage, contains chapters on the sentence, the paragraph, grammatical usage, and mechanics. The Handbook retains the commonsense approach that language is the tool of humans, not humans the tool of language, and that adherence to certain standards is not legislated but simply wise if one is to be understood.

Certain features of the text make it exceptionally teachable and student-oriented. Each chapter opens with clearly stated objectives, so that students know exactly what is expected of them. Then the topic is developed through discussion and examples, principles are explained, and a step-by-step procedure for writing or speaking is outlined. Plan Sheets, completed to serve as models, reinforce the need for careful preparation and show its result. Blank Plan Sheets corresponding to end-of-chapter assignments are provided for students to fill in. The pages are perforated so that the Plan Sheets may be torn out and handed in if the instructor wishes.

The readings in Part II have been selected to illustrate and amplify the writing assignments in Part I. The Handbook (Part III) is a resource that instructors and students can rely on according to individual needs and preferences. In its fifth edition, *Technical English* is once again a complete course.

Many individuals have had a part in the development of the fifth edition, some offering encouragement, some suggestions, and others materials for inclusion. Specifically, we are grateful to the following Hinds Community College personnel: McLendon Library staff, especially Nancy Baker; Jerry G. Carr, Faye Angelo, Jerry Agent, Ora Steele, and our students. We would also like to thank the following reviewers: Carolyn Boiarsky, Illinois State University; George Searles, Mohawk Valley Community College; and Thomas L. Warren, Oklahoma State University. To each individual, whatever the contribution, we say sincerely, "Thank you."

Nell Ann Pickett
Ann A. Laster

Part I
Forms of Communication

YOU MADE A WISE CHOICE when you decided to enter a technical education program. You are in good company, for you are a part of a group of several hundred thousand young men and women who have goals similar to your own. You realize that getting and keeping a good job and being promoted are not mere accidents but the result of good training, hard work, and the desire to be successful. You are in a division of higher education that is essential to the maintenance and progress of our technological civilization.

The courses in your particular technical program are required because you need them. You need them in order to be successful in your chosen field. Obviously, without sufficient technical knowledge and skill, the technician is out of step—or soon will be. Perhaps not so obvious to you, though, is your need for this English course, your need to develop skills in communication.

Need for Communication Skills

Think for a moment of some of the many different kinds of communication tasks technicians may be required to perform during a day. They may need to write a letter requesting information, ordering an item or a service, arranging a meeting, correcting an error, selling items or services; or they may need to transact such business orally via telephone conversation. They may be required to write reports on such subjects as the need for decreased or additional personnel; a new, more effective method for performing an operation; suggestions for revision, improvement, purchase; annual financial gain or loss; inventory of stock; or daily, weekly, or monthly activities. Technicians may need to demonstrate and explain a new process, technique, service, machine, or tool.

Think of some of the requirements that you as a communicator of technical and professional data must meet. You must be fact-minded, and you must be able to present information accurately, precisely, and clearly to audiences of varying abilities and degrees of understanding. You must be able to use many different forms for presenting information, such as reports, letters, memorandums. Some occasions may require written presentations; some, oral. Technicians indeed have a responsibility to themselves, their co-workers, and their professions to prepare themselves to handle communication tasks effectively.

Organization of Part I

Part I is carefully organized to assist you in each stage of the communication process. Each chapter begins with a list of objectives, that is, what you should be able to do when you complete the chapter. The remainder of the chapter is concerned with helping you learn these skills. First, a form, or technique, of communication is presented, as indicated in the chapter title. The form is discussed, and both student and professional examples are given. For student-prepared presentations, there are filled-in Plan Sheets so you can see how planning is essential to a successful presentation. Second, a list of general principles sums up the main points of the chapter. Then a procedure for presentation suggests a method of organization. This step-by-step guide shows what to include and where to include it. Finally, applications are given at the end of each chapter. These applications are exercises to help you attain the skills listed at the beginning of the chapter.

Throughout the entire communication process, therefore, you are accompanied by ample explanation, direction, and illustration. At no point are you left guessing what to do next.

Introduction

Communication is an active process with at least four parts: a sender, a message to be sent, a way to send the message, and a receiver (audience, reader) to comprehend, analyze, and respond or react to the message.

As a future communicator on the job, you must be very aware of this process, for you may be the person with a message or the person to receive a message. Making a sale, convincing management of needed change, giving employees instructions on operating new equipment, preparing a report on work in progress, proposing the purchase of new materials or equipment—a worker daily sends and receives many different types of messages to and from many different audiences.

Basically this text assumes that you are the sender in the communication process. You are introduced to some common types of messages—instructions and process, description of mechanisms, definitions, classification and partition, effect-cause, comparison-contrast, and summary—with explanations and suggestions on how to present these messages verbally and visually. Then you are asked to plan and present similar messages, making choices about subject, audience, purpose, content (verbal and visual), and layout and design.

For example, you might be assigned the following: You (sender) write (the way the message is to be sent) a set of instructions (message) for classmate Mary Jones, who was absent from class on the day the instructor gave the how-to explanation (receiver). As you study this text, you will learn to ask such questions as the following: Do I know enough about the subject to write a clear set of instructions for Mary? Will a review of notes be helpful? Should I practice using the tool or piece of equipment before I try to tell Mary how to use it? Are there questions I need to ask the instructor or another classmate? What exactly do I need to tell Mary so that she can use the tool or piece of equipment as required? How should I arrange the instructions on the page so that Mary can follow them easily and accurately? Would visuals help?

As these questions suggest, every decision you, the sender of a message, make about a message to be sent depends upon three factors:

- *Who* is to receive the message (the receiver, the audience, the reader)
- *Why* you are preparing the message (the purpose)
- *What* you want the receiver to do with the message (desired response)

Of course, the way an audience reads, comprehends, and responds to a message depends to a great extent on the content of the message and the reader's mental state, reason for reading, and special skills. The writer can, however, use various visual and verbal techniques to make a message easier to read and comprehend. Since these techniques may be new to you, the following pages introduce you to some of the frequently used visual and verbal techniques.

As you work through the various chapters in the text and prepare related assignments, you will practice selecting and adapting these techniques to enhance a particular message for a particular audience.

Layout

One of the simplest ways for a writer to make a message more readable is layout—that is, placement of words and sentences, paragraphs, lists, tables, graphs, and the like on the page. Layout makes an important visual impression on the reader.

For many years writers followed the nineteenth-century essay form, paragraph after paragraph after paragraph with only an occasional indentation to mark the beginning of a new paragraph. Sometimes no indentation appeared on an entire page; the page was literally filled with words with space only at the outer edges of the page. Look at Figure 1, Example A.

Today's writer uses layout to enhance writing and reading. Look at Example B in Figure 1. This example illustrates some techniques for making a page "come alive."

Basic layout techniques can be grouped into two general categories:

- □ White space providers
- □ Emphasis markers

White Space Providers. White space providers yield an uncluttered, easily readable, inviting page. Indenting for paragraphs, double spacing, and allowing ample margins (1½ inches minimum on all four perimeters) are the most usual ways of allowing for plenty of white space. Additional white space providers include headings (with triple spacing before a major heading and double spacing after the heading), vertical listing (placing the items up and down rather than across the page), and columns (setting up the page with several columns of short lines, as in a periodical, rather than a page of long lines).

Emphasis Markers. A number of techniques can be used in drawing attention, lending variety, or giving emphasis to material. Different sizes and styles of typefaces can be used, such as roman, italic, and boldface. Underlining gives emphasis, as do uppercase (capital) letters. Boxes can be used to set off material, such as a caution in a set of instructions, or to enclose an entire page (as used throughout this book to set off sample pieces of writing). Insets—small boxes that set off material—may provide a legend or an explanation of symbols for such visuals as maps, charts, and drawings; insets are often used to set off key thoughts, a summary, or supplementary material. Bars (heavy straight lines) typically separate a table or other visual from the text. Also, writers can employ multiple colors (for contrast) and shading (for depth) and a border (particularly in a paper requiring a title page).

Symbols can also be used as emphasis markers. Frequently used are the bullet (○), the square (□), other geometric shapes such as a triangle or diamond (△ ◇), the dash (—), the asterisk (*), and arrows (→ ↓). The various geometric shapes can be solid (●■▼) or open (○□△). Especially easy to make is the bullet: Use the letter "o" or the period on a typewriter, or hand insert the bullet with a pen.

All of these emphasis markers are fairly easy to produce. Readily available commercially prepared aids such as templates, transfer lettering sheets, transfer shading sheets, and charting tape can help to give your writing a professional look and to enhance readability and comprehension.

Design

Design concerns individual pages as well as the document as a whole. In designing a document, the writer makes careful, conscious decisions concerning matters that affect the appearance and the impact of the total document.

Design involves the elements of composition: contrast, balance, proportion,

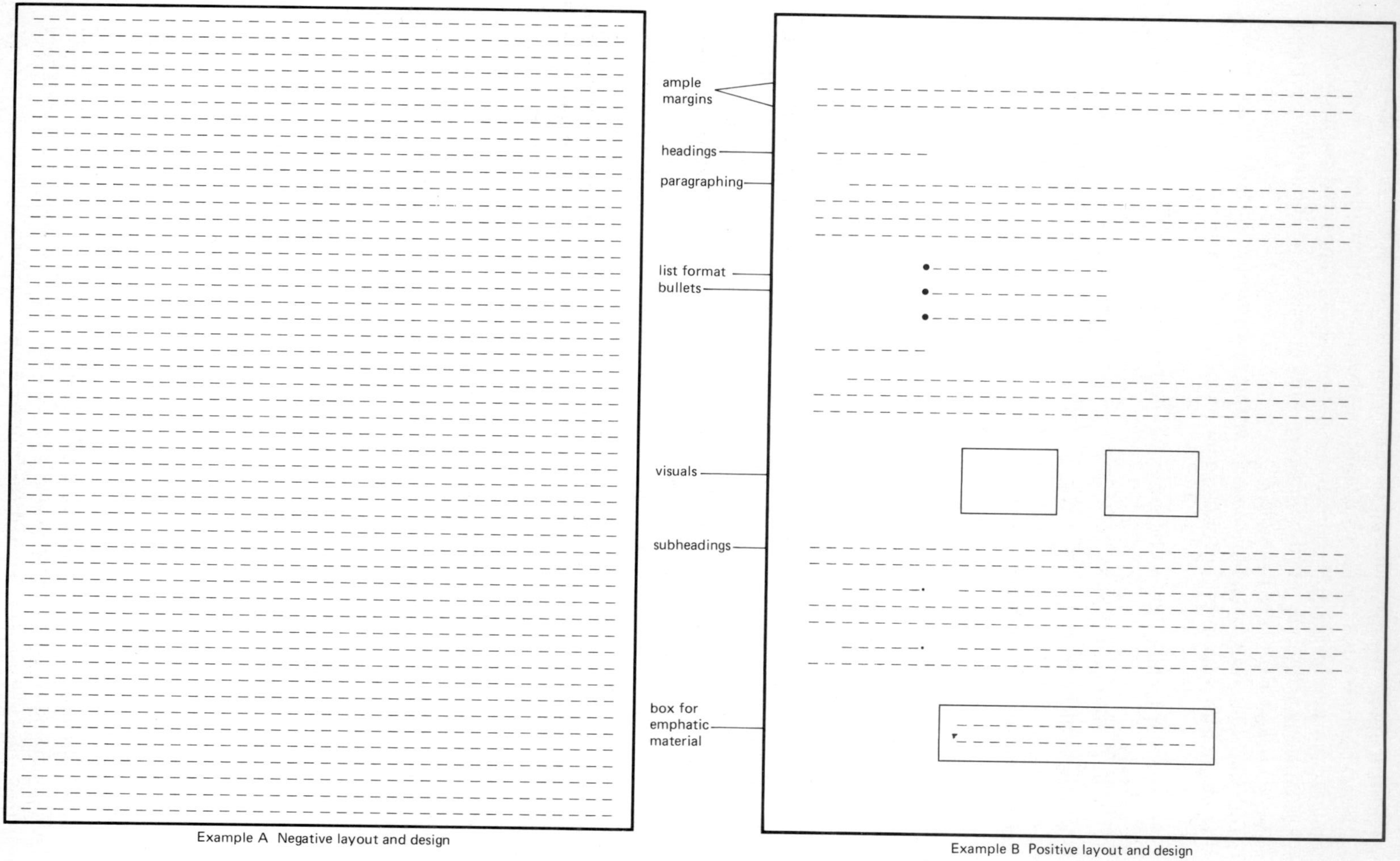

Example A Negative layout and design

ample
margins

headings

paragraphing

list format
bullets

visuals

subheadings

box for
emphatic
material

Example B Positive layout and design

Figure 1. Examples of negative and positive layout and design.

dominance, harmony, opposition, unity. These elements are used to direct the reader's attention and to establish hierarchies of emphasis.

Among the tools used in design are color, format, column width, visuals (including kind, size, and placement), spacing, texture, geometric shapes, and sizes and styles of typefaces (for text, headings, and special materials). Through the use of these tools, you can consciously make choices to achieve the desired intellectual and psychological impact on the reader.

Headings

A heading identifies the subject or topic written about in a block of information. Each heading should be meaningful, that is, clearly tell the reader what the block of information is about. The heading helps the writer to stay on the subject; the heading plus the white space marking the block of information helps the reader by making the pages of writing more inviting. Even a glance at a page of print without headings and one with headings (such as this page) will attest to the importance of headings. Headings give the reader a visual impression of major and minor topics and their relation to one another; headings reflect the organization of the material. Headings remind the reader of movement from one point to another. Headings help the reader interested only in particular sections, not the whole document, to locate these sections. And headings give life and interest to what otherwise would be a solid page of unbroken print.

For headings to show organization of material, the form and placement of each heading must indicate its rank or level. Use of capital letters and underlining and position of the headings on the page help to differentiate rank or level. Headings that identify higher levels of material should *look* more important than those that identify lower levels.

The writer should work out a system of headings to reflect the major points of a document and their supporting points. (Should a document contain an outline or table of contents, the headings in them should correspond exactly to the headings in the body.)

Following are suggested systems of headings.

2 LEVELS OF HEADINGS (THE UNDERLINING MAY BE OMITTED)

MAJOR HEADING (This heading may be centered)
Division of Major Heading

3 LEVELS OF HEADINGS (THE UNDERLINING MAY BE OMITTED)

MAJOR HEADING (This heading may be centered)
Division of Major Heading
   Subdivision of division of major heading. The paragraph begins here.

4 LEVELS OF HEADINGS

MAJOR HEADING (This heading may be centered)
Division of Major Heading
Subdivision of Division of Major Heading

Sub—subdivision of division of major heading. The paragraph begins here.

5 LEVELS OF HEADINGS

<u>MAJOR HEADING</u>
<u>DIVISION OF MAJOR HEADING</u>
<u>Subdivision of Division of Major Heading</u>
<u>Sub—Subdivision of Division of Major Heading</u>
  <u>Sub—sub—subdivision of division of major heading</u>. The paragraph begins here.

Headings must be visibly obvious if they are to be effective. Leave plenty of white space. As a general rule, triple-space above a heading and double-space below a heading (unless the heading is indented as a part of the paragraph).

Titles

The title of a document is the first thing a reader sees. It should, therefore, specifically identify the topic, suggest the writer's approach and coverage, and interest the reader. A title should indicate to the reader what to expect in the content.

A title such as "The Brentsville Police Department" is vague and broad, giving a reader no idea of the writer's approach to the topic or the coverage. A title such as "Brentsville Police Department: Organization and Responsibilities" clearly suggests the approach and coverage.

Be honest with your reader. A report entitled "Customizing a Van at Home" suggests the report covers everything an individual would need to know to customize a van. If the report discusses only special tools needed to customize a van at home, then the writer has misled the reader. An accurate title would be "Special Tools Needed for At-Home Van Customizing."

Titles may include subtitles. For example, a proposal to employ a commercial service for periodic cleaning of a swimming pool might be entitled "Employing Hill Brothers Pool Service: A Proposal." Such a title might also be worded "A Proposal to Employ Hill Brothers Pool Service." Or a report on the advantages of using a commercial service might be entitled "Advantages of a Commercial Pool Cleaning Service: A Report" or "A Report on the Advantages of a Commercial Pool Cleaning Service."

A title is usually a phrase rather than just one word or a complete sentence.

Word Choice

As a writer or speaker you must be very aware of words selected to convey a message to an audience. The English language is filled with a rich, diverse vocabulary. Thus, you have a wide choice of words to express an idea. To select the words that best convey intended meaning to an intended audience, use denotative and connotative words as needed, choose between specific and general words, avoid inappropriate jargon, practice conciseness where desirable, and select the appropriate word for the specific communication need.

Language could be used more easily and communication would be much simpler if words meant the same things to all people at all times. Unfortunately they do not. Meanings of words shift with the user, the situation, the section of the country, and the context (all the other words surrounding a particular word). Every word has at least two areas of meaning: denotation and connotation.

The denotative meaning is the physical referent the word identifies, that is, the thing or the concept: It is the dictionary definition. Words like *computer terminal*, *mannequin*, *T square*, *management*, *tractor*, *desk*, and *thermometer* have physical referents; *scheduling*, *production control*, *courage*, *recovery*, and *fear* refer to qualities or concepts.

The connotative meaning of a word is what an individual feels about that word because of past experiences in using, hearing, or seeing the word. Each person develops attitudes toward words because certain associations cause the words to suggest qualities either good or bad. For some people, such words as *Communist*, *Red*, *liberal*, *leftist*, *Democrat*, *Republican*, or *right-wing* are favorable; for others, they are not. The effects of words depend on the emotional reactions and attitudes that the words evoke.

Consider the words *fat*, *large*, *portly*, *obese*, *plump*, *corpulent*, *stout*, *chubby*, and *fleshy*. Each of these could be used to describe a person's size, some with stronger connotation than others. Some people might not object too much to being described as *plump*, but they might object strongly to being described as *fat*. On the other hand, they probably would not object to receiving a *fat* paycheck. So, some words may evoke an unfavorable attitude in one context, "*fat* person," but a favorable attitude in another, "*fat* paycheck."

Some words always seem to have pleasant connotations; for example, *craftsmanship*, *success*, *bravery*, *happiness*, *honor*, *intelligence*, and *beauty*. Other words, such as *defect*, *hate*, *spite*, *insanity*, *disease*, *rats*, *poverty*, and *evil*, usually have unpleasant connotations.

You should choose words that have the right denotation and the desired connotation to clarify your meaning and evoke the response you want from your reader. Compare the following sentences. The first illustrates a kind of writing in which words point to things (denotation) rather than attitudes; the words themselves call for no emotional response, favorable or unfavorable.

> Born in the Fourth Ward with its prevailing environment, John was separated from his working mother when he was one year of age.

The second sentence, through the use of words with strong connotative meaning, calls for an emotional response—an unfavorable one.

> Born in the squalor of a Fourth Ward ghetto, John was abandoned by his barroom-entertainer mother while he was still in diapers.

SPECIFIC AND GENERAL WORDS

A specific word identifies a particular person, object, place, quality, or occurrence; a general word identifies a group or a class. For example, *lieutenant* indicates a person of a particular rank; *officers* indicates a group. For any general word there are numerous specific words; the group identified by *officers* includes such specific terms as *lieutenant*, *general*, *colonel*, and *admiral*.

There are any number of levels of words that can take you from the general to the specific:

GENERAL   data processing   computer
SPECIFIC   IBM PC-XT
GENERAL   drafter   drafting pool   architectural drafting pool
SPECIFIC   Michael Baker Inc.

The more specific the word choice, the easier it is for the reader to know exactly the intended meaning.

GENERAL   The computer was expensive.
SPECIFIC   The IBM PC-XT cost $2195.
GENERAL   The instrument was on a shelf.
SPECIFIC   The thermometer lay on the topmost shelf of the medicine cabinet.
GENERAL   The place was damaged.
SPECIFIC   The tornado blew out all the windows in the administration building at Midwestern College.

Using general words is much easier, of course, than using specific words; the English language is filled with "umbrella" terms with broad meanings. These words come readily to mind, whereas the specific words that express exact meaning require thought. To write effectively, you must search until you find the right words to convey to your audience your intended meaning.

JARGON

Jargon is the specialized or technical language of a trade, profession, class, or group. This specialized language is often not understandable by persons outside those fields.

Jargon is appropriate (1) if it is used in a specialized occupational context and (2) if the intended audience understands the terminology. For example, a computer programmer communicating with persons knowledgeable about computer terminology and discussing a computer subject may use such terms as *input*, *interface*, *menu driven*, or *48K*. The computer programmer is using jargon appropriately.

If, however, these specialized words are applied to actions or ideas not associated with computers, jargon is used inappropriately. Or if such specialized words are used extensively—even in the specialized occupational context—with an audience who does not understand the terminology, jargon is used inappropriately.

The problem with inappropriate use of jargon is that the speaker or writer is not communicating clearly and effectively with the audience. Study these examples of inappropriate jargon:

The computer revolution has impacted on business.
The bottom line is a 10 percent salary increase.
Nurse to patient: "A myocardial infarction is contraindicated." (No heart attack)
TV repairperson to customer: "A shorted bypass capacitor removed the forward bias from the base-emitter junction of the audio transistor." (A capacitor shorted out and killed the sound.)

Gobbledygook. Jargon enmeshed in abstract pseudo-technical or pseudo-scientific words is called *gobbledygook* (from *gobble*, to sound like a turkey). Examine these sentences:

> The optimum operational capabilities and multiple interrelationships of the facilities are contiguous on the parameters of the support systems.
> Integrated output interface is the basis of the quantification.

The message in both sentences is unclear because of jargon and pseudo-technical and pseudo-scientific words—words that *seem* to be technical or scientific but in fact are not. Further, the words are all abstract; that is, they are general words that refer to ideas, qualities, or conditions. They are in contrast with concrete words that refer to specific persons, places, or things which one can see, feel, hear, or otherwise perceive through the senses. Examine again the two gobbledygook sentences above. Not one word in either sentence is a concrete word; not one word in either sentence creates an image in the mind of the reader.

CONCISENESS

In communication, clarity is of primary importance. One way to achieve clarity is to be concise, whenever possible. Conciseness—saying much in a few words—omits nonessential words, uses simple words and direct word patterns, and combines sentence elements.

It is important to remember that you can be concise without being brief and that what is short is not necessarily concise. The essential quality in conciseness is making every word count.

It is indeed misleading to suggest that all ideas can be expressed in simple, brief sentences. Some ideas by the very nature of their difficulty require more complex sentences. And frequently a longer sentence is necessary to show relationships between ideas.

Omitting Nonessential Words. Nonessential words weaken emphasis in a sentence by thoughtlessly repeating an idea or throwing in "deadwood" to fill up space. Note the improved effectiveness in the following sentences when unnecessary words are omitted.

WORDY	The train arrives at 2:30 P.M. in the afternoon.
REVISED	The train arrives at 2:30 P.M.
WORDY	He was inspired by the beautiful character of his surroundings.
REVISED	He was inspired by his surroundings.
WORDY	In the event that a rain comes up, close the windows.
REVISED	If it rains, close the windows.

Eliminating unnecessary words makes writing more exact, more easily understood, and more economical. Often, care in revision will weed out the clutter of deadwood and needless repetition.

Simple Words and Direct Word Patterns. An often-told story illustrates quite well the value of simple, unpretentious words stated directly to convey a message. A plumber who had found that hydrochloric acid was good for cleaning out pipes wrote a government agency about his discovery. The plumber received this reply: "The efficacy of hydrochloric acid is indisputable, but the corrosive residue is incompatible with metallic permanence." The plumber responded that he was glad his discovery was helpful. After several more garbled and misunderstood communications from the agency, the plumber finally received this clearly stated response: "Don't use hydrochloric acid. It eats the hell out of pipes." Much effort and time could have been saved if this had been the wording of the agency's *first* response.

Generally, use simple words instead of polysyllabic words and avoid giving too many details in needless modifiers. When necessary, show causal relationships or tie closely related ideas together by writing longer sentences.

WORDY AND OBSCURE	Feathered bipeds of similar plumage will live gregariously.
SIMPLE AND DIRECT	Birds of a feather flock together.
WORDY AND OBSCURE	Verbal contact with Mr. Jones regarding the attached notifications of promotion has elicited the attached representations intimating that he prefers to decline the assignment.
SIMPLE AND DIRECT	Mr. Jones does not want the job.
WORDY AND OBSCURE	Believing that the newer model air conditioning unit would be more effective in cooling the study area, I am of the opinion that it would be advisable for the community library to purchase a newer model air conditioning unit.
SIMPLE AND DIRECT	The community library should buy a new air conditioning unit.
IMPLIED RELATIONSHIP	The area was without rain for ten weeks. The corn stalks turned yellow and died.
STATED RELATIONSHIP	Because the area was without rain for ten weeks, the corn stalks turned yellow and died.

Combining Sentence Elements. Many sentences are complete and unified yet ineffective because they lack conciseness. Parts of sentences, or even entire sentences, often may be reduced or combined.

REDUCING SEVERAL WORDS TO ONE WORD	the registrar *of the college* the *college* registrar
REDUCING A CLAUSE TO A PHRASE OR TO A COMPOUND WORD	a house *that is shaped like a cube* a house *shaped like a cube* a *cube-shaped* house
REDUCING A COMPOUND SENTENCE	Mendel planted peas for experimental purposes, and from the peas he began to work out the universal laws of heredity.
TO A COMPLEX SENTENCE	As Mendel experimented with peas, he began to work out the laws of heredity.
OR TO A SIMPLE SENTENCE	Mendel, experimenting with peas, began to work out the laws of heredity.
COMBINING TWO SHORT SENTENCES	Many headaches are caused by emotional tension. Stress also causes a number of headaches
INTO ONE SENTENCE	Many headaches are caused by emotional tension and stress.

Active and Passive Voice Verbs

Generally use active voice verbs. Active voice verbs are effective because the reader knows immediately the subject of the discussion; the writer mentions first who or what is doing something.

The machinist *values* highly the rule depth gauge.
Roentgen *won* the Nobel Prize for his discovery of X rays.

Passive voice verbs are used when the who or what is not as significant as the action or the result and when the who or what is unknown, preferably unnamed, or relatively insignificant.

The lathe *has been broken* again. (More emphatic than "Someone has broken the lathe again.")
The blood sugar test *was made* yesterday.
The transplant operation *was performed* by an outstanding heart surgeon.
Light *is provided* for technical drawing classrooms by windows in the north wall.

A good rule is choose the voice of the verb that permits the desired emphasis.

ACTIVE VOICE   The ballistics experts *examined* the results of the tests. (Emphasis on *ballistics experts*)
PASSIVE VOICE   The results of the tests *were examined* by the ballistics experts. (Emphasis on *results of the tests*)
ACTIVE VOICE   Juan *gave* the report. (Emphasis on *Juan*)
PASSIVE VOICE   The report *was given* by Juan. (Emphasis on *report*)

Word Order

Word order plays a major role in readability. Writing sentences so that a reader is able to make sense out of word clusters improves readability. (Arranging groups of sentences or paragraphs developing one aspect of a topic into a block of information also improves readability.)

One desirable arrangement of words in sentences is placing the subject and the verb close together.

Also try to keep sentences within a maximum of four or five units of information. Look at the following two examples.

                Unit 1   Unit 2
The dimensions on a blueprint are called scale dimensions.

          Unit 1   Unit 2   Unit 3
Because the corners of the nut may be rounded or damaged and because the wrench

     Unit 4   Unit 5   Unit 6   Unit 7
may slip off the nut and cause an accident, choose a wrench with size and type suited

to the nut.

Notice that the second example becomes difficult to read and comprehend because of the many units of information.

Coordination and Subordination

Coordination and subordination are techniques used by the writer to combine ideas and to show the relationship between ideas. Showing the relationship between ideas helps the reader comprehend information more quickly and more accurately.

The ideas in sentences may be combined to make meaning clearer by adding a word, usually a subordinate conjunction or a coordinate conjunction, to show the relationship between the ideas.

From the two sentences

1. The company did not hire him.
2. He was not qualified.

a single sentence

1. The company did not hire him *because* he was not qualified.

makes the relationship of ideas clearer with the addition of the subordinate conjunction "because." The second sentence is changed to an adverb clause, "because he was not qualified," telling why the company did not hire him.

The two sentences

1. John applied for the job.
2. He did not get the job.

might be stated more clearly in

1. John applied for the job, but he did not get it.

The addition of the coordinate conjunction "but" indicates that the idea following is in contrast to the idea preceding it.

Study the following examples.

1. Magnetic lines of force can pass through any material.
2. They pass more readily through magnetic materials.
3. Some magnetic materials are iron, cobalt, and nickel.

The ideas in these three sentences might be combined into a single sentence.

1. Magnetic lines of force can pass through any material, but they pass more readily through magnetic materials such as iron, cobalt, and nickel.

The addition of the coordinate conjunction "but" indicates the contrasting relationship between the two main ideas; sentence 3 has been reduced to "such as iron, cobalt, and nickel." The three sentences might also be combined as follows:

1. Although magnetic lines of force can pass through any material, they pass more readily through magnetic materials: iron, cobalt, and nickel.

Using coordination and subordination, a writer can eliminate short, choppy sentences. The following five sentences

1. The mission's most important decision came.
2. It was early on December 24.
3. Apollo was approaching the moon.
4. Should the spacecraft simply circle the moon and head back toward earth?
5. Should it fire the Service Propulsion System engine and place the craft in orbit?

might be combined:

1. As Apollo was approaching the moon early on December 24, the mission's most important decision came.
2. Should the spacecraft simply circle the moon and head back toward earth or should it fire the Service Propulsion System engine and place the craft in orbit?

Combining sentences gives a flow of thought as well as makes relationships between ideas clearer.

Coordination and subordination are used to emphasize details; important details appear in independent clauses and less important details appear in dependent clauses and phrases. Almost any group of ideas can be combined in several ways. The writer chooses the arrangement of ideas that best "fits in with" preceding and following sentences. More importantly the writer arranges ideas to make important ideas stand out.

Consider the following group of sentences:

1. Carmen Diaz is the employee of the year.
2. She has been with the company only one year.
3. She has had no special training for the job she performs.
4. She is highly regarded by her colleagues.

The four sentences might be combined in several ways.

1. Although Carmen Diaz, the employee of the year, has been with the company only one year and has had no special training for the job she performs, *she is highly regarded by her colleagues.*
2. *Carmen Diaz is the employee of the year* although she has been with the company only one year and has had no special training for the job she performs; *she is highly regarded by her colleagues.*
3. Although Carmen Diaz, highly regarded by her colleagues, is the employee of the year, *she has been with the company only one year and has had no special training for the job she performs.*

Each sentence contains the same information; through coordination and subordination, however, different information is emphasized.

The way you arrange information, as well as what you say, affects meaning.

Positive Statements

Generally, statements worded positively are easier to read and comprehend. To understand negative wording, the reader reads the negative statement, mentally changes it to positive, and then changes it to negative, therefore taking a longer time to comprehend the statement.

Look at the following example.

NEGATIVE   If enrollment does not increase, the trustees will not vote to build new dormitories.

POSITIVE   The trustees will vote to build new dormitories only if enrollment increases.

The positively worded statement can be read and understood much more quickly.

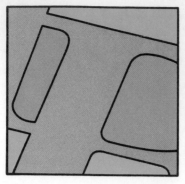

Chapter 1

Instructions and Process: Explaining a Procedure

Objectives

Upon completing this chapter, the student should be able to:

- Define instructions
- Classify instructions
- Plan a presentation giving instructions to two audiences
- Use visuals in giving instructions
- Plan the layout and design of instructions
- Explain the relationship between planning and giving instructions
- Give instructions in writing and orally
- Define process description
- Explain the difference between giving instructions and describing a process
- Give a process description directed to a general audience
- Give a process description directed to a specialized audience
- Use visuals in describing a process
- Give process descriptions in writing and orally

Introduction

You have been a giver and receiver of instructions practically from the beginning of your life. As a child, you were told how to drink from a cup, how to tie your shoes, how to tell time, and so on. As you matured, you became involved with more complex instructions: how to parallel park an automobile, how to throw a block in football, how to tune an electric guitar, how to stock grocery shelves. Since entering college, you have been confronted with even more complex and confusing instructions: how to register as an incoming freshman, how to write an effective report, how to get along with a roommate, how to spend money wisely, how to study.

Since all aspects of life are affected by instructions, every person needs to be able to give and to follow instructions. Frequently, clear, accurate, complete instructions save the reader or listener time, help do a job faster and more satisfactorily, or help get better service from a product. Being able to give and to follow instructions is essential for any employee. Certainly, in order to advance to supervisory positions, employees must be able to give intelligent, authoritative, specific, accurate instructions; and they must be able to follow the instructions of superiors.

Classifying Instructions

Giving instructions seems much simpler than following them. Telling someone how to study for a test, for instance, appears to be much easier than studying for it. But giving instructions—telling someone how to get somewhere or how to perform a particular operation—is deceptively simple.

You may find it helpful to look at two categories of instructions: locational and operational. As a student and a future worker you will have to give and receive both locational and operational instructions.

LOCATIONAL INSTRUCTIONS

Locational instructions, as the term suggests, help you locate a person, place, or thing. These instructions should clearly identify the starting point and the destination, the distance between the two, and the general direction. As with all instructions, giving them can be deceptively simple. Which of us at one time or another has not experienced the confusion of the delivery person in the following dialogue.

DELIVERY PERSON: Could you tell me where Mr. Sam Smith lives so I can deliver this load of lumber?

LOCAL INHABITANT: Go down the road a piece and turn left at the mailbox—the one just on this side of Mr. Jenkins's house. After you leave the main road and pass that bad curve, you should see the house you're looking for, not too far off up the road to the right.

The delivery person receiving these instructions may have difficulty in reaching the destination. Obviously, "down the road a piece" and "not too far" are, at best, indefinite; the inquirer has no idea where Mr. Jenkins lives; and what might be a "bad curve" to one person might not be to someone else. These instructions might have been more accurately and more clearly stated as follows:

LOCAL INHABITANT: Continue down this highway for about two miles. When you come to the second gravel road, turn left onto it. At this intersection, there is a large mailbox on a white wooden frame. When you take the gravel road, you will be about a half mile from where Mr. Smith lives. His is the second house on the right, the one that has a white paling fence around it.

In contrast to the lack of clarity and accuracy in the first instructions given to the delivery person, consider these instructions given to the driver of a moving van in an unfamiliar city:

DRIVER: After I took Exit 53 off the Interstate, I got confused somehow, and I can't seem to find Porter Drive, much less 4437 Porter Drive. It's supposed to be around here somewhere.

SERVICE STATION ATTENDANT: You took the right exit OK. As a matter of fact, you're only about three or four miles due south from where you want to be. If you knew where you were going and how to get there, it would take you ten minutes or less. Got a piece of paper to jot this down on? Now this street you are on, Charles Avenue, comes within three blocks of Porter Drive. Stay on Charles for about two-and-one-half or three miles until you get to Richard Street. You will turn left onto Richard. And Richard Street comes up just after you pass a fire station on the left. Got that? After you turn left on Richard from Charles, go three blocks. You run right into Porter Drive, which intersects with Richard Street. The address you are looking for is a few blocks on your left.

DRIVER: I think maybe I can find it now. Does this map I've sketched look right?

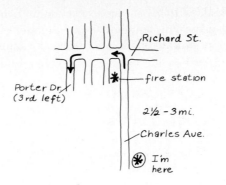

SERVICE STATION ATTENDANT: That looks perfect. Remember—just stay on this street till after you pass a fire station on your left, a couple of miles from here. Turn left onto Richard. Go three blocks. Take a left onto Porter.

Those directions are clear and easy to follow.

Or perhaps you have been in a situation similar to the following:

LOCKSMITH: I'm Jack Jones from National Lock and Key Company. I'm supposed to change a lock on Ms. Grady's desk.

RECEPTIONIST: Yes, I've been expecting you. Ms. Grady's office is on the third floor, Office 301. You'll notice identifying numbers over the door. After you enter the office, go to the desk on the left. As you face the desk, the drawer on the right is the one that needs to have the lock replaced.

Those directions are also clear and easy to follow.

The instructions above illustrate characteristics of effective locational instructions.

- When possible use "right" and "left" instead of "north," "south," "east," and "west."
- Be specific in giving distances and identifying places. Instead of "Go down the road a short distance," use "Continue down this road for two miles." Instead of "Turn left at the third street," use "Turn left at the third street, Richard Street."
- Identify any landmarks at, near, or on the way to the location. Identifying landmarks such as "a large mailbox on a white wooden fence," "house with a white paling fence around it," or "pass a fire station on the left" makes locational instructions easier to follow. Seeing such landmarks gives the person following the instructions confidence that he or she is moving in the proper direction.
- When possible include a map. Even a crudely drawn sketch can be helpful.

OPERATIONAL INSTRUCTIONS

Operational instructions tell how to carry out a procedure or an operation; for example, how to put a child's outdoor gym set together, how to run a lathe, how to prepare a blood smear, how to fill out an accident report, how to rescue a person from the tenth floor of a burning building, or any number of other "how to's."

The following student-written operational instructions explain how to draw blind contours.

HOW TO DRAW BLIND CONTOURS

Blind contour is a drawing technique by which the artist looking *only* at the subject and not at the drawing in progress produces a drawing. The artist tries to produce an outline of the subject. This technique teaches the student artist to concentrate on the subject rather than on the drawing and to develop hand-eye coordination.

REMEMBER! DO NOT LOOK AT THE DRAWING IN PROGRESS!

EQUIPMENT

Pen
Paper

PROCEDURE

1. Select a subject. For example, you might choose a tree, a chair, a person, a horse, a bicycle, or a hand.
2. Select a point on the subject as a beginning point.
3. Place the pen on the paper and begin outlining the subject, moving from the selected beginning point in whatever direction you choose. Concentrate on the subject. As your eyes move from the selected beginning point, move your pen as if your eyes and hand are connected. Imagine that your pen is touching the subject rather than the paper.
4. Use one continuous line to follow any lines, shapes, or shadows on the subject; draw what your eyes follow. Do not worry if your drawing little resembles the subject. See Figure 1.

Figure 1. A hand drawn by
the blind contour
technique.

CONCLUSION

Practicing blind contour drawing helps the artist focus on the subject. Also it helps the artist learn to avoid frequently looking from subject to painting.

Intended Audience

Giving instructions that can be followed successfully requires clear thinking and careful planning. One of the first things to be considered is the intended audience, that is, who will be hearing or reading and thus trying to follow the instructions. An explanation of how to operate the latest model X-ray machine would differ for an experienced X-ray technologist and for a student just being introduced to X-ray equipment. Instructions on how to freeze corn would differ for a food specialist at General Mills, for a homemaker who has frozen other vegetables but not corn, and for a seventh-grade student in a beginning home economics course.

Writing or speaking on a level that the intended audience will understand determines the kind and extent of details presented and the manner in which they are presented. Therefore, you must know who will be reading or hearing the explanation, and why. You must be fully aware whether a particular background, specialized knowledge, or certain skills are needed in order to understand the explanation.

Consider, for example, the excerpt from instructions included with a Black & Decker Dustbuster Model 9330 on page 22.*

For what readers were these instructions written? In what specific ways has the intended audience determined the kind and extent of details presented and the manner in which they are presented? Obviously the audience could be anyone able to purchase the Black & Decker Dustbuster Model 9330; the person might or might not be familiar with the procedure for removing dirt and cleaning the filter. Therefore, the instructions are kept simple and visuals are included to enhance the written explanation.

*Used by permission of the Black & Decker Mfg. Co.

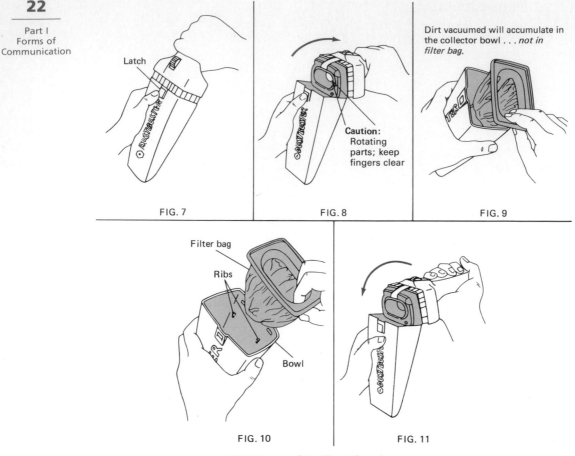

Latch

Caution:
Rotating
parts; keep
fingers clear

Dirt vacuumed will accumulate in
the collector bowl . . . *not in
filter bag.*

FIG. 7   FIG. 8   FIG. 9

Filter bag

Ribs

Bowl

FIG. 10   FIG. 11

Dirt Removal & Filter Cleaning

1. To remove dirt from the collector bowl, hold unit as shown in Figure 7.
2. Depress latch and swing collector bowl away and down from unit handle (Figure 8).
3. Place fingers into bowl and remove filter bag from collector bowl (Figure 9).
4. Empty dirt from collector bowl.
5. Clean filter bag by:
   a. Brushing the dust and dirt off the bag, or—
   b. Thoroughly shaking the bag clean into the waste can, or—
   c. Hand washing the filter bag with a mild detergent or dish washing soap. After washing, put a paper towel into the filter bag and squeeze the filter to remove the retained moisture. Be sure the filter bag is dry before replacing in the bowl.
6. Replace filter bag by pushing plastic ring into bowl until it rests on ribs at bottom surface of collector bowl (Figure 10).
7. Reassemble bowl to unit (Figure 11) by placing rear lower edge of bowl against lower ledge on front of unit. Line up so that front of unit fits into bowl, then snap together so that upper latch snaps into place holding bowl to unit.
8. Replacement bags are not required. Instead, DUSTBUSTER uses a long lasting Polyester reusable filter. If damaged, obtain new filter from your local dealer or B&D Service Center.

Oral Presentation

The content of oral instructions is very similar to that of written instructions; however, the delivery is different. In speaking, you don't have to be concerned with such things as spelling and punctuation, and you have the advantage of a visible audience with whom you can interact.

Your delivery of oral instructions will be improved if you follow these suggestions, whether addressing one or two persons or a large group:

1. *Look at your audience.* Use eye-to-eye contact with your audience, but without special attention to particular individuals. Avoid continuously looking at your notes, the floor, the ceiling, or a particular individual.
2. *Speak, don't read, to the audience.* Avoid memorizing or reading your presentation. Rather, have it carefully outlined on a note card.
3. *Repeat particularly significant points.* Remember that the audience is *listening*. Repeat main points and summarize frequently. The hearer cannot reread material as the reader can; therefore repetition is essential.
4. *Speak clearly, distinctly, and understandably.* Follow the natural pitches and stresses of the spoken language, and use acceptable pronunciation and grammar. Speak on a language level appropriate for the audience and the subject matter.
5. *Use bodily movements and gestures naturally.* Put some zest in your expression; be alive; show enthusiasm for your subject. Stand in an easy, natural position, with your weight distributed evenly on both feet. Let your movements be natural and well timed.
6. *Involve the audience, if practical.* Invite questions or ask someone to carry out the procedure or some part of it.
7. *Use visuals where needed.* Whenever visuals will make the explanation clearer, use them. The visuals should be large enough for everyone in the audience to see clearly.

For a detailed discussion of oral presentations, see Chapter 10, Oral Communication.

Visuals

To a person reading or hearing locational and operational instructions, it is often helpful if the instructions include such visuals as maps, drawings, photographs, diagrams, real objects, models, demonstrations, slides, and overhead-projected transparencies. Consider, for instance, the freehand map included with the locational instructions on page 19. Even this crudely drawn map with key points identified is extremely helpful to the individual in reaching the destination more quickly and easily. Such a map usually depicts pertinent streets, intersections, turns, landmarks, and distances.

Or consider the drawings included with the instructions on how to draw blind contours on pages 20–21, the drawings included with the instructions for removing the dirt from and cleaning the filter in a Black & Decker Dustbuster Model 9330 on page 22, and the drawings with the instructions for sharpening a ruling pen on pages 32–33. The drawings help to make the instructions clearer.

For a detailed discussion of visuals, see Chapter 11, Visuals.

Layout and Design

Especially in instructions, layout and design are of prime importance. The placement of material on the page should provide maximum ease in reading and comprehension. The overall design of the set of instructions must provide ready access to needed information, in a convenient format.

Consider, for example, the opening page of two sets of instructions for changing the oil and oil filter in a 1983 Cordoba (Figure 1).

Example A in Figure 1 gives the same information that Example B does. But which example is easier to read and comprehend? Example B, of course. Example B has plenty of *white space;* the material is uncluttered and so placed on the page that the eye easily and quickly sees the relationship of one part of the material to another. The *headings* provide key words for each section. Emphasis is achieved in several ways: through *uppercase letters* for the headings and for the precaution; the *box* to set off material about elevating the car; the *bullets* for a list in which sequence is not important; the *numbers* for the major steps (where sequence is very important); and the *letters of the alphabet* (for substeps in a sequence).

The main difference in the two presentations of the instructions for changing the oil and the oil filter in a 1983 Cordoba is the use of layout and design techniques in Example B that make reading and comprehension easier:

☐ White space
☐ Headings
☐ Uppercase letters
☐ Bullets
☐ Numbers
☐ Letters of the alphabet
☐ Boxed information

For a detailed discussion, see Layout and Design, pages 4–7.

General Principles in Giving Instructions

1. *Knowledge of the subject matter is essential.* To give instructions that can be followed, you must be knowledgeable about the subject. If need be, consult sources—knowledgeable people, textbooks, reference works in the library—to gain further understanding and information about the subject.
2. *The intended audience influences what information is to be presented and how it is to be presented.* Consider the audience's degree of knowledge and understanding of the subject. Avoid talking down to the audience as well as overestimating the audience's knowledge or skills.
3. *Effective instructions are accurate and complete.* The information must be correct. Instructions should adequately cover the subject, with no step or essential information omitted.

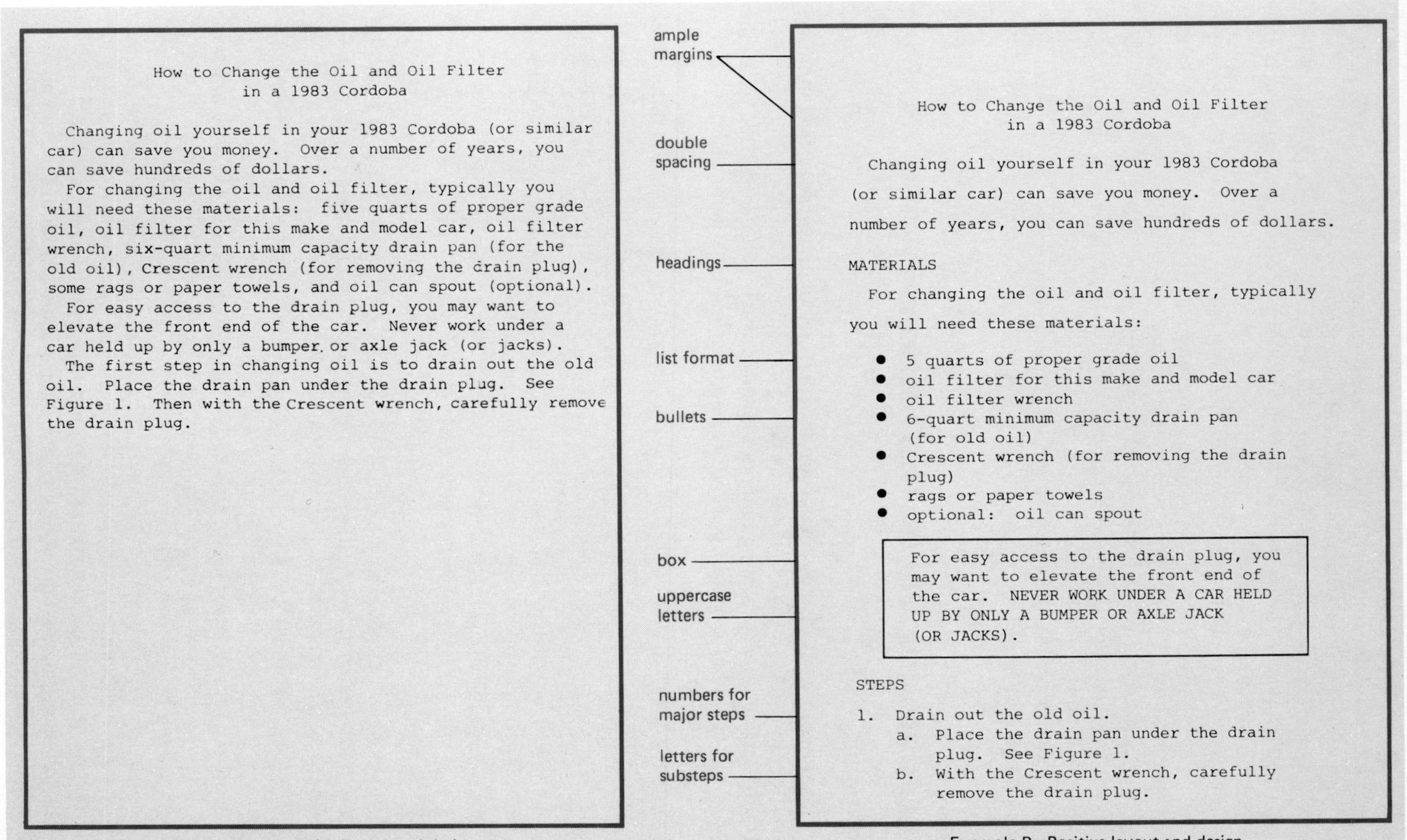

Example B (Positive layout and design):

How to Change the Oil and Oil Filter
in a 1983 Cordoba

Changing oil yourself in your 1983 Cordoba
(or similar car) can save you money. Over a
number of years, you can save hundreds of dollars.

MATERIALS

For changing the oil and oil filter, typically
you will need these materials:

- 5 quarts of proper grade oil
- oil filter for this make and model car
- oil filter wrench
- 6-quart minimum capacity drain pan
  (for old oil)
- Crescent wrench (for removing the drain
  plug)
- rags or paper towels
- optional: oil can spout

> For easy access to the drain plug, you
> may want to elevate the front end of
> the car. NEVER WORK UNDER A CAR HELD
> UP BY ONLY A BUMPER OR AXLE JACK
> (OR JACKS).

STEPS

1. Drain out the old oil.
   a. Place the drain pan under the drain
      plug. See Figure 1.
   b. With the Crescent wrench, carefully
      remove the drain plug.

Example B Positive layout and design

Labels:
- ample margins
- double spacing
- headings
- list format
- bullets
- box
- uppercase letters
- numbers for major steps
- letters for substeps

Example A (Negative layout and design):

How to Change the Oil and Oil Filter
in a 1983 Cordoba

Changing oil yourself in your 1983 Cordoba (or similar
car) can save you money. Over a number of years, you
can save hundreds of dollars.
For changing the oil and oil filter, typically you
will need these materials: five quarts of proper grade
oil, oil filter for this make and model car, oil filter
wrench, six-quart minimum capacity drain pan (for the
old oil), Crescent wrench (for removing the drain plug),
some rags or paper towels, and oil can spout (optional).
For easy access to the drain plug, you may want to
elevate the front end of the car. Never work under a
car held up by only a bumper or axle jack (or jacks).
The first step in changing oil is to drain out the old
oil. Place the drain pan under the drain plug. See
Figure 1. Then with the Crescent wrench, carefully remove
the drain plug.

Example A Negative layout and design

Figure 1. Examples of negative (poor) and positive (good) layout and design.

4. *Visuals help to clarify instructions.* Whenever instructions can be made clearer by the use of such visuals as maps, diagrams, graphs, pictures, drawings, slides, demonstrations, and real objects, use them.

5. *Instructions require careful layout and design.* An integral part of effective instructions is consideration of how material is placed on the page and of how the document as a whole is presented.

6. *Conciseness and directness contribute to effective communication.* An explanation that is stated in the simplest language with the fewest words is usually the clearest. If the instructions call for terms unfamiliar to the audience or for familiar terms with specialized meanings, explain them.

7. *Instructions that can be followed have no unexplained gaps in the procedure or vagueness about what to do next.* Well-stated instructions do not require the audience to make inferences, to make decisions, or to ask, "What does this mean?"

Procedure for Giving Instructions

FORM

The steps in a set of instructions are usually presented as a list, with numbers indicating the major steps. For nonsequential lists (such as a list of needed materials or equipment) and for indicating emphasis, symbols such as these may be used: the open or closed bullet ○ ● (see page 5), the open or closed square or box □ ■, the dash—, and the asterisk *.

In giving instructions, use the second person pronoun *you* (usually understood) and imperative or action verbs.

EXAMPLE

First, (*you* understood) *unplug* (imperative verb) the appliance. Then with a Phillips screwdriver (*you* understood) carefully *remove* (imperative verb) the back plate.

CONTENT

In giving instructions, be sure to tell your audience not only *what* to do but also *how* to do each activity, if there is any doubt that the audience might not know how. You may also add *why* the activity is necessary.

EXAMPLE

Remove the cap (tells *what*). Use the fingers to turn it counterclockwise until it can be removed (tells *how*). You can then insert the spout of the gasoline can to fill the tank with gasoline (tells *why*).

Use good judgment in selecting details. For example, if you are explaining how to pump gas, it is not necessary to say you must have a car and a gas tank.

The following suggestions may also be helpful. Generally include *a*, *an*, and *the*, unless space is limited. Notice the example above which includes *a*, *an*, and *the*. It is much easier to read than if it were written: "Remove cap. Use fingers to turn counterclockwise until it can be removed. You can then insert spout of gasoline can to fill tank with gasoline." Notice the example of the instructions for a Black & Decker Dustbuster, page 22. The *a, an,* and *the* have been omitted because all instructions were printed in a limited space.

When organizing material for giving instructions, whether in writing or orally, divide the presentation into two or three parts, as follows.

I. The identification of the subject is usually brief, depending on the complexity of the operation.
   A. State the operation to be explained.
   B. If applicable, give the purpose and significance of the instructions and indicate who uses them, when, where, and why.
   C. State any needed preparations, skills, equipment, or materials.

II. The development of the steps is the main part of the presentation and thus will be the lengthiest section.
   A. Clearly list each step and develop each step fully with sufficient detail to tell what to do and how to do it. Also tell why a step is necessary, if applicable.
   B. In a more complex operation, subdivide each major step, if necessary.
   C. Explain in clear detail exactly what is to be done to complete the operation.
   D. Sometimes it is helpful to emphasize particularly important points and to caution the reader where mistakes are most likely to be made.
   E. Plan the layout and design of the instructions, using headings and visuals wherever needed.

III. The closing may be
   A. The completion of the discussion of the last step.
   B. A summary of the main steps, especially in complex or lengthy instructions.
   C. A comment on the significance of the operation.
   D. Mention of other methods by which the operation is performed.

LENGTH OF PRESENTATION

The length of the presentation is determined by the complexity of the operation, the degree of knowledge of the reader, and the purpose of the presentation. In explaining how to do a simple operation, such as taking a temperature or changing a typewriter ribbon, the entire presentation may be very brief.

Planning and Giving Instructions

Planning instructions requires several steps. As with any planning, you need first to answer two questions:

1. Why am I giving the presentation? (Purpose)
2. For whom is the presentation intended? (Audience)

Although these two items do not necessarily appear in the actual presentation, purpose and audience directly affect the way you select and present details in a set of instructions, as illustrated by the three examples of instructions on pages 20–21, 22, and 32–33.

Review the General Principles in Giving Instructions on pages 24, 26. These principles summarize the major points of the chapter. Now review the Procedure for Giving Instructions on pages 26–27 on the form and the content for instructions. In the content outline, note that the suggestions under roman numeral I help you to know what kinds of information you may include in the introductory material; suggestions under II identify content for the body; and the suggestions under III give choices for the closing material. Then decide if including visuals will make the instructions clearer to the intended audience. Remember, if you use a visual, refer to the visual within the content of the instructions (see page 540).

The Plan Sheet (see the illustration on pages 29–30) includes parts of the content outline that are of major importance in planning the instructions. It serves as a guide to help you plan what you will write or tell; it helps you to clarify your thinking on a topic and helps you to select and organize details.

You may or may not use every bit of information on the Plan Sheet. Nevertheless, fill it in completely and fully so that once you begin to prepare a preliminary draft you will have thought through all major details needed in the presentation.

As you work, refer frequently to the Procedure for Giving Instructions and the Plan Sheet. These two guides should help you plan and develop an acceptable presentation.

Following is a sample presentation of instructions. The filled-in Plan Sheet, the rough copy showing revisions and corrections, and the final copy including marginal notes to outline the development of the instructions are included.

PLAN SHEET
FOR GIVING INSTRUCTIONS

Analysis of Situation Requiring Instructions

What is the procedure to be explained?
how to sharpen a ruling pen

For whom are the instructions intended?
a classmate in drafting

How will the instructions be used?
to sharpen a ruling pen

Will the instructions be written or oral?
written

Importance or Usefulness of the Instructions

Used regularly, a pen becomes dull. Anyone using one needs to know how to sharpen it. A sharp pen is required for neat, clean lines.

Organizing Content

Equipment and materials necessary:
3- or 4-inch sharpening stone (hard Arkansas knife piece; soaked in oil several days) crocus cloth (if desired)

Terms to be defined or explained:
ruling pen
nibs

Overview of the procedure (list as command verbs):
1. *Close the nibs.*
2. *Hold the sharpening stone correctly in the left hand.*
3. *Hold the ruling pen correctly in the right hand.*
4. *Round the nibs.*
5. *Sharpen the nibs.*
6. *Polish the nibs, if desired.*
7. *Test the pen.*

Major steps, with identifying information:

What to do	How to do	Why, if applicable
1. Close the nibs.	*Turn the adjustment screw (located above nibs) to the right until the nibs touch.*	*So the nibs can be sharpened to exactly the same shape and length*

Major steps, with identifying information, cont.:

What to do	How to do	Why, if applicable
2. Hold the sharpening stone correctly in the left hand.	Lay the stone across the left palm; grasp the stone with the thumb and fingers.	To give the best control of the stone
3. Hold the ruling pen correctly in the right hand.	Pick up the pen with the thumb and index finger of the right hand. Pick the pen up like a drawing crayon. Lightly rest the other three fingers on the pen handle.	To avoid an injured hand; to avoid a ruined pen
4. Round the nibs.	Stroke the pen back and forth on the stone. Start with the pen at a 30-degree angle and follow through to past a 90-degree angle. Usually 4 to 6 strokes.	Both the nibs must be the same length and shape so the tips of both will touch the paper.
5. Sharpen the nibs.	Open the blades slightly; turn the adjustment screw gently left. Sharpen the outside of each blade, one at a time. Hold the stone and the pen as in steps 2 and 3. Hold the pen at a slight angle with the stone. Rub the pen back and forth, 6 to 8 times. Rub the crocus cloth lightly over the nibs.	Rounding the nibs will leave them dull.
6. Polish the nibs. (optional)	Rub the crocus cloth lightly over the nibs.	To remove any rough places
7. Test the pen.	Add a small drop of ink between the nibs. Draw a line along the edge of the T square on a piece of paper.	To determine if the pen is properly sharpened

Precautions to emphasize (crucial steps, possible difficulties, dangers, places where errors are likely to occur):

Failure to hold the pen correctly and the stone correctly may cause an injured hand or a damaged pen.

Types and Subject Matter of Visuals to Be Included

drawings to show: 1. hand holding pen for sharpening
2. nibs to show correct shape

Sources of Information

30 class lecture in drafting, textbook in drafting, personal experience

Sharpening a Ruling Pen

A major instrument in making a mechanical drawing is the ruling pen.
which
~~It~~ is used to ink in straight lines and noncircular curves. *with a T square, a triangle, a curve, or a straightedge as a guide* The shape
(blades resembling tweezers)
of the nibs is the most important aspect of the pen. The nibs must be
create
rounded (elliptical) to ~~creat~~ an ink space between the nibs. *To assure neat, clean lines,* The ruling

pen must be kept sharp and in good condition. It must be sharpened from

time to time after extensive use because the nibs wear down.

Add heading: Equipment and materials   *List with visual marker:*

To sharpen the ruling pen, you need a 3- or 4-inch sharpening stone
for several days
(preferably a hard Arkansas knife piece that has been soaked in oil) and
seven   *Add*
a crocus cloth (optional). Sharpening a ruling pen involves ~~six~~ steps: *heading:*
Overview
close the nibs, hold the sharpening stone in the left hand, hold the ruling   *of procedure*

pen in the right hand, round the nibs, sharpen the nibs, polish the nibs

(optional), and test the pen.

Add
heading: *Steps*
1. Close the nibs. Turn the (adjustment) screw, located above the nibs,
then
to the right until the nibs touch. The nibs can be sharpened to exactly

the same shape and the same (lenght).

2. Hold the sharpening stone in the left hand in a usable position. With

the stone lying across the palm of the left hand, grasp it with the thumb and

fingers to give the best controll of the stone.

3. Hold the ruling pen in the right hand. Pick up the ruling pen with

the thumb and index finger of the right hand as if it were a drawing crayon.

The other three fingers should rest lightly on the pen handle. CAUTION: *Show*
hand
Place   *holding*
CAUTION Failure to hold the sharpening stone and ruling pen correctly may result in   *pen*
in a   *correctly.*
box.   an injured hand and a ruined ruling pen. (See Fig. 1)
Figure

4. Round the nibs. To round (actually to make elliptical) the nibs, stroke

the pen back and forth on the stone, starting with the pen at a 30-degree

angle to the stone and following through to past a 90-degree angle as the line
moves
across the stone (move) forward. Usually four to six strokes are needed.

Be sure that both nibs are the same length and shape so that the
Figure
tips of both nibs will touch the paper as the pen is used. (See Fig. 2)
Show correct shape.
(When the nibs are satisfactorily rounded, they will be left dull;)

5 Sharpen the nibs. Open the blades slightly by turning the
only the outside of
adjustment screw gently to the left. Sharpen each blade, one at a time

To sharpen, hold the stone and the ruling pen in the hands as
(see Step 3)
in rounding the nibs; the pen should be at a slight angle with the

stone. Rub the pen back and forth, *six to eight times, with a rocking,*
pendulum motion to restore the original shape *the nibs*   *by rubbing the cloth*
Make   If desired, ~~the nibs may be~~ polished with a crocus cloth, *lightly over the nibs, to remove*
this   *Make this Step 7.*
Step 6. any rough places. It is a good idea to test the ruling pen after
pen is properly sharpened   *will draw sharp,*
sharpening it. If ~~the job has been done~~ well, the pen ~~is capable of~~
clean
~~making~~ clean, sharp lines.

Add a small drop of ink between the nibs. Draw a line along the edge
of a T square placed on a piece of paper.

HOW TO SHARPEN A RULING PEN

Identification of ruling pen

A major instrument in making a mechanical drawing is the ruling pen, which is used to ink in straight lines and noncircular curves with a T square, a triangle, a curve, or a straightedge as a guide. The shape of the nibs (blades resembling tweezers) is the most important aspect of the pen. The nibs must be rounded (elliptical) to create an ink space between the nibs. To assure neat, clean lines, the ruling pen must be kept sharp and in good condition. It must be sharpened from time to time after extensive use because the nibs wear down.

Definition of nib

Reason for sharpening pen

Equipment and materials needed

Equipment and Materials

- 3- or 4-inch sharpening stone (preferably a hard Arkansas knife piece that has been soaked in oil for several days)
- crocus cloth (optional)

Listing of steps

Overview of Procedure

Sharpening a ruling pen involves seven steps: close the nibs, hold the sharpening stone in the left hand, hold the ruling pen in the right hand, round the nibs, sharpen the nibs, polish the nibs (optional), and test the pen.

Each step numbered and explained in detail: what to do, how to do it, and why (if needed)

Steps

1. Close the nibs. Turn the adjustment screw, located above the nibs, to the right until the nibs touch. The nibs can then be sharpened to exactly the same shape and the same length.
2. Hold the sharpening stone in the left hand in a usable position. With the stone lying across the palm of the left hand, grasp it with the thumb and fingers to give the best control of the stone.

Comparison with a familiar action

3. Hold the ruling pen in the right hand. Pick up the ruling pen with the thumb and index finger of the right hand as if it were a drawing crayon. The other three fingers should rest lightly on the pen handle. See Figure 1.

CAUTION! Failure to hold the sharpening stone and ruling pen correctly may result in an injured hand and a ruined ruling pen.

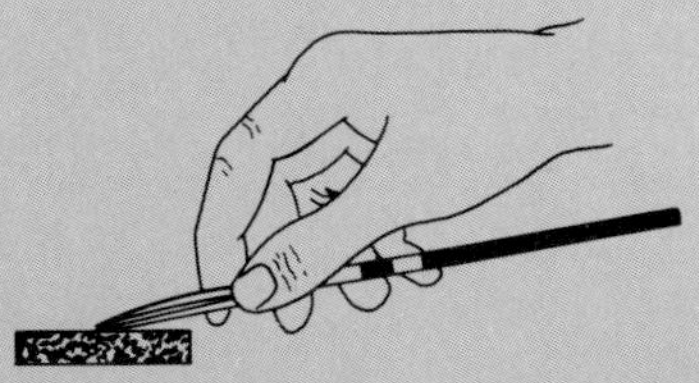

Figure 1. Holding the pen for sharpening.

4. Round the nibs. To round (actually to make elliptical) the nibs, stroke the pen back and forth on the stone, starting with the pen at a 30-degree angle to the stone and following through to past a 90-degree angle as the line across the stone moves forward. Usually four to six strokes are needed.

Particular emphasis

Be sure that both nibs are the same length and shape so that the tips of both nibs will touch the paper as the pen is used. See Figure 2.

**Figure 2.
Correct shape
of pen nibs.**

5. When the nibs are satisfactorily rounded, they will be left dull; sharpen the nibs. Open the blades slightly by turning the adjustment screw gently to the left. Sharpen only the outside of each blade, one at a time.

To sharpen, hold the stone and the ruling pen in the hands as in rounding the nibs (see step 3); the pen should be at a slight angle with the stone. Rub the pen back and forth, six to eight times, with a rocking, pendulum motion to restore the original shape.

Optional step

6. If desired, polish the nibs with a crocus cloth by rubbing the cloth lightly over the nibs to remove any rough places.
7. Test the ruling pen after sharpening it. Add a small drop of ink between the nibs. Draw a line along the edge of a T square placed on a piece of paper. If the pen is properly sharpened, the pen will draw sharp, clean lines.

Application 1   Giving Instructions

Make a list of five persons to interview about their jobs. After interviewing each of the five, make a list of examples showing how they use instructions on the job. From this experience, what can you speculate about the importance of instructions in your own future work?

Application 2   Giving Instructions

Explain to a new student how to get from the classroom to another campus building or location (the library, another classroom, the student center, etc.).

Application 3   Giving Instructions

Explain to a late-entering freshman in a lab how to find needed supplies and equipment.

Application 4   Giving Instructions

Explain to a visitor from out of town how to get from the airport, bus station, or train station to your home. Include a freehand map you have drawn.

Application 5   Giving Instructions

Find and attach to your paper a set of instructions that a manufacturer included with a product.

  a. Evaluate in a paragraph the layout and design of the instructions (see pages 4–7 and 24).
  b. Evaluate in a paragraph the clarity and completeness of the instructions by applying the General Principles in Giving Instructions, pages 24, 26.

Application 6   Giving Instructions

Assume that you are a foreman or a supervisor with a new employee on the job. In writing explain to the employee how to carry out some simple operation.

  a. Fill in the Plan Sheet on pages 37–38.
  b. Write a preliminary draft.
  c. Revise. See the inside back cover.
  d. Write the final draft.

Application 7   Giving Instructions

Give orally the instructions called for in Application 6 above. Remember to use visuals whenever they will be helpful. Ask your classmates to evaluate your speech by filling in an Evaluation of Oral Presentations from pages 527–535.

Application 8   Giving Instructions

From the following list, choose a topic for an assignment on giving instructions or choose a topic from your own experience. Consult whatever sources necessary for information.

a. Fill in the Plan Sheet on pages 39–40.
b. Write a preliminary draft.
c. Revise. See the inside back cover.
d. Write the final draft.

How to:

1. Copy a disk on a personal computer
2. Take a patient's blood pressure
3. Water ski
4. Install a room air conditioner
5. Sharpen a drill bit
6. Read the resistor color code
7. Clean a chimney
8. Produce a business letter—individualized to several people—on a word processor
9. Cure an animal hide
10. Cut a mat for a picture
11. Operate a piece of heavy-duty equipment
12. Rescreen a window
13. Open a checking account
14. Change a tire on a hill
15. Plant a garden
16. Change oil in a car
17. Read a micrometer or a dial caliper
18. Hang wallpaper
19. Set out a shrub
20. Replace a capacitor in a television set
21. Sterilize an instrument
22. Use a microfiche reader
23. Make a tack weld
24. Operate an office machine
25. Prepare a laboratory specimen for shipment
26. Input data into a computer
27. Fingerprint a suspect
28. Tune up a motor
29. Repair (or rebind) a book
30. Develop black and white film
31. Use a compass, architect's scale, divider, or French curve
32. Administer an intramuscular injection
33. Start an airplane engine
34. Set up a partnership
35. Topic of your choosing

Application 9 Giving Instructions

Give orally the instructions for one of the topics in Application 8 above. Remember to use visuals whenever they will be helpful. Ask your classmates to evaluate your speech by filling in an Evaluation of Oral Presentations from pages 527–535.

Application 10   Giving Instructions

Read the article "Clear Only If Known," by Edgar Dale, pages 568–571. List Dale's reasons why people have difficulty in giving and receiving instructions.

PLAN SHEET
FOR GIVING INSTRUCTIONS

Analysis of Situation Requiring Instructions

What is the procedure to be explained?

For whom are the instructions intended?

How will the instructions be used?

Will the instructions be written or oral?

Importance or Usefulness of the Instructions

Organizing Content

Equipment and materials necessary:

Terms to be defined or explained:

Overview of the procedure (list as command verbs):

Major steps, with identifying information:

What to do	How to do	Why, if applicable

Major steps, with identifying information, cont.:
What to do How to do Why, if applicable

Precautions to emphasize (crucial steps, possible difficulties, dangers, places where errors are likely to occur):

Types and Subject Matter of Visuals to Be Included

Sources of Information

40

Description of a Process

Describing a process—explaining how something is done—is similar to giving instructions. There are, however, two basic differences: a difference in the purpose and a difference in the procedure of presentation. The purpose in giving instructions is to enable an individual to perform a particular operation. The giver of the instructions expects the reader or hearer to *act*. In describing a process, however, the purpose is to describe a method or operation so that the intended audience will understand what is done. The presenter expects the reader or hearer to *understand* what happens.

Processes may be carried out by people, by machines, or by nature. Descriptions of processes carried out by people might be: how steel is made from iron, how glass is made, how diamonds are mined. Machine processes include how a clock works, how a Xerox copier works, or how a gasoline engine operates. Natural processes include how sound waves are transmitted, how rust is formed, how food is digested, and how mastitis is spread.

With instructions, you use commands (for example, *unplug, remove, insert*) so that the audience can act. You present and explain each step the reader or listener must carry out to perform the operation. In the description of a process, you emphasize the sequence of actions that is the procedure for an operation. The audience is unlikely to perform the operation. Notice below a comparison of possible procedures for giving instructions and for describing a process.

Giving instructions	*Describing a process*
"(you) Fasten the left strand . . ."	"The left strand is fastened . . ."
1. Imperative mood (orders or commands)	1. Indicative mood (statements of fact)
2. Active voice (subject does the action)	2. Passive (subject is acted upon)
3. Second person (person spoken to is subject). Subject is understood *you*.	3. Third person (thing spoken about is subject)

Through reading or hearing a description of a process, the audience develops an *understanding* of the operation. It would, in fact, be impossible to perform some processes, for example those carried out by nature. It is possible, however, to understand what happens as these natural processes occur; for instance, you can understand how sound is transmitted or how a tornado develops. Further, you can understand how bricks are made, but you would probably never make a brick.

Intended Audience

Just as in giving instructions, in giving a description of a process you must aim your presentation at a particular audience. You then write your explanation so that the intended audience will clearly understand. The intended audience determines the kind and extent of details you include in the description and the manner in which you present them. Audiences may be grouped into two broad categories: general audiences and specialized audiences.

The general adult audience requires a fairly inclusive description. The writer should assume that this audience has little, if any, of the particular background, knowledge, or skill necessary to understand a description of a technical process. Therefore, you need to describe that process as clearly and simply as possible, defining any terms that might have special meaning.

The following description of how the heart works is directed to a general audience. Note the use of drawings as well as simplified language in describing the process.

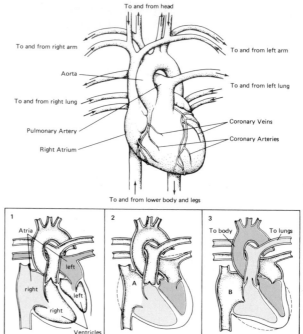

1. During the heart's relaxed stage (diastole), oxygen-depleted blood from body flows into right atrium, oxygenated blood from lungs into left. 2. Natural pacemaker, or sinoatrial node (A), fires electrical impulse and atria contract. Valves open and blood fills ventricles.

3. In pumping stage (systole), the electrical signal, relayed through atrioventricular node (B), causes ventricles to contract, forcing oxygen-poor blood to lungs, oxygen-rich blood to body.

The following description of how a common table mushroom grows illustrates an oral presentation for a general audience. A filled-in Plan Sheet and the description with directions for oral presentation are included.

In the description of how a common table mushroom grows, the speaker's purpose is to give the average adult listener a general view of the process. The speaker, being knowledgeable about the subject and understanding the process by which the mushroom grows sufficiently to give a general description of it, is successful in accomplishing that purpose. The speaker remembers the audience throughout the presentation by using a vocabulary and level of speaking that the average adult can easily understand.

The speaker is accurate and complete in the description, and the information is correct. All the necessary stages in the growth of the mushroom are given, and at no time is the mushroom unaccounted for, from the beginning to the finished product. Certainly, there is much more information about mushrooms that the speaker *could* have included, such as the kinds of mushrooms or differences in edible and inedible mushrooms. Or the speaker could have elaborated further about each step in the growth process. The important thing, however, is that the speaker included all *essential* information.

PLAN SHEET
FOR DESCRIBING A PROCESS

Analysis of Situation Requiring Description of a Process

What is the process to be described?
how a common table mushroom grows

For whom is the description intended?
a botany class

How will the description be used?
for general understanding of how an edible mushroom grows

Will the description be written or oral?
oral

Importance or Significance of the Process

Mushrooms can be cultivated to provide food.

Major Stages or Steps and Description of Each Step

(List the major stages or steps as -*ing* verb forms or in third person, present tense, active or passive voice.)

Stages or steps	What happens
1. spawn stage	Tiny webs that look like thread grow underground beneath a thin layer of compost and soil.
2. pinhead stage	After 7 to 8 weeks, if plenty of moisture is available, little knobs or pinheads show up on the surface of the soil. Look like tiny balls. If moisture is lacking, the pinheads may develop beneath the soil. Push through soil as they develop into buttons.
3. button stage	About one week later, the pinheads grow into buttons. About ½ inch across. Buttons are on the stalks or stems that push the button upward.
4. cap and gills stage	Buttons grow larger. Form caps or crowns at the upper part. Gills grow underneath the caps. At first gills are hidden by veillike covering; covering breaks away as the caps grow wider, exposing pink gills as thin ridges growing out from the center stalk to the edge of the cap. Gills bear spores from which new mushroom plants can grow.

Stages or steps, cont.	What happens, cont.
5. mature mushroom	Looks like a miniature open umbrella. Pink gills now dark brown. Mushroom ready to eat.

Important Points to Receive Special Emphasis

omit

Types and Subject Matter of Visuals to Be Included

a flowchart to outline major stages and show what each stage includes

*Put on chalkboard as
each step is mentioned:*

	spawn	*planting in compost* *covering with soil*
	pinhead	*knobs growing above or below soil*
	button	*knobs growing into buttons* *stalks or stems pushing buttons upward*
	cap/gills	*buttons growing larger* *forming caps and underneath caps* *forming gills* *gills changing from pink to brown*
	mature mushroom	*edible food*

two large glass bowls showing two stages: one containing compost and spawn with "threads" visible; one with mushrooms at pinhead stage

mature mushroom

Sources of Information

personal experience growing mushrooms
book on how to grow mushrooms for profit

ORAL PRESENTATION

HOW A COMMON TABLE MUSHROOM GROWS

When eating a mushroom, have you ever wondered how it grows into the umbrella shape? Mushrooms are fungi that grow in decaying vegetable matter. The common table mushroom grows in much the same way as other mushrooms. Belonging to a group of mushrooms called the agarics, the common mushroom grows wild, but it is also cultivated. The five stages of mushroom growth include the spawn, the pinhead, the button, the cap and gills, and the mature mushroom shaped like an opened umbrella.

As each major stage is listed, write on the chalkboard:

Spawn
   ↓
Pinhead
   ↓
Button
   ↓
Cap/Gills
   ↓
Mature Mushroom

STAGE 1 THE SPAWN

First a compost of decayed animal waste is prepared, usually in boxes. When the temperature is warm, pieces of mushroom spawn which look like tiny webs of thread are placed about a foot apart and just barely covered by the compost. The spawn, or mycelium, gets nourishment from the decayed compost matter. About one inch of soil is placed over the compost.

The spawn, the main part of the mushroom, grows underground. Its threads grow throughout the whole compost area and from them grow the pinheads.

As each aspect of the spawn stage is stated, jot it down on the chalkboard:

Spawn { *planting in compost*
         covering with soil
Show the glass bowl containing the spawn with "threads" visible through the sides.

STAGE 2 THE PINHEAD

In seven to eight weeks, if there is a lot of moisture, little knobs or pinheads appear on the surface of the soil. They look very much like tiny balls. If there is a lack of moisture, the pinheads may develop beneath the soil and then push through as they develop into buttons.

As each aspect of the pinhead stage is stated, jot it down on the chalkboard:
Pinhead {knobs growing above or below soil
Show the glass bowl containing mushrooms at the pinhead stage.

STAGE 3 THE BUTTON

In approximately one week the pinheads have grown into small buttons, about ½ inch across. The buttons are perched on top of tiny stalks or stems. The stalks or stems grow upward, pushing the buttons upward.

As each aspect of the button stage is stated, jot it down on the chalkboard:

Button $\begin{cases} \text{knobs growing into buttons} \\ \text{stalks or stems pushing buttons upward} \end{cases}$

STAGE 4 THE CAP AND GILLS

The buttons grow larger, forming caps or crowns at the upper part of the mushrooms. Underneath the caps, gills grow; at the early stage of growth the gills are hidden by a veillike covering which breaks away as the caps grow wider. Then the pink gills can be seen easily. They are thin ridges which grow out from the center stalks to the edges of the caps. The gills bear the spores, cells from which new mushroom plants may grow.

As each aspect of the cap and gills stage is stated, jot it down on the chalkboard:

Cap and Gills $\begin{cases} \text{buttons growing larger} \\ \text{forming caps and underneath caps forming gills} \\ \text{gills changing from pink to brown} \end{cases}$

STAGE 5 THE MATURE MUSHROOM

The mature mushroom looks like a miniature open umbrella. The pink gills have turned dark brown. Now the mushrooms can be harvested and sold for food.

Show mature mushroom.

PROCESS DESCRIPTION FOR A SPECIALIZED AUDIENCE

A specialized audience, as the term implies, has at least an interest in a particular subject and probably has the background, either from reading or from actual experience, to understand a description of a process related to that subject. For example, a person whose hobby is working on cars and reading about them would understand a relatively technical description of the operation of the Wankel engine. The specialized reader may even have a high degree of knowledge and skill. For example, sanitation engineers, fire chiefs, inhalation therapists, research analysts, programmers, or machinists would understand technical descriptions related to their fields of specialization. If, however, you have any doubt about the audience's level of knowledge, clarify any part of the description that you feel the reader might have difficulty understanding. (See also Accurate Terminology, pages 63–64.)

The following description of how a fire company officer sizes up a fire is directed to a person interested in fire science. The Plan Sheet used in preparing the description is included.

PLAN SHEET
FOR DESCRIBING A PROCESS

Analysis of Situation Requiring Description of a Process

What is the process to be described?
how fires are sized up

For whom is the description intended?
a fire science trainee

How will the description be used?
to learn the duties of a company officer

Will the description be written or oral?
a written explanation

Importance or Significance of the Process

Accurate sizing up of a fire determines fire-fighting techniques to be employed and the effectiveness of those techniques.

Major Steps and Description of Each Step

(List the major stages or steps as *-ing* verb forms or in third person, present tense, active or passive voice.)

1. *Determining the location of the fire:*
   Noticing the alarm code (alarm box report)
   Checking with caller (call-in report)
2. *Assessing the danger to life and materials:*
   Noticing time of day
   Noticing season of the year
3. *Judging what the fire can do:*
   Consulting water resources map
   Checking building and its surroundings
4. *Reviewing available resources:*
   Noting kinds of companies responding, including types and amounts of equipment, numbers of men, and the availability of additional help
   Estimating the amount of water needed
   Evaluating the adequacy of the water supply

5. *Checking progress of the fire:*
    Watching for drastic changes
    Considering any new information about building, occupants, contents

Important Points to Receive Special Emphasis

omit

Types and Subject Matter of Visuals to Be Included

omit

Sources of Information

lecture by company officer, observation of company officers at fires, training manual

HOW A COMPANY OFFICER SIZES UP A FIRE

A key person in a fire company is the fire company officer. When an alarm sounds, the officer must note several factors immediately and others at the site of the fire in order to direct other fire fighters in putting out the fire. Sizing up a fire involves these five factors: determining the location of the fire, assessing the danger to life and materials, judging what the fire can do, reviewing available resources, and checking the progress of the fire. The entire process of fighting a fire is largely determined by the decisions of the company officer.

Determining the Location of the Fire

Upon the sound of the alarm, the officer notices the alarm code to determine the location of the fire. If the alarm box is tied directly into the fire department's alarm system, the exact location of the fire may be determined. If not, only the area of the fire can be determined. If someone calls in to report a fire, the caller may or may not identify the exact location. Generally, the fire company responding must find the fire once the officer's unit reaches the general area.

Assessing the Danger to Life and Materials

Next the officer looks at the clock to determine whether danger to life is high, moderate, or low. For example, an early morning fire in a hotel or a hospital or any other place where people are sleeping is a high danger to life. The time also helps the officer to anticipate traffic problems and to decide whether police are needed to clear the streets or if the drivers of mobile fire fighting units should take an alternate route to reach the scene of the fire more quickly. Another time factor considered is the season of the year. If it is Christmas season and the fire is in a large department store, there will be more people in the store and extra-large stocks of merchandise.

Judging What the Fire Can Do

As the fire unit moves toward the fire, the officer consults the company water resources map to focus clearly in mind the location of hydrants or other water resources in the area. Radio communication may reveal the exact building on fire; more than likely, however, the exact building will not be identified until the unit reaches the scene of the fire.

Upon reaching the site of the fire, the officer notices the building and its surroundings: facts about smoke—the amount, its color, its location; visible flame, if any; facts about people involved, if any—at the windows, on the roof and the fire escapes, or in the streets. To assess hazards, the officer determines where within the building the fire is burning; notices the type of construction, the age, and the structural features of the building; and identifies any special hazards, such as explosive materials. An old building with open stairways and with wood used for much of its structure will burn rapidly.

Reviewing Available Resources

Next the officer reviews the resources available to handle the fire, noticing at least three factors. The first factor is the kinds of companies responding to the alarm (pump company, ladder company, and so on), the types and amount of equipment, the number of fire fighters, and the availability of additional help, if the severity of the fire indicates a need for such help. The second factor is the estimated amount of water needed to extinguish the fire. The third factor is the adequacy of the water supply.

Checking the Progress of the Fire
   Identifying resources available enables the officer to plan and direct tactics to fight the fire. During the course of the fire, the officer watches for any drastic changes. Any new facts about the building, its occupants, or its contents, whether observed by the officer or reported by another person, may mean new tactics and operations.

Conclusion
   The ability of a fire company officer to size up a fire determines fire-fighting effectiveness. Therefore, a high degree of skill in sizing up a situation would surely be a desirable trait for a company officer. Decisions may mean great economic loss or small economic loss, a building burned to the ground or a building partially burned but restorable, or, more significantly, life or death.

Oral Presentation and Visuals

For detailed discussions of oral presentation and visuals, see Chapter 10, Oral Communication, and Chapter 11, Visuals.

General Principles in Describing a Process

1. *The purpose of a description of a process is for audience understanding, not audience action.*
2. *The intended audience influences the kind and extent of details included and the manner in which they are presented.* Take into consideration the audience's degree of knowledge and understanding of the subject, and use a language level that will make the process description clear.
3. *Audiences may be grouped into two broad categories: general audiences and specialized audiences.*
4. *Accuracy and completeness are essential.* The process description should be correct in its information, and it should adequately cover all necessary aspects of the process.
5. *Visuals can enhance a process description.* Whenever a process description can be made clearer by such visuals as flowcharts, diagrams, drawings, real objects, demonstrations, and the like, include them.

Procedure for Describing a Process

FORM

In describing a process you may arrange the steps or stages of the process in a numbered list or in paragraphs. Remember, the purpose of the description is to make the reader understand *what happens* during a process. The usual way to describe what happens is to use the third person, present tense, active or passive voice.

EXAMPLES

An *officer* (third person) *fingerprints* (present tense, active voice) a suspect by . . .
Glass (third person) *is made* (present tense, passive voice) by . . .

Or, the description might use *-ing* verb forms.

EXAMPLES

> *Peeling* (*-ing* verb form) the thin layers . . .
> The final stage is *removing* (*-ing* verb form) the seeds . . .

Remember to be consistent with person, tense, and voice in a single presentation. For example, if you select third person, present tense, active voice for the major steps or stages, use the third person, present tense, active voice for all major steps or stages throughout the presentation. Do not needlessly shift to some other person, tense, or voice. (See pages 657–658, 697–698, 698–699, 701.)

CONTENT

When organizing a description of a process, divide the description into two or three parts, as shown below.

  I. The identification of the subject may be brief, perhaps only one or two sentences, depending on the complexity of the process.
- A. State the process to be explained.
- B. Identify or define the process.
- C. If applicable, give the purpose and significance of the process.
- D. Briefly list the main steps or stages of the process, preferably in one sentence.

 II. The development of the steps is the main part of the presentation and thus will be the lengthiest section.
- A. The guiding statement is a list of the main steps given in the introduction.
- B. Take up each step in turn, developing it fully with sufficient detail.
- C. Subdivide major steps, as needed.
- D. Insert headings for at least the main steps or stages.
- E. Use visuals whenever they will be helpful.

III. The closing is determined largely by the purpose of the presentation.
- A. If the purpose is simply to inform the audience of the specific procedure, the closing may be
  1. The completion of the discussion of the last step.
  2. A summary of the main steps.
  3. A comment on the significance of the process.
  4. Mention of other methods by which the process is performed.
- B. If the presentation serves a specific purpose, such as evaluation of economy or practicality, the closing may be a recommendation.

LENGTH OF PRESENTATION

The length of a presentation describing a process is determined by the complexity of the process, the degree of knowledge of the audience, and the purpose of the presentation. In a simple process, such as how a cat laps milk or how a stapler works, the steps may be adequately developed in a single paragraph, with perhaps a minimum of two or three sentences for each step. In a more complex process, the steps may be listed and numbered, and the description of each step may require a paragraph or more. The closing is usually brief.

Application 1   Describing a Process

Choose one of the following topics or another, similar topic of interest to you for describing a process for a general audience. Consult any sources needed.

    a. Fill in the Plan Sheet on pages 55–56.
    b. Write a preliminary draft.
    c. Revise. See the inside back cover.
    d. Write the final draft.

Processes carried out by people. How:

1. A synthetic heart is transplanted
2. A diamond is cut and polished
3. Ceramic tile, bricks, tires, sugar (or any other material) is (are) made
4. A person becomes a "star"
5. Community colleges and technical institutes are helping to alleviate the skilled labor shortage
6. A site is chosen for a business or industry
7. Flood prevention helps to eliminate soil erosion
8. Gold, silver, or coal is mined
9. An industrial plant works with community leaders
10. Penicillin (or any other "miracle" drug) was developed
11. An alcoholic beverage is produced
12. A buyer selects merchandise for a retail outlet
13. A television repairperson troubleshoots a television set
14. A college cafeteria dietician plans meals
15. A topic of your choosing

Application 2   Describing a Process

Give an oral process description of one of the topics in Application 1 above. Ask your classmates to evaluate your speech by filling in an Evaluation of Oral Presentations from pages 527–535.

Application 3   Describing a Process

Choose one of the following topics or another, similar topic for writing a process description for a specialized audience. Consult any sources needed.

    a. Fill in the Plan Sheet on pages 57–58.
    b. Write a preliminary draft.
    c. Revise. See the inside back cover.
    d. Write the final draft.

Processes carried out by machines. How:

1. A telephone answering service works
2. A space satellite works

3. A tape recorder works
4. A Xerox (or similar machine) reproduces copies
5. A computer copies a disk
6. An automatic icemaker works
7. A jet (diesel or gas turbine) engine works
8. Air brakes work
9. A mechanical cotton picker (or any other piece of mechanical farm or industrial machinery) operates
10. A lathe (or any other motorized piece of equipment in a shop) works
11. An autoclave works
12. A printing press prints material
13. An electronic calculator displays numbers
14. A dental unit works
15. A topic of your choosing

Application 4   Describing a Process

Give an oral process description of one of the topics in Application 3 above. Ask your classmates to evaluate your speech by filling in an Evaluation of Oral Presentations from pages 527–535.

Application 5   Describing a Process

Choose one of the following topics or another, similar topic for writing a process description. Consult any sources needed.

a. Fill in the Plan Sheet on pages 59–60.
b. Write a preliminary draft.
c. Revise. See the inside back cover.
d. Write the final draft.

Processes carried out by nature. How:

1. The human eye works
2. Oxidation occurs
3. Tornadoes are formed
4. Foods spoil
5. Aging affects the body
6. Gold, silver, granite, oil, peat (or any other substance) is formed
7. An acorn becomes an oak tree
8. Infants change during the first year
9. Microorganisms produce disease
10. Wood becomes petrified
11. A tadpole becomes a frog
12. Freshwater fish spawn
13. Sound is transmitted
14. An amoeba reproduces
15. A topic of your choosing

Application 6   Describing a Process

Give an oral process description of one of the topics in Application 5 above. Ask your classmates to evaluate your speech by filling in an Evaluation of Oral Presentations from pages 527–535.

USING PART II: SELECTED READINGS

Application 7   Describing a Process

Read the article "Videotex: Ushering in the Electronic Household," by John Tydeman, pages 574–582. In what ways are descriptions of processes used in the article? As directed by your instructor, answer the questions following the article.

Application 8   Describing a Process

Read "The World of Crumbling Plastics," by Stephen Budiansky, pages 622–623. Identify the parts of the article that describe process.

Application 9   Describing a Process

Read "Starch-Based Blown Films," by Felix Otey and others, pages 624–631. Under "Suggestions for Response and Reaction," page 631, do Number 2.

PLAN SHEET
FOR DESCRIBING A PROCESS

Analysis of Situation Requiring Description of a Process

What is the process to be described?

For whom is the description intended?

How will the description be used?

Will the description be written or oral?

Importance or Significance of the Process

Major Steps and Description of Each Step

(List the major stages or steps as -ing verb forms or in third person, present tense, active or passive voice.)

Important Points to Receive Special Emphasis

Types and Subject Matter of Visuals to Be Included

Sources of Information

PLAN SHEET
FOR DESCRIBING A PROCESS

Analysis of Situation Requiring Description of a Process

What is the process to be described?

For whom is the description intended?

How will the description be used?

Will the description be written or oral?

Importance or Significance of the Process

Major Steps and Description of Each Step

(List the major stages or steps as *-ing* verb forms or in third person, present tense, active or passive voice.)

Important Points to Receive Special Emphasis

Types and Subject Matter of Visuals to Be Included

Sources of Information

PLAN SHEET
FOR DESCRIBING A PROCESS

Analysis of Situation Requiring Description of a Process

What is the process to be described?

For whom is the description intended?

How will the description be used?

Will the description be written or oral?

Importance or Significance of the Process

Major Steps and Description of Each Step

(List the major stages or steps as -ing verb forms or in third person, present tense, active or passive voice.)

Important Points to Receive Special Emphasis

Types and Subject Matter of Visuals to Be Included

Sources of Information

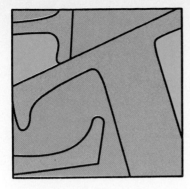

Chapter 2
Description of a Mechanism: Explaining How Something Works

Objectives

Upon completing this chapter, the student should be able to:

- Define mechanism
- Explain the difference between a general description and a specific description
- List and explain the three frames of reference (points of view) from which a mechanism can be described
- Use accurate and precise terminology in describing a mechanism
- Use visuals in a description of a mechanism
- Give a general description of a mechanism
- Give a specific description of a mechanism
- Describe orally or in writing a mechanism at rest and a mechanism in operation

Introduction

A mechanism is defined broadly as any object or system that has a functional part or parts. Thus, a mechanism is any item that performs a particular function. Items regarded as mechanisms are as diverse as a fingernail file, a diesel engine, a computer, a ballpoint pen, or a lawn mower. Another example of a mechanism is a heart, which receives and distributes blood through dilation and contraction. The human body could also be defined as a mechanism. And systems like the universe or a city, which are composed of parts that work together, could also be described as mechanisms. Most often, perhaps, the term *mechanism* suggests tools, instruments, and machines.

A technician constantly works with mechanisms and always needs to understand them: what they do, what they look like, what parts they have, and how these parts work together. At times an employee—when writing bid specifications, a memorandum for a repairperson, or purchase requests; when demonstrating a new piece of equipment; when learning how to perform heart catheterization or kidney dialysis; or when planning changes in a city department's activities—may need to describe a mechanism or a part of it, using a written or an oral presentation.

In a description of a mechanism, the mechanism may be described at rest or in operation or both, depending on the purpose of the description.

A description of a mechanism may be brief or it may be lengthy, depending upon audience and purpose. It may be a general description or a specific description.

General Description

General description focuses on describing a group or class of mechanisms, or on describing one mechanism as representative of the group or class. Examples are a word processor, a camera, a book, a cat, a pogo stick, or a rose.

The general description identifies and explains aspects *usually* associated with the mechanism. These aspects may include what the mechanism can do or can be used for, what it looks like, what its parts (components) are, what each part looks like, what each part does, and how these parts interact.

Logically there are three frames of reference, or points of view, from which a mechanism can be described generally: its function, its physical characteristics, and its parts.

Function. A mechanism is created to perform a particular function or task. The typewriter, for example, produces characters resembling printed ones, as a substitute for handwriting. The wedging board eliminates bubbles of air in clay. The fire hydrant provides a source of water for fire fighting. The hypodermic needle and syringe are used to inject medications under the skin. The microscope makes a very small object, such as a microorganism, appear larger so that it is clearly visible to the human eye. An automobile serves as a means of transportation. The kidney separates water and the waste products of metabolism from the blood. An elevator transports people, equipment, and goods vertically between floors.

Thus, the key element in a general description of a mechanism is an explanation of its function, that is, the answer to: What is it used for? This question automatically raises other questions: When is this mechanism used? By whom? How? These questions need answers if the general description is to be adequate.

Physical Characteristics. A general description points out the physical characteristics of the mechanism. The purpose is to help the audience "see," or visualize, the object, to give an overall impression of the appearance of the mechanism. Physical characteristics to consider include size, shape, weight, material, color, and texture.

Often, comparison with a familiar object is helpful too. The dividers, described on pages 67–68, are compared to a printed capital letter "A." Magnetic tape, described on pages 68–70, is compared to tape for a tape recorder and to movie film on a reel.

Sometimes a drawing of the parts of the mechanism is even more helpful. Notice the use of drawings with the descriptions, pages 67, 69, 70, 72, 75, 79, and 82; notice the use of cutaway diagrams on pages 69, 79, and 82.

Parts. A third frame of reference in describing a mechanism is its parts, or construction. The mechanism is divided into its parts, the purpose of each part is given, and the way that the parts fit together is explained. A typewriter, for example, would be described as having four major parts: frame, keyboard, ribbon, and carriage. When a key is struck against the ribbon, a character is printed on the paper in the carriage. In the part-by-part description, the frame, keyboard, ribbon, and carriage would each be described in turn. In essence, each of these parts becomes a new mechanism and is subsequently described according to its function, physical characteristics, and parts.

Identifying the parts is similar to partitioning (see pages 149–162), and explaining how they work together is similar to explaining a process (see pages 41–51).

Accurate Terminology

In describing a mechanism, use accurate and precise terminology. It is easier, of course, to use terms like *thing, good, large, narrow, tall,* and so forth. But your writing will be more effective if you use precise, specific words (see pages 9–10). Take time to think of or look up and use precise words.

Rather than "Magnetic tape is narrow" (not precise), use "Magnetic tape is ½ inch wide." Use "The chart shows manufactured goods representing 44 percent of the goods produced in Manitoba," not "The chart shows manufactured goods representing a large [not precise] percentage of goods produced in Manitoba." Or "Breaker points in use more than 10,000 miles need cleaning or replacing," not "Breaker points in use more than a few thousand miles [not precise] need cleaning or replacing."

Be careful to show variances. For example, rather than "Dividers may vary in size," use "Dividers may vary in size from 2 to 12 inches." Also be careful to refer to sizes by the standard method of measurement. The size of a bicycle, for instance, is referred to by the diameter of the wheels (such as 24-inch or 26-inch), not by the length of the frame or the minimum height of the saddle to the ground.

If you are not certain about precise words, many sources are available. Such sources include general dictionaries; general encyclopedias; textbooks; specialized dictionaries, handbooks, and encyclopedias; knowledgeable people; advertisements; mail order catalogs; and instruction manuals.

Purpose and Audience in Description

As always in planning a written or an oral presentation, you must identify the purpose of the presentation and the needs of the audience. For instance, descriptions of a hi-fi speaker would differ in emphasis and in detail, depending on whether the descriptions were given so a hi-fi buff could construct a similar speaker, or so a prospective buyer could compare the speaker to a similar one, or so the general public could simply understand what a hi-fi speaker is.

Sample General Descriptions

General descriptions are illustrated in the following two student-written descriptions. The first description includes the Plan Sheet that the student filled in before writing. Marginal notes have been included to help you analyze the descriptions.

PLAN SHEET
FOR A GENERAL DESCRIPTION OF A MECHANISM

Analysis of Situation Requiring a General Description of a Mechanism

What is the general mechanism to be described?
dividers

For whom is the description intended?
an apprentice drafter

How will the description be used?
to understand dividers as a mechanism used by drafters

Will the description be written or oral?
written

Analysis of the Mechanism

Definition, identification, and/or special features:
a tool commonly used by drafters, machinists, welders, and navigators to transfer measurements, to scribe circles, and to compare distances

Function, use, or purpose:
included in definition above

Who uses:
included in definition

When:
omit

Where:
omit

Physical characteristics:
  Size:
  2 to 12 inches
  3 to 18 inches
  (point range)

  Shape:
  printed capital letter "A" with a circle on top

  Weight:
  1 to 10 ounces

Material:
hardened steel

Other:
two adjustable legs or points joined at the top by a hinge and a circular spring

Major parts and a description of each:

Parts	Function	Physical characteristics
legs	*mark the points of measurement*	*thin rectangles of hardened steel; one end of each leg is sharpened to a point*
hinge	*allows the legs to pivot*	*small steel spool*
spring	*holds ends of legs together*	*flat piece of steel shaped into a circle*
adjustment screw	*makes it possible to change distance between the two legs*	*threaded rod with a nut, a round, knurled piece of metal with a center hole to fit onto the rod. The nut is threaded to match the rod. A small knob on the end of the rod stops the nut.*
handle	*for holding the divider*	*metal rod with knurled finger grip*

Operation Showing Parts Working Together

The handle is grasped between the thumb and first finger of one hand. Next, one point is placed on a scale. The other point is adjusted to the desired measurement by turning the adjustment screw.

Variations or Special Features

included under physical characteristics

Types and Subject Matter of Visuals to Be Included

figure showing hand holding dividers

Sources of Information

textbook, demonstration by instructor

A GENERAL DESCRIPTION OF DIVIDERS

Dividers defined; definition includes function

Control sentence naming topics to be covered

Physical characteristics: shape, material made from,

Dividers are a tool commonly used by drafters, machinists, welders, and navigators to transfer measurements, to scribe circles, and to compare distances. Dividers can be described by physical characteristics, parts, and use.

Physical Characteristics

The appearance of dividers is like a printed capital letter "A" with a small circle on top. Made of hardened steel, dividers have two adjustable legs or points joined at the top by a hinge and a circular spring. (See Figure 1.)

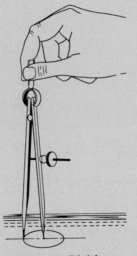

Figure 1. Dividers.

size,

weight
Control sentence renaming topics to be covered

Parts listed

Parts described by physical characteristics and function of each

Dividers may vary in size from 2 to 12 inches. They are classified, however, according to the maximum opening between the points, which ranges from 3 to 18 inches. Depending on the size, the weight may be 1 to 10 ounces. Although the dividers vary in size and weight, they have the same parts and are used in the same way.

Parts

The five parts of dividers are the legs, the hinge, the circular spring, the adjustment screw, and the handle.

Legs. The two legs, thin rectangles of hardened steel, are used to mark the points of measurement; the lower end of each leg is tapered to a point. At the ends opposite the points, the two legs are joined by a hinge and a spring.

Hinge. The hinge is a small steel spool that allows the legs to pivot.

Spring. The spring is a flat piece of steel shaped into a circle to hold the ends of the legs together.

Adjustment Screw. The adjustment screw, which fits below the hinge, makes it possible to change the distance between the two legs. It is a threaded rod with one end anchored to one leg of the dividers. On the opposite end of the rod is a nut, a round, knurled piece of metal with a center hole threaded to match the rod.

A small knob on the end of the rod stops the nut at the maximum opening of the dividers. Turning the nut moves the other leg back and forth along the threads from 0 to the maximum opening. The movable leg can be anchored at any point between 0 and the maximum opening by the nut.

Handle. Centered on top of the spring is a handle, a metal rod with a knurled finger grip for holding the dividers.

Using Dividers

Description of dividers "in use." A miniprocess explanation

Dividers are used in the following way. First, the handle is grasped between the thumb and first finger of one hand. Next, one point is placed on a scale. The other point is adjusted to the desired measurement by turning the adjustment screw. The measurement can then be transferred to another drawing, a circle can be drawn, or distances can be compared.

Note how the organization and headings in this second general description differ from those in the preceding general description of dividers.

DESCRIPTION OF MAGNETIC TAPE

OVERVIEW

Magnetic tape defined

Magnetic tape is a data representation medium that looks very much like tape recorder tape. It is used in a computer system for both input and output information.

Use of magnetic tape

To use magnetic tape requires a tape drive, a machine that reads and records in the magnetized areas. The tape drive also makes possible rewinding, backspacing, and skipping ahead.

Control sentence naming topics to be covered

Magnetic tape can be further described by explaining its standard characteristics and its parts.

STANDARD CHARACTERISTICS

Magnetic tape can be described by standard widths of the tape and typical lengths of the reels of tape and by the presence of tape markers.

Width

Most often, magnetic tape is ½ inch wide, although it may be ¾ inch or 1 inch wide. The tape is wound on a reel very similarly to the way movie film is wound on a reel (as shown in Figure 1).

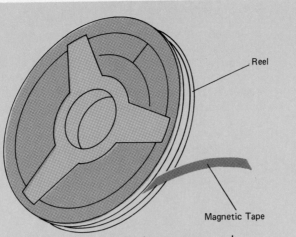

Figure 1. Magnetic tape on a reel.

Length

      The length of a reel of tape may range from 50 to 3600 feet, depending on the purpose of the tape. The most common length is 2400 feet and requires a 10½-inch-diameter reel.

Tape markers

      At the beginning and at the end of each reel of tape is a small strip of reflecting material that "marks" for the computer where usable information begins and when the end of the tape is near (to keep the tape from running off the reel). The strip at the beginning is appropriately called the load-pointer marker; the strip at the end, the end-of-reel marker.

PARTS

Two parts listed

      Magnetic tape has two basic parts: the base and the magnetic coating. See Figure 2.

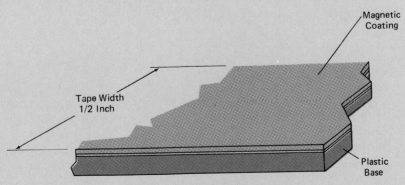

Figure 2. Parts of magnetic tape.

First part described
by material made
from and use

      The base comprises the major portion of the tape. Made of a type of flexible plastic, it provides a place on which the magnetic coating can be applied.

Second part described by material made from and use

Composed of ferric oxide, the magnetic coating is the most important part of the tape. On its surface in magnetized areas coded data representing numbers, letters, and special characters can be recorded.

Magnetic tape may have a seven-track or a nine-track coding area. A track is simply a horizontal row on the tape. The code is the computer language; short vertical lines are the coded language on the magnetized areas.

The magnetized areas containing data are called blocks. Blocks can carry tremendous amounts of information in a very small area. For example, more than 1000 letters and numbers can be recorded on 1 inch of tape.

Each block is separated by a space called an interblock gap (IBG). The interlock gaps allow starting and stopping and speeding and slowing as the tape is used. See Figure 3.

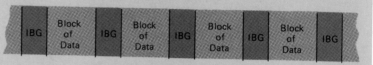

Figure 3. Tape showing data blocks and IBGs.

Specific Description

In contrast with the general description, the specific description of a mechanism emphasizes particular characteristics, aspects, qualities, or features. It focuses on particular characteristics of an identified model, style, or brand of mechanism. It tries to show what sets this mechanism apart from all similar mechanisms.

For example, a specific description of Hewlett-Packard (brand) ColorPro 7440A (model) might include facts about special features and unique construction features. See the sample description, pages 75–77.

USES OF SPECIFIC DESCRIPTION

Specific description is used in bid specifications, in advertisements for products, or in any description for an audience that wants to know particular features of a mechanism.

One example of specific description is bid specifications. Bid specifications name a product and describe the unique characteristics, qualities, or features the product must have to meet a need. A company or business that sells the product looks at the specifications, determines if the company can offer the product as it is described, and then offers to sell it at a stated price.

While there is no standard form for recording or presenting bid specifications, they always identify the required qualities clearly and precisely.

The following is an example of bid specifications for a particular brand name of lawn mower to be used in cutting a golf course. (See also page 81.)

Bid No: 519
Opening Date: April 23, 1988
Name: 7 Gang Blitzer Mower

Item No.	Quantity	Description
1	1	7 unit Gang Blitzer for cutting golf course roughs, 30" or wider cut per unit
		Reel to be a five-blade, 10" diameter unit with pneumatic or semipneumatic tires. Wheel shaft to rotate on tapered roller bearings. Internal gears to rotate on ball or roller bearings (rotation of gear drive train on brass/bronze bushing not acceptable). Reels to be 10" diameter and with hardened steel blades
		Separate drive train required for wheel of the unit
		Quick disabling of reel drive for transportation required
		Reversible bed knife preferred but not mandatory. (Bed knife must be hardened steel)
		Frames readily attachable or detachable to expand or reduce number of reel units in use
		Heavy duty frames equal to or better than Jacobsen Blitzer Frames or Toro "Aero-Frames"

The following is an example of a bid sheet without a specified brand name.

Bid No: 771
Opening Date: October 10, 1988
Bid Name: Insulation of Ceiling—Raymond Campus Cafeteria   Page _____

Item No.	Quantity	Description	Unit Price	Total Price
		The Hinds Junior College District is now accepting bids in the District Purchasing Office, Raymond Campus for the insulating of the ceiling area of Raymond Campus Cafeteria Building.		
		Specifications:		
		Area—13,400 sq. ft.		
		Material—Fiber Glass Blowing Wool installed with manufacturer's recommendations		
		R-value—R-38		
		Quantity—540 bags at a rate of 40 bags for each 1000 sq. ft.		
		Thickness—Minimum thickness 7½ in.		
		Job-site visit—Mandatory for all contractors or subcontractors. Contact Doug Fowler at 857-3370.		
		Completion requirements—15 working days from the date of the purchase order		
		Total Price:		

Another use of specific description is in magazines and journals in sections advertising products and in catalogs. The following description appeared in a section on new products in an issue of *The Office*.

Marvel Metal Products Co., 3843 West 43rd St., Chicago, Ill. 60632, introduces its Model 11-3030 utility stand for copiers, duplicators, audiovisual, and other office equipment. Stand is 30″ wide, 20″ deep, 30″ high, with top and shelf of skid-resistant textured steel. Stud-welded 1″ sq. steel legs and welded shelf corners provide rigid nonsway strength. Hooded ball casters, two with locks, allow mobile or stationary use. Stand can support up to 300 lbs.

Specific descriptions of mechanisms also appear in all types of merchandise catalogs. The description of hedge trimmers appeared in a Sears catalog.

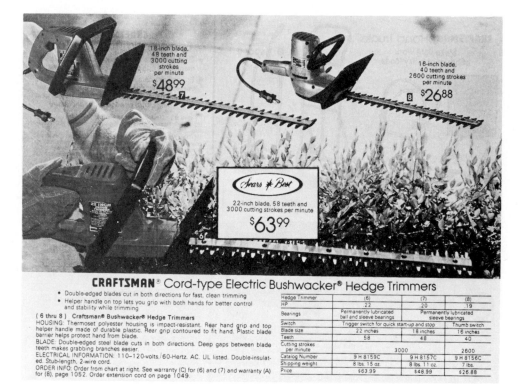

CRAFTSMAN® Cord-type Electric Bushwacker® Hedge Trimmers

- Double-edged blades cut in both directions for fast, clean trimming
- Helper handle on top lets you grip with both hands for better control and stability while trimming

(6 thru 8) **Craftsman® Bushwacker® Hedge Trimmers**

HOUSING: Thermoset polyester housing is impact-resistant. Rear hand grip and top helper handle made of durable plastic. Rear grip contoured to fit hand. Plastic blade barrier helps protect hand from blade.
BLADE: Double-edged steel blade cuts in both directions. Deep gaps between blade teeth makes grabbing branches easier.
ELECTRICAL INFORMATION: 110–120-volts, 60-Hertz. AC. UL listed. Double-insulated. Stub-length, 2-wire cord.
ORDER INFO: Order from chart at right. See warranty (C) for (6) and (7) and warranty (A) for (8), page 1052. Order extension cord on page 1049.

Hedge Trimmer	(6)	(7)	(8)
HP	.22	.20	.19
Bearings	Permanently lubricated ball and sleeve bearings		Permanently lubricated sleeve bearings
Switch	Trigger switch for quick start-up and stop		Thumb switch
Blade size	22 inches	18 inches	16 inches
Teeth	58	48	40
Cutting strokes per minute	3000		2600
Catalog Number	9 H 8159C	9 H 8157C	9 H 8156C
Shipping weight	8 lbs. 15 oz.	8 lbs. 11 oz.	7 lbs.
Price	$63.99	$48.99	$26.88

Following is a specific description of a mechanism prepared by a student. The description includes the Plan Sheet the speaker filled in; marginal notes have been added to outline the development of the description.

The description of the ColorPro plotter, to be given as an oral presentation, emphasizes the features that make the plotter desirable to use. The function, physical characteristics, and parts of this specific plotter are very similar to those of any other plotter; therefore, treatment of the *usual* aspects is kept to a minimum, and *unusual* aspects are emphasized.

Try reading the description aloud to approximate an oral presentation.

PLAN SHEET
FOR A SPECIFIC DESCRIPTION OF A MECHANISM

Analysis of Situation Requiring a Specific Description of a Mechanism

What is the specific mechanism to be described?
Hewlett-Packard ColorPro plotter 7440A

For whom is the description intended?
students majoring in Commercial Design and Advertising

How will the description be used?
to promote interest in a plotter used by professionals

Will the description be written or oral?
oral

Analysis of the Mechanism

Who uses:
businesspeople, writers, and presenters of reports with numerical data

When:
in making charts and graphs and transparencies of them

Where:
office, business meetings, reports, various publications

Physical characteristics:
    Size:
    4.9 in. high, 18.1 in. wide, 12.1 in. deep

    Shape:
    omit

    Weight:
    12 lb.

    Material:
    omit

    Other:
    cost — $1295

Special features to be emphasized:
1. *High-quality output*
   resolution of 0.025 mm (0.001 in.) for producing smooth circles, straight diagonal lines, crisp characters
   repeatability of 0.1 mm (0.004 in.) for closed circles and square corners
2. *Eight-pen carousel*
   spectrum of colors
   two widths of pens
3. *Programming capabilities*
   HP Graphics Language instructions
4. *Compatibility*
   with virtually all personal computers
   with a variety of HP and non-HP minicomputers and mainframes
5. *ROM cartridge slot*
   Graphics Enhancement cartridge: an accessory that extends such possibilities as additional HP-GL instructions (for drawing arcs, circles, polygons) and a buffer
6. *Graphics software*
   programs for personal computers
   programs for large systems

Variations available:
Cartridge Enhanced ColorPro plotter

Types and Subject Matter of Visuals to Be Included

picture of the plotter, presented on a transparency for overhead projector

example of a column chart in color, produced by the plotter

a poster listing each of the six distinguishing features of the plotter

Sources of Information

brochures from Hewlett-Packard

a Hewlett-Packard sales representative

ORAL PRESENTATION

DESCRIPTION:
THE HEWLETT-PACKARD COLORPRO PLOTTER 7440A

Identification of
mechanism

If you are responsible for making drawings and charts for professional presentations or reports, you should be aware of the capabilities of the Hewlett-Packard ColorPro plotter 7440A. This is what the plotter looks like.

Using an overhead projector, show a transparency of a picture of the plotter.

Figure 1. Picture of the plotter (to be presented as a transparency) (1987 Hewlett-Packard catalog).

Definition of mechanism

The HP ColorPro plotter is an eight-pen, A4/A-paper size plotter designed to provide quality color graphics for business and scientific applications.

Here is an example of a column chart—in color, of course—produced on the plotter.

Hold up the column chart, making sure that each person in the audience can see it.

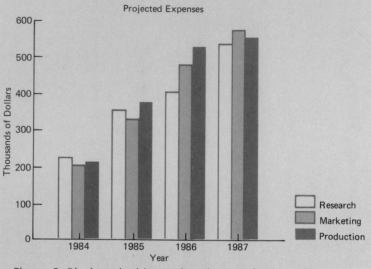

Figure 2. Black and white, reduced copy of color column chart (Hewlett-Packard brochure).

The HP ColorPro plotter makes impressive, colorful presentation graphics on regular and glossy paper and on overhead transparency film. This plotter does an exceptionally good job with lines. Curves and circles are smooth; diagonal lines are straight; letters are crisp.

The HP ColorPro Plotter is distinguished by these six features: high-quality output, an eight-pen carousel, programming capabilities, compatibility, ROM cartridge slot, and graphics software.

Show poster listing each feature. As each feature is explained in detail, point to its listing on the poster.

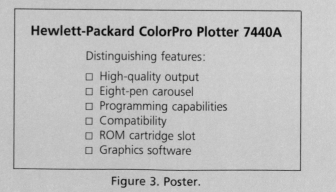

Figure 3. Poster.

First feature listed

The first distinguishing feature is high-quality output.

First feature
explained

Resolution and repeatability are the two major components of line quality. The plotter has resolution of 0.025 millimeter (0.001 inch) for producing smooth circles, straight diagonal lines, and crisp characters. Repeatability refers to the plotter's capability of returning to a given starting point. The plotter has a repeatability of 0.1 millimeter (0.004 inch) for closed circles and square corners.

The plotter can draw up to 1000 points in a 1-inch line. This precision helps ensure that circles are closed and that bar charts and pie charts are properly aligned.

Second feature listed

The second distinguishing feature is the eight-pen carousel.

Second feature
explained

When the eight-pen carousel is fully loaded, there is access to a spectrum of colors in two different widths—thick pens for headings, thin pens for details. The pens are capped automatically when not in use, to prolong pen life.

Third feature listed

The third distinguishing feature is the programming capabilities.

Third feature
explained

The Hewlett-Packard Graphics Language (HP-GL) instructions provide many capabilities. Their use can create many personalized graphics programs. HP-GL is a simple but powerful program that controls such plotting functions as pen movement, labeling, character set selection, and axis placement.

Fourth feature listed;
reminder of previously discussed
features

Now let us consider a fourth feature, compatibility. This fourth feature complements the previously explained features: high-quality output, an eight-pen carousel, and programming capabilities.

Fourth feature
explained

The HP plotter is compatible with many other computers. It works with virtually all personal computers. Such personal computers range from the Apple IIe to the IBM PC, the Company Deskpro to the HP Touchscreen, the AT&T PC 6300 to the HP Vector PC. It can also be connected to a variety of HP and non-HP minicomputers or mainframes.

Fifth feature listed

A fifth feature is the handy ROM cartridge slot.

Fifth feature
explained

To keep up with constantly expanding graphics needs (such as emerging graphics standards), this plotter has an ROM (reader only memory) cartridge slot. The Graphics Enhancement cartridge is an accessory that extends possibilities, such as additional HP-GL instructions to draw arcs, circles, and polygons, and a large RS-232-C buffer (1024 bytes).

Sixth feature listed

The last of the six features of the Hewlett-Packard ColorPro Plotter is the availability of graphics software.

Sixth feature
discussed

Graphics software is available for HP as well as non-HP computer systems. Among the graphics software packages for personal computers are these: Lotus 1-2-3, Energraphics, Key Chart, MacPlots II, SlideWrite, and Symphony.

Also available are graphics software packages for large systems. These include DISSPLA, TELL-A-GRAPH, DSG/3000, and Chart-Master.

Quality graphics help convey messages with immediacy and with impact. Graphics help the audience see relationships, discern trends, and draw conclusions.

Closing

The cost of the ColorPro plotter—$1295—is very reasonable, when one considers the possibilities of the income it can influence.

Graphics—well done—make a positive difference. The Hewlett-Packard ColorPro plotter 7440A can facilitate that positive difference.

Comparing General and Specific Descriptions

The general description of a mechanism emphasizes what the mechanism does or can be used for and what it looks like. It includes a description of the parts, what each part looks like, and what each does. Then the general description describes the parts as they relate to one another, that is, the mechanism in operation. This section of the description describes a process (see pages 41–51).

The general description explains a type of mechanism or a mechanism representing a class. It describes what all types of a stated mechanism or all mechanisms in a class have in common, even though variations may exist for different brands and different models.

A general description of a mechanism is the kind of description given in an encyclopedia. The following description of a microscope appears in *The World Book Encyclopedia.**

MICROSCOPE is an instrument that magnifies extremely small objects so they can be seen easily. It produces an image much larger than the original object. Scientists use the term *specimen* for any object studied with a microscope.

The microscope ranks as one of the most important tools of science. With it, researchers first saw the tiny germs that cause disease. The microscope reveals an entire world of organisms too small to be seen by the unaided eye. Physicians and other scientists use microscopes to examine such specimens as bacteria and blood cells. Biology students use microscopes to learn about algae, protozoa, and other one-celled plants and animals. The details of nonliving things, such as crystals in metals, can also be seen with a microscope.

There are three basic kinds of microscopes: (1) *optical*, or *light;* (2) *electron;* and (3) *ion*. This article discusses optical microscopes. For information on the other types, see the WORLD BOOK articles on ELECTRON MICROSCOPE and ION MICROSCOPE.

How a Microscope Works. An optical microscope has one or more lenses that bend the light rays shining through the specimen (see LENS). The bent light rays join and form an enlarged image of the specimen.

The simplest optical microscope is a magnifying glass (see MAGNIFYING GLASS). The best magnifying glasses can magnify an object by 10 to 20 times. A magnifying glass cannot be used to magnify an object any further because the image becomes fuzzy. Scientists use a number and the abbreviation X to indicate (1) the image of an object magnified by a certain number of times or (2) a lens that magnifies by that number of times. For example, a 10X lens magnifies an object by 10 times. The magnification of a microscope may also be expressed in units called *diameters*. A 10X magnification enlarges the image by 10 times the diameter of the object.

Greater magnification can be achieved by using a *compound* microscope. Such an instrument has two lenses, an *objective lens* and an *ocular*, or *eyepiece, lens*. The objective lens, often called simply the *objective*, produces a magnified image of the specimen, just as an ordinary magnifying glass does. The ocular lens, also called the *ocular*, then magnifies this image, producing an even larger image. Many microscopes have three standard objective lenses that magnify by 4X, 10X, and 40X. When these objective lenses are used with a 10X ocular lens, the compound microscope magnifies a specimen by 40X, 100X, or 400X. Some microscopes have *zoom* objective lenses that can smoothly increase the magnification of the specimen from 100X to 500X.

In addition to magnifying a specimen, a microscope must produce a clear image of the closely spaced parts of the object. The ability to provide such an image is called the

* Excerpted from *The World Book Encyclopedia*. © 1986 World Book, Inc.

resolving power of a microscope. The best optical microscopes cannot resolve parts of objects that are closer together than about 0.000008 of an inch (0.0002 of a millimeter). Anything smaller in the specimen—such as atoms, molecules, and viruses—cannot be seen with an optical microscope.

Parts of a Microscope. The microscopes used in most schools and colleges for teaching have three parts: (1) the *foot*, (2) the *tube*, and (3) the *body*. The foot is the base on which the instrument stands. The tube contains the lenses, and the body is the upright support that holds the tube.

The body, which is hinged to the foot so that it may be tilted, has a mirror at the lower end. The object lies on the *stage*, a platform attached above the mirror. The mirror reflects light through an opening in the stage to illuminate the object. The upper part of the body is a slide that holds the tube and permits the operator to move it up and down with a *coarse-adjustment* knob. This movement focuses the microscope. Most microscopes also have a *fine-adjustment* knob, which moves the tube a small distance for final focusing of a high-power lens.

The lower part of the tube contains the objective lens. In most microscopes, this lens is mounted on a revolving *nosepiece* that the operator can rotate to bring the desired lens into place. The upper end of the tube holds the ocular lens.

Using a Microscope. A microscope is an expensive instrument and can be damaged easily. When moving one, be sure to hold it with both hands and set it down gently on a firm surface.

Parts of a Microscope

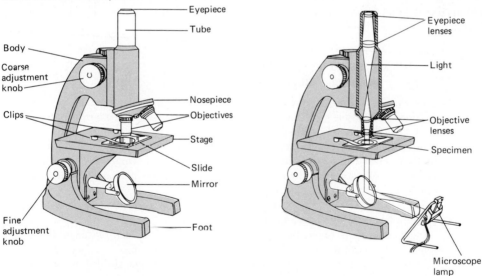

The diagram at the left shows the external parts of a microscope. A person adjusts these parts to view a specimen. The cutaway diagram at the right shows the path that light follows when passing through the specimen and then through the lenses and tubes of the microscope.

To prepare a microscope for use, turn the nosepiece so that the objective with the lowest power is in viewing position. Lower the tube and lens by turning the coarse-adjustment knob until the lens is just above the opening in the stage. Next, look through the eyepiece and adjust the mirror so a bright circle of light appears in the eyepiece.

The microscope is now ready for viewing a specimen. Most people keep both eyes open when looking into the eyepiece. They concentrate on what they see through the microscope and ignore anything seen with the other eye.

Most specimens viewed through a microscope are transparent, or have been made transparent, so that light can shine through them. Objects to be viewed are mounted on glass slides that measure 3 inches long and 1 inch wide (76 by 25 millimeters). The technique of preparing specimens for microscopic viewing is called *microtomy* (see MICROTOMY). See also the *Science Project* with this article.

To view a slide, place it on the stage with the specimen directly over the opening. Hold the slide in place with the clips on the stage. Look through the eyepiece and turn the coarse-adjustment knob to raise the lens up from the slide until the specimen comes into focus. Never lower the lens when a slide is on the stage. The lens could press against the slide, breaking both the slide and the lens.

After the specimen has been brought into focus, turn the nosepiece to an objective lens with higher power. This lens will reveal more details of the specimen. If necessary, focus the stronger objective with the fine-adjustment knob. A zoom microscope is changed to a higher power by turning a part of the zoom lens. Different parts of the specimen can be brought into view by moving the slide on the stage.

The specific description of a mechanism, on the other hand, describes the outstanding or unique features or characteristics of a stated brand or model. It emphasizes the aspects of the stated brand or model that set it apart from other brands or models of the same mechanism. The specific description below appears in a merchandise catalog from Arthur H. Thomas Company (note their omission of noncontent words to save space).

6545-H10   MICROSCOPES, Teaching
6545-H30   American Optical Series Sixty

Designed to withstand classroom handling. Advanced features include unique, spring-loaded focusing mechanism which protects slide and objective, avoids conventional rack and pinion. Equipped with in-base illuminator.

Revolving ball-bearing nosepiece is raised or lowered to effect focus, by means of coaxial coarse and fine adjustments, low-positioned for convenience. Adjustable stop (Autofocus) limits objective travel, permitting rapid focus while safeguarding objective and slide.

Stage, with clips, is 125 × 135 mm. Integral in-stage condenser with adjustable diaphragm has reference ring for centering specimen.

Inclined body is reversible for viewing from front or back of stage. Huygenian eyepiece, 10×, with pointer and measuring scale, is locked in.

Optics on both models have Americote magnesium fluoride coating on air-glass surfaces to reduce internal reflection and improve image contrast.

6545-H10 Microscope has 5-aperture disc diaphragm, double revolving nosepiece, 10× and 43× achromatic objectives, providing 100× and 430× magnification.

6545-H30 Microscope has iris diaphragm; triple revolving nosepiece; 4×, 10×, and 43× achromatic objectives, providing 40×, 100×, and 430× magnification.

With switch, cord, and plug, for 120 volts. For replacement bulb, see 6625-D30.

Still another type of specific description is illustrated in the following bid specifications for a microscope.

Bid No. 215
Opening Date: May 10, 1988
Name: Microscope, AO Series 10

Item No.	Quantity	Description	Unit Price	Total Price
	1	MICROSCOPE, AO Series 10 Microscope to permit simultaneous viewing of the same image by two persons in side-by-side position. To have dual viewing adapter binocular body, 10×, widefield eyepieces. To have illuminated green arrow, to be powered by microscope transformer, to be controlled by lever at base of adapter, and positioned anywhere in the field of view. To be equipped with ungraduated mechanical stage, Aspheric, N.A. 1.25, condenser with auxiliary swing-in lens, 2 pair 10× widefield eyepieces. SPECIFICATIONS: Viewing position—side by side Planachromatic 4×, 40×, 100×		

Mechanism at Rest and in Operation

In either a general or a specific description of a mechanism, the mechanism may be described at rest or in operation or both, depending on the purpose of the description.

A description of a mechanism at rest would perhaps mention the function and the physical appearance, but it would emphasize the parts of the mechanism, particularly their location in relation to one another. A description of a mechanism in operation would perhaps mention the function and the parts, but it would emphasize these parts working together to perform the function.

Occasionally a technician might want to emphasize only the operation of a mechanism. Such a description of a mechanism in operation is closely related to a description of a process (see pages 41–51).

The example below describes the four-cycle engine in operation.

FOUR-CYCLE ENGINE IN OPERATION

The four-cycle engine depends for its operation on a *carburetor* which provides a proper mixture of fuel and air, a *piston* contained within a cylinder, an *inlet valve* and an *outlet valve*, a *spark plug*, and a *crankshaft* to which the lower end of the piston is connected. The name of the engine derives from the four strokes through which the piston passes during engine operation.

1. On the *intake* stroke, the inlet valve is open and the outlet valve is closed. The mixture of air and fuel enters the cylinder through the inlet valve and the piston is at its lowest position.
2. During the *compression* stroke, the piston, impelled by the crankshaft, rises and compresses the fuel mixture. At this time both valves are closed.
3. As the piston nears its top dead center, the *power* stroke occurs and the fuel mixture is ignited, or "fired," by the spark plug. Both valves are still closed.

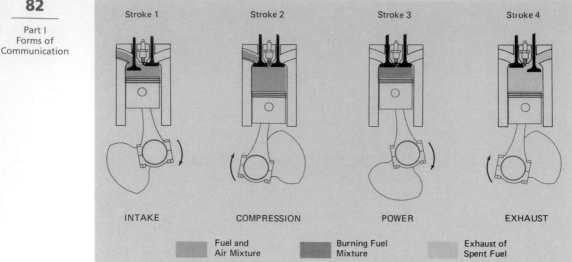

| Stroke 1 | Stroke 2 | Stroke 3 | Stroke 4 |

| INTAKE | COMPRESSION | POWER | EXHAUST |

Fuel and
Air Mixture

Burning Fuel
Mixture

Exhaust of
Spent Fuel

4. The force of the exploding fuel forces the piston downward as the inlet valve closes, the outlet valve opens and the *exhaust* stroke occurs. The spent gases are pushed out of the cylinder through the outlet valve by the rising piston.

This completes one cycle of operation and leaves the engine ready for a repetition of the four strokes. Two revolutions of the crankshaft are required for the above described operation to take place.

Oral Description

As a worker employed in such areas as sales, supervision, or training you may find occasions when you must describe a mechanism orally. This description might be given to a large group of potential buyers, to several newly hired workers, to students in a training program, or to a single individual. As a student, you may also be asked to give oral descriptions of mechanisms.

The content of an oral description might be organized following the procedures outlined within this chapter. Presenting the material, however, might require other considerations. For a detailed discussion on oral presentations, see Chapter 10, Oral Communication.

Visuals

Visuals are especially valuable in describing a mechanism. Generally in describing a mechanism, the goal is to enable the audience to see what the mechanism looks like and what its parts are and to understand how these parts work together to allow the mechanism to function.

Pictorial illustrations, such as photographs, drawings, and diagrams, that are included with a verbal description accomplish the goal more readily. The illustration above describes the four-cycle engine in operation; the illustration plus the verbal explanation provides a clear description, easily understood, of a fairly complex operation.

Photographs, drawings, and diagrams are useful in describing mechanisms. Photographs can provide a clear external or internal view; they can show the size of a mechanism; they can show color and texture. Drawings, ranging from simple, freehand sketches done with pencil, pen, crayon, or brush to engineers' or architects' minutely detailed drawings done with drafting instruments, machines, and computers, can do the same. Diagrams can outline an object or system, showing such things as its parts, its operation, its assembly. These outlines may be picture, schematic, or block diagrams. Drawings and diagrams can show the entire exterior or interior of a mechanism or only a part; they can show the shapes of parts, the relationships between parts, and the functions of parts as they work together; they can show cross sections. Or they can be exploded views that show the parts disassembled but arranged in sequence of assembly. Obviously, visuals such as these are of great value in describing mechanisms.

For a detailed discussion, see Chapter 11, Visuals.

General Principles in Describing a Mechanism

1. *The description of a mechanism may be a general description.* In a general description, emphasis is on giving an overall view of what the mechanism can do or can be used for.
2. *The three frames of reference in a general description are function, physical characteristics, and parts.* Although these three frames of reference are closely related, ordinarily they should be kept separate and in logical order.
3. *Accurate and precise terminology should be used.* Sources for such terminology are dictionaries, encyclopedias, textbooks, knowledgeable people, instruction manuals, and merchandise catalogs.
4. *The purpose of the description and the intended audience must be clear.* These two basic considerations determine the extent and the kind of details given and the manner in which they are given.
5. *The description may be a specific description of a particular mechanism.* In a specific description, emphasis is on particular characteristics or aspects of a mechanism identifiable by brand name, model number, and the like.
6. *The description may be of a mechanism at rest or in operation.* In an at-rest description, emphasis is on parts of the mechanism and their location in relationship one to another. In an in-operation description, emphasis is on the parts and the way these parts work together to perform the intended function.

Procedure for a General Description of a Mechanism

I. The identification of the mechanism is usually simple and requires only a few sentences.
   A. Define or identify the mechanism.
   B. Indicate why this description is important, if appropriate.

C. In a sentence, list the points (frames of reference) about the mechanism to be described.

II. The explanation of the function, physical characteristics, and parts is the lengthiest section of the presentation.
   A. Give the function, use, or purpose of the mechanism.
      1. If the mechanism is part of a larger whole, show the relationship between the part and the whole.
      2. If applicable, state who uses the mechanism, when, where, and why.
   B. Give the physical characteristics of the mechanism.
      1. Try to make the reader "see" the mechanism.
      2. Describe, as applicable, such physical characteristics as size, shape, weight, material, color, texture, and so on.
   C. Give the parts of the mechanism.
      1. List the major parts of the mechanism in the order in which they will be described.
      2. Identify each part.
      3. State what each part is used for—its function.
      4. Tell what each part looks like—its physical characteristics.
      5. Give the relationship of each part to the other parts.
      6. If necessary, divide the part into its parts and give their functions, physical characteristics, and parts.
   D. Use headings and visuals, when appropriate, to make meaning clear.

III. The closing, usually brief, emphasizes particular aspects of the mechanism.
   A. Show how the individual parts work together.
   B. If applicable, mention variations of the mechanism, such as optional features, other types, and other sizes.
   C. If applicable, comment on the importance or significance of the mechanism.

Procedure for a Specific Description of a Mechanism

I. The identification of the mechanism is usually simple and requires only a few sentences.
   A. Identify the mechanism by giving the brand name or model number.
   B. Tell the function or purpose of the mechanism.
   C. Tell who uses the mechanism, when it is used, and where.
   D. In a sentence, list the features or characteristics of the mechanism, or in a sentence, state that the mechanism possesses "unique features," "five specific characteristics," "the following desirable features," or some similar general phrase to introduce the main section.

II. The features or characteristics of the particular brand or model make up the main section of the description.
   A. Identify each feature or characteristic.
      1. Select information about the mechanism to set it apart from other similar mechanisms.
      2. Consider such features as size, shape, weight, material, overall appearance, available colors, cost, guarantee.

B. Describe each feature or characteristic in detail.

C. Use headings and visuals, when appropriate, to make meaning clear.

III. The closing, if included, is usually brief.

    A. The description may end after the discussion of the last feature or characteristic.

    B. The features or characteristics discussed may be summarized.

    C. Any variations of the mechanism, such as other models, different colors, other prices, or different designs, may be mentioned.

Application 1   Describing a Mechanism

Sort the sentences below into three general groups under the following three headings: Function, Physical Characteristics, and Parts. Write the numbers of the sentences under the proper headings.

1. The jaws are made of cast iron and have removable faces of hardened tool steel.

2. The machinist uses a small stationary holding device called "the machinist's bench vise" to grip the work securely when performing bench operations.

3. For a firm grip on heavy work, serrated faces are usually inserted.

4. In addition to the typical bench vise described here, there are many other varieties and sizes.

5. It is essential for holding work pieces when filing, sawing, and clipping.

6. A vise consists of a fixed jaw, a movable jaw, a screw, a nut fastened in the fixed jaw, and a handle by which the screw is turned to position the movable jaw.

7. To protect soft metal or finished surfaces from dents and scratches, false lining jaws are often set over the regular jaws.

8. This holding device is about the size of a small grinding wheel and is fastened to the work bench in a similar manner.

9. These lining jaws can be made from paper, leather, wood, brass, copper, or lead.

10. A smooth face is inserted to prevent marring the surface of certain work pieces.

After sorting the sentences into three groups, arrange the sentences in logical order to form a paragraph. The reorganized paragraph thus would read:

Sentence   <u>2</u> , <u>   </u> , <u>   </u> , <u>6</u> , <u>   </u> , <u>   </u> , <u>   </u> , <u>   </u> , <u>   </u> , <u>4</u>

Now write out the reorganized paragraph.

Application 2   Describing a Mechanism

Write a general description of one of the following mechanisms or of a mechanism in your major field.

a. Fill in the Plan Sheet on pages 89–90.
b. Write a preliminary draft.
c. Revise. See inside back cover.
d. Write the final draft.

Describe:

1. A disk drive
2. A human heart
3. A drill press
4. A power saw
5. A set of drawing instruments
6. A flashlight
7. A microphone
8. A wristwatch
9. A camera
10. A sprayer
11. A room air conditioner
12. A shotgun
13. A stapler
14. A washing machine
15. A lie detector
16. A diesel engine
17. A Geiger counter
18. An alternator
19. An antenna
20. A barometer
21. A sliding T bevel
22. A bicycle
23. An electric toaster or some other household appliance
24. A vaccinating needle
25. A calculator
26. A mechanical pencil
27. An autoclave
28. A dental unit
29. A stethoscope
30. An incubator

Application 3   Describing a Mechanism

Give an oral description of a mechanism in Application 2 above. Ask your classmates to evaluate your speech by filling in an Evaluation of Oral Presentations from pages 527–535.

Application 4   Describing a Mechanism

From the list in Application 2, choose a mechanism and write a specific description of it. This time, however, describe a *specific* model or make, such as a Swingline

90 stapler, a 20-gauge Remington automatic shotgun, or a Zenith portable radio, model number 461.

    a. Fill in the Plan Sheet on pages 91–92.
    b. Write a preliminary draft.
    c. Revise. See inside back cover.
    d. Write the final draft.

Application 5 Describing a Mechanism

Select a mechanism that you use in one of your lab courses. Follow this procedure:

    a. Fill in the Plan Sheet on pages 93–94.
    b. Write a preliminary draft.
    c. Revise. See inside back cover.
    d. Write the final draft.

Use appropriate information for audience and purpose to give the following three descriptions.

    1. Describe this mechanism so that an employee can find it and put an inventory number on it.
    2. Describe this mechanism for a new lab student who must use the mechanism.
    3. Describe this mechanism for a technician who will repair or replace a broken part.

Application 6 Describing a Mechanism

Choose a mechanism from the list in Application 2 or from your major field. Describe the mechanism at rest and in operation.

Application 7 Describing a Mechanism

Give orally a specific description of a mechanism. Plan to use a visual of the mechanism, either a sample mechanism, a picture, or a line drawing, or arrange for the class to visit your shop or laboratory and view the mechanism. Ask your classmates to evaluate your speech by filling in an Evaluation of Oral Presentations from pages 527–535.

Application 8 Describing a Mechanism

Research a mechanism to find out all possible options. Then "create" your own mechanism by adding your choice of options to the basic mechanism. After studying a sampling of brochures and fliers advertising mechanisms, plan and present a sales brochure to advertise your mechanism. Use appropriate visuals (see Chapter 11).

Application 9   Describing a Mechanism

Read the excerpt from *Zen and the Art of Motorcycle Maintenance*, by Robert M. Pirsig, pages 600–605. Identify a mechanism that you have had to cope with recently. Write a general description of the mechanism.

Application 10   Describing a Mechanism

Read "The Science of Deduction," by Sir Arthur Conan Doyle, pages 606–610. Under "Suggestions for Response and Reaction," page 610, write Number 5.

Application 11   Describing a Mechanism

Read "Holography: Changing the Way We See the World," pages 589–594. Under "Suggestions for Response and Reaction," page 594, write Number 1.

PLAN SHEET
FOR A GENERAL DESCRIPTION OF A MECHANISM

Analysis of Situation Requiring a General Description of a Mechanism

What is the general mechanism to be described?

For whom is the description intended?

How will the description be used?

Will the description be written or oral?

Analysis of the Mechanism

Definition, identification, and/or special features:

Function, use, or purpose:

Who uses:

When:

Where:

Physical characteristics:
    Size:

    Shape:

    Weight:

Material:

Other:

Major parts and a description of each:

Part	Function	Physical characteristics

Operation Showing Parts Working Together

Variations or Special Features

Types and Subject Matter of Visuals to Be Included

Sources of Information

Chapter 3

Definition:
Explaining What
Something Is

Objectives

Upon completing this chapter, the student should be able to:

- List the conditions under which a term should be defined
- Demonstrate the three steps in arriving at a sentence definition
- Give a sentence definition
- Give an extended definition
- State definitions according to their purpose and the knowledge level of the reader

Introduction

All too often in preparing a written or an oral presentation, the careless person dismisses the idea of including definitions in the presentation, apparently thinking, "Why bother with definitions? Aren't there dictionaries around for people who run into words they don't know?" Before evaluating this attitude, consider a common, frequently used word in the English language: *pitch*. Certainly, *pitch* is not a strange word to most people. A drafter writing about *pitch* for other drafters can be sure that they know the term refers to the slope of a roof, expressed by the ratio of its height to its span. An aeronautical technician, however, would automatically associate *pitch* with the distance advanced by a propeller in one revolution. A geologist would think of *pitch* as being the dip of a stratum or vein. A machinist thinks of *pitch* as the distance between corresponding points on two adjacent gear teeth or as the distance between corresponding points on two adjacent threads of a screw, measured along the axis. To the musician, *pitch* is that quality of a tone or sound determined by the frequency or vibration of the sound waves reaching the ear. To the construction worker, *pitch* is a black, sticky substance formed in the distillation of coal tar, wood tar, petroleum, and such, and used for such purposes as waterproofing, roofing, and paving. To the average layperson, *pitch* is an action word meaning "toss" or "throw," as in "*Pitch* the ball." Such expressions as "It's *pitch* dark" or "I see the neighborhood's chief con man has a new *pitch*" illustrate other uses of *pitch*.

This one word illustrates quite well the importance of knowing when to define for different readers, and it implies that the writer or speaker must know *how* to define.

Definition Adapted to Purpose

Because in business and industry many words have precise, specific meanings, the need for defining a term frequently arises. You must be able to give a clear, accurate definition that is appropriate for the situation, you must know how and when to write a sentence definition and an extended definition, and you must know how to word these definitions for a particular purpose and according to the knowledge level of the audience. A mechanical technician or a drafter, for instance, has to be able to adapt the length and simplicity of the definition of a micrometer caliper according to need. Consider each of the three following definitions and the different occasions on which each would best serve the writer's purpose.

1. A micrometer caliper is an instrument for measuring very small distances.
2. A micrometer caliper is a precision measuring instrument graduated to read up to one ten-thousandth of an inch and sized to measure stock from one to twenty-four inches. There are three kinds of micrometers—the inside, the depth gauge, and the outside. The last is the most common.
3. The micrometer caliper is a precision measuring instrument designed to use the decimal divisions of the inch. Some micrometers are graduated to read a thousandth part of an inch while others have verniers by which measurements of one ten-thousandth of an inch can be made.

   The micrometer has a U-shaped form to which are fastened an anvil and a barrel, or hub (see the drawing). Inside the barrel are very fine threads (40 per inch) that make a nut and screw for the moving spindle and thimble. On the front of the barrel is a long line divided by short crosslines; the inch of space along the barrel is divided into parts. The first line, starting from 0, is 0.025 inch; the second, 0.050; the third, 0.075; and the fourth (the long crossline), 0.100 inch. The spindle turns inside the barrel; the thimble, attached to the spindle, fits over the barrel. Turning the thimble to the right or the left moves the spindle toward or away from the anvil. A complete turn moves the spindle one-fortieth of an inch, or 0.025 inch. The thimble has a beveled edge divided into 25 equal parts or thousandths with each division representing 0.001 inch (one-thousandth of an inch). Turning the spindle one complete turn (0.025 inch) moves it 25 spaces on the beveled edge of the thimble. For every complete turn to the right, one more mark on the barrel is covered over.

   When the micrometer is closed, the zero line on the thimble and the zero line on the barrel coincide. Unscrewing the thimble one full turn will align the zero on the thimble with the horizontal line on the barrel and expose one space on the barrel. The micrometer has been opened one-fortieth of an inch, or 25 thousandths. Each complete turn of the thimble will expose one more mark on the barrel or an additional 25 thousandths. Also from a closed position the thimble can be turned so that the graduation next to zero on the beveled edge lines up with the horizontal line on the barrel, thus opening the micrometer one-thousandth of an inch. Every one-space turn of the thimble on the beveled edge opens the micrometer another one-thousandth.

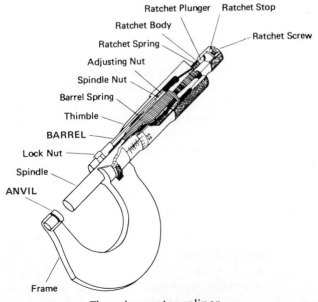

The micrometer caliper.

When the thimble has been turned 25 spaces along its beveled edge (one full revolution: 25 thousandths), the first line on the barrel is exposed.

The micrometer must be checked frequently for accuracy. To make the check, insert between the spindle and anvil a gauge block of known unit size, making sure both micrometer and gauge block are free from dirt and grit. The micrometer is accurate if the reading of the micrometer, with the block in place, is the same as the known size of the gauge block.

There are occasions on which each of the micrometer definitions is the most appropriate. The first definition, in its brevity, would be appropriate for a listing of precision measuring instruments and perhaps for a general desk dictionary. (A general desk dictionary does not include many of the specialized terms and meanings that technicians often need.) The second definition, which gives more detailed information, might be used in an introduction to a paper or an oral presentation describing the outside micrometer or in a paper or lecture describing the instruments a drafter should be familiar with. The extended definition would be appropriate for a textbook on machine shop tools, for a technical handbook or dictionary, or for a manufacturer to enclose with a micrometer.

You need to know when to define a term, to what extent to define it, and how to define it.

See also Jargon, page 10.

When to Define a Term

A writer or a speaker should define a term that (1) is unfamiliar to the audience, (2) has multiple meanings, or (3) is used in a special way in a presentation.

First, a term should be defined when the audience might not know its meaning but should. An audience might need such a definition simply because they are unfamiliar with it. A metallurgist, communicating with a general adult audience, would probably have to explain the use of *anneal* or *sherardize;* or an electronics technician, writing a repair memorandum for a customer's information, would have to explain the meaning of *zener diode* or *signal-to-noise ratio.*

Second, a term that has multiple meanings should be defined. A term may not be clear to an audience because it has taken on a meaning different from that which the audience associates with it. For instance, if the company nurse instructs the carpenter to put a new "2 by 4" and "4 by 8" on the cuts on his leg each morning, the carpenter may leave in disbelief—and certainly in misunderstanding. To the nurse, of course, "2 by 4" and "4 by 8" are common terms for *gauze squares* in those inch dimensions used in dressing wounds. But to the carpenter "2 by 4" and "4 by 8" mean *boards* in those inch dimensions. If there is any possibility that the intended meaning of a term may be misunderstood, define it.

Occasionally a writer or speaker gives a term or concept a special meaning within a presentation. The writer or speaker should then certainly let the reader know exactly how the term is being used.

Extent of Definition

The extent to which a term should be defined, thus the length of a definition, depends on the writer's or speaker's purpose and the knowledge level of the audience.

Sometimes merely a word or a phrase is sufficient explanation: "The optimum (most favorable) cutting speed of cast iron is 100 feet per minute." Notice the various ways in which the definitions are given in the following examples. "Consider, for instance, lexicographers—those who write dictionaries." "The city acquired the property by eminent domain, that is, the legal right of a government to take private property for public use."

At other times a definition may require a sentence or an entire paragraph. Occasionally a definition requires several paragraphs for a clear explanation, as when explaining an unfamiliar idea or item that is the central focus of a presentation.

Further, the extent to which a term should be defined depends on the complexity of the term. In a presentation for the general public, a definition of Ohm's law, for instance, undoubtedly would require more details than a definition of a Phillips screwdriver, because few readers or listeners would be familiar with the principles and terms of a specialized field like electronics and thus would have difficulty understanding the unfamiliar concept of Ohm's law. Most people, on the other hand, are familiar with a screwdriver; thus a definition of a Phillips screwdriver would be concerned only with how the Phillips is different from other screwdrivers. In addition, an abstract term like *Ohm's law* is usually more difficult to explain than a concrete item like *Phillips screwdriver*. The general knowledge and interest of the audience are certainly important in determining just how far to go in giving a definition. Most important of all, however, is the *purpose* for which the definition is given. In defining a term you must keep clearly in mind your purpose and a way to accomplish it. It may be that a definition is merely parenthetical. Or it may be that a definition is of major significance, as when an audience's understanding of a key idea hinges on the comprehension of one term or concept. And occasionally, though not often, an entire presentation is devoted to an extended definition.

How to Define a Term

Once you decide that a term needs defining and decide whether it should be a brief or an extended definition, you can then follow these suggested procedures to prepare an adequate definition.

Sentence Definition

A well-established, three-step method for giving a sentence definition includes stating:

1. The term (species)
2. The class (genus)
3. The distinguishing characteristics (differentiae)

The term is simply the word to be defined. The class is the group or category of similar items in which the term can be placed. For instance, a *chair* is a "piece of furniture"; a *stethoscope* is a "medical listening instrument." The distinguishing characteristics are the essential qualities that set the term apart from all other terms of the same class; the distinguishing characteristics make the definition accurate and complete. A *chair* (term) is a "piece of furniture" (class). But there are pieces of furniture that are not chairs; thus the chair as a piece of furniture must be distinguished from all other pieces of furniture. The characteristic "that has a frame, usually made of wood or metal, forming a seat, legs, and backrest and that is used for one person to sit in" differentiates the chair from other items in the class. The sentence definition would be: "A chair is a piece of furniture that has a frame, usually made of wood or metal, forming a seat, legs, and backrest and that is used for one person to sit in."

Study the following examples.

Term	Class	Distinguishing Characteristics
Program	Set of organized instructions	Composed of words and symbols and used to direct the performance of a computer
Rivet	Permanent metal fastener	Shaped like a cylinder with a head on one end; when placed in position, the opposite head is formed by impact
Photojournalist	Photographer	Works with newsworthy events, people, and places; works for newspapers, magazines, television; must possess skill in using cameras and in composing pictures

Thus, *program* (term) is a "set of organized instructions" (class) "composed of words and symbols and used to direct the performance of a computer" (distinguishing characteristics).

A *rivet* (term) is a "permanent metal fastener" (class) "that is shaped like a cylinder and has a head on one end. When placed in position, the opposite head is formed by impact" (distinguishing characteristics).

A *photojournalist* (term) is a "photographer" (class) who "works with newsworthy events, people, and places." This person "works for newspapers, magazines, and television" and "must possess skill in using cameras and in composing pictures" (distinguishing characteristics).

As these examples show, a sentence definition covers only one meaning of a word. Further, a sentence definition may actually be longer than just one sentence, as in the definition above of *rivet* and *photojournalist*. Two or more sentences may be needed in order to include all the essential distinguishing characteristics or for effective sentence structure.

When determining a class, be as precise and specific as possible. Placing a screwdriver in the class of "small hand tools" is much more specific than placing it in a class such as "objects, instruments, or pieces of hardware." The more specific the class is, the simpler it is to give the distinguishing characteristics, that is, the qualities that separate this term from all other terms in the same class. For instance, if, in a definition of *trowel*, the class is given simply as "a device," many distinguishing characteristics will have to be given to set a trowel apart from innumerable other devices (staple guns, rockets, tractors, phonographs, keys, pencil sharpeners, washing machines, etc.). If the class for *trowel* is given as "a hand-held, flat-bladed implement," then the distinguishing characteristics might be narrowed down to "having an offset handle and used to smooth plaster and mortar."

To avoid confusing peripheral information with the essential distinguishing characteristics, it is important to have an understanding of the *essence* of the term being defined. Essential to the nature of a brick, for instance, is that it is made out of baked clay. The color, methods of firing, size, shape, cost, and so on, would be peripheral information. Essential to the nature, the essence, of a refrigerator is that it preserves food by keeping it at a constant, cold temperature. Information that this large home appliance may be self-defrosting or may come in combination with a freezer would have no place, ordinarily, in a sentence definition.

Inadequate Sentence Definitions. The sentence definition, if it is to be adequate, must give the term, state the class as specifically as possible, and give characteristics that are really distinguishing. Such sentence definitions as "Scissors are things that cut" or "Slicing is something you ought not do in golf" or "Oxidization is the process of oxidizing" or "A stadium is where games are played" are inadequate if not completely useless. These definitions give little if any insight into what the term means.

The definition "Scissors are things that cut" equates scissors with razors, saws, drills, cookie cutters, plows, knives, sharp tongues, and everything else that cuts in any way. The definition is not specific in class ("things"), and the characteristic ("that cut") can hardly be called distinguishing. A more adequate definition might be "Scissors are a two-bladed cutting implement held in one hand. The pivoted blades are pressed against opposing edges to perform the cutting operation."

"Slicing is something you ought not do in golf" gives *no* indication of what slicing is. The so-called definition is completely negative. The class ("something") is about as nonspecific as it can be. Furthermore, there are no distinguishing characteristics ("you ought not do in golf" is meaningless—a golfer ought not slow down the players coming up behind, ought not bother other players' balls, ought not damage the green, ought not start at any hole except the first, ought not cheat on the score, etc.). A better definition is this: "Slicing is the stroke in golf that causes the ball to veer to the right."

"Oxidization is the process of oxidizing" as a definition is completely useless because the basic word oxidize is still unexplained. If a person knows what oxidizing means, he or she has a pretty good idea of what oxidization means; however, if someone has knowledge of neither word, the stated "definition" is only confusing. This kind of definition, which uses a form of the term as either the class or the distinguishing characteristic, is called a circular definition because it sends the reader in circles—the reader never reaches the point of learning what the word means. Oxidization might more satisfactorily be defined as the following: "Oxidization is a process of combining oxygen with a substance that reduces its strength."

"A stadium is where games are played" is inadequate as a sentence definition because no class is given and the distinguishing characteristic is not specific. The "is where" or "is when" construction does not denote a class. Furthermore, only certain types of games—sports events—are typically associated with a stadium. An adequate sentence definition is this: "A stadium is a large, usually unroofed building where spectator events, primarily sports events, are held."

Extended Definition

The extended definition gives information beyond stating the essence, or primary characteristic, of a term. The extended definition is concerned with giving enough information so the audience can gain a thorough understanding of the term. This definition in depth may contain such information as synonyms; origin of the term or item; data concerning its discovery and development; analysis of its parts;

physical description; necessary conditions, materials, equipment; description of how it functions; explanation of its uses; instructions for operating or using it, examples and illustrations; comparisons and contrasts; different styles, sizes, and methods; and data concerning its manufacture and sale. The central focus in an extended definition is on stating what something is by giving a full, detailed explanation of it.

The extended definition is closely related to other forms of explanation, particularly instructions and process description (see Chapter 1) and description of a mechanism (see Chapter 2), and often includes them.

Following is an example of a student-written extended definition (including the Plan Sheet that was filled in before the paper was written). Marginal notes have been added to help in analyzing the paper.

PLAN SHEET
FOR GIVING AN EXTENDED DEFINITION

Analysis of Situation Requiring Definition

What is the term to be defined?
tornado

For whom is the definition intended?
students involved in a tornado preparedness week

How will the definition be used?
for their knowledge and understanding

Will the definition be written or oral?
written

Sentence Definition

Term:	Class:	Distinguishing characteristics:
tornado	*severe storm*	*whirling winds (more than 200 miles an hour)*

As Applicable

Synonyms:
omit

Origin of the term or item:
omit

Information concerning its discovery and development:
omit

Analysis of parts:
omit

Physical description:
appears like a rotating funnel cloud; extends from thundercloud toward ground; gray or black

Necessary conditions, materials, equipment:
cold front; layers of air with contrasting temperature, moisture, density, and wind flow; cool, dry air from west or northwest; warm, moist air near earth's surface; narrow bands of strong wind

Description of how it functions:
omit

Explanation of its uses:
omit

Instructions for operating or using it:
omit

Examples and illustrations:
omit

Comparisons and contrasts:
appearance of funnel cloud and thunderstorm like huge mushroom; sounds like roar of hundreds of airplanes or several trains racing through the house

Different styles, sizes, methods:
omit

Data concerning its manufacture and sale:
omit

When and where used or occurring:
anywhere in the U.S.; central part of U.S. most common; in Southeast, most likely in March; in Midwest most likely in May, June; occur any time of day; most likely 3 to 7 P.M.; OK more tornadoes; TX more deaths

History:
March 18, 1925, 70 persons killed in MO, IN, KY, TN, and IL by a series of 8 tornadoes; 2000 injured; millions of dollars of property damage; March 21, 1932, 268 killed in AL by tornado series; 2000 injured; April 11, 1965, 257 killed in Midwest by series of 27 tornadoes; over 5000 injured

Other:
- □ *Speed and path: Average speed about 30 mph; winds may exceed 200 mph; average path 300–400 yds wide, 4 mi long; direction of movement: usually southwest to northeast. Example: May 26, 1917, tornado traveled 293 mi across IL and IN; lasted 7 hrs 23 mins.*
- □ *Destructive power: Combination of strong rotary winds and partial vacuum in center of vortex. Example: building has outside twisted and torn by winds; abrupt reduction of pressure in center of tornado causes building to "explode" outward; walls collapse or fall outward, windows explode, debris flies through air; train cars and airplanes moved.*

Types of Subject Matter of Visuals to Be Included

photograph of a tornado
table showing number of tornadoes, tornado days, and deaths by months, 1956–75

Sources of Information

pamphlets from the National Weather Service, <u>Weather Almanac</u>

AN EXTENDED DEFINITION OF TORNADOES

Term to be defined
Sentence definition
Kinds of information
to be given

Tornadoes are a common natural phenomenon in the central United States. A tornado is a severe storm with winds whirling at more than 200 miles an hour. Tornadoes can be further defined by a discussion of their formation, appearance, sound, speed and path, and destructive power, when and where tornadoes occur, and some of history's major tornadoes.

Formation
Conditions necessary
for formation

Formation of Tornadoes

Tornadoes usually form along a cold front, and they begin to form several thousand feet above the earth. Their formation requires layers of air with contrasting temperature, moisture, density, and wind flow. Cool, dry air from the west or northwest moves over warm, moist air near the earth's surface; if narrow bands of strong wind are present between these two layers of air, complex changes can produce a vortex or whirl.

Appearance

Comparison to a fa-
miliar object

Appearance of Tornadoes

Growing out of a thunderstorm, the tornado is a violently rotating column of air descending from a thundercloud system. Visible as a rotating funnel-shaped cloud extending from the thundercloud toward the ground, the funnel cloud's color may be gray or black. The funnel and cloud may look very much like a mushroom with the funnel extending to the ground resembling the stem or stalk and the thundercloud resembling the cap of the mushroom. See the photograph.

Figure 1. Photograph of a tornado.
(Courtesy of Bob Hodges)

The funnel may travel through the air or along a narrow path over land.

Sound

Comparison to a familiar object

Sound of Tornadoes

The sound of a tornado has been described as being like the roaring of hundreds of airplanes or like the sound of several freight trains racing through one's house. According to people who have heard the sound of a tornado, "It is a sound you never forget."

Speed and path

Speed

Speed and Path

The speed and path of tornadoes may vary widely. Most tornadoes move abut 30 miles an hour. However, some move very slowly and others may move as fast as 60 or more miles an hour. Winds may exceed 200 miles an hour.

Path

The average tornado is about 300 to 400 yards wide and about 4 miles long. However, tornadoes have been a mile or more wide and 300 miles long. On May 26, 1917, a tornado traveled 293 miles across Illinois and Indiana over a period of 7 hours and 23 minutes.

The direction of movement is usually southwest to northeast.

Destructive power

Cause of destructive power

Destructive Power

The destruction caused by a tornado occurs by the combination of the strong rotary winds and the partial vacuum in the center of its vortex. For example, as a tornado passes over a building, the winds twist and tear the outside and the abrupt reduction of pressure in the tornado's center causes the building to "explode" outward. Walls may collapse or fall outward, windows explode, and the debris from the building may be driven through the air.

The destructive power of a tornado is unmeasurable. Tornadoes have picked up cars from a train and airplanes and deposited them some distance away.

When and where tornadoes occur

When

When and Where Tornadoes Occur

Tornadoes may occur at any time of the day, but they are most likely to occur during the afternoon between 3 and 7 P.M.

While tornadoes may occur anywhere in the United States, the central part of the United States is the most likely place. For the Southeast, tornadoes occur most likely in March. Midwestern states, especially Oklahoma, Kansas, Iowa, and Nebraska, have a peak tornado season during May and June. Table 1 shows the number of tornadoes, tornado days, and deaths by month from 1956 to 1975.

Where

Oklahoma has the greatest number of tornadoes per 10,000 square miles, and Texas has had the most tornado-related deaths.

Major tornadoes

Some of History's Major Tornadoes

On March 18, 1925, a series of tornadoes, at least eight, killed 740 persons in Missouri, Indiana, Kentucky, Tennessee, and Illinois. Almost 2000 individuals were injured and millions of dollars of property damage occurred. On March 21, 1932, in Alabama, a series of tornadoes killed 268 and injured almost 2000. On April 11, 1965, a series of 37 tornadoes in the Midwest killed 257 and injured over 5000.

TABLE 1. Number of Tornadoes, Tornado Days, and Deaths by Months, 1956–1975

Year	January			February			March			April			May			June			July			August			September			October			November			December			Annual		
	Number	Days	Deaths	Number	Days	Deaths	Number	Days	Deaths	Number	Days	Deaths	Number	Days	Deaths	Number	Days	Deaths	Number	Days	Deaths	Number	Days	Deaths	Number	Days	Deaths	Number	Days	Deaths	Number	Days	Deaths	Number	Days	Deaths	Number	Days	Deaths
1956	2	2	0	47	12	8	31	7	1	85	15	67	79	24	4	65	21	0	91	26	1	43	20	2	16	10	0	29	8	0	7	6	0	9	4	0	504	155	83
1957	17	3	13	5	3	0	38	7	1	216	21	29	227	26	87	147	25	14	55	19	0	20	14	0	17	10	2	18	11	2	58	11	25	38	4	18	856	154	191
1958	12	7	0	20	5	13	15	10	0	76	19	4	68	21	0	127	27	42	121	30	1	46	20	1	24	14	1	18	6	4	45	6	4	1	1	0	564	166	66
1959	16	7	0	20	5	21	43	11	9	30	12	1	226	28	8	73	24	8	63	24	0	38	18	0	58	15	14	24	10	14	11	4	0	2	2	0	604	156	58
1960	9	4	0	28	10	0	28	10	0	70	20	7	201	26	34	124	27	3	43	22	0	47	23	1	22	13	0	18	10	1	25	6	1	1	1	0	616	172	46
1961	1	1	0	31	8	0	124	17	7	74	19	3	137	25	23	107	23	2	77	27	0	27	16	0	53	16	0	14	5	0	36	7	0	16	5	0	697	169	51
1962	12	3	1	25	7	0	37	9	17	41	8	17	200	21	3	171	29	0	78	26	0	51	16	0	24	11	6	24	10	0	4	4	0	2	2	0	657	152	28
1963	15	5	1	6	3	0	50	12	16	82	14	16	71	21	16	91	21	0	62	26	0	26	13	2	33	10	2	13	5	0	15	6	0	0	0	0	464	141	31
1964	14	3	10	2	2	0	36	11	6	157	23	15	135	20	15	136	24	0	63	23	0	79	23	2	25	10	2	22	4	22	17	8	8	18	5	2	704	156	73
1965	21	11	0	32	4	0	34	9	2	129	20	264	275	25	17	147	28	6	86	26	3	61	23	1	64	21	0	16	4	1	34	6	0	7	3	0	906	181	296
1966	1	1	0	28	5	0	12	6	58	80	20	12	98	17	0	126	28	9	100	27	3	58	21	3	22	13	0	29	6	6	20	3	3	11	3	0	585	150	98
1967	39	4	0	8	5	0	42	14	3	149	18	73	116	25	3	210	18	19	90	25	1	28	16	2	139	16	5	36	6	4	4	5	0	61	10	10	926	173	114
1968	5	3	0	7	3	0	28	8	0	102	15	40	145	26	72	136	27	0	56	22	2	66	23	2	25	14	0	14	9	0	44	12	3	32	9	1	660	171	131
1969	3	1	32	7	5	0	8	2	1	68	15	2	145	25	4	137	28	11	99	27	0	69	21	19	20	11	0	26	10	0	5	3	0	23	7	1	608	155	66
1970	9	5	0	16	3	0	25	16	2	117	16	29	88	19	26	134	24	6	81	26	3	55	21	3	54	20	0	50	13	6	10	4	0	14	8	0	653	171	72
1971	18	3	1	83	12	131	40	13	2	75	14	11	166	24	7	199	28	1	100	30	1	50	21	0	47	15	0	38	12	0	16	7	0	56	9	2	888	192	156
1972	33	10	5	5	4	0	69	17	0	96	20	16	140	27	0	114	25	2	115	29	0	59	23	3	49	19	0	34	10	0	17	7	12	8	6	0	741	194	27
1973	33	7	1	10	4	0	80	16	0	150	22	10	250	26	35	224	26	35	80	26	0	51	23	4	69	22	3	25	11	3	81	11	0	49	12	3	1102	206	87
1974	24	8	2	23	9	0	36	12	17	269	22	313	144	28	10	194	26	31	59	19	0	107	26	0	25	11	0	45	10	4	13	8	0	8	5	0	947	184	361
1975	53	7	12	45	12	7	84	16	12	108	20	13	188	30	5	196	28	6	76	26	2	60	25	2	34	17	0	12	7	0	40	8	0	22	8	1	918	204	60
1956–1975: Total	337	94	88	448	121	180	860	219	147	2174	353	926	3099	485	355	2858	520	160	1595	506	14	1041	411	46	820	291	43	483	168	50	507	129	48	378	105	38	14600	3402	2095
Mean	17	5	4	22	6	9	43	11	7	109	18	46	155	24	18	143	26	8	80	25	1	52	21	2	41	15	2	24	8	3	25	6	2	19	5	2	730	170	105

SOURCE: *Weather Almanac.*

Intended Audience and Purpose

Whether the definition involves one sentence, a paragraph, or several paragraphs, the writer or speaker must be aware of the audience to whom the communication is directed and the purpose of the communication.

DEFINITIONS IN GENERAL REFERENCE WORKS

Professional writers are keenly aware of who will be reading their material and why. Consider, for instance, lexicographers—those who write dictionaries. Lexicographers know that people of all ages and with varying backgrounds turn to the dictionary to discover or ascertain the meanings that most people attach to words. Lexicographers are aware that dictionary definitions must be concise, accurate, and understandable to the general reader. The following definition of *measles*, for instance, from a standard desk dictionary is adequate for its purpose.

> **mea·sles** / ′mē-zelz / *n pl but sing or pl in constr* [ME *meseles*, pl. of *mesel* measles, spot characteristic of measles; akin to MD *masel* spot characteristic of measles] (14c) **1a:** an acute contagious viral disease marked by an eruption of distinct red circular spots **b:** any of various eruptive diseases (as German measles) **2** [ME *mesel* infested with tapeworms, lit., leprous , fr. OF, fr. ML *misellus* leper, fr. L, wretch, fr. *misellus*, dim. of *miser* miserable]: infestation with or disease caused by larval tapeworms in the muscles and tissues*

DEFINITIONS IN SPECIALIZED REFERENCE WORKS

A specialized dictionary or encyclopedia, however, is not designed to give a concise definition for the general reader. The specialized reference book aims its information at a well-defined, select audience. The following entry, "Measles," from the *McGraw-Hill Encyclopedia of Science and Technology*, illustrates such a definition. Note the number of cross references suggested for the reader to see.

MEASLES

An acute, highly infectious viral disease, with cough, fever, and maculopapular rash. It is of worldwide endemicity. *See* ANIMAL VIRUS.

The infective particle is an RNA virus about 100—150 nm in diameter, measured by ultrafiltration, but the active core is only 65 nm as measured by inactivation after electron irradiation. Negative staining in the electron microscope shows the virus to have the helical structure of a paramyxovirus with the helix being 18 nm in diameter. Measles virus will infect monkeys easily and chick embryos with difficulty; in tissue cultures the virus may produce giant multinucleated cells and nuclear acidophilic inclusion bodies. The virus has not been shown to have the receptor-destroying enzyme associated with other paramyxoviruses. Measles, canine distemper, and bovine rinderpest viruses are antigenically related. *See* EMBRYONATED EGG CULTURE; MYXOVIRUS; PARAMYXOVIRUS; TISSUE CULTURE; VIRAL INCLUSION BODIES.

The virus enters the body via the respiratory system, multiplies there, and circulates in the blood. Prodromal cough, sneezing, conjunctivitis, photophobia, and fever occur, with Koplik's spots in the mouth. A rash appears after 14 days' incubation and persists

* By permission. *From Webster's Ninth New Collegiate Dictionary* © 1987, by Merriam-Webster, Inc., publisher of the Merriam-Webster® dictionaries.

5–10 days. Serious complications may occur in 1 of every 15 persons; these are mostly respiratory (bronchitis or pneumonia), but neurological complications are also found. Encephalomyelitis occurs, but it is rare. Permanent disabilities may ensue for a significant number of persons. Laboratory diagnosis (seldom needed since 95% of cases have the pathognomonic Koplik's spots) is by virus isolation in tissue culture from acute-phase blood or nasopharyngeal secretions, or by specific neutralizing, hemagglutination-inhibiting, or complement-fixing antibody responses.

In unvaccinated populations, immunizing infections occur in early childhood during epidemics which recur after 2–3 years' accumulation of susceptible children. Transmission is by coughing and sneezing. Measles is spread chiefly by children during the catarrhal prodromal period; it is infectious from the onset of symptoms until a few days after the rash has appeared. By the age of 20 over 80% of persons have had measles. Second attacks occur but are very rare. Treatment is symptomatic.

At one time, prevention was limited to use of gamma globulin, which protects for about 4 weeks, and can modify or prevent the disease. Prevention is advisable in infants 4–12 months of age or in sick children. However, if the disease has been prevented by administration of gamma globulin, the child develops no immunity, whereas the illness modified by gamma globulin may confer lasting immunity. *See* IMMUNOGLOBULIN.

Killed virus vaccine should not be used, as certain vaccinees become sensitized and develop local reactions when revaccinated with live attenuated virus, or a severe illness upon contracting natural measles.

Live attenuated virus vaccine can effectively prevent measles; vaccine-induced antibodies persist for years. Prior to the introduction of the vaccine, over 500,000 cases of measles occurred annually in the United States. Following mass immunization in 1966–1967, the number of cases decreased to 22,000 in 1968. Failure to immunize children from certain segments of the population resulted in 75,000 cases in 1971; however, with renewed emphasis on immunization, the number of cases declined to 32,000 in 1972, and by 1974 was again down to 22,000. In many areas of the United States, measles occurs in sporadic epidemics among nonimmunized children, in which the attack rate for immunized and nonimmunized children is about 2 and 34%, respectively. *See* BIOLOGICALS; HYPERSENSITIVITY; SKIN TEST.

Measles antibodies cross the placenta and protect the infant during the first 6 months of life. Vaccination with the live virus fails to take during this period; thus immunization is not recommended. Vaccination is not recommended also in persons with febrile illnesses, with allergies to eggs or other products used in production of the vaccine, and with congenital or acquired immune defects.

Measles virus appears to be responsible for subacute sclerosing panencephalitis (SSPE, Dawson's inclusion body encephalitis), a rare chronic degenerative brain disorder. The disease manifests itself in children and young adults by progressive mental deterioration, myoclonic jerks, and an abnormal EEG with periodic high-voltage complexes. The disease develops a number of years after the initial measles infection. A virus closely resembling measles virus, but not completely identical to it, has been isolated from brain tissue of patients. The virus is not localized only in brain tissues, since isolations have been made from lymph nodes. The presence of a latent intracellular measles virus in lymph nodes suggests a tolerant infection with defective cellular immunity.

Patients with SSPE have a functioning humoral attack system against cells which express surface measles virus antigens; cultured cells from the brain of a patient with SSPE have been lysed by the patient's own serum. Lysis occurs only when measles virus antigens are expressed on the cell surface, and is dependent on the presence of antibody to measles virus and complement. Lysis can also be induced by sera and cerebrospinal fluid from other SSPE patients, by sera from patients who have convalesced from normal measles virus infections, and by heterologous rabbit serum against measles virus. *See* VIRUS INFECTIONS, LATENT, PERSISTENT, SLOW.

A laboratory-produced, defective (temperature-sensitive) mutant of measles virus has caused hydrocephalus when inoculated intracranially into newborn hamsters; this finding shows the need for caution in use of experimentally induced virus variants.

[JOSEPH L. MELNICK]

Bibliography: P. Isacson and A. Stone, Allergic reactions associated with viral vaccines, *Progr. Med. Virol.*, 13:239–270, 1971; F. E. Payne, Measles virus associated with subacute sclerosing panencephalitis, *Progr. Med. Virol.*, vol. 22, 1976; J. J. Whitte, The epidemiology and control of measles, *Amer J. Epidemiol.*, 100:77–78, 1974.*

Definition as Part of a Longer Communication

Giving a definition may be the main purpose of a communication; perhaps more often, however, a definition is an integral part of a longer communication. Consider, for instance, the paragraph definition of the term *report* on page 347 in Chapter 8, Reports. While the chapter, of course, is concerned with much more than definition, the definition of the focal term *report* serves as a framework for the entire chapter.

Oral Definition

In giving a definition orally, you would generally follow the procedures suggested for a sentence or extended definition. Suggestions for effective oral presentation of information are made in Chapter 10, Oral Communication.

Visuals

Visuals are very helpful in definitions. For example, the definition of a micrometer on pages 97–98 is much clearer because of the included drawing of the micrometer; the drawing helps the reader to see what a micrometer looks like, to know its parts and where they are located, and to understand better how the micrometer is used. Similarly, the definition of a tornado on pages 105–107 is made clearer with the photograph.

In defining concepts, visuals can be especially helpful. Consider, for instance, the following sentence definition of *horsepower*. Note how each of the three distinguishing characteristics (raising 33,000 pounds, distance of 1 foot, in 1 minute) is visually illustrated.

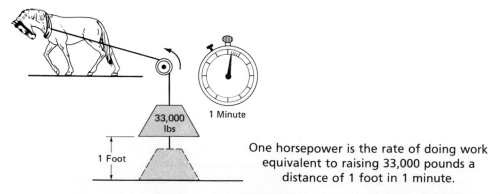

One horsepower is the rate of doing work equivalent to raising 33,000 pounds a distance of 1 foot in 1 minute.

* Reprinted by permission from *McGraw-Hill Encyclopedia of Science and Technology*, Vol. 8 (New York: McGraw-Hill, 1982).

Any time you can illustrate a term you are defining and thus help your audience understand it more easily and clearly, do so. For a discussion of types of visuals and guidance in preparing them, see Chapter 11, Visuals.

General Principles in Giving a Definition

1. *Knowing when to define a term is essential.* A term should be defined if the audience does not know the meaning of a word but should, a word is used in a meaning different from that which the audience ordinarily associates with it, or a term is given a special meaning within a presentation.
2. *The extent to which a term should be defined depends on several factors.* It depends on the complexity of the term, the general knowledge and interest of the audience, and, primarily, the purpose for which the definition is given.
3. *A sentence definition has three parts: the term, the class, and the distinguishing characteristics.* The term is simply the word to be defined. The class is the group or category of similar items in which the term can be placed. The distinguishing characteristics are the essential qualities that set the term apart from other terms in the same class.
4. *An extended definition gives information beyond stating the essence, or the primary characteristic, of a term.* The extended definition includes such information as origin, development, analysis of parts, physical description, function, and so on.
5. *The crucial factor in giving a definition is understanding the essence of the term being defined.* The writer or speaker must understand the nature of the object or concept to define it accurately.

Procedure for Giving an Extended Definition

An extended definition, whether it is one paragraph or several, or whether it is an independent presentation or part of a longer whole, usually has two distinct parts: identification of the term and additional information. A formal closing is unnecessary, although frequently the presentation ends with a comment or summarizing statement.

I. The identification of the term is usually brief.
   A. State the term to be defined.
   B. Give a sentence definition.
   C. Indicate the reason for giving a more detailed definition.
   D. State the kinds of additional information to be given.

II. The additional information forms the longest part of the presentation.
   A. Select additional information (as applicable): synonyms; origin of the term or item; information concerning its discovery and development; analysis of its parts; necessary conditions, materials, equipment; description of how it functions; explanation of its uses; instructions for operating or using it; examples and illustrations; comparisons and contrasts; different styles, sizes, methods; and data concerning its manufacture and sale.
   B. Organize the selected additional information.
   C. Give the additional information, including whatever details are needed to give the audience an adequate understanding of the term.

D. Use connecting words and phrases so that each sentence flows smoothly into the next and so that all the sentences in a paragraph and all paragraphs hang together as a unit.

E. Include visuals if their use will enhance understanding of the term defined.

III. Generally there is no formal closing, although a comment or summarizing statement is often included.

Application 1   Giving Definitions

By giving the distinguishing characteristics, complete the following to make general sentence definitions.

1. An orange is a citrus fruit. . . .
2. A speedometer is a gauge. . . .
3. A hammer is a hand tool. . . .
4. Measles is a disease. . . .
5. A printer is a data output device. . . .
6. An ambulance is a vehicle. . . .
7. A library is a building. . . .
8. A molar is a tooth. . . .
9. An anesthetic is a drug. . . .
10. Handcuffs are a restraining device. . . .

Application 2   Giving Definitions

Analyze the following definitions, noting their degree of accuracy and usefulness. Make whatever revisions are needed for adequate general sentence definitions.

1. A compass is for drawing circles.
2. A fire lane is where you should not park.
3. A T square, helpful in drawing lines, is a device used by drafters.
4. Immunity means to be immune to disease.
5. A crime is a violation of the law.
6. An anesthetic is a drug.
7. A board foot is a piece of material 1 inch thick, 12 inches long, and 12 inches wide.
8. Sterilization is the process of sterilizing.
9. A pictorial drawing is a drawing that is drawn from a drawing that was first drawn flat.
10. Airplanes without engines are called gliders.

Application 3   Giving Definitions

Write a sentence definition of five of the following terms.

1. Architects' scale
2. Diode
3. Solenoid valve
4. Database
5. Coupling
6. Helix

7. Brazing	20. Dividers	33. Lineup
8. Leader	21. Sawhorse	34. Disinfectant
9. Calipers	22. Depreciation	35. Arrest
10. Dial indicator	23. Terminal	36. Metabolism
11. Dowel	24. Thermometer	37. Anesthesia
12. Flashing	25. Osmosis	38. Tort
13. Feedback	26. Serum	39. Tourniquet
14. Inductance	27. Coagulation	40. Gland
15. Thermostat	28. Antibiotic	41. Disk
16. Bias	29. Conflagration	42. Airbrush
17. Traction	30. Nutrition	43. Flowchart
18. Herbicide	31. Enzyme	44. Management
19. Hydraulic lift	32. Canister mask	45. Fee

Application 4   Giving Definitions

Name your major field. Then make a list of ten words from that field that you should be thoroughly familiar with. Write an adequate sentence definition of each of the words.

Application 5   Giving Definitions

As an exercise in definition study, choose from your major field a term that can be found in each reference work specified below and note how it is treated in each of them. In giving the requested information, include the title of each reference work consulted. You may prefer photocopying a lengthy definition.

1. State the term.
2. Give the meaning of the term as stated in a standard desk dictionary.
3. Give the meaning as stated in a technical handbook or dictionary.
4. Give the meaning as stated in the *McGraw-Hill Encyclopedia of Science and Technology* and its yearbooks or a similar encyclopedia pertinent to your field. (Find the term in the index and then turn to the pages referred to.)

Application 6   Giving Definitions

Read the following excerpt, "Clinical Laboratory Technologists and Technicians," from the *Occupational Outlook Handbook*, 1986–1987 edition.

1. Write a sentence definition of (a) a medical technologist and (b) a medical laboratory technician. Use your own wording; do not merely copy phrases from the article.
2. Write a paragraph definition of a medical laboratory worker. Use your own wording; do not merely copy phrases and sentences from the article.

CLINICAL LABORATORY TECHNOLOGISTS AND TECHNICIANS

Nature of the Work

Laboratory tests play an important part in the detection, diagnosis, and treatment of disease. They are essential in detecting the presence of illnesses in which there are changes in the body fluids and tissues. Examples of such changes include chemical changes in the blood, urine, or other body fluids; increases or decreases in the count of various types of white or red blood cells; microscopic changes in the structure of the cells of a diseased tissue or organ; and the presence of parasites, viruses, or bacteria in the blood or tissue.

Although physicians use the results of laboratory evaluation and diagnosis, they do not perform the tests themselves. Instead, the tests are done by clinical laboratory personnel. These specialists provide laboratory services ranging from routine tests to highly complex analyses, and their level of skill and educational preparation vary accordingly. This section of the *Handbook* discusses the work of two levels of laboratory personnel: technologists and technicians.

Medical technologists have a bachelor's degree in science, as a rule. They perform complicated chemical, biological, hematological, microscopic, and bacteriological tests. These may include chemical tests to determine, for example, the blood cholesterol level, or microscopic examination of the blood to detect the presence of diseases such as leukemia. Technologists microscopically examine other body fluids; make cultures of body fluid or tissue samples to determine the presence of bacteria, parasites, or other micro-organisms; and analyze the samples for chemical content or reaction. They also type and cross-match blood samples for transfusions.

Technologists in small laboratories perform many types of tests, while those in large laboratories usually specialize. Among the areas in which they can specialize are biochemistry (the chemical analysis of body fluids), blood bank technology (the collection and preparation of blood products for transfusion), cytotechnology (the study of human body cells), hematology (the study of blood cells), histology (the study of human tissue), and microbiology (the study of bacteria and other micro-organisms).

Most medical technologists perform tests related to the examination and treatment of patients. Others do research, develop laboratory techniques, teach, or perform administrative duties.

Medical laboratory technicians generally have an associate degree or a diploma or certificate from a private postsecondary trade or technical school. They are midlevel laboratory workers who function under the supervision of a medical technologist or laboratory supervisor. They perform a wide range of routine tests and laboratory procedures which do not require the analytical knowledge of medical technologists. Like technologists, they may work in several areas or specialize in one field.

Application 7 Giving Definitions

Choose a term in your major field for an exercise in writing extended definitions that serve different purposes.

a. Fill in the Plan Sheet on pages 117–118.
b. Write preliminary drafts.
c. Revise. See the inside back cover.
d. Write the final drafts.

Write a paper to:

1. Define the term for a fifth-grade *Weekly Reader*.

2. Define the term for your English teacher, who wants to understand the term well enough to judge the accuracy and completeness of students' sentence definitions of the term.
3. Define the term as if you were having an examination in your technical field and you were asked to write a paragraph definition of it.

Application 8   Giving Definitions

Orally give the definitions you prepared for Application 7 above. Ask your classmates to evaluate your speech by filling in an Evaluation of Oral Presentations from pages 527–535.

Application 9   Giving Definitions

Write an extended definition (200–300 words) of a term or concept pertaining to your technical field.

a. Fill in the Plan Sheet on pages 119–120.
b. Write a preliminary draft.
c. Revise. See the inside back cover.
d. Write the final draft.

EXAMPLES

heat treating, tolerance, orthographic projection, laminating, square foot cost estimating, inert-gas welding, magnetism, radio, amplifier, integrated circuit, blueprint, metabolism, hematology, anemia, library, farm recreation, modeling, proteins, pH meter, radiograph, holography

Application 10   Giving Definitions

Give   an oral extended definition of a term in Application 9 above. Ask your classmates to evaluate your speech by filling in an Evaluation of Oral Presentations from pages 527–535.

USING PART II: SELECTED READINGS
Application 11   Giving Definitions

Read the poem "The Unknown Citizen," by W. H. Auden, pages 572–573.

a. In what ways is definition used in this selection?
b. Under "Suggestions for Response and Reaction," page 573, write Number 6.

Application 12 Giving Definitions

Read the excerpt from *Zen and the Art of Motorcycle Maintenance*, by Robert M. Pirsig, pages 600–605.

    a. In what ways is definition used in the selection?
    b. Under "Suggestions for Response and Reaction," page 605, write Number 2.

Application 13 Giving Definitions

Read "Holography," by Jonathan Back and Susan S. Lang, pages 589–594. Under "Suggestions for Response and Reaction," page 594, write Number 2.

Application 14 Giving Definitions

Read "Starch-Based Blown Films," by Felix Otey and others, pages 624–631. Under "Suggestions for Response and Reaction," page 631, write Number 3.

Application 15 Giving Definitions

Read "The World of Crumbling Plastics," by Stephen Budiansky, pages 622–623. Under "Suggestions for Response and Reaction," page 623, write Number 2.

PLAN SHEET
FOR GIVING AN EXTENDED DEFINITION

Analysis of Situation Requiring Definition

What is the term to be defined?

For whom is the definition intended?

How will the definition be used?

Will the definition be written or oral?

Sentence Definition

Term:　　　　　Class:　　　　　Distinguishing characteristics:

As Applicable

Synonyms:

Origin of the term or item:

Information concerning its discovery and development:

Analysis of its parts:

Physical description:

Necessary conditions, materials, equipment:

Description of how it functions:

Explanation of its uses:

Instructions for operating or using it:

Examples and illustrations:

Comparisons and contrasts:

Different styles, sizes, methods:

Data concerning its manufacture and sale:

When and where used or occurring:

History:

Other:

Types and Subject Matter of Visuals to Be Included

Sources of Information

Chapter 4

Analysis Through Classification and Partition: Putting Things in Order

Objectives

Upon completing this chapter, the student should be able to:

- Define classification
- State the basis of division into categories in a classification system
- Select for a classification system a basis of division that is useful and purposeful
- Set up a classification system whose categories are coordinate, mutually exclusive, nonoverlapping, and complete
- Present classification data in outlines, in verbal explanations, and in visuals
- Select an appropriate order for presentation of classification categories
- Give an analysis through classification
- Define partition
- State the basis of division in a partition system
- Select for a partition system a basis of division that is useful and purposeful
- Set up a partition system whose divisions are coordinate, mutually exclusive, nonoverlapping, and complete
- Present partition data in outlines, in verbal explanations, and in visuals
- Select an appropriate order for presentation of partition divisions
- Give an analysis through partition
- Give an analysis using both classification and partition
- Describe situations in which an employee would need to know how to give an analysis using classification or partition, or both

Introduction

Human beings try to make sense out of the world in which they live. They try to see how certain things are related to other things. They try to impose some kind of order on their environment. More specifically, the technical student tries to see how a skill in building trades or in engineering technology or health occupations is related to getting and keeping a good job and being able to support a family. Or perhaps the student is trying to devise a practical plan for getting enough money for college expenses.

Persons on the job try to make sense out of the industrial and business world in which they work. There are situations in which they need to put things in some kind of organized relationship. The situation may be a request from a superior for a list of parts that must be replaced in a machine. The situation may be a weekly report of the number of units produced in a particular department. The situation may be a memorandum to the division head on the problems encountered in a new manufacturing process. The situation may be a report to the doctor on the temperature changes of a patient.

All of these kinds of situations, whatever their nature and wherever they are experienced, call for analysis—looking at a subject closely so that it can be put into a useful, meaningful order. Establishing an order, or relationship, is the basic step in solving a problem, whether the problem is how to reduce air pollution, how to operate a blood bank more efficiently, or how to improve a variety of cotton.

The purpose of this chapter is to help an individual give order to information—to give order by classifying it into related groups or by partitioning it into its components.

Definition of Classification

Classification is a basic technique in organization, and thus in writing. It starts with the recognition that different items have similar characteristics and develops with the sorting of these items into related groups. In a letter of application, for instance, details about places of former employment, dates of employment, or names of supervisors all have similar characteristics in that they all have to do with work experience. In organizing the letter of application, items would be sorted into one group on the basis of work experience, into another group on the basis of education, and so on. Classification, then, is the grouping together, according to a specified basis, of items having similar characteristics.

For example, in preparing to explain how batteries generate power, batteries might first be broadly classified as primary and secondary types. The explanation might then be given as follows:

> Batteries are broadly classified as primary and secondary types. Primary batteries generate power by irreversible chemical reaction and require replacement of parts that are consumed during discharge (or, more commonly, the batteries are thrown away when discharged). Secondary batteries, on the other hand, involve reversible chemical reactions in which their reacting material will be restored to its original "charged" state by applying a reverse, or charging, current. In general, primary batteries provide higher energy density, specific energy, and specific power than do secondary batteries.

Classification is obviously the method used to organize the material in the paragraph.

Analysis Through Classification

Suppose that a newspaper reporter is visiting a campus to gather data from students for an informative article about the student body. During interviews with students she has jotted down some of their comments. The notes include the following:

A lot of the students live at home and commute to college.
Some students have real hangups, particularly when it comes to sex.
There are more freshmen than sophomores.
I'm a technical student in document processing. When I finish the two-year program here, I'll be able to get a good job in my specialized field.
I'm not taking a full number of courses this semester because I work eight hours a day to support my family.
Many students have a real money problem. Like my roommate. He's completely paying his own way.
I chose this college because it's close to home.
They say you gotta go to college to get anywhere. OK, here I am.
Somehow I didn't do too well on my ACT score, so I can't take all the courses I want to take.
Some students are lucky enough to have their own apartment.
The biggest thing I've had to cope with is finding enough time to study and to play—all in the same weekend.
This college is really growing. This year we have the biggest freshman class in the history of the college.

The newspaper reporter must put these items in order. Assist her in organizing the information into related groups so that she can write an article her readers can follow.

In sorting this information into meaningful categories, group together the items that are related in a specific way. In the sorting process, start with the first statement:

> A lot of the students live at home and commute to college.

This has to do with student residence. Find other statements that have to do with where students live while attending college.

> Some students are lucky enough to have their own apartment.

This is the one directly related item.

Further analysis of the reporter's notes shows that there are additional groupings of information. Several statements deal with problems of students, several with reasons for attending college, several with the status of the student according to the number of hours completed, and several with the status of the student according to the number of hours currently enrolled in, as shown below:

Residence of Students While Attending College:

A lot of the students live at home and commute to college.
Some students are lucky enough to have their own apartment.

Problems of Students:

Some students have real hangups, particularly when it comes to sex.
Many students have a real money problem. Like my roommate. He's completely paying his own way.
The biggest thing I've had to cope with is finding enough time to study and to play—all in the same weekend.

Reasons for Students' Attending College:

I'm a technical student in document processing. When I finish the two-year program here, I'll be able to get a good job in my specialized field.
I chose this college because it's close to home.
They say you gotta go to college to get anywhere. OK, here I am.

Status of Student According to Number of Hours Completed:

There are more freshmen than sophomores.
This college is really growing. This year we have the biggest freshman class in the history of the college.

Status of Student According to Number of Hours Currently Enrolled In:

I'm not taking a full number of courses this semester because I work eight hours a day to support my family.
Somehow I didn't do too well on my ACT score, so I can't take all the courses I want to take.

The process of organizing the newspaper reporter's notes into groups, or categories, is analysis through classification. This process of organization was not carried out haphazardly but logically. In the first place, it is clear that the subject being classified is the reporter's notes, not students or courses or factors affecting student achievement. Moreover, to help the newspaper reporter sort the notes into useful categories, you had to be knowledgeable about the various aspects of a student body.

Five different bases were used for sorting the information:

Residence of students while attending college
Problems of students
Reasons for students' attending college
Status of student according to number of hours completed
Status of student according to number of hours currently enrolled in

These bases are clear and logical. If, for instance, the last two bases had been considered as one basis, such as "Status of student," there would be confusion. The term *status of student* has two different meanings here, depending on whether reference is made to the student as a freshman, sophomore, or upper classman ("Status of student according to number of hours completed") or whether reference is made to the student as a full-time student, part-time student, probationary student, or whatever ("Status of student according to number of hours currently enrolled in").

In the example, the categories are coordinate; that is, they are all on the same level, with no confusion of major categories with subcategories. If, however, along with the five major categories there had been included "Students who have money problems," the major categories, or classes, would not be coordinate. They would no longer be on the same level because "Students who have money problems" is a subcategory, or subclass, of "Problems of students."

Each of the categories is mutually exclusive; that is, each of the five categories is composed of a clearly defined group that would still exist without the other categories. "Residence of students while attending college," for example, would be a valid category even if some or all of the other categories were unnecessary.

In the example, the categories do not overlap; an item can be placed in only one category. For instance, the item "I chose this college because it's close to home" can be placed in only one of the groups: "Reasons for students' attending college." The speaker of the "close to home" comment may live at home and commute to school, but the comment itself could not possibly be grouped under "Residence of students while attending college" or any of the groups except "Reasons for students' attending college."

Finally, in the classification of the reporter's notes into five categories, each item of information fits into a category; no item is left out.

ITEMS THAT CAN BE CLASSIFIED

Any group of items *plural* in meaning may be classified. These items may be objects, concepts, or processes; the items may be classified into a variety of categories. Objects such as men's beach coats might be classified by size, style, fabric, color, and so on. Concepts or ideas, such as the causes of the Industrial Revolution in England, may make use of the principles of classification; the causes, for instance, may be classified according to significance, origin, historical influences, and so forth.

Processes frequently employ classification—methods of desalinizing water may be classified according to cost, required time, materials needed, and so on, or particular steps may be classified as one phase of a desalinization method.

CHARACTERISTICS OF A CLASSIFICATION SYSTEM

The categories in a classification system, as illustrated in organizing the information about a student body, must be *coordinate, mutually exclusive, nonoverlapping,* and *complete*. A classification system must have all of these qualities if it is to be adequate.

Coordination. The categories in a classification system must be coordinate, or parallel. The groups, or categories, that items are sorted into must be on the same level in grammatical form and content. For instance, classification of the refrigerators in Mr. Lee's appliance store, according to source of power, into the following categories would be inadequate: natural gas, butane gas, and electric. The error is in the grammatical form of the word *electric*, which here is an adjective; *natural gas* and *butane gas* here are nouns. These categories, however, would be adequate: natural gas, butane gas, and electricity.

Categories must be coordinate in content as well as in grammatical form. The classification of automobile tires as whitewall, blackwall, red stripe, and tubeless is not coordinate because *tubeless* does not refer to the same content, or substance, that *whitewall, blackwall,* and *red stripe* do. These three terms have to do with decorative coloring; *tubeless* does not.

See also Parallelism in Sentences, pages 655–656, and Coordinating Conjunctions, page 692.

Mutual Exclusiveness. The categories in a classification system must be mutually exclusive; that is, each category must be independent of the other categories in existence. A category must be composed of a clearly defined group that would still be valid even if any or all of the other categories were unnecessary or nonexistent. For instance, if a shipment of coats were being classified according to the amount of reprocessed wool each coat contained and the categories were designated "Coats in Group A," "Coats with more reprocessed wool than Group A," and "Coats with less reprocessed wool than Group A," the categories would be inadequate. Two of the categories depend upon "Group A" for their meaning, or existence. A more logical classification could be "Coats containing 40 to 60 percent reprocessed wool," "Coats containing more than 60 percent reprocessed wool," and "Coats containing less than 40 percent reprocessed wool." In this grouping, each category is independent, that is, mutually exclusive, of the other categories.

Nonoverlapping. The categories in a classification system must be nonoverlapping. It should be possible to place an item into only *one* category. If an item, however, can reasonably be placed under more than one category, the categories should be renamed and perhaps narrowed. For instance, if the fabrics in a sewing shop were being classified as natural fibers, synthetics, or blends, how would a fabric that is part cotton and part wool be classified? The fabric is of natural fibers but it is also a blend. Thus the given categories are inadequate. More satisfactory would be categories such as these: fabrics of a pure natural fiber, fabrics of blended natural fibers, synthetic fabrics, and fabrics that have a blend of synthetic and natural fibers.

Completeness. The categories in a classification system must be complete. Every item to be classified must have a category into which it logically fits, with no item left out. For instance, the categories in a classification of American-made automobiles according to the number of cylinders would not be sufficient unless the categories were these: four cylinders, six cylinders, and eight cylinders. The omission of any one of the categories would make the classification system incomplete, for a number of automobiles then would have no category into which they would logically fit.

BASES FOR CLASSIFICATION

Usefulness. The bases on which classifications are made should be useful. Classification of trees according to bark texture or according to leaf structure would be of significance to one interested in botany. It would take a stretch of the imagination, however, to understand the usefulness of classifying trees according to the number of leaves produced and shed over a 50-year span.

Purpose. The usefulness of a classification system depends on its purpose. And according to this purpose, the writer or speaker, as classifier, will emphasize certain aspects of the subject. Suppose, for instance, that a salesperson in a garden center has three pieces of mail regarding lawn mowers that must be answered. There is a letter from a woman who wishes to purchase a lawn mower that she can start easily. She wants suggestions for such models. There is a letter from a meticulous gardener who gives certain motor, cutting blade, and attachment specifications. He would like suggestions for several models that most nearly meet his requirements. There is a memorandum from the store manager requesting a list of the best-selling lawn mowers. Obviously, each of the three persons is seeking quite different information for different purposes—but the information concerns the same group of lawn mowers, the lawn mowers that the garden center stocks. In order to meet each request, the salesperson must classify the lawn mowers according to at least three different bases: ease in starting (for the woman); motor, cutting blade, and attachment specifications (for the gardener); and popularity of sale (for the store manager).

ORDER OF DATA PRESENTATION

The categories, or classes, in a classification system should be presented in the order that will best help to accomplish your purpose. In classifying steps in a process, probably a *chronological*, or *time*, *order* would be used. In classifying items such as breeds of cattle in the United States, it might be wise to use an *order of familiarity*, that is, to start with the best known or most familiar and go to the least known or least familiar. In classifying the qualities a nurse should have, the qualities might be listed in *order of importance*. At times, *order according to complexity*, that is, movement from the simple to the more difficult, might be best to use, as in classifying swimming strokes or in classifying casserole recipes. A *spatial order*, movement from one physical point to another (from top to bottom, inside to outside, left to right, etc.), may be the most practical arrangement, as in classifying the parts of an automobile engine or the furnishings in a house. If the categories are such that order is not important, an *alphabetical* or a *random presentation* may be used. The important thing to remember is this: Use whatever order is appropriate for whatever you want to accomplish in your presentation.

For additional discussion on order, see Part III: Handbook, pages 675–677.

FORMS OF DATA PRESENTATION

Classification may be the major organizing principle in a presentation or in a section of a presentation, or it may be the major organizing principle within a paragraph or in a part of a paragraph. In whichever way classification is used, there are several forms in which the data may be presented: in outlines, in verbal explanations, and in visuals.

Outline. Outlines are particularly helpful for grouping items in an orderly, systematic arrangement. The outline may be purely for personal use in preparing information for presentation in a verbal explanation or in a visual. Or the outline may be the form selected for presentation of the information. The outline that others are to read should be clear, consistent, and logical. If the outline is to meet these requirements, certain accepted standards should be followed:

1. Choose either the traditional number-letter outline form or the decimal outline form. (Illustrations of both outline forms follow.)
2. Make the outline either a topic outline (illustrations follow) or a sentence outline (illustrated on pages 138–139). In a topic outline each heading is a word, phrase, or clause. In a sentence outline each heading is a complete sentence. *Do not combine both topics and sentences in an outline.* (Such a combination usually confuses the reader.)
3. Headings should be given an appropriate number or letter. In the traditional number-letter outline form, first-level, or major, headings are designated with roman numerals (I, II, III, etc.); second-level headings have uppercase letters (A, B, C, etc.); third-level headings have Arabic numerals (1, 2, 3, etc.); and fourth-level headings have lowercase letters (a, b, c, etc.). In the decimal outline form a system of decimal points is used to designate the various levels of headings.
4. Headings should be indented to indicate the degree of subclassification.
5. Headings on a given level should keep the same grammatical structure. Nouns should be used with nouns, gerund phrases with gerund phrases, etc.
6. There should be at least two headings on each level. Each level is composed of divisions of the preceding heading. (If, however, a heading calls for a *list*, it is possible to have only one item in the list.)

TOPIC OUTLINES: CLASSIFICATION OF SAWS

Two Bases: Uses and Types

Traditional Number-Letter
Outline Form

    I. Basic uses
       A. Crosscutting
       B. Ripping

Decimal Outline Form

    1. Basic uses
       1.1 Crosscutting
       1.2 Ripping

II. Types
    A. Hand-operated saws
        1. Handsaw
        2. Backsaw
        3. Keyhole saw
        4. Coping saw
        5. Hacksaw
    B. Power saws
        1. Portable power saws
           a. Circular saw
           b. Saber saw
           c. Reciprocating saw
        2. Stationary power saws
           a. Radial arm saw
           b. Table saw
           c. Motorized miter saw

2. Types
  2.1 Hand-operated saws
      2.1.1 Handsaw
      2.1.2 Backsaw
      2.1.3 Keyhole saw
      2.1.4 Coping saw
      2.1.5 Hacksaw
  2.2 Power saws
      2.2.1 Portable power saws
          2.2.1.1 Circular saw
          2.2.1.2 Saber saw
          2.2.1.3 Reciprocating saw
      2.2.2 Stationary power saws
          2.2.2.1 Radial arm saw
          2.2.2.2 Table saw
          2.2.2.3 Motorized miter saw

A well-organized classification presentation can be easily outlined. For example, the student-written classification of southern lawn grasses on pages 133–134 might be outlined as follows:

SOUTHERN LAWN GRASSES

I. Warm-season lawn grasses
    A. Bermuda grasses
    B. Zoysia grasses
    C. St. Augustine grass
    D. Centipede grass

II. Cool-season lawn grasses
    A. Bluegrasses
    B. Tall fescue grasses

Verbal Explanation. Although the outline, as a presentation form of a classification system, may be used alone, it is more frequently accompanied by verbal explanation; sometimes it is used with a visual. Most often, however, a classification system is presented as a verbal explanation, as in the following classification system of southern lawn grasses. Note that in the explanation, lawn grasses are divided into two broad categories. These two broad categories are divided into subcategories, which are still further divided. Note also the interrelationship of classification and definition. The Plan Sheet that the student filled in before writing the report is included, and marginal notes have been added to indicate the development of the classification system.

PLAN SHEET
FOR GIVING AN ANALYSIS THROUGH CLASSIFICATION

Analysis of Situation Requiring Classification

What is the subject to be classified?
southern lawn grasses

For whom is the classification intended?
new home owners

How will the classification be used?
to decide on a lawn grass

Will the classification be written or oral?
written

Setting up the Classification System

Definition or identification of subject:
selection of lawn grass a major decision before preparing lawn

Basis (or bases) for classification:
planting season, varieties of grasses, characteristics, planting method, care

Significance or purpose of basis (or bases):
An attractive lawn increases the value of a house.

Categories of the subject, with identification of each category:
Warm-season lawn grasses — planted in spring (March–July)
   Bermuda
      Types — Tifgreen, Tiflawn, Tifway, Tidwarf, common
      Characteristics — bright emerald to dark green; fine leaf texture; does not grow
         well in shade
      Planting method — common seeded (1 lb. per 1000 sq. ft.); other sprigs or plugs
         (10–12 inches apart, 10–12 inches between rows)
      Care — 2–3 lbs. nitrogen per 1000 sq. ft. in April–May, June–July, Aug.–Sept.;
         mowed regularly
   Zoysia
      Types — Emerald, Japonica, Matrella, Meyer
      Characteristics — grows well in shade; tolerant of low temperatures, frost
      Planting method — sprigs or plugs; plants and rows 6–8 inches apart; plugs
         (2 inches apart) and rows (6 inches apart)
      Care — 2 lbs. nitrogen per 1000 sq. ft., April–May, June–July, Aug.–Sept.

St. Augustine
 One type
 Characteristics—dark green, coarse textured; susceptible to insects, disease, weather. Grows in shade; tolerates salt spray.
 Planting method—no seed; sprigs or plugs, plants and rows 10–12 inches apart
 Care—2 lbs. nitrogen per 1000 sq. ft. April–May and June–July; 1 lb. Aug.–Sept.
Centipede
 One type
 Characteristics—light green; coarse leaf texture; needs warm temperatures; light shade. Tolerates insects and disease. Prefers slightly acid soil.
 Planting method—seed (¼ to ½ lb. per 1000 sq. ft.; very expensive); sprigs, plants and rows 10–12 inches apart
 Care—light fertilization, 1 lb. nitrogen per 1000 sq. ft. in April–May, June–July, Aug.–Sept.; little mowing
Cool-season lawn grasses—planted in fall (Sept.–Nov.)
 Bluegrasses
  Types—Fylking, Kenblue, Park, Windsor
  Characteristics—medium green; fine leaf; reasonable growth in shade
  Planting method—seed, 2–3 lbs. per 1000 sq. ft; solid sodding
  Care—1–2 lbs. nitrogen per 1000 sq. ft. in March; 2–3 lbs. in Sept.
 Tall Fescue
  Types—Kentucky 31, Kenwell
  Characteristics—dark green; coarse texture; fair shade growth
  Planting method—seed, 5–10 lbs. per 1000 sq. ft.
  Care—2 lbs. nitrogen per 1000 sq. ft. in March, 2–3 lbs. in Sept.

Presentation of the System

Most logical order for presenting the categories:
random listing

Types and subject matter of visuals to be included:
omit

Sources of Information

The Pasture Book, World Book Encyclopedia, Cooperative Extension Service, county agent

CLASSIFICATION OF SOUTHERN LAWN GRASSES
BY PLANTING SEASON

Subject identified
Importance of subject

An attractive lawn increases the value of a house. But an attractive lawn does not develop by accident; it is carefully planned, prepared, and maintained. A major decision in the planning stage for a southern lawn is the selection of the right grass. Basically the selection of grass depends upon the season when planting will be done.

Subject divided into two broad categories on basis of planting season

There are two major planting seasons for southern lawn grasses: warm-season planting and cool-season planting. Warm-season grasses should be planted in the spring (March–July). Cool-season grasses should be planted in the fall (September–November).

First broad category divided on basis of selection of grasses

WARM-SEASON LAWN GRASSES

The possible selections of warm-season grasses include Bermuda grasses, zoysia, St. Augustine, and centipede. The choice of one of these grasses can be made after considering the available varieties, characteristics, planting method, and basic care.

Each selection described by varieties, characteristics, planting method, care

First selection described by varieties

Bermuda Grasses

The Bermuda grasses recommended for home lawn use in the South include Tifgreen, Tiflawn, Tifway, Tidwarf, and common Bermuda. The color of these grasses varies from a bright emerald green to a dark green. Generally Bermuda grasses have a fine leaf texture. Bermuda grasses do not grow well in the shade. Common Bermuda grasses can be seeded; all other varieties require vegetative planting, either by sprigs or plugs. Common Bermuda should be seeded 1 pound per 1000 square feet; other varieties should have sprigs or plugs placed 10–12 inches apart in rows 10–12 inches apart.

by characteristics

by planting method

by care

All varieties must be highly fertilized. Recommended fertilizer includes 2–3 pounds of nitrogen per 1000 square feet in April–May, June–July, and August–September. All varieties must be mowed regularly to look their very best.

Second selection described by varieties

Zoysia Grasses

Zoysia grasses used for southern lawns are Emerald, Japonica, Matrella, and Meyer.

by characteristics
by planting method

These grasses grow well in the shade and are tolerant of low temperatures, including even frost. They should be planted by sprigs or plugs. Sprigs should be placed 6–8 inches apart in rows 6–8 inches apart; 2-inch plugs should be placed 6 inches apart in rows 6 inches apart. Zoysia should be fertilized with 2 pounds of nitrogen per 1000 square feet in April–May, June–July, and August–September.

by care

Third selection described by characteristics

St. Augustine Grass

St. Augustine, a single variety, is dark green and coarse textured. It does not tolerate insects, diseases, or weather well. St. Augustine does grow in shade and is not affected by salt spray, making it a good choice for lawns in coastal areas.

by planting method

No St. Augustine seeds are available; the grass is established vegetatively by sprigs or plugs. Sprigs or plugs should be planted by care | 10–12 inches apart in rows 10–12 inches apart. Nitrogen amounts to maintain healthy green grass are 2 pounds of nitrogen per 1000 square feet in April–May and June–July with 1 pound per 1000 square feet in August–September.

Centipede Grass

Fourth selection described by characteristics

by care | Centipede grass, a single variety, is light green with a coarse leaf texture. It prefers warm temperatures and will grow in lightly shaded areas. Not susceptible to insects or disease, centipede grows in most soils, although it prefers a slightly acid soil. This grass can be started from seed or sprigs, but the seeds are quite expensive. Each 1000 square feet requires the spreading of ¼ to ½ pound of seed; sprigs should be set 10–12 inches apart in rows 10–12 inches apart. Centipede requires little fertilization, 1 pound of nitrogen per 1000 square feet in April–May, June–July, and August–September. It also needs little mowing.

COOL-SEASON LAWN GRASSES

Second broad category divided on basis of selection of grasses

Two common cool-season lawn grasses are bluegrass and tall fescue.

Bluegrasses

First selection described by varieties

by characteristics
by care | Improved varieties of Kentucky bluegrass are Fylking, Kenblue, Park, and Windsor. Bluegrass has a medium green color and a fine leaf texture. It will grow reasonably well in shaded areas. It can be grown from seed or by solid sodding. Seeding rate is 2–3 pounds per 1000 square feet. Nitrogen amounts recommended are 1–2 pounds per 1000 square feet in March and 2–3 pounds per 1000 in September.

Tall Fescue Grasses

Second selection described by varieties

by characteristics
by care | Varieties of tall fescue include Kentucky 31 and Kenwell.

Tall fescue has a very coarse texture and is a dark green. This grass will grow fairly well in the shade. Tall fescue is established by seeding 5–10 pounds of seed per 1000 square feet. Nitrogen should be applied in March, 2 pounds per 1000 square feet, and in September, 2–3 pounds per 1000 square feet.

Closing comment | The person planning a lawn in the South has a wide choice of grasses among the warm-season and the cool-season lawn grasses. For information concerning lawn care and maintenance, an excellent source is the Cooperative Extension Service personnel in each county.

Visuals. In addition to presenting data in a classification system as an outline or as a verbal explanation, visuals may be used. Such visuals as charts, diagrams, maps, photographs, drawings, graphs, and tables frequently make a mass of information understandable. Study the line graph/bar chart on page 135. Determine the data being organized; then determine the bases on which the data are organized. Why are the data presented in this form?

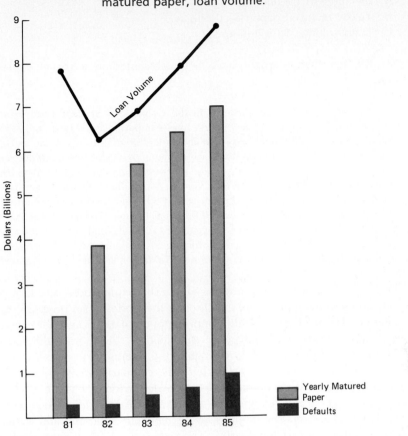

Guaranteed Student Loan Program, FY81–FY85 defaults, matured paper, loan volume.

Of course, the information in the chart might have been presented in outline form, as follows:

I. Fiscal year 1981
   A. Defaults: $0.3 billion
   B. Yearly matured paper: $2.3 billion
   C. Loan volume: $7.8 billion

II. Fiscal year 1982
   A. Defaults: $0.3 billion
   B. Yearly matured paper: $3.8 billion
   C. Loan volume: $6.3 billion

III. Fiscal year 1983
   A. Defaults: $0.5 billion
   B. Yearly matured paper: $5.7 billion
   C. Loan volume: $6.9 billion

IV. Fiscal year 1984
   A. Defaults: $0.7 billion
   B. Yearly matured paper: $6.4 billion
   C. Loan volume: $7.9 billion

V. Fiscal year 1985
   A. Defaults: $1.0 billion
   B. Yearly matured paper: $7.0 billion
   C. Loan volume: $8.9 billion

Or the information might have been presented as a verbal explanation, as follows:

> One of the important concerns in the Guaranteed Student Loan Program is the escalating default costs. A view of the five-year period from 1981 to 1985 shows that defaults have steadily increased (with annual additions to loan volume and increases in yearly matured paper—matured paper is the cumulative dollar amount of loans that have ever entered repayment). In fiscal year 1981, defaults totaled $0.3 billion; the matured paper was $2.3 billion and the loan volume $7.8 billion. In fiscal year 1982, defaults totaled $0.3 billion; the matured paper was $3.8 billion and the loan volume $6.3 billion. In fiscal year 1983, defaults totaled $0.5 billion; the matured paper was $5.7 billion and the loan volume $6.9 billion. In fiscal year 1984, defaults totaled $0.7 billion; the matured paper was $6.4 billion and the loan volume $7.9 billion. In fiscal year 1985, defaults totaled $1.0 billion; the matured paper was $7.0 billion and the loan volume $8.9 billion.

Of the three forms—chart, outline, and verbal explanation—used for presenting the data concerning money paid out through the student loan program, the chart is the clearest. The reader can easily comprehend the data because for each of the two items charted, yearly matured paper and defaults, there are light and dark bars representing each year; for the loan volume, the amounts are plotted as a line graph; and there is a monetary scale on the left. The comparison is visibly evident in the chart. In the outline, the comparison is somewhat evident, but the reader has to do more work to grasp the relationships that are so plainly reflected in the chart. In the verbal explanation, the comparison of monetary amounts and dates becomes wearisome after the first few figures. The reader has to do a great deal of work to comprehend the information and understand the relationships that the writer is trying to show.

Thus the data regarding student loan claims were presented in a chart because this form clearly communicates the information.

For further discussion of visuals, see Chapter 11, Visuals.

Combination. Frequently a combination of outline, visual, and verbal explanation is used in presenting data. The following example classifying flowchart symbols effectively combines verbal explanation and visuals.

CLASSIFICATION OF DATA PROCESSING FLOWCHART SYMBOLS

Data processing flowchart symbols may be classified into three groups, depending on their use: basic symbols, symbols related to programming, and symbols related to systems.

Basic Symbols

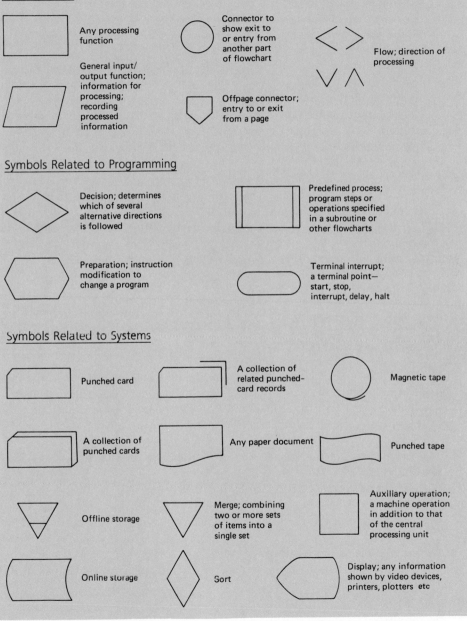

Any processing function

Connector to show exit to or entry from another part of flowchart

Flow; direction of processing

General input/ output function; information for processing; recording processed information

Offpage connector; entry to or exit from a page

Symbols Related to Programming

Decision; determines which of several alternative directions is followed

Predefined process; program steps or operations specified in a subroutine or other flowcharts

Preparation; instruction modification to change a program

Terminal interrupt; a terminal point— start, stop, interrupt, delay, halt

Symbols Related to Systems

Punched card

A collection of related punched- card records

Magnetic tape

A collection of punched cards

Any paper document

Punched tape

Offline storage

Merge; combining two or more sets of items into a single set

Auxiliary operation; a machine operation in addition to that of the central processing unit

Online storage

Sort

Display; any information shown by video devices, printers, plotters etc

General Principles in Giving an Analysis Through Classification

1. *Classification is a basic approach in analysis.* It places related items into categories, or groups.
2. *Only a plural subject or a subject whose meaning is plural can be classified.* If a subject is singular, it can be partitioned but not classified.
3. *The categories in classification must be coordinate, or parallel.* All categories on the same level must be of the same rank in grammatical form and in content.
4. *The categories must be mutually exclusive.* Each category should be composed of a clearly defined group that would still exist without the other categories on the same level.
5. *The categories must not overlap.* An item can have a place in only one category.
6. *The categories must be complete.* There should be a category for every item, with no item left out.
7. *The data in a classification analysis may be presented in outlines, in verbal explanation, and in visuals.* The form or combination of forms that presents the data most clearly should be used.
8. *The basis on which classification is made should be clear, useful, and purposeful.*
9. *The order of presentation of categories depends on their purpose.* Among the possible orders are time, familiarity, importance, complexity, space, and alphabetical and random listing.

Procedure for Giving an Analysis Through Classification

This kind of analysis is generally a part of a larger whole, as have been the other forms of communication discussed in the preceding chapters. Regardless of whether the classification analysis is a dependent or an independent communication, however, the structure of such an analysis is as follows:

I. The presentation of the subject and of the bases of division into categories is usually brief.
   A. State the subject that is to be divided into categories.
   B. Identify or define the subject.
   C. If applicable, list various bases by which the subject can be divided into categories.
   D. State explicitly the bases of the categories.
   E. Point out the reasons the categories are significant and what purpose they serve.
II. The listing and the discussion of the categories are the longest parts of the presentation.
   A. List the categories and, if any, the subcategories.
   B. Give sufficient explanation to clarify and differentiate among the given categories.

C. Present the categories in whatever order best serves the purpose of the analysis.

D. Divide the categories into subcategories wherever needed.

E. Use outlines and visuals whenever they will help clarify the explanation.

III. The closing (usually brief) depends on the purpose of the presentation.

A. The closing may be the completion of the last point in the analysis.

B. The closing may be a comment on the analysis or a summary of the main points.

Application 1 Giving an Analysis Through Classification

Some of the following classifications are satisfactory and some are not. Indicate the sentences containing unsatisfactory classifications. Rewrite them to make them satisfactory.

1. The students on our campus can be classified either as blonds or as brunets.
2. Automobiles serve two distinct functions: They are used either for pleasure or for business.
3. Cattle may be classified as purebred, crossbred, or unbred.
4. The courses a technical student takes at this college can be divided into technical courses, related courses, and general education courses.
5. Structural material may be wood, brick, aluminum, steel, concrete blocks, or metal.

Application 2 Giving an Analysis Through Classification

Each of the following classification systems is lacking in *coordination, mutual exclusiveness, nonoverlapping,* and/or *completeness*. Point out the specific errors. Then make whatever changes are necessary for the outlines to be more adequate.

I. Classification of Watches According to Shape
   A. Round
   B. Oblong
   C. Thin
   D. Oval

II. Classification of Phonograph Records According to Type of Music
   A. Classical
   B. Jazz
   C. Michael Jackson
   D. Country and Western
   E. Instrumental

III. Classification of Mail According to Collection Time
   A. Day
      1. Weekdays
      2. Saturdays
      3. Holidays
   B. Hour
      1. A.M.
      2. Afternoons

IV. Classification of Television Shows
   A. Westerns
   B. Situation comedies
   C. Today Show
   D. NBC
V. Classification of College Students
   A. Academic classification
      1. Freshmen
      2. Sophomores
      3. Upperclassmen
   B. Residence
      1. Dormitory
      2. Local
      3. In state
      4. Out of state

Application 3 Giving an Analysis Through Classification

For each of the following groups of items, suggest at least three bases of classification.

EXAMPLE

Group of items: Books
Bases of classification: Subject matter, nationality of author, cost, date of writing, publisher, alphabetical listing by author

1. Metals
2. Office machines
3. Diseases
4. Buildings
5. Clothing

Application 4 Giving an Analysis Through Classification

For Application 3 above, choose one group of items and one basis of classification. For that basis of classification write an explanatory paragraph.

Application 5 Giving an Analysis Through Classification

Joe received the following injuries in an automobile accident: black eyes, broken nose, fractured ribs, crushed pelvis, broken thumb, twisted ankle, cuts on the forehead, teeth knocked out, dislocated knee, gash on the right leg, bruised shoulder, and tip of little finger cut off. Organize Joe's injuries into related groups. Use a formal outline.

Application 6   Giving an Analysis Through Classification

From the following list of words pick out at least four groups of related items and identify the relationship within the group. Use your dictionary to look up any unfamiliar words.

EXAMPLE

List of Words	Word Group	Relationship
Electrical pressure	Electrical pressure	Basic electric quantities
Magnets	Ohm	
Ohm	Ampere	
Ampere	Coulomb	
Copper wire		
Coulomb		

1. Gothic
2. Propeller
3. Airbrush
4. Dry cells
5. FORTRAN
6. Blood pressure
7. Horn
8. Nonphotographic pencil
9. Drawing board
10. Classifying
11. Drill press
12. Keyboard
13. Tractor
14. Dissecting microscope
15. Storing
16. Micrometer
17. COBOL
18. X-acto knife
19. Dial indicator
20. Plaintiff
21. Breathing
22. Ruling pen
23. Planer
24. Sorting
25. Text
26. Wing
27. Stirrup
28. Primary cells
29. Punched cards
30. RPG
31. Preserved specimen
32. Screen
33. Pulse
34. Arrest
35. Lathe
36. T square
37. Central processing unit
38. Slide
39. Magnetic tape
40. Updating
41. Dial caliper
42. Roman
43. Lead-acid storage cells
44. Reflexes
45. BASIC
46. Flexible disk
47. Printer
48. Witness
49. Strap
50. Shaper
51. Hay baler
52. Compass
53. Triangles
54. Seat
55. Italic
56. Vernier bevel protractor
57. PASCAL
58. Program
59. Drawing set
60. Combine

Application 7   Giving an Analysis Through Classification

Select one of the following subjects. Divide the subject into classes in accordance with at least four different bases; subdivide wherever necessary. Use outline form to show the relationship of divisions.

EXAMPLE

Automobiles

I. Body style
  A. Sedan
  B. Coupe
  C. Convertible
  D. Hard top

II. Body size
  A. Regular
  B. Intermediate
  C. Compact
  D. Subcompact

III. Kind of fuel
  A. Gasoline
  B. Diesel
  C. Gasohol
  D. Butane
  E. Propane

IV. Cost
  A. Inexpensive—under $10,000
  B. Average—$10,000–$18,000
  C. Expensive—over $18,000

1. Computers
2. Farm machinery
3. Cattle
4. Hand tools
5. Drawing instruments
6. Scales
7. Precision measuring instruments
8. Building materials
9. Transistors
10. Muscles
11. Home air conditioners
12. Commercial aircraft
13. Typewriters
14. Clocks
15. Clothing
16. Food
17. Transportation
18. Tires
19. Trees
20. Meters
21. Furniture
22. Wrenches
23. Books
24. Radios
25. Paint
26. Flexible disks

Application 8   Giving an Analysis Through Classification

For the subject you chose in the preceding application (or another of the listed subjects), present the data in a written explanation.

  a. Fill in the Plan Sheet on pages 145–146.
  b. Write a preliminary draft.
  c. Revise. See inside back cover.
  d. Write the final draft.

Application 9   Giving an Analysis Through Classification

Give an oral classification analysis of one of the topics in Application 7 above. Ask your classmates to evaluate your speech by filling in an Evaluation of Oral Presentations from pages 527–535.

Application 10 Giving an Analysis Through Classification

Modern automobiles can be bought with numerous options. List these options, and then group them into categories. Use outline form to show the relationship between divisions.

Application 11 Giving an Analysis Through Classification

Present the data from Application 10 in a written explanation.

a. Fill in the Plan Sheet on pages 147–148.
b. Write a preliminary draft.
c. Revise. See inside back cover.
d. Write the final draft.

Application 12 Giving an Analysis Through Classification

Assume that an insurance company will give you an especially good rate if you will insure all your possessions with it. Prepare an *organized* list of your belongings so that the insurance company can suggest the amount of coverage you need.

Application 13 Giving an Analysis Through Classification

From the statistics in Table 1, select information and present it in another visual form, such as a pie chart (see pages 544–545) or bar chart (see pages 545–546).

TABLE 1 Carraway Community College
Enrollment Summary 1987–1988

Session	Campuses			Total
Fall	Sanders	Mayville	Rock Springs	
Academic	2840	367	291	3498
Technical	1937	680	130	2747
Vocational	682	357	250	1289
Other	54	34	1196	1284
				8818
Spring				
Academic	2576	347	268	3191
Technical	1881	742	72	2695
Vocational	636	359	201	1196
Other	36	16	1147	1199
				8281
Summer				
Academic	1066	129	220	1415
Technical	538	156	58	752
Vocational	266	418	138	822
Other	27	66	58	151
				3140

Application 14   Giving an Analysis
Through Classification

Select a subject suitable for classification. Present the classification system in a visual.

USING PART II: SELECTED READINGS

Application 15   Giving an Analysis
Through Classification

Read "Are You Alive?" by Stuart Chase, pages 611–615.

a. Outline the article to show its use of classification.
b. Under "Suggestions for Response and Reaction," write Number 1.

Application 16   Giving an Analysis
Through Classification

Read "Videotex: Ushering in the Electronic Household," by John Tydeman, pages 574–582. Under "Suggestions for Response and Reaction," write Number 2.

PLAN SHEET
FOR GIVING AN ANALYSIS THROUGH
CLASSIFICATION

Analysis of Situation Requiring Classification

What is the subject to be classified?

For whom is the classification intended?

How will the classification be used?

Will the classification be written or oral?

Setting Up the Classification System

Definition or identification of subject:

Basis (or bases) for classification:

Significance or purpose of basis (or bases):

Categories of the subject, with identification of each category:

Presentation of the System

Most logical order for presenting the categories:

Types and subject matter of visuals to be included:

Sources of Information

PLAN SHEET
FOR GIVING AN ANALYSIS THROUGH
CLASSIFICATION

Analysis of Situation Requiring Classification

What is the subject to be classified?

For whom is the classification intended?

How will the classification be used?

Will the classification be written or oral?

Setting Up the Classification System

Definition or identification of subject:

Basis (or bases) for classification:

Significance or purpose of basis (or bases):

Categories of the subject, with identification of each category:

Presentation of the System

Most logical order for presenting the categories:

Types and subject matter of visuals to be included:

Sources of Information

Analysis Through Partition

DEFINITION OF PARTITION

Partition is analysis that divides a singular item into parts, steps, or aspects. Only *singular* subjects can be partitioned; plural subjects are classified. Partition breaks down into its components a concrete subject, such as a tree (parts: roots, trunk, branches, and leaves), or an abstract subject, such as how to build a herd of cattle (steps: select good stocker cows, select a good herd bull, breed the cows at the right time, keep the heifers, sell the bull calves, and sell the old cows), or such as inflation (aspects: causes, effects on consumers, etc.).

A partition system must have certain characteristics if it is to be adequate; a partition system may be presented in various forms; divisions may be made on various bases; and data may be presented in several orders. All these aspects of analysis through partition are similar to those of analysis through classification.

CHARACTERISTICS OF A PARTITION SYSTEM

If a partition system is to be adequate, that is, if it is to fulfill its purpose, the divisions must have certain characteristics:

1. The divisions must be coordinate.
2. The divisions must be mutually exclusive.
3. The divisions must not overlap.
4. The divisions must be complete.

Consider the following partition of a concrete subject, a lamp socket, on the basis of its construction.

PARTITION OF A LAMP SOCKET

For the do-it-yourself person, replacing the base in a lamp socket can be simple. The parts of a lamp socket are few and uncomplicated as shown in the following drawing.

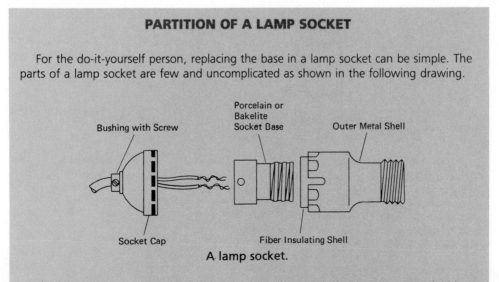

A lamp socket.

The lamp socket has an outer metal covering with an interlocking cap topped with a bushing and screw. Lodged inside in a fiber insulating shell is the socket base, usually made of porcelain or Bakelite. To replace the socket base, the repairer has only to disconnect the wires from the old base, remove that base, insert a new socket base, reconnect the wires, and pull the socket cap over the outer metal shell so that the lamp socket is one piece again.

The partition of the lamp socket is adequate. The divisions are coordinate: They are of equal rank in grammatical form and in content, as shown below:

Lamp Socket Partition

Outer parts } *First level of partition*
  Outer metal shell ⎤
  Socket cap ⎬   *Second level of partition*
  Bushing with screw ⎦

Inner parts } *First level of partition*
  Fiber insulating shell ⎤
  Socket base ⎦   *Second level of partition*

On each level, the divisions are mutually exclusive; that is, each division could exist without the other divisions. For instance, the outer metal shell could still exist even if there were no socket cap. The divisions do not overlap; a part has a place in only *one* division. The divisions are complete; every part of the lamp socket is accounted for and no part is left out.

Or, consider the claw hammer. It can be partitioned into two main parts: the handle and the head, as illustrated in the outline and the visual following.

  I. Handle
 II. Head
    A. Neck
    B. Poll
    C. Face
    D. Cheek
    E. Claw
    F. Adze eye

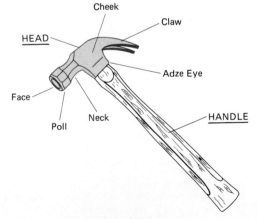

Figure 1. A claw hammer.

A carpenter would be concerned about the main parts and the subparts as they function in various ways when the hammer is in use.

The following student-written report is an analysis through partition of an abstract subject, a police department. The subject is given careful examination by separating it into its constituent parts so that, through individual consideration of each part, a better understanding of the whole can be achieved. The Plan Sheet that the student filled in before writing is included. The notes in the left-hand margin of the report have been added as an aid in studying the organization of the material.

PLAN SHEET
FOR GIVING AN ANALYSIS THROUGH PARTITION

Analysis of Situation Requiring Partition

What is the subject to be partitioned?
Brentsville Police Department

For whom is the partition intended?
a new police officer in Brentsville

How will the partition be used?
for the reader's general understanding

Will the partition be written or oral?
written

Setting Up the Partition

Definition or identification of subject:
efficient, effective police department

Basis (or bases) for partition:
administrative organization, responsibilities

Significance or purpose of basis (or bases):
All the divisions and bureaus must work together if the police department is to function easily and effectively.

Divisions of the subject, with identification of each division:

Chief of Police
  Chaplain's Office
    —Ministers to individuals under stress
  Internal Affairs
    —"Polices" dept. members
    —Enforces policies
  Intelligence Div.
    —Uncovers & analyzes info (organized crime, radical groups)

Assistant Chief of Police
  Administrative Services Bureau
    —Performs tasks beneficial to entire dept.
    —Includes: Crime Prevention, ICAP, License & Permits, Personnel, Public Information, Training
  Detective Bureau
    —Takes care of daily operations
    —Includes: Burglary/Auto, Crime Lab, Homicide/Forgery, Evidence, Identification, Vice & Narcotics, Youth

Administrative Asst.
  —Screens info
  —Directs certain items to Chief
Research & Dev.
  —Anticipates & plans for future needs
Legal Office
  —Aids in prosecuting criminals
  —Gives legal advice

Operations Bureau
  —Oversees activities that directly assist the public
  —Includes: Precincts, Traffic, Tactical Air Patrol, Reserves
Technical Services Bureau
  —Supplies needs and services to other divisions
  —Includes: Detention, Communications, Records, Supply, Vehicle Mgt.

Presentation of the System

Most logical order for presenting the divisions:
random listing

Types and subject matter of visuals to be included:
organization chart to depict visually the police department personnel

Sources of Information

interview with Administrative Assistant to the Chief of Police

154

BRENTSVILLE POLICE DEPARTMENT:
ORGANIZATION AND RESPONSIBILITIES

Partition of police department on basis of administrative organization

Two major administrators

The city of Brentsville has an efficient, effective police department. This is possible largely because of the administrative organization. (See the accompanying organization chart.) The police department is administered by two persons: the Chief of Police and the Assistant Chief of Police. Various divisions and bureaus support one another as well as the police department as a whole by fulfilling specific responsibilities.

All divisions within the department ultimately are responsible to the Chief of Police. However, only six divisions report directly to the Chief of Police, while four bureaus report directly to the Assistant Chief of Police.

DIVISIONS REPORTING TO THE CHIEF OF POLICE

First major administrator

Divisions reporting directly to the Chief

The following divisions report directly to the Chief of Police: Chaplain's Office, Internal Affairs, Intelligence Division, Administrative Assistant, Research and Development, and Legal Office. Their duties are varied.

Chaplain's Office

First division explained

Personnel in the Chaplain's Office are in contact with all police department employees. They minister to individuals in any stress situation, either personal or professional.

Internal Affairs

Second division explained

The Internal Affairs Division has as its major responsibility "policing" each member of the department. This division investigates any charges or questions from the public, other police officers, or independently developed sources about the activities or action(s) of any department member as to misconduct and criminality. This division enforces written policies, procedures, and rules to maintain discipline within the department.

Intelligence Division

Third division explained

Finding and analyzing information is the task of the Intelligence Division. This information has to do with such matters as the activities of organized crime and radical groups. Usually this information is gathered from informants and by undercover personnel. Officers in this division are usually department veterans, experienced in investigation, with unquestionable integrity.

Administrative Assistant

Fourth division explained

The Administrative Assistant is a go-between who listens to complaints, problems, concerns, and suggestions. The assistant screens all information, handles routine situations, and decides what information to pass on to the Chief of Police for handling.

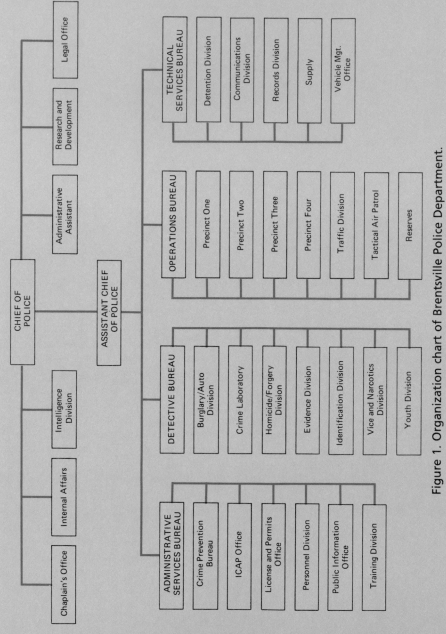

Figure 1. Organization chart of Brentsville Police Department.

Fifth division
explained

Research and Development

The Research and Development group are planners for the police department. They anticipate future needs and prepare to meet those needs. For example, they would be aware of city growth and population shifts, determining when to increase or decrease the number of beats. Or they might determine that within 12 months the department will need to add six additional squad cars.

Sixth division
explained

Legal Office

The Legal Office aids officers in the prosecution of criminal cases. This office serves in any capacity where legal assistance is needed; advises on policies, planning, internal hearings, and problems in criminal cases; and maintains liaison with the courts.

Second major admin-
istrator

Bureaus reporting to
the Assistant Chief

BUREAUS REPORTING TO THE ASSISTANT CHIEF

Reporting to the Assistant Chief of Police are four bureaus: the Administrative Services Bureau, the Detective Bureau, the Operations Bureau, and the Technical Services Bureau. Within each of these bureaus are numerous divisions to facilitate the responsibilities of each bureau.

First bureau
explained

Administrative Services Bureau

The Administrative Services Bureau's tasks have long-term application; they are tasks performed for the benefit of the department as a whole.

Divisions within the bureau include the Crime Prevention Bureau, ICAP Office, License and Permits Office, Personnel Division, Public Information Office, and Training Division.

Crime Prevention Bureau. The Crime Prevention Bureau controls, reduces, and prevents crime. Personnel concentrate on informing the public on ways to make crimes more difficult to carry out, such as improved security, better lighting, locks, and alarms.

ICAP Office. The Integrated Criminal Apprehension Program (ICAP) is concerned with crime analysis. Personnel collect reports, analyze the reports over a period of time, and determine needs. For example, a study of accident reports might reveal a high-risk accident area during specific hours; additional patrol persons or traffic officers could then be assigned to the area during these hours.

License and Permits Office. Any individual wishing to apply for licenses and permits for such activities as operating a dance hall, selling beer, or driving a taxi must make application through the License and Permits Office. Office personnel screen both the applicant and the location of the desired activity and review city ordinances to be sure carrying out the activity would not violate any ordinance.

Personnel Division. Personnel Division employees oversee all tasks related to employees—recruitment, selection, assignment, transfer, promotion, termination, and labor relations.

Public Information Office. The Public Information Office handles public relations and press relations; its major task is to keep the public informed on police activities through the various media—radio, television, newspapers, public lectures.

Training Division. The Training Division meets the professional needs of police service. It provides, for example, in-class and on-the-job training, physical and mental training, theory and practice, recruit training, and in-service training.

Second bureau
explained

Detective Bureau

The Detective Bureau is a collection of divisions that are directly involved in the daily operation of the department. These divisions include Burglary/Auto, Crime Laboratory, Homicide/Forgery, Evidence, Identification, Vice and Narcotics, and Youth.

Burglary/Auto Division. Personnel in Burglary/Auto are skilled in investigating crimes of theft. A special division is set aside to handle such matters since theft is a very common crime.

Crime Laboratory. Crime Laboratory personnel provide examination and classification of concrete evidence such as tire tracks, bloodstains, fingerprints, and fibers. Physical evidence has a major role in the prosecution of cases.

Homicide/Forgery Division. This division's personnel are especially trained or experienced in investigating all crimes related to homicide and forgery. Often they must testify in follow-up court sessions.

Evidence Division. Employees in the Evidence Division are responsible for the safekeeping of all evidence from a crime, accident, or other such occurrence. They must catalog the evidence and store it for easy access and make sure no one tampers with it.

Identification Division. Identification Division personnel fingerprint and photograph suspects and prisoners. They must be skilled since they provide permanent records and investigate major crimes.

Vice and Narcotics Division. The Vice and Narcotics Division enforces vice and drug laws. These laws have to do with such activities as illegal gambling, selling liquor illegally, obscene conduct, pornography, and sale of drugs.

Youth Division. The Youth Division has the task of dealing with juveniles. The personnel must follow special legal and practical guidelines that pertain to minors. Frequently personnel become involved in such matters as child abuse, neglect, and runaways. The focus is on social welfare rather than on crime.

Third bureau
explained

Operations Bureau

The Operations Bureau oversees activities carried out to directly assist the public. This bureau is the division of police work that the average person knows about. Its work includes such tasks as patrol, traffic, community relations, vice, and crime prevention. Precincts in strategic geographical locations throughout the city are "mini police departments" that oversee the carrying out of these tasks.

Three special divisions—Traffic Division, Tactical Air Patrol, and Reserves—also aid in implementing the tasks of the Operations Bureau.

Fourth bureau
explained

Technical Services Bureau

Activities within the Technical Services Bureau directly support the other divisions within the police department. This bureau usually operates every hour, every day of the year. The divisions include Detention, Communications, Records, Supply, and Vehicle Management.

<u>Detention Division.</u> Detention personnel handle the confinement, usually temporary, of persons arrested and the needs of the confined persons.

<u>Communications Division.</u> Communications employees are the link between the public and the police. They answer telephone calls from persons seeking help and they dispatch help.

<u>Records Division.</u> Records personnel collect, organize, and store data on wanted persons, on traffic accidents, on parking tickets, on arrests, and so forth. Using various report forms, records personnel keep reports on file on all types of information that would be useful to the department. Many records systems are on computer for quick, easy access. Record information is available 24 hours a day, every day.

<u>Supply.</u> Supply employees make sure that the department has everything necessary to function, everything from flashlight batteries, to parking ticket forms, to bullets. This division must keep an accurate inventory of supplies and make purchases as needed.

<u>Vehicle Management Office.</u> The Vehicle Management Office insures that all vehicles are maintained and "ready to roll."

Closing comment Obviously, the Brentsville Police Department is a complex organization. Its structure, however, allows it to function easily and effectively.

BASIS OF PARTITION

Partition, like classification, must be done on a useful, purposeful basis. A carpenter's partitioning, or dividing, the tasks in the construction of a proposed porch on the basis of needed materials would be useful and purposeful.

ORDER OF DATA PRESENTATION

The order of data presentation in a partition system is determined by your purpose. As in classification, among the possible orders are time, familiarity, importance, complexity, space, and alphabetical and random listing. For example, the work of a general duty nurse can be partitioned, or divided, into what she does the first hour, second hour, and so on (time order), if the purpose of partition is to show how she spends each working hour. Or a light socket can be partitioned according to the outer and the inner parts (spatial order), if the purpose of the partition is to show the construction of the light socket.

FORMS OF DATA PRESENTATION

The forms of presentation of analyses through partition are the same as those for classification. Logical methods are outlines, verbal explanations, and visuals.

Look at the following floor plan of the nation's capitol (Figure 1). It partitions the building to show the design and layout and to show space utilization.

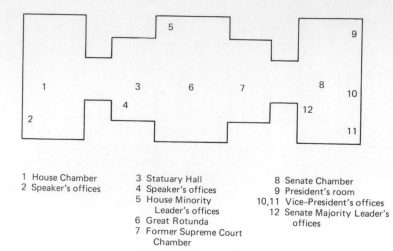

1 House Chamber
2 Speaker's offices

3 Statuary Hall
4 Speaker's offices
5 House Minority
  Leader's offices
6 Great Rotunda
7 Former Supreme Court
  Chamber

8 Senate Chamber
9 President's room
10,11 Vice-President's offices
12 Senate Majority Leader's
  offices

Figure 1. Floor plan of the U.S. Capitol.

The same information could be shown in outline form:

Floor Plan of the U.S. Capitol

1. House of Representatives wing }   *(First level of partition)*
  1.1 House Chamber
  1.2 Speaker's offices   *(Second level of partition)*
2. Great Rotunda and center wings } *(First level of partition)*
  2.1 Great Rotunda
  2.2 Statuary Hall
  2.3 Speaker's offices   *(Second level of partition)*
  2.4 House Minority Leader's offices
  2.5 Former Supreme Court Chamber
3. Senate wing }   *(First level of partition)*
  3.1 President's room
  3.2 Senate Chamber
  3.3 Vice-President's offices   *(Second level of partition)*
  3.4 Senate Majority Leader's offices

Visuals are frequently used in verbal partition presentations, such as the organization chart in the preceding explanation of a police department. Other often used visuals include diagrams, flowcharts, and drawings. (See Chapter 11 for a fuller discussion of visuals.)

Contrast of Classification and Partition

Both classification and partition are approaches to analyzing a subject, though for each the approach is from a different direction. Classification and partition differ in the number of the subject being analyzed, in the relationship of divisions, and in the overall result.

Classification divides plural subjects into kinds or classes. The subject may be singular in form, but if the meaning is plural (such as mail, furniture, food) the subject is classifiable. Partition, on the other hand, divides a singular subject, or item, into its component parts.

In partition, the parts do not necessarily have anything in common, other than being parts of the same item. For instance, the handle, shank, and blade of a screwdriver share no relationship beyond their being parts of the same item. In classification, however, all the items in a division have a significant characteristic in common. For example, in classifying typewriters as manual or electric, all manual typewriters use the typist as the source of energy, and all electric typewriters use electricity as the source of energy.

Classification and partition differ also in their end results. Since their purposes are different—the purpose of classification being to sort related items into groups and the purpose of partition being to divide an item into its parts—their outcomes are different. For instance, regardless of the category into which screwdrivers are classified (common, Phillips, spiral ratchet, powered, etc.), they are still screwdrivers. In the partition of a screwdriver, however, a part—whether the handle, blade, or shank—is still only a part of the whole.

General Principles in Giving an Analysis Through Partition

1. *Partition is a basic approach in analysis.* It divides a subject into parts, steps, or aspects so that through individual consideration of these, a better understanding of the whole can be achieved.
2. *Only a singular subject can be partitioned.* If a subject is plural it can be classified but not partitioned.
3. *The divisions in partition must be coordinate, or parallel.* All divisions on the same level must be of the same rank in grammatical form and in content.
4. *The divisions must be mutually exclusive.* Each division should be composed of a clearly defined group that would still exist without the other divisions on the same level.
5. *The divisions must not overlap.* A part can have a place in only *one* division.
6. *The divisions must be complete.* Every part must be accounted for, with no part left out.
7. *The data in a partition analysis may be presented in outlines, in verbal explanation, and in visuals.* The form or combination of forms that presents the data most clearly should be used.
8. *The basis on which partition is made should be clear, useful, and purposeful.*
9. *The order of presentation of divisions depends on their purpose.* Among the possible orders are time, familiarity, importance, complexity, space, and alphabetical and random listing.
10. *Classification and partition may be used together or separately.*

Procedure for Giving an Analysis Through Partition

Analysis through partition is generally a part of a larger whole, as have been the other forms of communication discussed in the preceding chapters. Regardless of whether the partition analysis is a dependent or an independent communication, however, the structure of such an analysis is as follows:

I. The presentation of the subject and of the bases of partition is usually brief.
   A. State the subject that is to be partitioned.
   B. Identify or define the subject.
   C. If applicable, list various bases by which the subject can be partitioned.
   D. State explicitly the bases of the divisions.
   E. List the divisions and, if any, the subdivisions.

II. The listing and discussion of the divisions are the longest parts of the presentation.
   A. Point out the reasons why the divisions are significant and what purpose they serve.
   B. Give sufficient explanation to clarify and differentiate among the given divisions.
   C. Present the divisions in whatever order best serves the purpose of the analysis.
   D. Use outlines and visuals whenever they will help clarify the explanation.

III. The closing (usually brief) depends on the purpose of the presentation.
   A. The closing may be the completion of the last point in the analysis.
   B. The closing may be a comment on the analysis or a summary of the main points.

Application 1   Giving an Analysis Through Partition

Select one of the following for partition. First fill in the Plan Sheet on pages 165–166; then give the partition in outline form.

1. Disk or magnetic tape
2. Resistor
3. House
4. A store (department, grocery, etc.)
5. Typewriter
6. Stapler
7. Clock
8. Telephone
9. Egg
10. A part of the body (ear, eye, heart, etc.)
11. Cash register terminal
12. Musical instrument (piano, trumpet, guitar, etc.)
13. Pocket calculator
14. Drafting table
15. Term of your own choosing from your major field

Application 2   Giving an Analysis Through Partition

For Application 1 above, give the partition as a verbal explanation.

Application 3 Giving an Analysis Through Partition

For Application 1 above, give the partition as a visual.

Application 4 Giving an Analysis Through Partition

Take the material you have prepared in Applications 1, 2, and 3 above and integrate them into one written presentation.

a. Add or delete material, as needed, in the Plan Sheet filled out in Application 1 above.
b. Write a preliminary draft.
c. Revise. See inside back cover.
d. Write the final draft.

Application 5 Giving an Analysis Through Partition

Give an oral partition analysis of the topic from Application 4 above. Ask your classmates to evaluate your speech by filling in an Evaluation of Oral Presentations from pages 527–535.

Application 6 Giving an Analysis Through Partition

Select a subject suitable for partition. Present the partition system in a visual.

Application 7 Giving an Analysis Through Partition

Select a business or organization (store, industry, farm, etc.) that you are familiar with (or can become familiar with). Show how this organization functions by dividing it into departments or areas. Include an organization chart (review pages 547–548 in Chapter 11, Visuals). Your presentation will be an explanation of a process; within this framework you will be relying heavily on analysis through classification and through partition. (One way of attaining information is interviewing knowledgeable people. For suggestions for informational interviews, see pages 522–523 in Chapter 10, Oral Communication.)

a. Fill in the Plan Sheet on pages 167–168.
b. Write a preliminary draft.
c. Revise. See inside back cover.
d. Write the final draft.

Application 8   Giving an Analysis Through Partition

Give an oral partition analysis of the topic from Application 7 above. Ask your classmates to evaluate your speech by filling in an Evaluation of Oral Presentations from pages 527–535.

USING PART II: SELECTED READINGS

Application 9   Giving an Analysis Through Partition

Read "Are You Alive?" by Stuart Chase, pages 611–615.

a. Show the structure of the essay by a partition analysis.
b. Under "Suggestions for Response and Reaction," write Number 3.

Application 10   Giving an Analysis Through Partition

Read "Videotex: Ushering in the Electronic Household," by John Tydeman, pages 574–582. In what ways is partition used in the article?

PLAN SHEET
FOR GIVING AN ANALYSIS THROUGH PARTITION

Analysis of Situation Requiring Partition

What is the subject to be partitioned?

For whom is the partition intended?

How will the partition be used?

Will the partition be written or oral?

Setting Up the Partition

Definition or identification of subject:

Basis (or bases) for partition:

Significance or purpose of basis (or bases):

Divisions of the subject, with identification of each division:

Presentation of the System

Most logical order for presenting the divisions:

Types and subject matter of visuals to be included:

Sources of Information

PLAN SHEET
FOR GIVING AN ANALYSIS THROUGH PARTITION

Analysis of Situation Requiring Partition

What is the subject to be partitioned?

For whom is the partition intended?

How will the partition be used?

Will the partition be written or oral?

Setting Up the Partition

Definition or identification of subject:

Basis (or bases) for partition:

Significance or purpose of basis (or bases):

Divisions of the subject, with identification of each division:

Presentation of the System

Most logical order for presenting the divisions:

Types and subject matter of visuals to be included:

Sources of Information

Chapter 5

Analysis Through Effect-Cause and Comparison-Contrast: Looking at Details

169

Objectives

Upon completing this chapter, the student should be able to:

- Establish the cause of an effect
- Differentiate between an actual cause and a probable cause
- Differentiate between logical and illogical or insufficient causes of an effect
- List and explain the steps in problem solving
- Solve a problem
- Give an analysis through effect and cause
- List and explain three organizational patterns for a comparison and contrast analysis
- Give an analysis through comparison and contrast

Introduction

Analysis through effect and cause and through comparison and contrast involves an intense examination of details. This examination may be as extraordinary as Sherlock Holmes's solving a case or as ordinary as a technician troubleshooting an air conditioning unit. Both Mr. Holmes and the technician have the same problem: discovering the cause of a particular effect.

Similarly, a comparison-contrast analysis may be as complex as a scientist's comparison of life in Hiroshima before and after the atom bomb. Or a comparison-contrast analysis may be as pointed as a technician's explaining that a chip in a computer is like a note in a piece of music: Remove one chip or one note and the whole is disrupted but not destroyed.

The purpose of this chapter is to help you develop skill in analyzing details through effect-cause and comparison-contrast and in reporting the results of these analyses.

See also cause and effect as a method of paragraph development, pages 673–674.

Analysis Through Effect and Cause

Trying to find the cause of a situation (given the effect) is a common problem. Not at all unusual are such questions as these: Why was production down last week? Why does this piece of machinery give off an unusual humming sound after it has been in operation a few minutes? Why didn't this design sell? Why does this hairspray make some customers' hair brittle? Why does this automobile engine go dead at a red light when the air conditioner is on? Why does this patient have recurring headaches?

When persons begin to answer these questions, they are using an investigative, analytical approach; they are seeking a logical reason for a situation, event, or condition. Explaining the cause of an effect is a kind of process; in this analysis, however, the concern is with answering *why* rather than *how*. In explaining the process of making a print by the diazo method, for example, the emphasis is on *how* the operation is done. In analyzing the effect-cause relationship in the diazo process, the concern might be *why* this particular process was used in reproducing a print and not another method, or the concern might be *why* exposed diazo film must be subjected to ammonia gas for development.

In the following report, condensed from a student's paper, the writer is analyzing the reasons for the loss of 1500 lives aboard the *Titanic*. The student wants to know *why* so many people perished in the tragedy. (Marginal notes have been added.)

WHEN THAT GREAT SHIP WENT DOWN

Effect investigated

On the evening of 14 April 1912, on its maiden trip, the "un-sinkable" luxury liner the *Titanic* struck an iceberg in the freezing waters of the North Atlantic. Within a few hours the ship sank, taking with it some 1500 lives. Why did so many people perish, particularly when there was another ship within ten miles? Subsequent investi-

Actual causes (con-clusions) established

gation by national and international agencies showed that the *Titanic* crew was small and insufficiently trained, that the ship did not have sufficient lifesaving equipment, and that international radio service was inadequate.

Causes discussed in order of least to most important

Small and Insufficiently Trained Crew

First cause explained in detail

One reason so many people perished in the *Titanic* tragedy was that the crew was small and insufficiently trained for such an emergency. In the face of grave danger, evidently the crew of the *Titanic* was simply not able to meet the situation. Many of the lifeboats left the ship only half full and others could have taken on several more people. The passengers were not informed of their imminent danger upon impact with the iceberg, investigation suggests, and the officers deliberately withheld their knowledge of the certain sinking of the ocean liner. Therefore, when lifesaving maneuvers were finally be-gun, many passengers were unprepared for the seriousness of the moment. Further indication of poor crew leadership is that when the *Titanic* disappeared into the sea, only one lifeboat went back to pick up survivors.

Insufficient Lifesaving Equipment

Second cause ex-plained in detail

Another reason so many people lost their lives was that the ship was not properly equipped with lifeboats. Although millions of dollars had been spent in decorating the ship with palm gardens, Turkish baths, and even squash courts, this most luxurious liner afloat was lacking in the vitally important essentials of lifesaving equipment. Records show that the *Titanic* carried only 16 wooden lifeboats, ca-pable of carrying fewer than 1200 persons. With a passenger and crew list exceeding 2200, there were at least 1000 individuals un-provided for in a possible sea disaster—even if every lifeboat were filled to capacity.

Inadequate International Radio Service

Third cause explained
in detail

    Undoubtedly, however, the primary reason that so many of the persons aboard the *Titanic* were lost is that international radio service was inadequate. Within ten miles of the *Titanic* was the *Californian,* which could have reached the *Titanic* before she sank and could have taken on all of her passengers. When the *Titanic* sent out emergency calls, near midnight, the radio operator of the *Californian* had already gone to bed. A crew member, however, was still in the radio room and picked up the signals. Not realizing that they were distress signals from a ship in the immediate vicinity, the crew member decided not to wake up the radio operator to receive the message. At that time there were no international maritime regulations requiring a radio operator to be on duty around the clock.

Comment on signifi-
cance of causes

    Had there been a more adequate international radio system, re-gardless of the *Titanic*'s small and insufficiently trained crew and its insufficient lifesaving equipment, probably those 1500 lives could have been saved when that great ship went down.

In "When That Great Ship Went Down," the writer begins by identifying the topic and stating the circumstances. Next, the question is posed, "Why did so many people perish," with the crucial modification, "particularly when there was another ship within ten miles." Then the writer summarizes in one sentence the three main causes. In the following three paragraphs, each cause is explained in turn, with sufficient supporting details. The writer discusses the causes in the order of their significance, leading up to the most critical cause. Then the writer closes the analysis with a brief comment that includes a relisting of the three main causes and emphasis on the most critical cause.

Effect-to-Effect Reasoning

Analysis through effect and cause may involve a chain of reasoning in which the cause of an effect is the effect of another cause. In the following paragraph (from the *Occupational Outlook Handbook*) dealing with technological progress and labor, for instance, industrial applications of scientific knowledge and invention (cause) result in increased automation (effect). Increased automation (cause) calls for skilled machine operators and service people (effect). This (cause), in turn, results in occupational changes in labor (effect).

    Technological progress is causing major changes in the occupational makeup of the nation's labor force. Rapid advances in the industrial applications of scientific knowledge and invention are making possible increasing use of automatic devices that operate the machinery and equipment used in manufacturing. Nonetheless, the number of skilled and semiskilled workers is expected to continue to increase through the 1980s, despite this rapid mechanization and automation of production processes. It is expected that our increasingly complex technology generally will require higher levels of skill to operate and service this machinery and related equipment.

Intended Audience and Purpose

173

Analysis Through
Effect-Cause and
Comparison-
Contrast: Looking
at Details

You must be aware of who will read the effect-cause analysis and why, for these two basic considerations determine the extent of the investigation and the manner in which the investigation will be reported. For instance, the analysis may be presented for the general adult reader who wants a general understanding of the topic or the analysis may be for a division manager who will base decisions and actions on the analysis.

Relationship to Other Forms of Communication

Analysis through effect and cause is a method of thinking and of organizing and presenting material. The effect-cause analysis frequently includes other forms of communication, such as definition, description, process explanation, classification, and partition, as well as oral presentation and the use of visuals. (For discussion of oral communication, see Chapter 10; for discussion of visuals, see Chapter 11.) Several of these forms are found in the following student-written effect-cause analysis. Given first is the Plan Sheet that the student filled in before writing the analysis.

PLAN SHEET
FOR GIVING AN ANALYSIS THROUGH EFFECT
AND CAUSE

Analysis of Situation Requiring Effect-Cause Analysis

What is the subject to be analyzed?
why a flashlight produces light

For whom is the analysis intended?
the general public

How will the analysis be used?
for a basic understanding of why a flashlight produces light

Will the analysis be written or oral?
written

Effect to Be Investigated

why, when the switch is ON, a flashlight produces light

Significance of the Effect

omit

Causes (as applicable)

Possible causes:
omit

Probable cause (or causes):
omit

Actual cause (or causes):

four parts of flashlight involved in producing light—battery, switch, bulb, crown

☐ *battery supplies the electricity by means of a chemical reaction*

☐ *switch when ON closes the circuit*
current flows from the negative side of the battery through the circuit to the positive side

☐ *filament inside the bulb resists electricity*
filament heats up, glows, and gives off light. Protected from burning up by the glass covering which keeps air out and contains a mixture of gases to prolong life

☐ *crown houses the bulb and the reflector*
reflector directs light

Supporting Evidence or Information for the Probable or Actual Cause (or Causes)

see above

Organization of the Analysis

I will explain the battery, the switch, the bulb, and the crown as parts of the flashlight that cause the light.

Types and Subject Matter of Visuals to Be Included

outline of battery showing positive and negative sides
flow of energy
crown (reflector, bulb, glass covering)

Sources of Information

physics textbook
World Book Encyclopedia

WHY A FLASHLIGHT PRODUCES LIGHT

177

Analysis Through
Effect-Cause and
Comparison-
Contrast: Looking
at Details

A flashlight is a common household item used often as a light source which requires no external energy source. Four parts of the flashlight work together to produce light: the battery, the switch, the bulb, and the crown.

The Battery

The source of energy is a battery or batteries that fit inside the flashlight's casing. The battery which produces electricity by means of chemical action has a negative and a positive side. See Figure 1.

Positive + Side   — Negative Side

Figure 1. Battery.

The Switch

The switch controls the circuit that carries the energy from the battery to the bulb. When the switch is moved to the ON position, current flows from the negative side of the battery through the circuit to the positive side of the battery. See Figure 2.

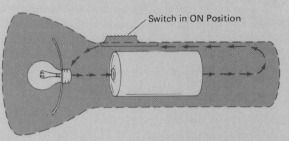

Switch in ON Position

Figure 2. Flow of current.

The Bulb

Inside the bulb is a filament, a coil of thin wire. The filament resists the flow of electricity, causing the electricity to heat the filament to a very high temperature. The heated filament glows and gives off light. The filament does not burn up because the bulb keeps air away from the filament. Also, the bulb may contain a mixture of gases to make the filament last longer.

The Crown

The crown of the flashlight houses the bulb and a reflector. Opposite the end of the bulb is a glass covering. When the heated filament gives off light, the light is reflected through the glass covering to direct a beam of light in whatever direction the flashlight is pointed. See Figure 3.

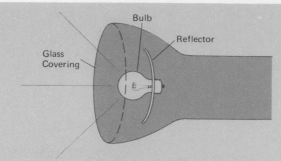

Figure 3. Crown and its parts.

Conclusion

When the battery, the switch, the bulb, and the crown function properly, the flashlight becomes a useful light source.

Establishing the Cause of an Effect

A common problem is to discover the cause, or causes, of a given effect. Through reading, observation, consultation with knowledgeable people, and so on, an answer is sought. Why was there a high rate of absenteeism last week? Investigation shows that 97 percent of the absenteeism was due to a highly contagious virus. Why has the new paint already begun to peel on a car that was repainted only two months ago? Investigation shows that the cause is inferior paint, not an improper method of paint application or improper conditioning of the old finish, as at first suspected. In these two situations, it was possible to establish a definite or *actual cause*—one that evidence proves beyond question is the true cause.

Often, however, a definite cause cannot be established. A possible or *probable cause* must suffice. Why was the Edsel car a flop? Why is this patient antagonistic toward the nurses? Why is red hair associated with a quick temper? Why have sales this year more than tripled those of last year? Why is my roommate so popular? Although such questions as these cannot be answered with complete certainty, causes that answer them, nevertheless, need to be established.

In the process of establishing a probable cause, all the possible causes must be considered. After examining each possible cause, eliminate it entirely, keep it as merely a possibility, or decide that—all things considered—it is a probable cause. In the investigation of possible causes, the reasoning must be logical and relevant. Each possible cause must be logical and relevant. Each possible cause must be examined on the basis of reason, not on the basis of emotions or preconceived ideas; and each possible cause must *really* be a possible cause.

ESTABLISHING THE CAUSE: WHY SAM DIDN'T GET 100 BUCKS

Consider the plight of Sad Sam, who asks, "Why didn't I get the 100 bucks I asked my parents for to get new front tires on my car?" Perhaps—not on paper but at least mentally—Sam thinks of these possible reasons:

179

Analysis Through
Effect-Cause and
Comparison-
Contrast: Looking
at Details

1. My parents don't love me.
2. They don't have the money.
3. They didn't receive my letter.
4. They mailed me a check but something happened to the letter.
5. They just don't realize what bad shape those front tires are in.
6. They think I should pay my car expenses out of the money I earn from my part-time job.

Sam turns over each possibility in his mind. He immediately dismisses possibility 1 (My parents don't love me). He knows this was only a childish reaction and that love cannot be equated with money. Sometimes parents who don't love their children send them money; and sometimes, for very good reasons, parents who love their children don't send them money.

There is a real possibility Sam's parents didn't have 100 dollars to spare (possibility 2). But Sam reasons if that is the cause, they would have asked him to wait a while longer or at least made some response.

Perhaps his parents did not receive his request (possibility 3). Maybe they mailed him a check but something happened to the letter (possibility 4). Either of the letters could have gotten lost in the mail. To his knowledge, however, none of his other letters has ever gotten lost.

Possibility 5 (They just don't realize what bad shape those front tires are in) could certainly be a true statement. But Sam realizes that his parents could be aware of the condition of the tires and still have reason not to send the money.

The more Sam considers possibility 6 (They think I should pay my car expenses out of the money I earn from my part-time job), the more this seems the likely reason. After all, Sam had promised his parents that if they would help him buy a car, he would keep it up. Furthermore, in his excitement of owning a car at long last, he had cautioned his parents to ignore any requests for car money, no matter how desperate they sounded.

So, as Sam carefully considers each possibility, he comes to the conclusion that the probable reason he did not receive the money is that his parents expect him to pay for his car expenses.

Whether the conclusion that is reached is the definite cause or the most probable cause of a situation, it must be arrived at through *logical* investigation, as in Sam's case. A conclusion, to be logical, must be based on reliable evidence, on relevant evidence, on sufficient evidence, and on an intelligent analysis of the evidence. Reliable evidence is the *proof* that can be gathered from trustworthy sources: personal experience and knowledge, knowledgeable individuals, textbooks, encyclopedias, and the like. Relevant evidence is information that directly influences the situation. Sufficient evidence means enough or adequate information with no significant facts that would alter the situation omitted. Evidence that is reliable, relevant, and sufficient must be analyzed intelligently. The meaning and significance of each individual piece of information and of all the pieces of information together must be considered if the most plausible conclusion is to be reached.

Illogical and Insufficient Causes

In an investigation of the "why" of a situation, especially guard against four pitfalls in reasoning, lest you arrive at an illogical or insufficient cause.

A FOLLOWING EVENT CAUSED BY A PRECEDING EVENT

One pitfall in reasoning assumes that a preceding event causes a following event, simply because of the time sequence. This fallacy in logic is called *post hoc, ergo propter hoc*, which literally means "after this, therefore because of this." Many superstitions are based on this false assumption. For instance, if a black cat runs across the street in front of Tom and later his automobile has a flat tire, blaming the cat is illogical. The two events (black cat running across street and flat tire) are not related in any way, except that they both concern the same person and they happened one after the other.

HASTY CONCLUSION

A common pitfall in effect-cause reasoning is jumping to a conclusion before all the facts are in. This results from not thoroughly investigating all aspects of a situation before announcing or indicating the cause. Consider this telephone call from a teenager to her father:

"Hello. Dad? I'm calling from City Drug Store. I've been in a slight automobile accident."
"Automobile accident? Just how fast were you driving?"
"I had stopped at a traffic light and this guy just rammed me from the rear."

Note the hasty conclusion that the daughter had caused the accident by driving too fast. The father had jumped to a conclusion before all the facts were made known.

OVERSIMPLIFICATION

Avoid the pitfall of oversimplifying a situation by thinking the situation has only one cause when it has several causes. If May says the reason she did not get a job she wanted is that she was nervous during the interview, she is oversimplifying the situation. She is not taking into consideration such possibilities as her training was inadequate, she lacked experience in the type of work, another applicant was better qualified, or she was not recommended highly enough.

SWEEPING GENERALIZATION

All-encompassing general statements that are not supportable reflect illogical cause-effect thinking. Consider these statements:

Labor unions are never concerned with consumer safety.
Our products always completely satisfy the customer.

The two preceding statements are sweeping generalizations. They are broad statements that are difficult if not impossible to support with sufficient evidence. Be wary of using all-inclusive terms such as *never, always, completely, everyone,* and the like.

Cause-to-Effect Analysis

181

Analysis Through
Effect-Cause and
Comparison-
Contrast: Looking
at Details

Thus far the discussion has centered around movement from a known or evident effect or situation to reasons explaining that effect or situation. Frequently there are occasions, however, when people must approach this problem from the opposite end. That is, they are confronted with a cause and want to know what the effects are, or will be. For instance, if a piece of stock is not perfectly centered in a lathe (cause), what will happen (effect)? If a television antenna is not properly installed (cause), what will be the consequences (effect)? If the swelling of an injured foot (cause) is not attended to, what will happen to the patient (effect)? The following sample paragraph explains that aspirin (cause) has many advantages as well as some disadvantages (effects).

Since its introduction in Germany 85 years ago, aspirin (the acetyl derivative of salicylic acid) has proved to be a miracle drug. Aspirin banishes headache, reduces fever, and eases pain. Not without its dangers, however, aspirin taken in excess will cause stomach upset, bleeding ulcer, and even death. Nevertheless, aspirin is responsible for more good than harm. It may even prevent heart attacks. According to current research, it is possible that aspirin contains anticoagulant properties that prevent thrombi, the blood clots that clog the coronary arteries, causing heart attacks.

Whether the occasion warrants approaching a problem by moving from effect to cause or by moving from cause to effect, the principles of logic are the same.

Problem Solving

Closely related to answering the "why"—the effect/cause—of a situation is answering the "what should I do" in a situation. What courses should I take next semester? Should I drop out of school for a while until I can make some money? Should I continue in a technical program or should I switch to a baccalaureate degree program? Should I take another job that has been offered to me? Which tires should I buy?

All these questions indicate problems that people are trying to solve. Everyone has problems—some of them insignificant, some very significant—that must be faced and dealt with. Such problems as which shoes to wear today or which vegetables to choose at lunch or which movie to go to tonight require only a few moments consideration and, viewed a few days later, are quite insignificant. But other problems, other decisions, require thoughtful consideration and will still be significant a few days or a few years from now. How does one go about solving such problems?

Problem solving is not easy; several suggestions, however, may help in reaching a decision:

1. Recognize the problem for what it is.
2. Realize that there are *various* possibilities for solution.
3. Consider *each* possibility on its own merits.
4. Decide which possibility is *really* best.
5. Determine to make the chosen course of action successful.

RECOGNITION OF THE REAL PROBLEM

The first step in solving any problem is recognizing the problem for what it is. The student who wrote only three pages on a test and who thinks a classmate received an A because the classmate filled up ten pages is avoiding the *real* problem: what the pages contain, not how many pages there are. In looking at a problem for what it is, reason and logical thinking must prevail. Glorifying oneself as a martyr or exaggerating the circumstances is avoiding the real issue. Sometimes emotions and pressures from others cloud reason and thus camouflage the real problem.

VARIOUS POSSIBLE SOLUTIONS

Various choices, or possibilities, or answers are available for solving problems. Sometimes, in a hastily posed problem, there seem to be only two alternatives, thus an either-or situation. An example is the problem of the student who does not have money for college and says, "I'll either have to go so far in debt it will take me ten years to get a clear start in life, or I'll have to postpone college until I have enough money in the bank to see me through two years." This student is certainly not considering *all* the alternatives available. The student has not considered that many colleges have an agreement with various industries whereby the student goes to college for a semester and works for a semester. Or the possibility of evening school, through which the student can attend courses one or more evenings per week after work and earn college credit. Or the possibility of taking a part-time job to help defray most or all of the college expenses. Or the possibility of taking a reduced college load and working full time. Thus a problem that seemed to have only two possible solutions really has a number of possible solutions.

MERITS OF EACH POSSIBLE SOLUTION

The third step in solving a problem is considering each possible choice on its own merits. Each possible solution must be analyzed and weighed carefully from every angle. This may involve investigation: consulting with knowledgeable people, drawing upon the experiences of family and friends, turning to reference materials, and so forth.

CHOOSING THE BEST SOLUTION

Once each of the possible choices has been given an honest, thoughtful consideration, it is time to decide which *one* choice is best. This is the most difficult step in problem solving, but if it is preceded by a mature analysis of the problem and all of its possible solutions, the final decision is more likely to be satisfactory even in a few weeks or in several years. Sometimes it may seem easier to let someone else make the decision; in that way, if things do not turn out as wished, someone else gets the blame. But shunning responsibility by avoiding making a decision goes back to the first step in problem solving: recognizing the problem for what it is.

DETERMINATION TO SUCCEED

Finally, in problem solving there must be a determination to make the chosen course of action successful. This may involve acquainting others with the decision and the reasons for it and perhaps even persuading them to accept the decision. A

183

Analysis Through
Effect-Cause and
Comparison-
Contrast: Looking
at Details

positive, open-minded attitude toward a well-thought-out decision is the best assurance that the decision, after all, was the best one.

General Principles in Giving an Analysis Through Effect and Cause

1. *The precise effect (situation, event, or conditions) being investigated must be made clear.* The focus is on gathering information that relates directly to that one effect.
2. *Awareness of who will read the analysis, and why, is essential to the writer.* These two considerations—the who and the why—determine the extent of the investigation and the manner of reporting it.
3. *Additional forms of communication may be needed.* Definition, description, process explanation, classification, and partition—plus visuals—may help clarify the effect-cause analysis.
4. *If the actual cause can be established, sufficient supporting evidence should be given.* The evidence must be reliable, relevant, and sufficient.
5. *If the actual cause cannot be established, adequate support for the probable cause should be given.* The support must be reliable, relevant, and as sufficient as possible.
6. *A preceding event may or may not cause a particular following event.* It is the task of the effect-cause analyst to determine whether there is a causal relationship between two events.
7. *A hasty conclusion indicates lack of thorough analysis.* All aspects of a situation should be investigated and weighed carefully before reaching a conclusion.
8. *A situation may be oversimplified by attributing to it only one cause when it may have several causes.* Significant situations are usually more complex than they at first seem.
9. *Sweeping generalizations weaken a presentation because they are misleading and are often untrue.* Caution should guide the use of such all-inclusive terms as *never, always,* and *every person.*
10. *Problem solving is a logical process.* The steps in problem solving are recognizing the problem for what it is, realizing that there are various possible solutions, considering each possible solution on its own merits, deciding which possibility is really best, and determining to make the chosen course of action successful.

Procedure for Giving an Analysis Through Effect and Cause

The analysis may be an independent presentation, or it may be a paragraph or a part of a presentation. Whichever the case, the procedure is the same and presents no unusual problems. (Although the procedure suggested here is for analyzing from effect to cause, the particular communication situation may call for analyzing from cause to effect.) Generally, the analysis should be divided into three sections:

I. Stating the problem usually requires only a few sentences.
   A. State the effect (situation, event, or condition) that is being analyzed.
   B. If applicable, give the scope and limitations of the investigation (as in reporting the causes of traffic accidents involving fatalities at the Highway 76–Justice Avenue intersection from January 1 to July 1).
   C. Give the assumptions, if any, on which the analysis or interpretation of facts is based (as in assuming that the new machine which is not working properly today was correctly installed and serviced last month).
   D. If applicable, give the methods of investigation used (reading, observation, consultation with knowledgeable people, etc.).
   E. Unless the order of the presentation requires otherwise, state the conclusions that have been reached.

II. Reporting the investigation of the problem is the longest section of the presentation.
   A. Consider in sufficient detail possible causes; eliminate improbable causes.
   B. If the actual cause can be established, give the cause and sufficient supporting evidence.
   C. If the actual cause cannot be established, give the probable cause and sufficient supporting evidence.
   D. Interpret facts and other information when necessary.
   E. If necessary, especially in a longer presentation, divide the subject into parts and analyze each part individually.
   F. Organize the information around a logical pattern or order.
      1. The topic may suggest moving from the less important to the more important, or vice versa.
      2. The topic may suggest using a time order of what happened first, second, and so on.
      3. The topic may suggest moving from the more obvious to the less obvious.
      4. The topic may suggest going from the less probable to the more probable.
      5. The nature of the presentation may suggest some other order.

III. The conclusion (usually brief) reflects the purpose of the analysis.
   A. If the purpose of the analysis is to give the reader a general knowledge of an effect-cause situation, summarize the main points or comment on the significance of the situation.
   B. If the analysis is a basis for decisions and actions, summarize the main points, comment on the significance of the analysis, *and* make recommendations.

Application 1   Giving an Analysis Through Effect and Cause

Each of the following statements contains an illogical or insufficient cause for an effect. Point out why the stated or implied cause is inadequate.

1. The reason my cow died is that I didn't go to church last Sunday.
2. Strikes occur because people are selfish.
3. Mr. Smith's television appearance on election eve won him the election.

185

Analysis Through
Effect-Cause and
Comparison-
Contrast: Looking
at Details

4. Automobile insurance rates are higher for teenagers than for adults because teenagers are poorer drivers.
5. Susan isn't dependable. She was supposed to turn in a report today but she hasn't shown up.
6. The atom bomb won World War II.
7. If I were you, I wouldn't switch kinds of drinks at a party. I did so last night, and this morning I have a terrible headache.
8. I can't stand pizza. I ate some once and it was as soggy as a wet dishrag.
9. Most people who are successful in business have large vocabularies. If I develop a large vocabulary, I'll be successful in business.
10. "I joined the Confederacy for two weeks. Then I deserted. The Confederacy fell."—Mark Twain

Application 2   Giving an Analysis Through Effect and Cause

In each of the following situations, indicate whether a probable cause or an actual cause is the more likely to be established. Point out why.

1. I thought that I had a B for sure in history, but the grade sheet shows a D. Why did I get a D?
2. This is the fourth time that the same jaw tooth has been filled. Why won't the filling stay in?
3. For the third night in a row I have asked my neighbors to please be a little quieter. Tonight they have the stereo turned up even higher. Why won't they be quieter?
4. In the 1984 presidential election the Republican candidate won. Why did he receive more votes than the Democratic candidate?
5. Every time my mother eats a lot of tomatoes, she breaks out in a rash. Why?

Application 3   Giving an Analysis Through Effect and Cause

Make a list of five problems regarding your schoolwork, job, or vocation that require thoughtful consideration. Choose one of these problems and solve it by listing:

1. The real problem
2. Various possibilities for solution
3. Consideration of each possible solution
4. The solution
5. Ways to assure that the chosen course of action will be successful

Application 4   Giving an Analysis Through Effect and Cause

Choose one of the general subjects below (or one that is similar). Restrict the subject to a specific topic.

EXAMPLE OF SUBJECT RESTRICTION

General subject: Decrease or increase of students in certain majors

Specific topic: Increase in students majoring in building trades

a. Fill in the Plan Sheet on pages 189–190.
b. Write a preliminary draft.
c. Revise. See the inside back cover.
d. Write the final draft.

Cause (or causes) of:

1. Improper functioning of a mechanism
2. A mishap of national importance
3. A decrease or increase in jobs in your city, county, or state
4. Variations in fringe benefits with different companies
5. A decrease or increase in sales during a particular period of time
6. Factors in a company's consideration of a plant site, a new piece of equipment, decrease or increase of employees
7. A change in attitude toward a decision
8. Increased cost of an item or service

Application 5   Giving an Analysis Through Effect and Cause

Give an oral effect-to-cause analysis of a topic from Application 4 above. Ask your classmates to evaluate your speech by filling in an Evaluation of Oral Presentations from pages 527–539.

Application 6   Giving an Analysis Through Effect and Cause

Write an explanation of the cause (or causes) in one of the topics below.

a. Fill in the Plan Sheet on pages 191–192.
b. Write a preliminary draft.
c. Revise. See the inside back cover.
d. Write the final draft.

Explain:

1. Why an automobile engine starts when the ignition is turned on
2. Why a light comes on when the switch is turned on

187

Analysis Through
Effect-Cause and
Comparison-
Contrast: Looking
at Details

3. Why water and oil do not mix
4. Why computer literacy is important
5. Why two houses on adjacent lots and with the same floor space may require different amounts of electricity for air conditioning or for heating
6. Why an earthquake occurs
7. Why a diesel engine is more economical to operate than a gasoline engine
8. Why artificial breeding has become a popular practice among ranchers
9. Why extreme care should be taken in moving an accident victim
10. Why different fabrics (such as silk, cotton, wool, linen) supposedly dyed the same color may turn out to be different shades
11. Why proper insulation decreases heating and cooling costs
12. Why an appliance wired for alternating current will not operate on direct current
13. Why a dull saw blade should be sharpened
14. Why a stain should not be applied over a varnish
15. Why the Rh factor is important in blood transfusions and in pregnancy
16. Why the blood bank insists on having complete identification in a patient sample of blood
17. Why a new book in the library does not appear on the shelf immediately after it is received
18. Why computer documentation is essential
19. Why vitamins are essential in human growth
20. A topic of your own choosing

Application 7 Giving an Analysis Through Effect and Cause

Give an oral effect-to-cause explanation of a topic from Application 6 above. Ask your classmates to evaluate your speech by filling in an Evaluation of Oral Presentations from pages 527–535.

USING PART II: SELECTED READINGS

Application 8 Giving an Analysis Through Effect and Cause

Read the short story "Quality," by John Galsworthy, pages 595–599. In what ways are effect-cause relationships used in the story? As directed by your instructor, answer the questions following the selection.

Application 9 Giving an Analysis Through Effect and Cause

Read "The Science of Deduction," by Sir Arthur Conan Doyle, pages 606–610. In what ways are effect and cause used in this article? As directed by your instructor, answer the questions following the selection.

Application 10 Giving an Analysis Through Effect and Cause

Read "American Labor at the Crossroads," by Steven M. Bloom and David E. Bloom, pages 583–588. Under "Suggestions for Response and Reaction," pages 587–588, write Numbers 1, 2, and 3.

Application 11 Giving an Analysis Through Effect and Cause

Read "American Labor at the Crossroads," by Steven M. Bloom and David E. Bloom. In an outline (see pages 128–129) show the reasons given in the article why American labor is at a crossroads.

PLAN SHEET
FOR GIVING AN ANALYSIS THROUGH EFFECT AND CAUSE

Analysis of Situation Requiring Effect-Cause Analysis

What is the subject to be analyzed?

For whom is the analysis intended?

How will the analysis be used?

Will the analysis be written or oral?

Effect to Be Investigated

Significance of the Effect

Causes (as applicable)

Possible causes:

Probable cause (or causes):

Actual cause (or causes):

**Supporting Evidence or Information for the Probable
or Actual Cause (or Causes)**

Organization of the Analysis

Types and Subject Matter of Visuals to Be Included

Sources of Information

PLAN SHEET
FOR GIVING AN ANALYSIS THROUGH EFFECT
AND CAUSE

Analysis of Situation Requiring Effect-Cause Analysis

What is the subject to be analyzed?

For whom is the analysis intended?

How will the analysis be used?

Will the analysis be written or oral?

Effect to Be Investigated

Significance of the Effect

Causes (as applicable)

Possible causes:

Probable cause (or causes):

Actual cause (or causes):

Supporting Evidence or Information for the Probable or Actual Cause (or Causes)

Organization of the Analysis

Types and Subject Matter of Visuals to Be Included

Sources of Information

Analysis Through Comparison and Contrast

193

Analysis Through
Effect-Cause and
Comparison-
Contrast: Looking
at Details

A basic method of looking at things closely is comparison and contrast, that is, showing how two or more things are alike and how they are different. Comparison or contrast may be used singly, of course, but most often they are used together; further the term *comparison* is commonly used to encompass both likenesses and differences.

Frequently, comparison-contrast is used to explain the unfamiliar. For instance, an airbrush looks like a pencil and is held like one; a computer terminal resembles a typewriter keyboard.

Often, too, comparison-contrast is used to highlight specific details. Llamas, for instance, are related to camels but are smaller and have no hump. *Approve* and *endorse*, though synonyms, have slightly different meanings: *approve* means simply to express a favorable opinion of, and *endorse* means to express assent publicly and definitely.

Analysis through comparison-contrast is a part of the life of every consumer: Should I buy Brand A or Brand B tires; which is the better buy in a color TV set—Brand X or Brand Y; which of these hospital insurance plans is better? Comparison of products is such an important aspect of the economy that some periodicals, such as *Consumers' Research, Consumer Reports,* and *Changing Times,* are devoted to test reports and in-depth comparisons.

See also comparison and contrast as a method of paragraph development, pages 672–673.

Organizational Patterns of Comparison-Contrast Analysis

An analysis developed by comparison or contrast must be carefully organized. The writer must decide in what order the details will be presented. Although various organizational patterns are possible, three are particularly useful.

POINT BY POINT

In the point-by-point pattern, sometimes called "comparison of the parts," a point (characteristic, quality, part) of each subject is analyzed, then another point of each subject is analyzed, and so on. The report below on a comparison of two electronics technology programs is organized point by point. The first point of comparison is cost. Cost details are given for each subject—Stevens College and Tinnin Community College. Then the next point, lab equipment, is analyzed for each program; and then the last point, instruction, is analyzed for each program. The point-by-point pattern can be delineated in this way:

Point A (cost) of Subject I (Stevens College)
Point A (cost) of Subject II (Tinnin Community College)

Point B (lab equipment) of Subject I (Stevens College)
Point B (lab equipment) of Subject II (Tinnin Community College)

Point C (instruction) of Subject I (Stevens College)
Point C (instruction) of Subject II (Tinnin Community College)

The point-by-point organizational pattern is usually preferred if the entire analysis is complex, that is, if the comparison is longer than a paragraph or two or if the analysis treats in depth more than two or three points.

SUBJECT BY SUBJECT

The subject-by-subject organizational pattern, sometimes referred to as "comparison of the whole," gives all the details concerning the first subject, then all the details concerning the next subject, and so on. The report below on a comparison of two electronics technology programs could be reorganized around the subject-by-subject pattern: All the details concerning the first subject, Stevens College (cost, lab equipment, and instruction), could be given first. Then would follow all the details concerning the second subject, Tinnin Community College (cost, lab equipment, and instruction). The subject-by-subject pattern can be delineated in this way:

Subject I (Stevens College)
   Point A (cost)
   Point B (lab equipment)
   Point C (instruction)
Subject II (Tinnin Community College)
   Point A (cost)
   Point B (lab equipment)
   Point C (instruction)

SIMILARITIES/DIFFERENCES

In the similarities/differences pattern, a comparison of the similarities of the subjects is given first, followed by a contrast of differences between the subjects. Again, the report below on a comparison of two electronics technology programs could be reorganized. If the similarities/differences pattern were used, the similarities between Stevens College and Tinnin Community College would be given, and then the differences between the two colleges would be given. The similarities/differences pattern can be delineated in this way:

Similarities of Subjects I and II (Stevens College and Tinnin Community College)
   Point A (cost)
   Point B (lab equipment)
   Point C (instruction)
Differences between Subjects I and II (Stevens College and Tinnin Community College)
   Point A (cost)
   Point B (lab equipment)
   Point C (instruction)

The similarities/differences (or differences/similarities) organizational pattern is particularly effective when positive and negative emphases are desired. The subject or emphasis presented last makes the greatest impression.

THREE PATTERNS COMPARED

195

Analysis Through
Effect-Cause and
Comparison-
Contrast: Looking
at Details

These three organizational patterns can be delineated as follows (of course, there can be any number of points and any number of subjects; here it is assumed that two subjects are being compared/contrasted on three points):

Point by Point	*Subject by Subject*	*Similarities/Differences*
Point A of Subject I	Subject I	Similarities between
Point A of Subject II	Point A	Subjects I and II
Point B of Subject I	Point B	Point A
Point B of Subject II	Point C	Point B
Point C of Subject I	Subject II	Point C
Point C of Subject II	Point A	Differences between
	Point B	Subjects I and II
	Point C	Point A
		Point B
		Point C

The following student-written report is an analysis through comparison-contrast. Note that the report is organized around three points of comparison: cost, lab equipment, and instruction. The Plan Sheet that the student filled in before writing is included. Marginal notes have been added to show how the student developed and organized the paper.

PLAN SHEET
FOR GIVING AN ANALYSIS THROUGH
COMPARISON AND CONTRAST

Analysis of Situation Requiring Comparison-Contrast Analysis

What are the subjects to be analyzed?
the electronics technology program at Stevens College and the electronics technology program at Tinnin Community College

For whom is the analysis intended?
prospective electronics students

How will the analysis be used?
to decide which college the student should attend

Will the analysis be written or oral?
a written report

Major Points (Characteristics, Qualities, Parts) to Be Compared/Contrasted

cost, lab equipment, instruction

Details Concerning the Major Points of Each Subject

Stevens College   *Tinnin Community College*

I. Cost

Tuition $6100.00   *Tuition ($300 a semester) . . . $1200*
No charge for books   *Books (after reselling for ½ price) $300*

II. Lab Equipment

15	*Power supply*	*50*
12	*Digital multimeter*	*40*
4	*Microprocessor trainer*	*12*
6	*Sine-square wave generator*	*25*
0	*Transistor curve tracer*	*1*
0	*Decade boxes*	*40*
4	*Digital trainer*	*20*
4	*Function generator*	*30*
10	*Oscilloscope*	*34*
1	*Personal computer system*	*4*

Value: *Value:*
$50,000   *over $150,000*

III. Instruction

1 Instructor
A.A.S. from Stevens, taught 3 yrs at Stevens, worked 2 yrs in industry (IBM)

3 Instructors
Instructor A: B.S., taught at 1 other college, taught 5 yrs, worked in industry 3 yrs
Instructor B: B.S., taught in 1 other college, taught 7 yrs, worked in industry 4 yrs and alternate summers
Instructor C: M.S., taught in 2 other colleges, taught 18 yrs, worked in industry 7 yrs

Classes 8–12, 1–2, M–Th
Electronics lectures 8—10:30, lab 10:30–12
A.A.S. degree: Eng I & II, Bus Law I & II, Soc, Math for Electronics, Digital Math, Computer Math, Basic Electronics, Fund of Elec, Digital Circuits, Semiconductors, Transistor Circuits

Variable schedules 8–3, M–F
Electronics—usually 2-hr block
A.A.S. degree: Tech Writing I & II, Soc SC, P.E., Ind Psy, Tech Math I & II, Tech Physics I & II, Fund of Drafting, Elec for Electronics, Electron Devices & Circuits, 4 sophomore-level electronics courses, 1 elective

Organization of the Analysis

Point by point. First I'll discuss the cost of attending each college. Then I'll talk about the lab equipment in each, and finally, the instruction in each.

Types and Subject Matter of Visuals to Be Included

table showing the cost of tuition and books for each college
table showing the major pieces of lab equipment for each college

Sources of Information

catalog from each college; chairman of the electronics technology dept. at each college

199

Analysis Through
Effect-Cause and
Comparison-
Contrast: Looking
at Details

A COMPARISON OF TWO ELECTRONICS TECHNOLOGY PROGRAMS: STEVENS COLLEGE AND TINNIN COMMUNITY COLLEGE

Subjects compared

General conclusion reached

Sources of information

Specific conclusions and points of comparison

A comparison of the electronics technology program at Stevens College and at Tinnin Community College indicates that the Tinnin program has more to offer the student. This conclusion is based on a thorough examination of the catalog from each college and a visit to each college for a personal interview with the chairman of the electronics technology department. This investigation shows that the Tinnin program costs less, has more lab equipment, and possibly has better instruction.

COST

First point of comparison for each subject

The cost of tuition and books for each college is as follows:

Table visually depicts supporting information

Stevens College

Tuition (18-month program)	$6100.00
Books—included with tuition	0
	$6100.00

Tinnin Community College

Tuition (4-semester program) $300.00 per semester	$1200.00
Books (about $150.00 per semester) 4 semesters × $150.00 = $600.00 resale value = 300.00 $300.00	300.00
	$1500.00

Interpretation of information

The Stevens program costs $4600.00 more than the Tinnin program. This large difference in cost is due primarily to Stevens being a private college and Tinnin being a public community college.

LAB EQUIPMENT

Second point of comparison for each subject

Each college has these major pieces of lab equipment:

Table visually depicts supporting information

Equipment	Stevens	Tinnin
Power supply	15	50
Digital multimeter	12	40
Microprocessor trainer	4	12
Sine-square wave generator	6	25
Transistor curve tracer	0	1
Decade resistance/capacitance boxes	0	40
Digital trainer	4	20
Function generator	2	30
Dual-trace oscilloscope	5	34
Personal computer system	1	4

The total value (approximate retail value) of the Stevens equipment is approximately $50,000. The total value of the Tinnin equipment is well over $150,000.

The program at Tinnin Community College has more kinds and more pieces of each kind of lab equipment than does Stevens College.

INSTRUCTION

The programs in both colleges differ in three aspects of instruction: qualifications of the instructors, daily class schedule, and required courses. Stevens College has one electronics instructor. He has an Associate in Applied Science degree from Stevens, has taught three years (all at Stevens), and has worked two years in industry (at IBM). Tinnin Community College has three electronics instructors. Two of them have Bachelor of Science degrees and the other has a Master of Science degree. Each has taught in at least one other college, has taught at least five years, and has worked in industry at least three years. One instructor works in industry every other summer.

The daily class schedule is set up differently at each school. At Stevens, the students are in class from 8:00 A.M. until 12:00 noon and from 1:00 P.M. to 2:00 P.M. Monday through Thursday (there are no classes on Friday). For the electronics courses, the students have lecture from 8:00 to 10:30 and lab from 10:30 to 12:00. The afternoon hour is for a nonelectronics course. At Tinnin, the students have variable schedules from 8:00 A.M. to 3:00 P.M. five days a week, with electronics courses typically in two-hour blocks. The distribution of lecture and lab time is at the discretion of the instructor.

Both Stevens and Tinnin offer an Associate in Applied Science degree upon successful completion of their programs. Requirements for the associate degree differ slightly at the two schools. At Stevens, the required courses are College English I and II, Business Law I and II, Sociology, Math for Electronics, Digital Mathematics, Computer Mathematics, Basic Electronics, Fundamentals of Electricity, Digital Circuits, Semiconductors, and Transistor Circuits. At Tinnin the required courses are Technical Writing I and II, any course in social science, Physical Education, Industrial Psychology, Technical Mathematics I and II, Technical Physics I and II, Fundamentals of Drafting, Electricity for Electronics, Electron Devices and Circuits, any four additional sophomore-level electronics courses, and one elective. Both programs offer the same amount of classroom instruction—18 months—but the Stevens program does not have a three-month summer break.

Thus, the instructors at Tinnin may be better qualified than the instructor at Stevens, the daily class schedule is more flexible at Tinnin, but the requirements for graduation are similar in both colleges.

CONCLUSIONS

A person planning to specialize in electronics technology should thoroughly explore the programs in prospective colleges. This report shows that two colleges (both in the same city) vary widely in their electronics programs. The electronics technology program at Tinnin Community College costs less, has more lab equipment, and possibly has better instruction than does the electronics technology program at Stevens College.

Relationship to Other Forms of Communication

Comparison-contrast analyses are closely related to other forms of communication, especially to description, definition, classification, and partition, and often involve them. Often, too, the comparison-contrast analysis is given orally. (For a discussion of oral communication, see Chapter 10.)

Various kinds of visuals—drawings, diagrams, tables, graphs, charts—are particularly helpful in showing comparison-contrast. (For a discussion of visuals, see Chapter 11). Table 1, for instance, shows the average annual energy use and annual cost of home appliances in a typical American home. Note how the arrangement of information makes it easy for the reader to see similarities and differences.

TABLE 1 Home Appliance Annual Energy Use and Cost

Use of electricity is stated in the number of kilowatt-hours (kwh) used per year. Cost is based on an average charge by utilities of 9¢ per kwh.

Electronic Appliance	Annual Energy Use	Annual Cost
Air conditioner	2000 kwh	$180.00
Can opener	1 kwh	.09
Clock	17 kwh	1.53
Clothes dryer	1200 kwh	108.00
Coffee maker	100 kwh	9.00
Dishwasher	350 kwh	31.50
Floor furnace	480 kwh	43.20
Freezer (16 cu. ft.)	1200 kwh	108.00
Frying pan	240 kwh	21.60
Hair dryer	15 kwh	1.35
Hot plate	100 kwh	9.00
Iron	150 kwh	13.50
Lighting	2000 kwh	180.00
Radio	20 kwh	1.80
Radio-phonograph	40 kwh	3.60
Range	1500 kwh	135.00
Refrigerator (12 cu. ft.)	750 kwh	67.50
Shaver	1 kwh	.09
Television (black and white)	400 kwh	36.00
Television (color)	550 kwh	49.50
Toaster	40 kwh	3.60
Vacuum cleaner	45 kwh	4.05
Washer (clothes)	100 kwh	9.00
Water heater (all electric)	4200 kwh	378.00

General Principles in Giving an Analysis Through Comparison and Contrast

1. *Comparison-contrast can be used to explain the unfamiliar or to highlight specific details.*
2. *The comparison-contrast analysis must be carefully organized.* Various organizational patterns are possible.
3. *The point-by-point organizational pattern can be used.* In this pattern, a point of each subject is analyzed, then another point of each subject, and so on.
4. *The subject-by-subject pattern can be used.* In this pattern, all the details concerning the first subject are presented, then all the details about the next subject, and so on.
5. *The similarities/differences pattern can be used.* In this pattern, a comparison of similarities of the subjects is given, followed by a contrast of differences between the subjects.

Procedure for Giving an Analysis Through Comparison and Contrast

The comparison-contrast analysis may be independent or it may be part of a longer presentation. Whichever the case, the analysis is typically divided into three sections.

I. Stating the subjects and points to be compared and contrasted requires only a few sentences.
   A. State the subjects that are being compared and contrasted.
   B. State the points (characteristics, qualities, parts) by which the subjects are being compared and contrasted.
   C. Give any needed background information.
   D. If applicable, give the source of information or methods of investigation used (reading, observation, consultation with knowledgeable people, etc.).
   E. Unless the order of the presentation requires otherwise, state the conclusions that have been reached.

II. Reporting the details is the longest section of the presentation.
   A. Give sufficient supporting details for each point of each subject.
   B. When needed, subdivide the major points.
   C. Interpret facts and other information when necessary.
   D. Use visuals if they will help clarify the analysis.
   E. Organize the information around a logical pattern or order.

III. The conclusion (usually brief) reflects the purpose of the analysis.
   A. If the purpose of the analysis is to give the reader a general knowledge of the subject, summarize the main points or comment on the significance of the comparison.
   B. If the analysis is a basis for decisions and actions, summarize the main points, comment on the significance of the analysis, *and* make recommendations.

203

Analysis Through
Effect-Cause and
Comparison-
Contrast: Looking
at Details

Application 1   Giving an Analysis Through Comparison and Contrast

Compare and contrast your program of study at your institution with a similar program in another institution.

a. Fill in the Plan Sheet on pages 205–206.
b. Write a preliminary draft.
c. Revise. See the inside back cover.
d. Write the final draft.

Application 2   Giving an Analysis Through Comparison and Contrast

Choose one of the topics below for a comparison-contrast analysis.

a. Fill in the Plan Sheet on pages 207–208.
b. Write a preliminary draft.
c. Revise. See the inside back cover.
d. Write the final draft.

Analyze:

1. Two brands of a product (Examples: an RCA and a Zenith color television set, a John Deere and a Massey-Ferguson tractor, Levi and Dickey blue jeans, a Chrysler Cordoba and a Ford Thunderbird)
2. Two synonyms (Examples: job-career, technician-technologist, education-training, friend-acquaintance)
3. Before-and-after situations (Examples: a piece of furniture or a building before and after renovation, an engine before and after overhauling, efficiency before and after using a new procedure)
4. A topic of your own choosing

Application 3   Giving an Analysis Through Comparison and Contrast

Give an oral comparison-contrast explanation of Application 1 or of a topic from Application 2 above. Ask your classmates to evaluate your speech by filling in an Evaluation of Oral Presentations from pages 527–535.

USING PART II: SELECTED READINGS

Application 4   Giving an Analysis Through Comparison and Contrast

Read "Are You Alive?" by Stuart Chase, pages 611–615. In what ways are comparison and contrast used in this article? As directed by your instructor, answer the questions following the selection.

Application 5   Giving an Analysis Through Comparison and Contrast

Read "To Serve the Nation: Life Is More Than a Career," by Jeffrey R. Holland, pages 632–638. In what ways are comparison and contrast used in this article? As directed by your instructor, answer the questions following the selection.

Application 6   Giving an Analysis Through Comparison and Contrast

Read "American Labor at the Crossroads," by Steven M. Bloom and David E. Bloom, pages 583–588. Under "Suggestions for Response and Reaction," pages 587–588, do Numbers 1, 2, and 3 as directed by your instructor.

Application 7   Giving an Analysis Through Comparison and Contrast

Read "The World of Crumbling Plastics," by Stephen Budiansky, pages 622–623. Under "Suggestions for Response and Reaction," page 623, write Number 3.

PLAN SHEET
FOR GIVING AN ANALYSIS THROUGH
COMPARISON AND CONTRAST

Analysis of Situation Requiring Comparison-Contrast Analysis

What are the subjects to be analyzed?

For whom is the analysis intended?

How will the analysis be used?

Will the analysis be written or oral?

Major Points (Characteristics, Qualities, Parts) to Be Compared/Contrasted

Details Concerning the Major Points of Each Subject

Organization of the Analysis

Types and Subject Matter of Visuals to Be Included

Sources of Information

PLAN SHEET
FOR GIVING AN ANALYSIS THROUGH
COMPARISON AND CONTRAST

Analysis of Situation Requiring Comparison-Contrast Analysis

What are the subjects to be analyzed?

For whom is the analysis intended?

How will the analysis be used?

Will the analysis be written or oral?

Major Points (Characteristics, Qualities, Parts) to Be Compared/Contrasted

Details Concerning the Major Points of Each Subject

Organization of the Analysis

Types and Subject Matter of Visuals to Be Included

Sources of Information

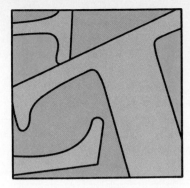

Chapter 6

The Summary: Getting to the Heart of the Matter

Objectives

Upon completing this chapter, the student should be able to:

- Define the terms *descriptive summary*, *informative summary*, and *evaluative summary*
- List the specific purposes that each form of summary may serve
- Prepare each form of summary
- Give bibliographical information for a summary
- Describe specific situations in which a student and an employee would need to know and use the preceding skills

Introduction

Why do you, a student, need to know how to give a summary, a digest of the main points of an original presentation? Of what importance is this to you? First, because you are a *student*, you should be able to give a summary orally and in writing. As a student, for example, you must be able to grasp the main points in a reading assignment for a course. One very effective way of studying textbook material is to write down the significant points, the important facts, in the lesson. Writing summaries of class lectures and of textbook assignments helps to clarify information and makes it stick in the mind longer.

This practice is quite helpful not only in day-to-day preparations but also in reviewing for tests. Pulling the main points and ideas from pages and pages of textbook material and class notes and then writing them down coherently and understandably is a sound approach to studying for tests. Since tests cannot possibly cover everything a student is supposed to learn in a course, test questions usually concentrate on major points and aspects of the subject.

Often a class requirement is to make a report or review (a summary) of reading assignments. Usually the only directions that are given have to do with what sources are to be read and how long the report should be. Rarely is the student told *how* the report should be presented.

Thus, for very practical reasons of immediate importance, you need to learn how to give a summary.

There are occasions, too, in which giving summaries is part of a person's work, such as condensing reports, articles, speeches, or discussions for a superior too busy to read or hear the original. Further, people *read* summaries, and thus should be able to recognize a summary for what it is.

Purpose and Intended Audience

The first consideration in giving a summary is *why* it is being written, what purpose it will serve. The purpose may be to *describe* very briefly, perhaps in only a sentence or two, what the article or situation is about. The purpose may be to *inform*, to present, in general, the principal facts and conclusions given in the original work. Or the purpose may be to *evaluate*, to judge the accuracy, completeness, and usefulness of the work.

Closely related to the purpose is the intended audience. If it is a study aid, the audience may be the student writing the summary. If the summary is an assignment, the audience may be an instructor. Or the audience may be a business executive who wants only the main points and conclusions of an original work. Or the audience may be a student trying to determine what sources would be most helpful for a research paper.

In other words, the direction that a summary takes is not haphazard; you must have clearly in mind the purpose of the summary and the intended audience.

Understanding the Material to Be Summarized

You must have a thorough understanding of the material before you can summarize it. Since summary concerns main ideas, you must be able to distinguish main ideas from minor ideas and supporting details. To do this, you have to understand the material.

The first step in understanding the material to be summarized is to scan it. In looking over the material, note the title, introductory matter, opening and closing paragraphs, headings, marginal notes, major topics, organization, and method of presentation. This scant familiarity with the material is preparation for reading it closely.

Now read the material very carefully. Underline or jot down on a separate sheet of paper (particularly if the work belongs to someone else) the key ideas. Think through the relationship of these key ideas to get the overall significance of the material.

To illustrate the procedure for the reading of material to be summarized, an article about fire ants follows. Assume that a summary of the main points is to be written for a general adult reader. In scanning the article, note that the opening paragraph and the closing paragraph focus on the central idea of the article: fire ants won't be eradicated, only temporarily controlled.

Key points and ideas are printed in boldface italic type, and marginal notes explain how to determine what information to include and what to omit in a summary.

Bring Mounds of Trouble

FIRE ANTS: LOOKS LIKE THEY'LL STAY*

OMIT IN A SUMMARY

by **Preston Smith**

INCLUDE IN A SUMMARY

Controlling idea (main point) of the article

It's official. *Fire ants are here to stay in the South. Despite years of research and millions of dollars, a dozen or so chemicals for their control and some successes, the hopes of eradicating the pest are over.* Resignation is all that remains.

Supporting quotation omitted

"Having fire ants is like living with ticks or fleas—we all have to learn to coexist with them," laments Miles Karner, area extension entomologist in Altus, Oklahoma. Within the last year, *fire ants have been spotted for the first time in southern Oklahoma.* Only four areas have had the ants, but Karner believes it is only a matter of time before they spread where the climate will allow.

Important fact: Oklahoma affected

Supporting details omitted

Mark Trostle of the Texas Department of Agriculture fire ant program believes *eliminating the fire*

Important observation: lost cause

* Reprinted with permission from the October 1986 issue of *Successful Farming*.

ant from the 230 million southern acres it already infests is a lost cause.

"There's no way we'll ever get rid of the fire ant, not with our present tools. There's not enough funds in the world to control the fire ant. It would take the national budget to control them," he observes.

In 1985 fire ants were found in Lubbock, says Trostle, to have not only survived two previous winters but also have thrived. *Lubbock, at 3,281 feet elevation, is the highest and coldest place fire ants have ever been found in this country.*

"I used to think they wouldn't get to Lubbock. I'm not saying if they'll get to Amarillo," says Trostle.

Ants on the Move

Fire ants have moved farther north than anyone would have guessed. Currently the ants are found in 11 southern states. Only Maryland and Virginia have been spared (so far) any permanent infestation of the pest. Clifford Lofgren, research leader at the largest USDA fire ant research lab in the country, expects the fire ants will spread only 50–100 miles farther north before the colder weather will beat them back. Some species of fire ants in South America have been known to adapt to higher altitudes and much cooler temperatures. If the imported fire ants in the U.S. adapt as well as their southern cousins, *they could move as far as Washington, D.C.*

Ant infestation to the West is a different story. *The ants could easily move up the California coast and into other arid western states.*

New Ways to Control the Pest

Even though fire ants offer some benefits to row-crop farmers, mostly in the form of some pest suppression, *they are an economic problem.* An average infestation of 30 mounds an acre *can reduce soybean yields* 5.3 bushels an acre. *Mounds also slow harvest, affect livestock and have been known to kill calves.*

One out of every 100 persons is highly allergic to fire ant stings. More reports of allergic reactions will occur as more people are exposed.

Currently, *no chemical fire ant control* is registered for broadcast use on cropland. EPA also has resisted registration of chemicals having residual killing effects. [Some 65 products are labeled as fire ant killers, but all of the products only temporarily control the ants.]†

You Can Kill Too Many Ants

Believe it or not, *you can kill too many fire ants.* Sherman Phillips, Texas Tech entomologist, has observed that *if* a field has 80–90 mounds per acre, and

† Several paragraphs detailing specific products are here omitted.

OMIT IN A
SUMMARY

INCLUDE IN A
SUMMARY

213

The Summary:
Getting to the
Heart of the Matter

all the fire ants and beneficial ants are killed back, within a year many more fire ants move into the area and there could be 100 or 200 mounds in a short time. A more correct control method might be to kill only a few of the 80–90 mounds. The remaining workers would prey on new queens and prevent new colonies.

Texas A&M entomologist Brad Vinson reports that new, multi-queen colonies have sprung up. These fire ants live in super colonies with perhaps 200–300 mounds per acre and 50–500 queens per mound. The best chemical controls today offer only temporary relief.

All entomologists stress that fire ants will always be in the South and that any control is only temporary. There's no magic cure to wiping out fire ants. We'll just have to learn to grimace and bear them.

Reminder of main
idea: control of
fire ants is only
temporary

Ways to Shorten the Original

A summary is a shorter, briefer version of an original work. It may be a condensation or it may be an abridgment of the main points in the original work. The summary omits introductory matter, details, examples, and illustrations—everything except the major idea and facts.

SUMMARY BY CONDENSATION

A summary that results from the condensation of the original material presents the main content (a digest) of the original, rephrased. That is, the writer or speaker has carefully read the original and then restated the details in his or her own words.

The summary may sometimes include direct quotations from the original work. The quoted material may be a phrase, a sentence, a paragraph, or perhaps an even longer section. Quoted material should be put in quotation marks. Sometimes page or chapter numbers for quotations are helpful for reference. Footnotes are unnecessary, since only one source is involved and that source is identified.

Below is a paragraph from "Health Technologists and Technicians" in the *Occupational Outlook Handbook*, 1986–1987 Edition, page 182. Following the original paragraph is a paragraph summarized through condensation.

ORIGINAL PARAGRAPH

Many jobs in the health field owe their existence to the development of new laboratory procedures, diagnostic techniques, and treatment methods. Quite a few of these involve clinical applications of the computer made possible by the microchip. Clinical laboratories have been transformed by the installation of automated instruments that offer low-cost analyses in minutes. Elsewhere in the hospital, new kinds of equipment—computed tomography (CT) scanners, dialysis machines, and ultrasound scanners, for example—have made possible earlier and more accurate diagnoses and more effective treatment. Technologies that are candidates for widespread use in the years ahead include magnetic resonance imaging, brain wave mapping, laser surgery, and organ transplantation. However, the field of medicine is changing so rapidly that it is impossible to predict what the next generation of devices will bring.

PARAGRAPH SUMMARIZED BY CONDENSATION

Many health-field jobs result from new and ever-changing medical advances. Many of these developments, made possible by the computer, involve automated instruments and new kinds of equipment, thus giving rise to such new technologies as magnetic resonance imaging, brain wave mapping, laser surgery, and organ transplantation.

SUMMARY BY ABRIDGMENT

A summary produced by abridgment retains the essential content of a work in the original wording. Supporting details and supplementary matter are omitted.

Below is the paragraph from "Health Technologists and Technicians"; lines mark out the supporting matter, leaving the essential content of the paragraph.

ORIGINAL PARAGRAPH WITH SUPPORTING MATTER MARKED OUT

Many jobs in the health field owe their existence to the development of new ~~laboratory~~ procedures, ~~diagnostic techniques~~, and ~~treatment~~ methods. Quite a few of these involve clinical applications of the computer ~~made possible by the microchip. Clinical laboratories have been transformed by the installation of automated instruments that offer low cost analyses in minutes. Elsewhere in the hospital,~~ new kinds of equipment—~~computed tomography (CT) scanners, dialysis machines, and ultrasound scanners, for example~~—have made possible earlier and more accurate diagnoses and more effective treatment. Technologies ~~that are candidates for widespread use~~ in the years ahead include magnetic resonance imaging, brain wave mapping, laser surgery, and organ transplantation. ~~However, the field of medicine is changing so rapidly that it is impossible to predict what the next generation of devices will bring.~~

PARAGRAPH SUMMARIZED BY ABRIDGMENT

Many jobs in the health field owe their existence to the development of new procedures and methods. Quite a few of these involve clinical applications of the computer. New kinds of equipment have made possible earlier and more accurate diagnoses and more effective treatment. Technologies in the years ahead include magnetic resonance imaging, brain wave mapping, laser surgery, and organ transplantation.

Forms of Summaries

Depending on the writer's primary purpose, the summary can take various forms. It may be:

- □ A descriptive summary
- □ An informative summary
- □ An evaluative summary

DESCRIPTIVE SUMMARY

A summary may describe; it may state what a work is about in a very general way. The descriptive summary simply indicates the main topics that are discussed. It is usually very brief, sometimes only one or two sentences in length.

215

The Summary:
Getting to the
Heart of the Matter

Descriptive summaries frequently occur in the headings to chapters in books, in the headings to magazine articles, on the table of contents pages in magazines, in publicity for new books and pamphlets, and in annotated bibliographies.

The following excerpt from a table of contents page of *FDA Consumer* (July–August 1986) illustrates a common use of the descriptive summary.

Stroke: Fighting Back Against America's No. 3 Killer   6
Each year, about 400,000 Americans suffer strokes. About 160,000 die; about half the survivors are seriously disabled. But there are steps you can take to reduce your risk of stroke.

The Growing Use of Irradiation to Preserve Food   12
Fruits and vegetables join the list of foods that may be irradiated to preserve freshness and kill insects. Labeling information, including a special symbol, will alert shoppers to products that have been processed with this newest form of food preservation.

Through the Bureaucratic Jungle: A Guide for the Confused Consumer   16
A consumer who's looking for help with a problem about some product and finds the right government agency on the first try is lucky, indeed. Knowing who does what before you pick up the phone or write a letter can help improve the odds.

Another use of the descriptive summary is the *descriptive abstract*. An abstract is a brief presentation of the essential contents of a document. The descriptive abstract frequently serves as an indicator of the usefulness of the original material.

Descriptive abstracts are very important to technicians who want to keep abreast of what is going on in their field. So invaluable are these descriptive abstracts in determining whether or not some particular article is pertinent to a specific interest that entire books and periodicals—for instance, *Chemical Abstracts, Nuclear Science Abstracts,* and *Biological Abstracts*—are devoted to descriptive abstracts of what is being written in particular fields. Persons interested in the field can read the abstracts to determine whether they want or need to obtain and read copies of the original works.

Other professional journals include within each issue a section for abstracts. These may be abstracts of papers presented at meetings of professional groups or they may be of articles appearing in related literature. Again, the abstract serves as an indicator of usefulness, directing persons to articles of concern and interest.

The first example below is a descriptive abstract of a report prepared for the U.S. Department of Agriculture, Economic Research Service.

"The Electronic Scanner Checkout and Item Price Removal." Charlene C. Price and Charles Handy. (202)447-6363.

Electronic scanner checkout is here to stay. This report focuses on the benefits of scanning, its effect on item pricing, the pros and cons of item price removal, and state legislation affecting price marking. The conclusion reached is that the question to be answered is how should item price information be provided.

The following sample abstract is from *Fire Technology*.* Note the headings, particularly Reference, which gives complete bibliographical information.

* Reprinted from the November 1982 *Fire Technology* (Vol. 18, No. 54), copyright © 1982 National Fire Protection Association, Quincy, Massachusetts 02269. Reprinted by permission.

KEY WORDS: abrasion resistance, burn injury, clothing, fabrics, fire fighting, heat protection, insulation, physical properties, tensile strength, turnout coats.

ABSTRACT: Several fabrics commonly used in the outer shells of structural fire fighters' turnout coats were subject to a variety of laboratory tests. These included breaking and tearing strength tests as well as several kinds of abrasion tests. The fabrics were also subjected to radiant heat (8.4 kJ/m²s = 2.0 kcal/m²s) and flames (84 kJ/m²s), and their resistance to heat deterioration and insulative characteristics were measured. The flame exposure tests were also conducted on assemblies of outer shell fabric, vapor barrier and innerliner. It should be emphasized that the results were obtained on one piece of each material and that they cannot be generally applied to all fabrics of the same or similar description.

REFERENCE: "Performance Evaluation of Fabrics Used in Fire Fighters' Turnout Coats," J. F. Krasny, R. W. Singleton, and J. Pettengill, *Fire Technology*, Vol. 18, No. 4 (November 1982), pp. 309–318.

The descriptive summary or descriptive abstract is an essential part of a report. It brings together in one section the major points discussed in the report. (For a detailed discussion of the summary or abstract in reports, see Chapter 8, Reports, page 353. See also the abstract included with a library research report on page 479 and the abstract included with a government report on page 624.)

The summary that follows appeared in a student-written report, "Police Work as an Occupation."

The continued need for police officers is obvious. The opportunities are not overwhelming, but they can be found by the qualified applicant. It is an occupation with a long history and tradition and is continuously being upgraded. It provides a secure job even though the salaries are not as high as those in other occupations requiring similar qualifications.

In the classroom, instructors frequently mention standard resource materials in particular fields and suggest that students become familiar with them. The student who examines the material and jots down several facts about the kind and manner of presentation of information is writing an annotation, a kind of descriptive summary, as in the following:

The McGraw-Hill Encyclopedia of Science and Technology, 1987 edition, from McGraw-Hill, 20 volumes, including an index. Updated by yearbooks. Specialized encyclopedia written for knowledgeable persons with a high reading level. Includes drawings, diagrams, and schematics. Long, detailed, highly technical articles.

Descriptive summaries are often used to publicize books, pamphlets, brochures, catalogs, and the like, as in this excerpt from *U.S. Government Books*, Catalog Number 4–6.

217

The Summary:
Getting to the
Heart of the Matter

United States Women in Aviation, 1930–1939. The 1930's marked a positive change for aviation in general and for women in aviation in particular. This era saw the entrance of women into the highly competitive spheres of air racing and commercial air travel. Stories and photos of the outstanding women pilots of that period, including Amelia Earhart and her ill-fated attempt at an around-the-world flight. 1985.
74 p. il.
S/N 047-002-00028-1
$3.50 (SI 1.42:6)

INFORMATIVE SUMMARY

The primary purpose of an informative summary is to inform; that is, it is designed to present the principal facts and conclusions given in the original work. When most people use the word *summary*, they are probably referring to the informative summary. No personal feelings or thoughts are injected; the main points of the material are presented objectively. Unlike the descriptive summary that tells *about* the work in only a few sentences, the informative summary tells what is *in* the work in a paragraph to several pages, depending on the length of the original.

Given below are two examples of the informative summary. The first is from the *American Journal of Public Health* (March–April 1986), a publication that includes with each report a synopsis (informative summary).

THE CONTRIBUTION OF SMOKING TO SEX DIFFERENCES IN MORTALITY

Ingrid Waldron, PhD

Synopsis

The contribution of smoking to sex differences in mortality is estimated on the basis of data from 12 studies of the mortality of nonsmoking men and women, together with mortality data for comparable general population samples. Most of the data are for samples drawn from the U.S. population from the late 1950s to 1980. The findings from different studies are generally consistent, once methodological factors are taken into account.

The findings indicate that, for total mortality, the proportion of sex differences attributable to smoking decreases from about two-thirds at age 40 to about one-quarter at age 80. Over the adult age span, it appears that about half of the sex difference in total mortality is attributable to smoking. Findings for ischemic heart disease mortality show a similar pattern. For lung cancer, it appears that about 90 percent of the sex difference in mortality is attributable to smoking.

The estimated contributions of smoking include both the effects of sex differences in smoking habits and the effects of sex differences in the increase in mortality caused by smoking. The quantitative results should be interpreted with caution, since several lines of argument suggest that multivariate analyses controlling for other relevant factors would produce lower estimates of the contribution of smoking to sex differences in mortality. Despite this limitation, the findings analyzed in this review, together with additional evidence from related research, strongly support the conclusion that cigarette smoking makes a major contribution to men's higher mortality, but other factors also play an important role.

The following student-written summary (with filled-in Plan Sheet) is informative. The length of the original article is approximately 1500 words; the length of the summary is approximately 300 words. The original article, "Videotex: Ushering in the Electronic Household," appears on pages 574–582.

PLAN SHEET
FOR GIVING AN INFORMATIVE SUMMARY

Analysis of Situation Requiring Summary

What type of summary is this to be?
informative

For whom is the summary intended?
students in communications class

How will the summary be used?
for conveying main points of original article

Will the summary be written or oral?
written

Identification of Work (Bibliographical Information) as Applicable

Title:
"Videotex: Ushering in the Electronic Household"

Author:
John Tydeman

Date:
February 1982

Type of material (magazine article, book, report, etc.):
magazine article

Title of publication:
The Futurist

City of publication and publishing company:
omit

Volume number:
omit

Page numbers:
54–61

Other:

Key Facts and Ideas in the Work

Change through electronics technology by home information services

Two systems: teletext (a one-way flow of information)
videotex (a two-way flow of information)

Uses of videotex: news, advertising/shopping, ordering

Future possibilities: added services to existing TV coverage, entertainment and game potential, retrieval of information by libraries, telephone directories, message sending

Most popular use of teletext: information retrieval

Most popular use of videotex: probably information retrieval, computer-based gaming, and transactions

Projected users: 30–40% of all households by 2000

Effects on society: living facilities, places of employment, home shopping, electronically controlled schooling, new careers in information management

Relatively slow development because of no infrastructure

Move toward electronic household to continue

221

The Summary:
Getting to the
Heart of the Matter

Centered
Bold
underline

Source
info

AN INFORMATIVE SUMMARY OF "VIDEOTEX: USHERING IN THE ELECTRONIC HOUSEHOLD"

John Tydeman. "Videotex: Ushering in the Electronic Household." *The Futurist*, February 1982, pp. 54–61.

Electronics technology has brought about tremendous change in the world. One aspect of this electronic technology change is in the home information service, teletext and videotex, "systems that primarily disseminate verbal and pictorial information by wholly electronic means, for visual display or printing under the control of the user."

Teletext is a one-way flow of information that utilizes television signals. With the necessary equipment, a keypad, an individual can choose information to appear on the TV screen. In contrast, videotex is a two-way flow of information through such existing equipment as telephone networks, some cable TV systems, or "a one-way cable into the home with a normal telephone link out."

Home Applications

Videotex is in the trial stage in the United States. It offers such things as news, weather, and sports information from newspapers and wire services. The systems also provide advertising and shopping information; the user of some systems can not only see the advertised products but can also order the products through the system.

Possible future uses of videotex systems include adding services to video programs covering such events as elections or sports, or the system could provide services for the hearing impaired or multilingual television audience.

In addition to providing "pages" of information to the home user, "both teletext and videotex have entertainment and game potential as well." Also retrieval of information is made possible through electronic retrieval services offered by specialized libraries and information banks. Telephone directories could be delivered electronically with a more frequent update on yellow pages than the current annual update. Another possible use is message sending, such as community bulletin board information access; health, gardening, and weather tips; and financial information.

The Household Market

Of the potential uses for teletext, market surveys indicate it would probably be used for information retrieval. Videotex systems are in such a state of evolution, it is difficult to identify the most popular applications; more than likely "information retrieval, computer-based gaming, and transactions (including home-based shopping) will be the driving applications." The many potential uses may be classified into five areas of application: information retrieval, transaction, messaging, computing, and telemonitoring.

Projected users of these systems include 30 to 40 percent of all households by 2000. The systems' use will develop because bankers, retailers, manufacturers, and wholesalers are realizing that these systems can offer their services more economically than can conventional ways.

Where to Next?

The systems today are in very early stages. Those of the future will little resemble current systems. As they evolve, the systems will also bring about many changes in society. These include:

☐ Living facilities (home, apartment) will become the place of employment, changing the structure and the location of these facilities.

□ Shopping from the home; consumers' orders will result in "production on demand" thus controlling manufacturing processes.
□ Schooling will be influenced, if not controlled, electronically.
□ The need for information management will create new careers.

Videotex: Ready or Not?

Compared to the speed of development of radio and television, videotex perhaps more closely resembles the development of radio. "There is no infrastructure in place." Further, the product(s) of videotex is not clearly defined. "Is it information retrieval, is it transactions, is it messaging, or is it all of these things?" The one thing that can be stated with certainty is that "the move toward the electronic household is irrevocable."

EVALUATIVE SUMMARY

A summary may evaluate; it may be the primary purpose of the summary to analyze the accuracy, completeness, usefulness, appeal, and readability of a piece of writing. In the evaluative summary you include your own comments and reactions, your thoughts and feelings regarding the material. The evaluative summary emphasizes assessment of the original material.

The evaluative summary should contain sufficient information to show that you thoroughly understand the work being evaluated. Include specific main points from the original and make specific references to the original work.

In the following student-written evaluative summary, evaluative comments are underlined. Read the summary, skipping the underlined comments; then read the entire summary.

EVALUATIVE SUMMARY: "CLEAR ONLY IF KNOWN"

The well-written, clearly illustrated article "Clear Only If Known," by Edgar Dale,* answers the question, "Why do people give directions poorly and sometimes follow excellent directions inadequately?"

In answering the first half of the question, "Why do people give directions poorly?" Dale states six specific reasons.

1. People do not always understand the complexity of the directions they attempt to give.
2. People give overly complex directions that include unnecessary elements.
3. People overestimate the experience of the person asking directions.
4. People make explanations more technical than necessary.
5. People are unwilling to say, "I don't know."
6. People use the wrong medium for giving directions.

Each reason is easily identified because Dale uses words like "First of all," "Another frequent reason," and "Another difficulty in communicating directions" as introductory phrases. Further, each reason is explained in detail, as Dale gives at least one excellent illustration for each reason.

---

* Edgar Dale, "Clear Only If Known," reprinted in Nell Ann Pickett and Ann Laster, *Technical English,* 5th ed. (New York: Harper & Row, 1988), pp. 568–571.

223

The Summary:
Getting to the
Heart of the Matter

The last half of the article <u>briefly</u> answers the second part of the question, "Why do people sometimes follow excellent directions inadequately?" Dale gives two reasons:

1. People don't understand directions but think they do.
2. People often are in too big a hurry when they ask for directions.

In closing, Dale emphasizes the need for clear communication, saying that "clarity in the presentation of ideas is a necessity."
<u>Anyone reading this article will have a clear understanding of why the communication process often breaks down.</u>

The book review is a popular form of evaluative summary. (See also Reading Report, pages 355–363.) The following book review appeared in *Monthly Labor Review*, May 1986. Compare the length of the original work, 354 pages, to the length of the evaluative summary, some 400 words. Clearly the purpose of this review is to evaluate the usefulness of the book as well as to tell what it contains.

> *Labor Market Economics*. By Saul D. Hoffman. Englewood
> Cliffs, N.J., Prentice-Hall, Inc., 1986. 354 pp. $28.95.

The first interesting thing about this book appears on its cover—the title. Although the author may assume that his readership would pay no particular attention to the word "market," that is precisely the word that piques the interest of the classical economist. Has a modern text finally been written which begins with the premise that labor is, among other things, a commodity which is bought and sold? Yes, and Saul D. Hoffman is its author.

Indeed, a major strength of this book is its sound economic analysis, which is based on the understanding that labor is not magically exempt from the economic laws and pressures which operate in a market economy. (Anticipating the objections of those who would say that today's economy is too far removed from the free-market ideal to justify such an approach, Hoffman has included a convincing defense.) The use of the traditional economic tools of analysis, avoiding the need to reinvent the wheel in order to examine the subject, contributes to the effectiveness of this text, which uses a "neoclassical" perspective.

This comprehensive study is well written and well organized. There are separate chapters on demand for labor, labor supply, human capital, discrimination, unions, income, and unemployment, as well as one on "further topics." "Applications sections" are employed, many of which delve into real-world cases. Much of the author's analysis is quite enlightening, while imprecisions are mostly minimal. The exception is the statement: " . . . it should be obvious that the wage income of all workers is exactly equal to the labor costs of all firms." The author should have added a section on benefits as an aspect of compensation, or included a disclaimer stating that the term "wages" is used synonymously with "compensation." Also, because the idea of labor cost is not useful without considering productivity, it cannot be considered in such absolute terms. In a lesser violation, Puerto Rico is listed as one of the "countries" (sic) which has minimum wage legislation.

Although some equations and graphs merely quantify the obvious, most are helpful. Hoffman makes good use of new ideas and research. Divergent opinions are presented in a balanced, unbiased manner. Overall, *Labor Market Economics* is an excellent choice for the introductory college course and for those interested in the subject.

—MICHAEL WEINERT
Chicago Regional Office
Bureau of Labor Statistics

Another form of evaluative summary—closely related to the book review—is the video review. The following video review is from the September–October 1986 issue of *The Conservationist*.

Introduction to Fly Fishing, L. L. Bean
Outdoor Video Library, $59.95, rebate of $40 on return.

There was six inches of snow on the ground and flakes as big as quarters were still falling as I sat looking out the window on a cold December day. With the push of a button I was suddenly on a trout stream in Maine and my fishing companion via the TV screen was none other than Dave Whitlock: author, artist, flytier, teacher and fisherman.

Whitlock is the host for *Introduction to Fly Fishing*, one of a series of videos in the L. L. Bean Outdoor Video Library.

Many books have been written about fly fishing for the beginner and expert and most are excellent, but having someone such as Dave Whitlock with his folksy, down-to-earth style at your side makes the printed word come alive with the sounds and action of fly fishing.

The photography is very good and there are some interesting underwater and surface level sequences showing how a fly should be presented and how a fly is taken by a cruising trout. Whitlock covers all the basics including selecting tackle, knots, flies, clothing and safety.

By watching Whitlock work with several students on a stream to improve their casting techniques you get the feeling that fly fishing is really not all that difficult, which it is not.

Introduction to Fly Fishing, which runs for 70 minutes, is instructional and at the same time entertaining. The tape comes with the L. L. Bean *Fly-Fishing Handbook* written and illustrated by Dave Whitlock. Although I do not know him personally, the author appears to be the kind of fellow I would enjoy meeting on a stream some day.

Introduction to Fly Fishing is available for rental or purchase in either Beta or VHS formats and can be obtained from L. L. Bean, Freeport, Maine 04033.

—Arthur Woldt

Accuracy of Fact and Emphasis

A summary must agree with the original material in two ways: in the presentation of factual information and in the emphasis given the information.

The misrepresentation of factual information may occur through careless error, as in omitting a digit or a decimal point in a number. Through the process of cutting out words, misleading information may be given. Consider this sentence: "Wood posts can be expected to last at least 30 years *if they have been pressure treated and handled properly.*" The last part of the sentence is essential; omission of the *if* qualification changes the meaning of the entire sentence. The sentence might be condensed as follows: "Pressure-treated and properly handled wood posts last at least 30 years."

Misrepresenting the emphasis placed on information in the original material may occur because of overemphasizing or underemphasizing information. If a one-page summary of a ten-page report on a new weed killer stresses the possible danger to animals when this was only a minor point in the original report, the summary has misrepresented the emphasis. A summary must agree in emphasis with the original; the intended audience of the summary should get the same impression that the intended audience of the original should get.

Documentation

225

The Summary:
Getting to the
Heart of the Matter

Material being summarized should be cited, that is, be completely identified. Such an identification is referred to as documentation. Documentation is complete bibliographical information on the original material. In addition to author and title, it includes facts about publication (place, publisher, and date). See pages 473–474 for examples of bibliographical forms. Depending on the purpose of the summary, documentation may also include such information as price, number of pages, and availability in paper or cloth binding. See pages 469–474 for a discussion of documentation.

PLACEMENT OF BIBLIOGRAPHY

The placement of bibliographical information may vary. Although the bibliographical information may be given within the summary, it usually appears as a formal bibliographical entry at the beginning of the summary.

For examples of bibliography given as a separate entry at the beginning (or the end) of a summary, see pages 216, 221, 222, 223, and 224.

Also, the arrangement of information and the punctuation in an entry may vary, although the information itself does not vary.

General Principles in Giving a Summary

1. *The emphasis of a summary is determined by the purpose and the intended audience.* The primary purpose may be to describe, to inform, or to evaluate.
2. *Summarizing requires a thorough understanding of the material.* Scanning, reading carefully, and underlining or taking notes are important in comprehending the material and thus in distinguishing which information to include in the summary and which to omit.
3. *A descriptive summary states what the material is about in a very general way.* Frequently only a few sentences are required to indicate the major concern of the material.
4. *An informative summary states objectively what is in the material.* The principal facts and conclusions given in the original are presented in a paragraph to several pages, depending on the length of the original.
5. *An evaluative summary emphasizes assessment of the material.* The accuracy, completeness, and usefulness of the information are judged. Include your own reactions, thoughts, and feelings along with a report of the main facts and ideas in the original.
6. *A summary is accurate in fact and in emphasis.* Unintentional distortion of the original may occur through careless error, through omission of essential information, or through unequal treatment of ideas presented as equally significant in the original.
7. *The material being summarized should be identified.* The identification may be given as a formal bibliographical entry at the beginning (or end) of the summary or within the summary.

Procedure for Giving a Summary

1. Decide whether the primary purpose of the summary is to describe what the work is about; to inform by presenting the principal facts and conclusions given in the original work; or to evaluate the accuracy, completeness, and usefulness of the work.
2. Look over the work. Note the opening and closing paragraphs, the major concern of the work, the organization, and the method of presentation.
3. Read the work very carefully. Underline or jot down on a separate sheet of paper (particularly if the work does not belong to you) the key ideas. Think through the relationship of these key ideas; be sure that you understand what the writer is trying to communicate.
4. Identify the original material specifically by author, work, and date. This information may be given within the summary or as a formal bibliographical entry at the beginning or end of the summary. (See pages 473–474 for bibliographical forms.)
5. Write a draft of the summary, giving the author's main ideas in the order presented.
6. Reread the original work; then check the summary for accuracy of fact and emphasis and for completeness.

Application 1   Giving a Summary

Bring to class two *descriptive* summaries, as explained on pages 214–217, written by others. State the specific sources.

Application 2   Giving a Summary

Write a two-to-three sentence *descriptive* summary of "Fire Ants" on pages 211–213.

Application 3   Giving a Summary

Write a *descriptive* summary of any five works (specialized encyclopedias, dictionaries, etc.) related to your major field of study. Before you can write the summary, you need to become familiar with each book. Examine its physical characteristics (size, type of binding, number of volumes), organization (table of contents, index, illustrations), content (chapter or entry headings, kind and extent of information, intended reader), and publishing information (publisher, year of publication, method of updating information).

Application 4   Giving a Summary

In the library find two books, two periodicals, and two other sources (pamphlets, reports, bulletins, audiovisual materials, etc.) that relate to your major field. Write a *descriptive* summary of each.

Application 5   Giving a Summary

227

The Summary:
Getting to the
Heart of the Matter

Give orally the descriptive summaries you wrote in Application 2, 3, or 4 above. Ask your classmates to evaluate your speech by filling in an Evaluation of Oral Presentations from pages 527–535.

Application 6   Giving a Summary

Write two *descriptive* summaries of a book of your selection. First, write a descriptive summary for a librarian who will order the book; second, write a descriptive summary for a student who will use the book for reference.

Application 7   Giving a Summary

In a paragraph of approximately 75 words, write an *informative* summary of "Fire Ants" on pages 211–213, or write an *informative* summary of this chapter in your textbook.

a. Fill in the Plan Sheet on pages 233–234.
b. Write a preliminary draft.
c. Revise. See inside back cover.
d. Write the final draft.

Application 8   Giving a Summary

Write an *informative* summary of the following excerpt from the *Occupational Outlook Handbook*.

a. Fill in the Plan Sheet on pages 235–236.
b. Write a preliminary draft.
c. Revise. See inside back cover.
d. Write the final draft.

ELECTRICAL AND ELECTRONICS TECHNICIANS

(D.O.T. 003.161; .261; 019.281; 194.381; 726.261; .281-010; and 828.261-018)

Nature of the Work

Knowledge of science, mathematics, and principles of electricity and electronics enables electrical and electronics technicians to work in all phases of business and government—from research and design to manufacturing, sales, and customer service. Although their jobs are more limited in scope and more practically oriented than those of engineers or scientists, they often apply the theoretical knowledge developed by engineers and scientists to actual situations. Electrical and electronics technicians develop, manufacture, and service equipment and systems such as radios, radar, sonar, television, industrial and medical measuring or control devices, navigational equipment, and computers. Because the field is so broad, these technicians often specialize in one area, such as automatic control devices or electronic amplifiers.

Technicians working in design, production, or customer service use sophisticated measuring and diagnostic devices to test, adjust, and repair equipment. In many cases, they must understand the purpose for which the electronic device is being used. To design equipment for space exploration, for example, a technician must consider the need for minimum weight and volume and maximum resistance to shock, extreme temperature, and pressure.

In research and development, one of the largest areas of employment, technicians set up equipment, prepare experiments, and calculate the results, sometimes with the aid of computers. They also assist engineers and scientists by making prototype versions of newly designed equipment and, frequently, by doing routine design work.

In production, technicians usually follow the general directions of engineers and scientists, but often without close supervision. They may prepare specifications for components and devise and conduct tests to ensure product quality. They often supervise production workers to make sure they follow prescribed procedures.

As sales workers or field representatives for manufacturers, wholesalers or retailers, technicians give advice on the installation, operation, and maintenance of complex equipment and may write specifications and technical manuals.

Although electrical and electronics technicians often repair equipment, workers whose main job is the repair of electronic equipment are included elsewhere in the *Handbook* under electrical and electronic equipment repairers.

Working Conditions

Most electrical and electronics technicians work regular hours in laboratories, offices, electronics shops, or industrial plants. Those in sales and service usually work much of the time in customers' facilities. Some occasionally are exposed to electric shock hazards from equipment.

Employment

Electrical and electronics technicians held about 404,000 jobs in 1984, over one-third of which were in manufacturing. The largest manufacturing employers were the electrical equipment, machinery, and professional and scientific equipment industries. Over one-fourth worked in wholesale trade, mainly for distributors of machinery, equipment, and electrical goods. Large numbers also worked for public utilities and service and repair companies.

The Federal Government employed about 20,000 electrical and electronics technicians, mainly in the Departments of Defense and Transportation. State and local governments employed about 6,000.

Training, Other Qualifications, and Advancement

Although persons can qualify for electrical or electronics technician jobs through many combinations of work experience and education, most employers prefer applicants who have had some specialized technical training. Specialized training is available at technical institutes, junior and community colleges, extension divisions of colleges and universities, and public and private vocational-technical schools. Persons with college courses in engineering, science, and mathematics may qualify for some positions but additional specialized training or experience may be needed.

Many electrical and electronics technicians qualify for their jobs through on-the-job training or on the basis of experience gained in the Armed Forces. Some qualify through correspondence schools.

Some of the types of schools that provide electrical and electronic training are technical institutes, junior and community colleges, and area vocational-technical schools.

Some large corporations conduct training programs and operate private schools to meet their needs for electrical and electronics technicians trained in specific areas; such

229

The Summary:
Getting to the
Heart of the Matter

training rarely includes general studies. The Armed Forces also train many types of electrical and electronics technicians. Although military job requirements generally differ from those in the civilian economy, military technicians often find private or civilian government jobs with only minimal additional training.

Many private technical and correspondence schools specialize in electronics. Some of these schools are owned and operated by large corporations that have the resources to provide up-to-date training in a technical field.

Persons interested in a career as an electrical or electronics technician should have an aptitude for mathematics and science and enjoy technical work. An ability to do detailed work with a high degree of accuracy is necessary; for design work, creative talent also is desirable. Electrical and electronics technicians are part of a scientific team, and often work closely with engineers and scientists as well as other technicians and skilled workers. Technicians in service and sales should be able to work independently and deal effectively with customers.

Electrical and electronics technicians usually begin work as trainees in routine positions under the close supervision of an experienced technician, scientist, or engineer. As they gain experience, they carry out more difficult assignments under only general supervision. Some eventually become supervisors and, a few, electrical engineers.

Job Outlook

Employment of electrical and electronics technicians is expected to increase much faster than the average for all occupations through the mid-1990's due to anticipated continued strong demand for computers, communications equipment, and electric products for military, industrial, and consumer use. More technicians will be needed to help develop, produce, and service these products. Opportunities will be best for graduates of 2-year postsecondary school technical training programs.

Although a relatively small proportion of electrical and electronics technicians leave the occupation each year, most job openings will be to replace those who transfer to other occupations, retire, or leave the labor force.

Since many jobs for electrical and electronics technicians are defense related, cutbacks in defense spending could result in layoffs in defense-related industries.

Earnings

Median annual earnings of full-time electrical and electronics technicians were about $21,800 in 1984; the middle 50 percent earned between $17,300 and $28,000. Ten percent earned below $13,700 and 10 percent earned over $34,700.

In the Federal Government, electrical and electronics technicians could start at $11,458, $12,862, or $14,390 in 1985, depending on their education and experience.

Related Occupations

Electrical and electronics technicians apply scientific principles in their work. Others whose work involves the application of scientific principles include engineering technicians, science technicians, broadcast technicians, drafters, surveyors, and health technicians and technologists. Others who repair electrical and electronic equipment are communications equipment mechanics, data processing equipment repairers, electronic home entertainment equipment repairers, and office machine repairers.

Application 9   Giving a Summary

Write an *informative* summary of a reading assignment in one of your other courses or of an article in a recent issue of a periodical relating to your major field.

a. Fill in the Plan Sheet on pages 237–238.
b. Write a preliminary draft.
c. Revise. See inside back cover.
d. Write the final draft.

Application 10   Giving a Summary

Give orally the informative summary you wrote in Application 7, 8, or 9 above. Ask your classmates to evaluate your speech by filling in an Evaluation of Oral Presentations from pages 527–533.

Application 11   Giving a Summary

In a paragraph of approximately 75 words, write an *evaluative* summary of "Fire Ants" on pages 211–213.

a. Fill in the Plan Sheet on pages 239–240.
b. Write a preliminary draft.
c. Revise. See inside back cover.
d. Write the final draft.

Application 12   Giving a Summary

Write an *evaluative* summary of a reading assignment in one of your other courses, or of an article in a recent issue of a periodical relating to your major field.

a. Fill in the Plan Sheet on pages 241–242.
b. Write a preliminary draft.
c. Revise. See inside back cover.
d. Write the final draft.

Application 13   Giving a Summary

Give orally the evaluative summary you wrote in Application 11 or 12 above. Ask your classmates to evaluate your speech by filling in an Evaluation of Oral Presentations from pages 527–535.

Application 14   Giving a Summary

Find and attach to your paper an example of a descriptive summary, of an informative summary, and of an evaluative summary. Write a one-page report on

231

The Summary:
Getting to the
Heart of the Matter

what is accomplished in the summaries. Be sure to include in your report specific identification on the source of each summary, the form of each summary, the purpose of each, and any other pertinent information.

Application 15 Giving a Summary

Select a piece of writing, such as a magazine article, pamphlet, or chapter in a book, relating to your major field. For the selection, write:

□ A descriptive summary
□ An informative summary
□ An evaluative summary

a. Fill in the Plan Sheet on pages 243–244.
b. Write the preliminary drafts.
c. Revise. See inside back cover.
d. Write the final drafts.

USING PART II: SELECTED READINGS

Application 16 Giving a Summary

Read "Quality," by John Galsworthy, pages 595–599. Write a descriptive summary of the story.

Application 17 Giving a Summary

Read "Clear Only If Known," by Edgar Dale, on pages 568–571. Under "Suggestions for Response and Reaction," page 571, write Number 3.

Application 18 Giving a Summary

Read "Are You Alive?" by Stuart Chase, pages 611–615. Under "Suggestions for Response and Reaction," page 615, write Number 4.

Application 19 Giving a Summary

Read "To Serve the Nation: Life Is More Than a Career," by Jeffrey R. Holland, pages 632–638. Under "Suggestions for Response and Reaction," page 638, write Number 6.

Application 20   Giving a Summary

Give descriptive, informative, and evaluative summaries of other selections from Part II as directed by your instructor.

PLAN SHEET
FOR GIVING AN INFORMATIVE SUMMARY

Analysis of Situation Requiring Summary

What type of summary is this to be?

For whom is the summary intended?

How will the summary be used?

Will the summary be written or oral?

Identification of Work (Bibliographical Information), as Applicable

Title:

Author:

Date:

Type of material (magazine article, book, report, etc.):

Title of publication:

City of publication and publishing company:

Volume number:

Page numbers:

Other:

Key Facts and Ideas in the Work

PLAN SHEET
FOR GIVING AN INFORMATIVE SUMMARY

Analysis of Situation Requiring Summary

What type of summary is this to be?

For whom is the summary intended?

How will the summary be used?

Will the summary be written or oral?

Identification of Work (Bibliographical Information), as Applicable

Title:

Author:

Date:

Type of material (magazine article, book, report, etc.):

Title of publication:

City of publication and publishing company:

Volume number:

Page numbers:

Other:

Key Facts and Ideas in the Work

PLAN SHEET
FOR GIVING AN EVALUATIVE SUMMARY

Analysis of Situation Requiring Summary

What type of summary is this to be?

For whom is the summary intended?

How will the summary be used?

Will the summary be written or oral?

Identification of Work (Bibliographical Information), as Applicable

Title:

Author:

Date:

Type of material (magazine article, book, report, etc.):

Title of publication:

City of publication and publishing company:

Volume number:

Page numbers:

Other:

Key Facts and Ideas in the Work

My Evaluative Comments

PLAN SHEET
FOR GIVING AN EVALUATIVE SUMMARY

Analysis of Situation Requiring Summary

What type of summary is this to be?

For whom is the summary intended?

How will the summary be used?

Will the summary be written or oral?

Identification of Work (Bibliographical Information), as Applicable

Title:

Author:

Date:

Type of material (magazine article, book, report, etc.):

Title of publication:

City of publication and publishing company:

Volume number:

Page numbers:

Other:

Key Facts and Ideas in the Work

My Evaluative Comments

PLAN SHEET
FOR GIVING AN EVALUATIVE SUMMARY

Analysis of Situation Requiring Summary

What type of summary is this to be?

For whom is the summary intended?

How will the summary be used?

Will the summary be written or oral?

Identification of Work (Bibliographical Information), as Applicable

Title:

Author:

Date:

Type of material (magazine article, book, report, etc.):

Title of publication:

City of publication and publishing company:

Volume number:

Page numbers:

Other:

Key Facts and Ideas in the Work

My Evaluative Comments

Chapter 7

Memorandums and Letters: Sending Messages

Objectives

Upon completing this chapter, the student should be able to:

- List and define the two regular parts of a memorandum
- Set up a heading for a memorandum
- Write a memorandum
- Analyze the form and content of a letter
- Show and identify by label the three most often used layout forms
- Head a second page properly
- List, define, label, and write an example of the six regular parts of a letter
- List, define, label, and write an example of seven special parts of a letter
- List, define, label, and write an example of the two regular parts on the envelope
- Write a letter of inquiry or request
- Write an order letter
- Write a sales letter
- Write a claim letter and an adjustment letter
- Write a collection series
- Write a job application letter
- Write a resumé
- Fill out a job application form
- Write application follow-up letters

Introduction

Being able to write effective memorandums and letters, according to business people and industrialists, is one of the major writing skills that an employee needs. These employers say that the employee must be able to handle the aspects of work that involve correspondence, such as communicating with fellow employees, making inquiries about processes and equipment, requesting specifications, making purchases, answering complaints, and promoting products. Over half of all business is conducted in part or wholly by correspondence. Thus it is impossible to overemphasize the importance of effective business correspondence.

Too, memorandums and letters are business records; copies show to whom messages have been written and why. Since more than one message frequently is required in settling a matter, copies become a necessity for maintaining continuity in the correspondence. Even now you should begin to develop the habit of keeping a copy of every memorandum and letter you send.

Whether or not you get a job may be determined by your application materials (letter of application, resumé, application form). Employers, particularly personnel managers, are concerned that often a potentially good worker does not get the job desired because the application materials do not make a good impression.

Memorandums

Memorandums are communications typically between persons within the same company. The correspondents may be in the same building or in different branch offices of the company. Memorandums are used to convey or confirm information.

They serve as written records for transmittal of documents, policy statements, instructions, meetings, and the like. Although the term *memorandum* formerly was associated with a communication of only a temporary nature, usage of the term has changed. Now a memorandum is regarded as a communication that makes needed information immediately available or that clarifies information.

The memorandum, unlike the business letter, has only two regular parts: the heading and the body. The formalities of an inside address, salutation, complimentary close, and signature usually are omitted. (Some companies, however, prefer the practice of including a signature, as in the memorandum on page 251.) The memorandum may be initialed in handwriting following the typed "From" name if there is no signature at the end of the memo, as illustrated in the memorandum on page 253. This initialing (or signature at the end) indicates official verification of the sender. As in business letters, an identification line (see page 260) and a copy line (see pages 260–261) may be used, if appropriate.

See also Memorandum in Chapter 8, Reports, pages 351 and 369.

HEADING

The heading in a memorandum is a concise listing of:

- □ *To* whom the message is sent
- □ *From* whom the message comes
- □ The *subject* of the message
- □ The *date*

For ease in reading, the guide words To, From, Subject, and Date usually appear on the memorandum. (These guide words are not standardized as to capitalization, order, or placement at the top of the page.) Most companies use a memorandum form, printed with the guide words and the name of the company or the department or both. It is perfectly permissible, however, to make you own memorandum form by simply typing in the guide words, as illustrated in the memo on page 251.

BODY

The body of the memorandum is the message. It is written in the same manner as any other business communication. The message should be clear, concise, complete, and courteous. If internal headings will make the memo easier to read, insert them (see Headings, pages 7–8 and 354–355).

In the following three examples of memorandums, note how the messages serve different purposes. The first memorandum, with a filled-in Plan Sheet, explains the reason for an industrial manufacturing error and sets forth a new procedure to eliminate the problem; the second memorandum explains a change in company policy concerning annual leave; the third announces a sales increase and plans for sales strategy.

PLAN SHEET
FOR WRITING A MEMORANDUM

Heading

To:
Anna Shulermann, Quality Control Manager

From:
Henry J. McCord, Fabrication Superintendent

Subject:
Weld failure on pull rod

Date:
March 23, 1987

Body

The purpose of this memorandum is:
to explain why the pull rod failed on the Charlesworth Power & Light circuit breaker and to set forth a procedure to eliminate the problem

I need to include this information:
Cause of pull rod failure; inadequate weld penetration because of incorrect machine setting
New procedure: weld current and voltage to be recorded by shop operator on routing card
Routing card to be checked by Quality Control before the pull rod goes to next work station
Correct machine settings to be sent to Quality Control

Special Parts

This memorandum requires these special parts:
signature—Henry J. McCord

TO: Anna Shulermann, Quality Control Manager

FROM: Henry J. McCord, Fabrication Superintendent

SUBJECT: Weld failure on pull rod

DATE: March 23, 1987

Manufacturing Engineering has determined that the pull rod
failure on the Charlesworth Power and Light circuit
breakers was due to inadequate weld penetration, caused by
incorrect machine setting.

To provide a check on future pull rod welds, Manufacturing
Engineering has recommended that the weld current and
voltage used on pull rod welds be recorded by the shop
operator on the shop routing card.

This card will be checked by the Quality Control inspector
before the pull rod goes to the next work station.

Correct machine settings established by Manufacturing
Engineering will be provided the Quality Control
inspector.

Henry J. McCord

Memorandum on Printed Form with Letterhead

Mt. Vernon, NY 10553

DISCOUNT MERCHANDISE

To: Board of Directors

Subject: Spring sales are UP!

From: Thomas Bane, President

Date: 10 April 1987

The figures are in for sales in March, and as we had
anticipated, sales are up. Total sales for March were
$31,976, up 10% over February sales and up 14% over sales
for March 1986.

Much of the increase in sales is no doubt due to the
opening of the Hunter Corporation factory which has
brought 300 new families into the immediate area. We are
not yet sure of other reasons that may account for the
increase.

A more comprehensive report will be available at the next
board meeting when we will also have the April sales
total. After studying that report, we will plan strategy
for late summer sales promotion.

THE NATIONAL BANK OF AMERICA
IN JEFFERSON PARISH (LOUISIANA)

MEMORANDUM   DATE: July 1, 1987

FOR   All employees

FROM   Donald S. Milton, President

Effective January 1, 1988, requests for annual leave of
three or more consecutive days must be presented in
writing to your immediate supervisor at least three weeks
in advance.

Business Letters

The basic principles of composition apply to letters as they do to any other form of writing. The writer should have the purpose of the letter and its intended reader clearly in mind and should carefully organize and write the sentences and paragraphs so that they say what the writer wants them to say.

Business letters are usually not very long; therefore, sentences and paragraphs tend to be short. Often the first and last paragraphs each contain only a single sentence and seldom more than two sentences. Other paragraphs may contain from two to five sentences. In letters that are longer than average, paragraphs are likely to be longer. The longer paragraph should stay on a single topic, and the sentences should be clearly related one to another.

Study the sample letters throughout this chapter, noting sentence and paragraph length and development.

One of the main differences between the business letter and other forms of writing discussed in previous chapters is that in the business letter there is a specific reader who is named and to whom the communication is directly addressed.

THE "YOU" EMPHASIS

Since the business letter is directed to a named reader, it becomes more personal; courtesy becomes important. To achieve a personal, courteous tone, stress the "you."

Focus on qualities of the addressee and minimize references to "I," "me," "my," "we," and "us." For example,

> We were pleased to receive your order for ten microscopes. We have forwarded it to the warehouse for shipment.

can easily be rephrased to:

> Thank you for your order for ten microscopes. You should receive the shipment from our warehouse within two weeks.

The first sentence is writer centered; the rephrased sentence is reader centered.

THE POSITIVE APPROACH

The "you" emphasis helps the writer attain a positive approach. For example, the negative statement

> We regret that we cannot fill the order for ten microscopes by December 1. It is impossible to get the shipment out of our warehouse because of a rush of Christmas orders.

can be rephrased to:

> Your order for ten microscopes should reach you by December 10. Your bill for the microscopes will reflect a 10% discount to say thank you for accepting a delayed shipment caused by a backlog of Christmas orders.

The negative statement rewritten to emphasize "you" becomes a positive statement. The letter writer should follow the admonition of a once-familiar popular song: "Accentuate the positive; eliminate the negative."

NATURAL WORDING

The wording of a letter should be as natural and normal as possible. Jot down thoughts or organize them in your mind so that they can be presented clearly and naturally. Consider the following wording:

> Pursuant to your request of November 10 that I be in charge of the program at the December DECA meeting, I regret to inform you that my impending departure for a holiday tour of Europe will prevent a positive response.

This sentence probably resulted from a writer trying to sound impressive. The result is stilted, awkward wording. The same information might be clearly and naturally worded:

> I am sorry I cannot be in charge of the DECA program for December. I will be leaving on December 5 for a holiday tour of Europe.

As you write a business letter, keep your reader in mind. How will the letter sound to the reader? Is the information presented in a natural, normal way so that the content is clear?

Form in Letters

Learning about the form of a letter includes becoming familiar with standard formats and with the parts of a letter and their arrangement.

FORMAT

There are standard practices concerning paper, typewriting, appearance, and layout. Failure to follow these standards shows poor taste, reflects the ignorance of the writer, and invites an unfavorable response. In business correspondence, there is no place for unusual or "cute" stylistic practices. (A specialized type of business letter, the sales letter, sometimes uses attention-getting gimmicks. Follow-up sales letters, however, tend to conform to standard practices.)

Paper. Use unruled, good-quality, standard-size paper. Do not use notebook paper.

Handwriting or Typewriting. Letters should be typewritten or produced on a word processor because the printed word is neater and easier to read. If a letter is handwritten, only black, blue, or blue-black ink should be used.

Appearance. The general appearance of a letter is very important. A letter that is neat and pleasing to the eye invites reading and consideration much more readily than one that is unbalanced or has noticeable erasures. The letter should be like a picture, framed on the page with margins in proportion to the length of the letter. Allow at least a 1½-inch margin at the top and bottom, and a 1-inch margin on the sides. Short letters should have wide margins and be appropriately centered on the page. As a general rule, single space within the parts of the letter; double space between the parts of the letter and between paragraphs. It is always a wise investment to spend the extra time to retype a letter if there is even the slightest doubt about its making a favorable impression on the reader. (See the letter on page 259 for proper spacing.)

Layout Forms. Although there are several standardized layout forms for letters, two seem to be preferred by most firms: the modified block (pages 259, 272, 281, 282, 297, 317), and the block (pages 267, 277, 284, 287, 293). Another layout form gradually gaining favor is the simplified block form (page 271).

Modified Block Form. In the modified block form, the inside address, salutation, and paragraphs are flush, or even, with the left margin. The heading, complimentary close, and signature are to the right half of the page. Paragraphs may or may not be indented. Open punctuation is sometimes used; that is, no punctuation follows the salutation and the complimentary close.

Block Form. In the block form, *all* parts of the letter are even with the left margin. Open punctuation is sometimes used with the full block form; that is, no punctuation follows the salutation and the complimentary close.

Simplified Block Form. The parts of the letter in the simplified block form, or the AMS (American Management Society) form, are the heading, the inside address, the subject line, the body, and the signature. The salutation and the complimentary close are omitted. A subject line is always used. Like the block, all parts begin at the left margin.

PARTS OF A BUSINESS LETTER

The parts of a business letter follow a standard sequence and arrangement. The six regular parts in the letter include: heading, inside address, salutation, body, complimentary close, and signature (these are illustrated in the letter on page 259). In addition, there may be several special parts in the business letter. On the envelope, the two regular parts are the outside address and the return address (these are illustrated on page 262).

Regular Parts

1. *Heading*. Located at the top of the page, the heading includes the writer's complete mailing address and the date, in that order, as shown below. As elsewhere in standard writing, abbreviations should generally be avoided. (Note that the heading does *not* include the writer's name.)

Route 12, Box 758   704 South Pecan Circle
Elmhurst, IL 60126   Hanover, PA 17331
July 25, 1987   9 April 1987

In writing the state name, use the two-letter abbreviation suggested by the postal service.

TWO-LETTER ABBREVIATIONS FOR STATES

Alabama	AL	Maryland	MD
Alaska	AK	Massachusetts	MA
Arizona	AZ	Michigan	MI
Arkansas	AR	Minnesota	MN
California	CA	Mississippi	MS
Colorado	CO	Missouri	MO
Connecticut	CT	Montana	MT
Delaware	DE	Nebraska	NE
District of Columbia	DC	Nevada	NV
Florida	FL	New Hampshire	NH
Georgia	GA	New Jersey	NJ
Hawaii	HI	New Mexico	NM
Idaho	ID	New York	NY
Illinois	IL	North Carolina	NC
Indiana	IN	North Dakota	ND
Iowa	IA	Ohio	OH
Kansas	KS	Oklahoma	OK
Kentucky	KY	Oregon	OR
Louisiana	LA	Pennsylvania	PA
Maine	ME	Rhode Island	RI

South Carolina	SC	Virginia	VA
South Dakota	SD	Washington	WA
Tennessee	TN	West Virginia	WV
Texas	TX	Wisconsin	WI
Utah	UT	Wyoming	WY
Vermont	VT		

Note that each abbreviation is written in all capitals and that no periods are used.

Letterhead Stationery. Many firms use stationery that has been especially printed for them with their name and address at the top of the page. Some firms have other information added to this letterhead, such as the names of officers, a telephone number, or a slogan. The letterhead on stationery is always put there by a printer. Thus, the writer of a business letter, whether a student or an employee, never makes his or her own letterhead by typing, writing, or drawing one in.

If letterhead stationery is used, the address is already printed on the paper; only the date must be added. Letterhead paper is used for the first page only. (See the letters on pages 272, 277, 281, 282, 284, 287.)

2. *Inside Address.* The inside address is placed even with the left margin and is usually three spaces below the heading. It contains the full name of the person—including a proper title such as *Mr., Ms., Dr.,* and the like—or firm being written to and the complete mailing address, as in the following illustrations:

Mr. Ronald M. Benrey
Electronics Editor, *Popular Science*
355 Lexington Avenue
New York, NY 10017-0127

Kipling Corporation
Department 40A
P.O. Box 127
Beverly Hills, CA 90210

Preface a person's name with a title of respect if you prefer; and when addressing an official of a firm, follow the name with a title or position. Write a firm's name in exactly the same form that the firm itself uses. Although it may be difficult to find out the name of the person to whom a letter should be addressed, it is always better to address a letter to a specific person rather than to a title, office, or firm. In giving the street address, be sure to include the word *Street, Avenue, Circle,* and so on. Remember: No abbreviations.

3. *Salutation.* The salutation, or greeting, is two spaces below the inside address and is even with the left margin. The salutation typically includes the word *Dear* followed by a title of respect plus the person's last name or by the person's full name: "Dear Ms. Badya:" or "Dear Maron Badya." In addressing a company, acceptable forms include "Dear Davidson, Inc.:" or simply "Davidson, Inc." Some writers use "Dear Sir or Madam."

Usually the salutation is followed by a colon. Other practices include using a comma if the letter is a combination business-social letter, and, in the modified and full block forms, the option of omitting the mark of punctuation after both the salutation and the complimentary close.

4. *Body.* The body, or the message, of the letter begins two spaces below the salutation. Like any other composition, the body is structured in paragraphs.

Generally it is single spaced within paragraphs and double spaced between paragraphs. The paragraphs may or may not be indented, depending on the layout form used (see pages 255–256).

Second Page. For letters longer than one page, observe the same margins as used for the body of the first page. The second-page top margin should be the same as that of the sides of the second page. Be sure to carry over a substantial amount of the body of the letter (at least two lines) to the second page.

Although there is no one conventional form for the second-page heading, it should contain (*a*) the name of the addressee (the person to whom the letter is written), (*b*) the page number, and (*c*) the date. The following illustrate two widely used forms:

Mr. Thomas R. Racy Page 2 May 22, 1987

or

Mr. Thomas R. Racy
Page 2
May 22, 1987

See the resumés on pages 300–301 and 305–306 and the report letter on pages 391–392.

5. *Complimentary Close.* The complimentary close, or closing, is two spaces below the body. It is a conventional expression, indicating the formal close of the letter. "Sincerely" is the commonly used closing. "Cordially" may be used when the writer knows the addressee well. "Respectfully" or "Respectfully yours" indicates that the writer views the addressee as an honored individual or that the addressee is of high rank.

Capitalize only the first word, and follow the complimentary close with a comma. (In using open punctuation, the comma is omitted after both the salutation and the complimentary close.)

6. *Signature.* Every letter should have a legible, handwritten signature in ink. Below this is the typewritten signature. If the entire letter is handwritten, of course there is only one signature.

If a person's given name (such as Dale, Carol, Jerry) does not indicate whether the person is male or female, the person may want to include a title of respect (Ms., Miss, Mrs., Mr.) in the parentheses to the left of the typewritten signature (see the letter on page 271). In a business letter, a married woman uses her own first name, not her husband's first name. Thus, the wife of Jacob C. Andrews signs her name as Thelma S. Andrews. In addition, she may type her married name in parentheses (Mrs. Jacob C. Andrews) below her own name (see page 272).

The name of a firm as well as the name of the individual writing the letter may appear as the signature (see the letters on pages 272, 282). In this case, responsibility for the letter rests with the name that appears first.

Following the typewritten signature there may be an identifying title indicating the position of the person signing the letter; for example: Estimator; Buyer, Ladies Apparel; Assistant to the Manager, Food Catering Division. (See pages 277, 281, 282, 284, 287.)

At least 1½-inch margin at top

HEADING

Route 12, Box 75B
Elmhurst, IL 60126
25 July 1987

Begin address
at center

Four typewriter carriage returns

INSIDE
ADDRESS

Mr. Ronald M. Benrey
Electronics Editor, Popular Science
355 Lexington Avenue
New York, NY 10017-0127

Same as address
on envelope

Double space

SALUTATION

Dear Mr. Benrey: Followed by colon

Double space

BODY OF
LETTER

In an electronics laboratory I am taking as a part of my second-year training at Midwestern Technical College, Elmhurst, Illinois, I have developed a six-sided hi-fi speaker system. The speaker system is inexpensive (the materials cost less than $50), lightweight, and quite simple to construct. The sound reproduction is excellent.

Double space

My electronics instructor believes that other hi-fi enthusiasts may be interested in building such a speaker system. If you think the readers of Popular Science would like to look at the plans, I will be happy to send them to you for reprint in your magazine.

Double space

COMPLIMENTARY
CLOSE

Closing and
signature
begin at center

Sincerely yours,

Thomas G. Stein

Thomas G. Stein

Capitalize first word;
comma after close

Four typewriter
carriage returns

TYPED NAME

At least 1½-inch margin at bottom

Special Parts. In addition to the six regular parts of the business letter, sometimes special, or optional, parts are necessary. The main ones, in the order in which they would appear in the letter, are the following.

1. *Attention Line.* When a letter is addressed to a company or organization rather than to an individual, an attention line may be given to help in mail delivery. An attention line is never used when the inside address contains a person's name. Typical are attention lines directed to: Sales Division, Per-

sonnel Manager, Billing Department, Circulation Manager; the attention line may also be an individual's name. The attention line contains the word *Attention* (capitalized and sometimes abbreviated) followed by a colon and name of the office, department, or individual.

Attention: Personnel Manager   *or*   Attn: Mr. Robbin Carmichael

The attention line appears both on the envelope and in the letter. On the envelope, the attention line appears directly underneath the first line of the address. In the letter, the attention line is even with the left margin and it may be in one of two places: (1) two spaces below the inside address or (2) directly underneath the first line of the inside address. (See page 293.) Since a letter addressed to an individual is usually more effective than one addressed simply to a company, the attention line should be used sparingly.

2. *Subject, or Reference, Line.* The subject, or reference, line saves time and space. It consists of the word *Subject* or *Re* (a Latin word meaning "concerning") followed by a colon and a word or phrase of specific information, such as a policy number, account number, or model number.

Subject: Policy No. 10473A   *or*   Re: Latham Stereo Tape Deck Model 926

The position of the subject line is not standardized. It may appear to the right of the inside address or salutation; it may be centered on the page several spaces below the inside address; it may be even with the left margin and several spaces below the inside address; it may even be several spaces below the salutation. (See pages 271, 282.)

3. *Identification Line.* When the person whose signature appears on the letter is not the person who typed the letter, there is an identification line. Current practice is to include in the line only the initials (in lowercase) of the typist. The identification line is two spaces below the signature and is even with the left margin of the letter. (See pages 272, 281, 282.)

4. *Enclosure.* When an item (pamphlet, report, check, etc.) is enclosed with the letter, an enclosure line is usually typed two spaces below the identification line and even with the left margin. If there is no identification line, the enclosure line is two spaces below the signature and even with the left margin. The enclosure line may be written in various ways and may give varying amounts of information.

Enclosure
Enclosures: Inventory of supplies, furniture, and equipment
            Monthly report of absenteeism, sick leave, and vacation leave
Encl: Application of employment form
Encl. (2)

(See pages 271, 272, 284, 293, 297.)

5. *Copy.* When a copy of a letter is sent to another person, the letter *c* (usually lowercase) or the word *copy* followed by a colon and the name of the person or persons to whom the copy is being sent is typed one space below the identification line and even with the left margin of the letter. (See page 282.)

If there is no identification line, the copy notation is two spaces below the signature and even with the left margin.

c: Mr. Jay Longman
copy: Joy Minor

6. *Personal Line*. The word *Personal* or *Confidential* (capitalized and usually underlined) indicates that only the addressee is to read the letter; obviously, this line should appear on the envelope. It usually appears to the left of the last line of the outside address. The personal line may be included in the letter itself, two spaces above the inside address and even with the left margin.
7. *Mailing Line*. If the letter is sent by means other than first class mail, a notation may be made on both the letter and the envelope. The mailing line in the letter is even with the left margin and appears after all other notations.

Delivered by Messenger
Certified Mail

THE ENVELOPE

The U.S. Postal Service has established guidelines for sizes of and addresses on envelopes. These guidelines allow an Optical Character Reader (OCR), an automated device that reads the address and sorts the mail by ZIP code, to operate efficiently.

Regular business envelopes should be a minimum of 3½ inches high and 5 inches long and a maximum of 6⅛ × 11½ inches. Mailable thickness is 0.007 to 0.25 inch.

Envelope sizes regularly used by businesses are classified as All-Purpose (3⅝ × 6¼ inches), Executive (3⅞ × 7½ inches), and Standard (4⅛ × 9½ inches). Larger manila envelopes are available in three standard sizes: 6½ × 9½ inches, 9 × 12 inches, and 11½ × 14⅝ inches.

Regular Parts. The two regular parts on the envelope, the outside address and the return address, are explained and illustrated on page 262.

1. *Outside Address*. The outside address on the envelope is identical with the inside address. The postal service prefers single spacing, all uppercase letters, and no punctuation marks for ease in sorting mail by an OCR. For obvious reasons, the address should be accurate and complete. For large-volume mailing the postal service encourages using the nine-digit zip code to facilitate mail delivery.
2. *Return Address*. Located in the upper left-hand corner of the envelope, not on the back flap, the return address includes the writer's name (without "Mr.," etc.) plus the address as it appears in the heading. The ZIP code should be included.

Special Parts. In addition to the two regular parts on the envelope, sometimes a special part is needed. The main ones are the attention line, the personal line, and mailing directions.

1. *Attention Line*. An attention line may be used when a letter is addressed to a company rather than to an individual. The wording of the attention line on the envelope is the same as that of the attention line in the letter. On the envelope, the attention line is written directly under the first line of the address.

2. *Personal Line*. The word *Personal* or *Confidential* (capitalized and usually underlined) indicates that only the addressee is to read the letter. The personal line, aligned with the left margin of the return address, appears three spaces below the return address.

3. *Special Delivery or Registered Mail*. Mailing directions such as SPECIAL DELIVERY or REGISTERED MAIL are typed in all capital letters below the stamp.

Regular Parts of the Envelope and Their Spacing

RETURN ADDRESS	Thomas G. Stein Route 12, Box 75B Elmhurst, IL 60126	Except for name and date, same as heading in letter
		Begin outside address slightly left of center
OUTSIDE ADDRESS	Content same as in inside address in letter	MR RONALD M BENREY ELECTRONICS EDITOR POPULAR SCIENCE 355 LEXINGTON AVENUE NEW YORK NY 10017-0127

Types of Letters

There are many, many types of letters. There are numerous books devoted wholly to discussions of letters. This chapter discusses some of the common types of letters written by businesses and individuals: inquiry, order, sales, claim and adjustment, collection, job application, and application follow-up. (See also Chapter 8, Reports, for further discussion of letters, page 351, and the letter of transmittal, pages 352, 393, and 475.)

LETTER OF INQUIRY

A letter to the college registrar asking for entrance information, a letter to a firm asking for a copy of its catalog, a letter to a manufacturing plant requesting information on a particular product—each is a letter of inquiry, or request. Such a letter is simple to write if these directions are followed:

1. State clearly and specifically what is wanted. If asking for more than two items or bits of information, use an itemized list.

2. Give the reason for the inquiry, if practical. Remember: If you can show clearly a direct benefit to the company or person addressed, you increase your chances of a reply.

3. Include an expression of appreciation for the addressee's consideration of the inquiry. Usually a simple "Thank you" is adequate.

4. Include a self-addressed, stamped envelope with inquiries sent to individuals who would have to pay the postage themselves to send a reply.

On page 265 is a Plan Sheet filled in by a student preparing to write a letter of inquiry. The letter follows.

PLAN SHEET
FOR WRITING A LETTER OF INQUIRY

Layout Form (modified block, block, simplified block)

I will use:
block

Heading

My mailing address is:
3505 Joy Hill Drive
North Liberty, IN 46554

The date is:
10 June 1987

Inside Address

The person (or firm) to whom I am writing (including any identifying title after the name) is:
ABC Art Supply

The complete mailing address is:
1510 West Street
Buffalo, NY 14200

Salutation

I will use:
ABC Art Supply:

Body

Through this letter I am seeking:
information on texture and cost of brushes

The reason for my inquiry is:
art major; like to be aware of new materials that might be useful

To make my inquiry clear and specific, I need to include this information:
saw advertisement in <u>Modern Art</u>, May 1987

Complimentary Close

I will use:
Sincerely,

Signature (including, if applicable, the name of the firm and identifying title)

I will use:
Teresa Bridges

Special Parts

This letter requires the following special parts:
omit

```
3505 Joy Hill Drive
North Liberty, IN 46554
10 June 1987

ABC Art Supply
1510 West Street
Buffalo, NY 14200

ABC Art Supply:

The May 1987 issue of Modern Art contained an
advertisement showing your watercolor brushes. The article
indicated that further information was available upon
request.

Would you please send information on the texture and cost
of these brushes.

I am an art major and I like to be aware of new materials
that might be useful to me.

Thank you.

Sincerely,

Teresa Bridges
Teresa Bridges
```

ORDER LETTER

The order letter, as the term implies, is a written communication to a seller from a buyer who wishes to make a purchase. For the transaction to be satisfactory, the terms of the sale must be absolutely clear to both. Since in the order letter it is the writer (purchaser) who is requesting certain merchandise, it is the writer's responsibility to state clearly, completely, and accurately exactly what is wanted, how it will be paid for, and how it is to be delivered.

In writing an order letter:

1. State clearly, completely, and accurately what is wanted. If ordering two or more different items, use an itemized list format. Include the exact name of the item, the quantity wanted, and other identifying information such as model number, catalog number, size, color, weight, and finish.
2. Give the price of the merchandise and the method of payment: check, money order, credit card (give name, number, and expiration date), COD, charge to account if credit is established (give account name and number).
3. Include shipping instructions with the order. Mention desired method for shipment of the merchandise: parcel post, truck freight, railway express, air express, or the like. If the date of shipment is important, say so.

As the two following order letters indicate, the general principles for writing an order letter are the same, whether the order be for a small or a large amount, or from an individual or a company. The first order letter has a filled-in Plan Sheet, which shows the thinking that the student did before writing the letter.

PLAN SHEET
FOR WRITING AN ORDER LETTER

Layout Form (modified block, block, simplified block)

I will use:
simplified block

Heading

My mailing address is:
602 Joyce Street
Whiteoak, MO 63880

The date is:
16 April 1987

Inside Address

The person (or firm) to whom I am writing (including any identifying title after the name) is:
Sports Illustrated

The complete mailing address is:
541 North Fairbanks Court
Chicago, IL 60611

Salutation

I will use:
not used in simplified block

Body

The items I am ordering and their identification (model number, catalog number, size, color, weight, finish, etc.) are:

Item	Quantity	Identification	Cost
ankle weights set	*1*	*advertised in March Sports Illustrated, page 41*	*$16.95*

The manner of payment (check, money order, credit card, COD, charge, etc.) is:
check

The method of shipping (parcel post, truck freight, railway freight, air express, etc.) is:
truck freight; cost includes shipping

Other information or instructions:
omit

Complimentary Close

I will use:
not used in simplified block

Signature (including, if applicable, the name of the firm and identifying title)

I will use:
Lynn Reed

Special Parts

The letter requires the following special parts:
Subject: ankle weights (Subject line required in simplified block form)
Enclosure: Check for $16.95

602 Joyce Street
Whiteoak, MO 63880
16 April 1987

Sports Illustrated
541 North Fairbanks Court
Chicago, IL 60611

Subject: ankle weights

Please mail one set of the ankle weights advertised in
Sports Illustrated, March issue, page 41. Enclosed is a
check for $16.95.

Please send the weights by truck freight.

Lynn Reed

(Ms.) Lynn Reed

Enclosure: check for $16.95

Do not use

the ANDREWS COMPANY

432 FIELDING AVENUE
YOUNGSTOWN, OREGON 97386

May 1, 1987

Woodfield Wholesale Company
300 N. State Street
Braxton, CA 90318

Woodfield Wholesale:

Please send by railway express the following items
advertised in your spring sale catalog:

Quantity	Catalog No.	Item	Unit Price	Amount
5	77X1628	Chain Saw, 5 H.P. engine	$195.00	$ 975.00
10	32W5602	Bow and Arrow Set	22.00	220.00
100	11R7487	Golf Club Pac	36.00	3,600.00
			Total	$4,795.00

I am enclosing a check for one-fourth of the amount
($1,198.75), and the balance ($3,596.25) plus shipping
charges will be paid within 30 days after the merchandise
arrives.

Sincerely,

THE ANDREWS COMPANY

Thelma S. Andrews

Thelma S. Andrews
(Mrs. Jacob C. Andrews)

msw

Enclosure: Check for $1,198.75

Printed Order Forms. Many companies provide a printed order form for
ordering merchandise. These forms indicate specific information needed and provide
spaces to fill it in, and they usually give directions for filling in the form. An example
of a filled-in order form is shown on page 273.

You can count on

Sears

Satisfaction Guaranteed
or Your Money Back

	DAY	ORDERS	# LINES	SOURCE OF-SALE	TYPE SALE	METHOD SHIPMT	CASH	DISCT.	TAX EXMT	SPEC CODE	SPECIAL INFORMATION (DO NOT TRANSMIT DASHES)
				5							

	TERMS TABLE	NO. OF MONTHS

My SearsCharge number is:

Direct Mail Order Blank

PLEASE DO NOT WRITE IN THIS SPACE

NAME AND PRESENT ADDRESS

NAME (first, middle initial, last) Please use the same name on ALL orders from your household

M I K E S G I T A N O

MAILING ADDRESS | APT. NO.

3 7 1 R O S E S T

CITY STATE | ZIP CODE 3 5 9 0 3

G A D S D E N A L

AREA CODE PHONE NUMBER TODAY'S DATE

2 0 5 9 3 8 2 0 0 6 2 2 4 8 7

METHOD OF PAYMENT

☐ Add to my SearsCharge account

My SearsCharge signature: _____

☒ CASH: (check or money order payable to Sears, Roebuck and Co.)

☐ PLEASE OPEN AN ACCOUNT. Completed credit application enclosed.

☐ Amount enclosed to be applied to my Sears Credit Account $_____

NOTE:
✔ PLEASE GIVE COMPLETE DELIVERY INFORMATION
Be sure to give complete mailing address at left filling in the correct information, on the lines provided, including telephone number.

✔ PLEASE MAIL ALL INQUIRIES NOT DIRECTLY RELATED TO THIS ORDER UNDER SEPARATE COVER

✔ C.O.D. ORDERS NOT ACCEPTED BY DIRECT MAIL

IF YOUR ADDRESS HAS CHANGED since your last order, please give your old mailing address here:

FORMER ADDRESS | APT. NO.

CITY/STATE | ZIP CODE

SHIP TO ANOTHER ADDRESS? If you want this order shipped to another person or to a different address, freight or express station, give address here:

NAME (first, middle initial, last)

MAILING ADDRESS | APT. NO.

CITY/STATE | ZIP CODE

See Yellow Pages in General Catalog for shipping information

If the merchandise you are ordering exceeds postal limitations, it is usually more economical to ship your order to our store nearest your home for pickup. Please indicate the Sears store most convenient for you

Store Name and Location: _____

	CATALOG NUMBER	HOW MANY	Color NO.	SIZE	Name of Item	PRICE EACH	TOTAL PRICE Dollars	Cents	Please do not write below Code	Special Instructions	Weight Lbs.	Oz.	PAGE NO.
1	34 E 68928	1	—	—	CORN POPPER	17.99	& 17	99	—	CR LF	6	2	426
2	33 E 56515 F	2 PR		XL	GLOVES	15.00	& 30	00	—	CR LF		10	194
3	96 BR 3796 H	3	—	542	TOWELS	9.99	& 29	97	—	CR LF	3	6	79
4		—	—				&			CR LF			
5		—	—				&			CR LF			
6		—	—				&			CR LF			
7		—	—				&			CR LF			

To order additional items, use other side

TAX: Please be sure to add correct state, county, city and local taxes applicable.

*SHIPPING AND HANDLING: Figure separate charges for weight of each individual item ending in a "C" or "L" as each of these items must be shipped separately. Do not add the weight of multiples of same item together, figure each separately. Calculate on reverse side. See Big Book for information on items ending with an "N"

F6015 3.82 Printed in U.S.A. Sears, Roebuck and Co.

Fill in spaces below on CASH ORDERS only

			Total Pounds	Total Ounces
TOTAL FOR GOODS	77	96		
SHIPPING, HANDLING	3	45	9	18
TAX (See at left)	3	12	Total Weight* in Pounds	
Amount I owe Sears on previous order		—		11
TOTAL CASH PRICE	84	53	thank you for shopping at Sears	
AMOUNT ENCLOSED	Sears Checks			
	Money Order or Check	84	53	

Send Mail Orders to the nearest Sears Catalog Distribution Center listed below

ATLANTA, GA 30395
BOSTON, MA 02215
CHICAGO, IL 60607
COLUMBUS, OH 43228
DALLAS, TX 75295
GREENSBORO, NC 27480
JACKSONVILLE, FL 32297
KANSAS CITY, MO 64127
LOS ANGELES, CA 90051
MEMPHIS, TN 38140
MINNEAPOLIS, MN 55440
PHILADELPHIA, PA 19132
SEATTLE, WA 98184

Reprinted by permission of Sears, Roebuck and Co.

The sales letter is a specialized kind of business letter requiring careful planning if it is to serve its purpose: to convince the reader to buy a particular product or service. (See also "Basic Considerations in Persuasion," pages 507–508.)

A successful letter gets the reader's attention and arouses interest. The "you" attitude is especially important and should be particularly emphasized in the sales letter. Since the letter is directed toward the *reader's* interests and needs, references to what *"I"* the writer can do or think are kept to a minimum. Emphasis is on the reader and the benefits from the product.

The good sales letter is positive and sincere in approach. The writer must be thoroughly familiar with the product and its capabilities and limitations. Rather than trying to sell by downgrading a competitor, it is wiser to present the product on its own merits. The description of the product should be truthful and should avoid misleading and sensational promises.

The reader must be convinced of personal need for the product. Tactfully, yet concisely and forcefully, the reader must be shown the relationship between the product and good business and the ways in which the product will be of profit or benefit. Finally, the effective sales letter leads to the purchase of the product. The sales letter is successful if the reader immediately begins (whether by a plan presented in the letter or by one devised by the reader) to make arrangements to use the product.

In writing a sales letter:

1. Appeal to the reader through something that is important: home, family, business, community involvement, prestige, or some other area of importance.
2. Identify and describe, accurately and honestly, the product or service offered.
3. Make clear the reader's need for the product or service.
4. State confidently the action you want the reader to take.

On the following pages is a filled-in Plan Sheet for a sales letter, followed by the letter.

PLAN SHEET
FOR WRITING A SALES LETTER

Layout Form

The layout form (modified block, block, simplified block) that I will use is:
block

Heading

My mailing address is:
letterhead

The date is:
February 22, 1987

Inside Address

The person (or firm) to whom I am writing (including any identifying title after the name) is:
Mr. W. D. Adams

The complete mailing address is:
Adams Service Center
4102 Green Street
Jackson, CO 80172

Salutation

I will use:
Dear Mr. Adams:

Body

To get the reader's attention and arouse interest, I will:
mention that automobiles need special care in the spring and summer and then mention particular products

The product (or service) I am selling is:
a new consignment program for automotive supplies

The product can be described in this way:
We will stock your shelves with a complete supply of Autoright supplies. You pay us for only what you sell. After four months if you are not satisfied with the plan, you may discontinue it.

The product will benefit or profit the reader in these ways:
It will help you give even better service to your customers. A participating station cannot lose money.

The plan that I will present whereby the reader can immediately begin to use the product is:
I will call on you at your station on March 2.

Complimentary Close

I will use:
Sincerely,

Signature (including, if applicable, the name of the firm and identifying title)

I will use:
E. C. Pace
Jackson Area Distributor

Special Parts

This letter requires the following special parts:
omit

AUTORIGHT

2829 CHARLOTTE DRIVE
DENVER, COLORADO 80213
TEL. 303-269-3313

February 22, 1987

Mr. W. D. Adams
Adams Service Center
4102 Green Street
Jackson, CO 80172

Dear Mr. Adams:

With spring approaching, you and your customers will be
thinking of special automotive needs for the spring and
summer. And you will be checking your stock of radiator
hoses, summer coolants, plugs, points, condensers, and fan
belts.

As the Autoright distributor for the Jackson area, may I
introduce you to our new consignment program for
automotive supplies. We will stock your shelves with
Autoright parts, at no cost to you. You will have on hand
a complete supply of parts for automobiles and trucks of
all makes and models. You will not pay for the parts until
you sell them. If, after four months, you are not entirely
satisfied with the program, I will pick up the parts and
issue full credit. With a program of this kind you have
all to gain and nothing to lose.

You are already familiar with the high quality of
Autoright products and their competitive prices. We
believe our new consignment plan will help you not only
maintain but increase the efficient service you want to
give your customers.

I would like to call on you at your station on March 4. If
you have any questions, I will be more than glad to answer
them.

Sincerely,

E. C. Pace

E. C. Pace
Jackson Area Distributor

WE CARRY EVERYTHING AUTOMOTIVE

CLAIM AND ADJUSTMENT LETTERS

Claim (complaint) and adjustment letters are in some ways the most difficult letters of all to write. Frequently the writer is angry or annoyed or extremely dissatisfied, and the first impulse is to express those feelings in a harsh, angry, sarcastic letter. But the purpose of the letter is to bring about positive action that satisfies the complaint. A rude letter that antagonizes the reader is not likely to result in such positive action. Thus, above all in writing a claim or an adjustment letter, be calm, courteous, and businesslike. Assume that the reader is fair and reasonable. Include only factual information, not opinions; and keep the focus on the real issue, not on personalities.

In writing a claim letter:

1. Identify the transaction (what, when, where, etc.). Include copies (not the originals) of substantiating documents—sales receipts, canceled checks, invoices, and the like.
2. Explain specifically what is wrong.
3. State the adjustment or action that you think should be made.
4. Remember that reputable companies are eager to have satisfied customers and that most respond favorably to justifiable complaints.

In writing an adjustment letter:

1. Respond to the claim letter promptly and courteously.
2. Refer to the claim letter, identifying the transaction.
3. State clearly what action will be taken. If the action differs from that requested in the claim letter, explain why.
4. Remember to be fair, friendly, and firm.

The following claim letter, with its filled-in Plan Sheet, illustrates a common complaint: The purchaser feels there has simply been an oversight or error in the shipment of goods he has ordered. The second letter is an adjustment letter, replying to the first.

PLAN SHEET
FOR WRITING A CLAIM OR AN
ADJUSTMENT LETTER

Layout Form (modified block, block, simplified block)

I will use:
modified block

Heading

My mailing address is:
letterhead

The date is:
February 20, 1987

Inside Address

The person (or firm) to whom I am writing (including any identifying title after the name) is:
Middleton Manufacturing Company

The complete mailing address is:
Alton, IL 60139

Salutation

I will use:
Middleton Manufacturing Company

Body

Identification of the transaction (items purchased, date, invoice number, model number, style, size, etc.):
ten refrigerators, invoice 479320

Statement of the problem or complaint:
Supposed to include five ice trays for each refrigerator. No trays received.

Desired action or action that will be taken:
Send trays by air express.

Complimentary Close

I will use:
Sincerely,

Signature (including, if applicable, the name of the firm and identifying title)

I will use:
John A. Manuel, Manager
Appliance Department

Special Parts

This letter requires the following special parts:
Identification line — jph

appleton's, inc.

4636 MOCKINGBIRD EXPRESSWAY
BATON ROUGE, LOUISIANA 70806

February 20, 1987

Middleton Manufacturing Company
Alton
Illinois 60139

Middleton Manufacturing Company:

Thank you for your prompt delivery of the ten
refrigerators, invoice #479320.

Your catalog indicated that each refrigerator contains
five ice trays. None of these was included in the order we
received. Since your invoice does not show a back order or
a separate delivery of the trays, we believe there may
have been an error in filling the order.

Since we are having to delay our sales promotion until we
receive the trays, please ship them by air express.

Sincerely,

John A. Manuel

John A. Manuel, Manager
Appliance Department

jph

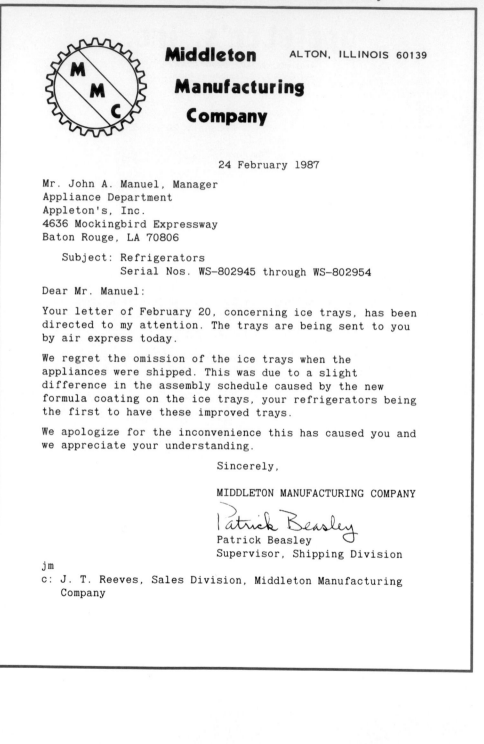

Middleton

Manufacturing

Company

ALTON, ILLINOIS 60139

24 February 1987

Mr. John A. Manuel, Manager
Appliance Department
Appleton's, Inc.
4636 Mockingbird Expressway
Baton Rouge, LA 70806

    Subject: Refrigerators
          Serial Nos. WS-802945 through WS-802954

Dear Mr. Manuel:

Your letter of February 20, concerning ice trays, has been
directed to my attention. The trays are being sent to you
by air express today.

We regret the omission of the ice trays when the
appliances were shipped. This was due to a slight
difference in the assembly schedule caused by the new
formula coating on the ice trays, your refrigerators being
the first to have these improved trays.

We apologize for the inconvenience this has caused you and
we appreciate your understanding.

        Sincerely,

        MIDDLETON MANUFACTURING COMPANY

        Patrick Beasley
        Patrick Beasley
        Supervisor, Shipping Division

jm
c: J. T. Reeves, Sales Division, Middleton Manufacturing
   Company

COLLECTION LETTERS

Collection letters are used to collect overdue accounts. They must be firm yet friendly to cause the customer to pay up yet keep the customer's goodwill.

Often collection letters are form letters and often they are a series of letters, starting with a reminder letter and moving through various appeal letters to an ultimatum letter, depending on the response of the customer.

A reminder letter, written early in a collection series, might mention that the unpaid balance is overdue and appeal to the customer's pride. It might include the sentence: "If you have already mailed your payment, please disregard this reminder."

An appeal letter, written later in a collection series, might

1. Ask for payment.
2. Ask about possible problems. Suggest the customer call (include a toll-free number) to discuss any such problems.
3. Remind the customer of the value of good credit.
4. Enclose a return envelope to make payment easier.

A sample letter on page 284 illustrates this type of appeal letter.

The ultimatum letter is the last resort. It may

1. State that payment must be received, often by stating a deadline.
2. Mention previous notices sent.
3. State what action will be taken if payment is not received by the deadline, such as turning the account over to a collection agency or a lawyer.

The sample letter on page 287 is an example of the ultimatum letter. The Plan Sheet used to organize the letter is included.

M&P APPLIANCES
HWY 15S
ALBANY, GA 31707
(912)601-0043

February 8, 1987

Mr. Walter O. Casey
5701 Honda Lane
Forsyth, GA 39129

Dear Mr. Casey:

We are concerned about your overdue account of $161.
During the past 90 days we have sent three statements,
reminding you of this problem. Since you have always paid
bills promptly, we believe some special circumstances may
have caused this delay.

If you need to make special arrangements for paying your
account balance, we will be happy to work out a
satisfactory payment plan. Or, by sending us a check today
for the $161, you can preserve your excellent credit
rating. A good credit rating is of inestimable value.

Please use the enclosed postpaid envelope to mail us your
check or call us at 800-601-0043 to discuss your account.

Sincerely,

James W. Malley

James W. Malley
Credit Manager

Enc: postpaid envelope

PLAN SHEET
FOR WRITING A COLLECTION LETTER

Layout Form (modified block, block, simplified block)

I will use:
block

Heading

My mailing address is:
letterhead

The date is:
March 11, 1987

Inside Address

The person (or firm) to whom I am writing (including any identifying title after the name) is:
Walter O. Casey

The complete mailing address is:
5701 Honda Lane
Forsyth, GA 39129

Salutation

I will use:
Dear Mr. Casey:

Body

Appeal to customer:
Why not preserve your heretofore excellent credit rating and avoid an unpleasant situation by mailing your check for $161 today.

Reference to balance due and payment:
You have been given every opportunity to pay your overdue balance of $161. We do not feel we can permit any further delay in payment.

Problem(s) preventing payment:
omit

Action the company owed has taken:
omit

Action the company owed will take:
If payment is not received immediately, we will be forced to turn your account over to a collection agency.

Complimentary Close

I will use:
Sincerely,

Signature (including, if applicable, the name of the firm and identifying title)

I will use:
James W. Malley
Credit Manager

Special parts

This letter requires the following special parts:
omit

APPLIANCES
HWY 15S
ALBANY, GA 31707
(912)601-0043

March 11, 1987

Mr. Walter O. Casey
5701 Honda Lane
Forsyth, GA 39129

Dear Mr. Casey:

You have been given every opportunity to pay your overdue
balance of $161. We do not feel that we can permit any
further delay in payment.

If payment is not received immediately, we will be forced
to turn your account over to a collection agency.

Why not preserve your heretofore excellent credit rating
and avoid an unpleasant situation by mailing your check
for $161 today.

Sincerely,

James W. Malley
Credit Manager

LETTER OF APPLICATION

The most important letter you ever write may well be a letter of application for a job. As a college student, you may want to apply for a part-time job or a summer job. As a family breadwinner, you may be seeking employment. Since there are usually several applicants for a position, the letter may be the decisive factor, particularly in determining whether or not an interview is granted; the letter is successful if it results in an interview.

The letter of application has three sections: purpose of the letter, selected background information that relates directly to the job sought, and request for an interview.

Purpose of the Letter. In the first paragraph, state that you are applying for a position (better to apply for a specific position). Tell how you found out about the job (if in a periodical or newspaper, give name and date; if from a person, give name), and explain the reason for wanting it. If you have special qualifications for the job, state those qualifications.

Background Information. From the resumé, select those facts that qualify you for the job. State that a resumé with more complete information is included, and, of course, make the proper "enclosure" notation. If you refer to yourself, do so modestly. Don't undersell yourself, however. You want to convince the potential employer that you are the person to hire, that your skills and knowledge will be an asset to the company.

Request for an Interview. Any firm interested in employing an applicant will want a personal interview. In the closing paragraph of the letter of application, therefore, request an interview at the prospective employer's convenience. Should there be restrictions regarding time, however, such as classes or work, say so. If distance makes an interview impractical (living in Virginia and applying for a summer job in Yellowstone National Park), suggest some alternative, such as an interview with a local representative or sending a tape recording if speaking ability is essential to the job.

Be sure to include in the closing paragraph how and when you may be reached. (For suggestions on the job interview—preparing for an interview, holding an interview, and following up an interview—see Chapter 10, Oral Communication, pages 516–522.)

RESUMÉ (DATA SHEET OR VITA)

Background information is given on a separate sheet of paper headed "Resumé," "Data Sheet," or "Vita." (See pages 299, 300–301, and 305–306 for examples of resumés.) By whatever name, this organized listing of background information on a separate page(s) keeps the letter from being overly long and makes locating specific details easier. The resumé is a *full* listing of information concerning the applicant. The accompanying letter, or the cover letter with an application form, focuses attention on facts relevant to the prospective job.

It is a good idea to keep the resumé as short as possible since busy employers are interested in necessary information only. If possible, keep the resumé to one page. After you have had considerable work experience, the resumé may be longer.

Information in a Resumé. The information in a resumé may include the following:

1. *Job you are seeking.* State your immediate employment objective. For example, Computer Programmer. You may also state your ultimate career goal. For example, Manager Trainee with future opportunity to become a buyer.

2. *Heading.* Include name, address, telephone number, and you may include salary expected. Some experts advise against mentioning a salary; you might have been paid more, or if you mention too high a figure, you may not even get an interview.

3. *Education.* College education and high school education may be sufficient. After gaining considerable experience, you may want to include your highest level of college education only. If you include several levels of education, begin with your most recent education and work backward.

4. *Experience.* Experience on a job of almost any kind is valuable; do not hesitate to mention any job you have had. If you have a listing of several job experiences, you might divide the information into two sections: experience related to your job objective and general or other experience. In listing experience, begin with current or most recent employment and work backward. If your resumé focuses on skills, you will list skills and related employment experience(s) that helped you to develop the skills.

5. *Personal data.* You may want to include statistics on age, height, weight, health, and so on, *obsolete* particularly if these statistics are important to the job objective. Also you may mention hobbies or leisure interest activities as well as memberships in organizations. You cannot be required to give data on race, sex, color, religion, age, or national origin unless such data are bona fide occupational qualifications, are required by national security laws, or are required for other legally permissible reasons.

6. *References.* Select three to five individuals who can speak knowledgeably about your qualifications and character. A student with little or no experience might select a friend of long standing, a teacher, or a community leader. Give the name, occupation, business address, and telephone number of each individual. If not obvious, explain your relationship to the individual. If you do not wish to list references, you may state: "References supplied upon request."

7. *Special qualifications.* List any special qualifications, awards, honors, and the like. Include these in a properly headed section.

Types of Resumés. A resumé follows no standard form. Once you have jotted down facts under the topics listed above, you can then decide on the headings you will use and the arrangement of the information. The sections of a resumé may be arranged in several different ways with different headings.

One way is to include the sections listed above, with the headings "Education," "Experience," "Personal Data," and "References," as illustrated in the sample resumés on pages 299, 300–301. Arrange the headings so that the most impressive details will appear near the top of the page. This type of resumé is called a *traditional* or *chronological resumé*.

Another arrangement of sections of the resumé emphasizes functions you can perform or skills you have. For example, you might use headings such as "Communication Skills," "Supervisory Skills," "Budgetary Skills," "Machine Operating Skills," "Personnel Skills," "Research Skills," and the like. Following the headings give a summary of activities that helped you become competent at performing a function or helped you develop a skill.

You may also include traditional headings such as "Education," "Personal Data," and "References." The heading "Experience" may also be included if you do not cover experience in the function or skills section, or the heading "Experience" may be used to introduce the function or skills section.

This type of resumé may be called a *skills* or a *functional resumé*. See the sample resumé on pages 305–306.

A good rule of thumb: Select the type of resumé and arrange the information on the page to show your qualifications as a potential employee in the best possible way.

Examples of Letters of Application. On the following pages are two letters of application. The first letter, written by a student, includes a filled-in Plan Sheet. The letter has an accompanying traditional resumé. (See page 299 for the traditional resumé.)

The second letter, written by an experienced technician, has an accompanying skills resumé. The letter has a filled-in Plan Sheet. (See pages 305–306 for the skills resumé with filled-in Plan Sheet.)

PLAN SHEET
FOR WRITING A LETTER OF APPLICATION WITH A RESUMÉ

Layout Form (modified block, block, simplified block)

I will use:
block

Heading

My mailing address is:
621 Green Cove
Sheridan, MI 48885

The date is:
12 December 1986

Inside Address

The person (or firm) to whom I am writing (including any identifying title after the name) is:
Carter Insurance Agency
Attention: Mr. Jim Garrett

The complete mailing address is:
1531 Jayne Avenue
Big Rapids, MI 49307

Salutation

I will use:
Dear Mr. Garrett:

Body

Job applied for is:
programmer

I learned about it:
computer programming instructor

The reason I want the job is:
begin work in field trained for by education and limited experience

From the resumé I want to emphasize:

Education:
will complete degree in computer programming on May 31
ranked one of top three students in computer-related classes
recognized as Outstanding First-year Student in Programming

Experience:
student worker for Mr. Lewis, Director of Data Processing, at John Williams
Community College
night supervisor in Midwestern Bell Telephone's data processing department

Personal data:
omit

Other:
omit

I would like an interview on:
Any weekday afternoon; all day Saturday

I can be reached by:
313-373-5057 home

Complimentary Close

I will use:
Sincerely,

Signature (including, if applicable, the name of the firm and identifying title)

I will use:
Alice M. Rydel

Special Parts

This letter requires the following special parts:
Attention line
Enclosure notation

621 Green Cove
Sheridan, MI 48885
1 May 1987

Carter Insurance Agency
Attention: Mr. Jim Garrett
1531 Jayne Avenue
Big Rapids, MI 49307

Dear Mr. Garrett:

Through my computer programming instructor, Mr. Tom Lewis,
I have learned that you have an opening in your department
for a programmer.

On May 31 I will complete my degree in Computer
Programming. While a student I have worked for Mr. Lewis,
Director of Data Processing at John Williams Community
College, and as a night supervisor in Midwestern Bell
Telephone's data processing department. Now I am eager to
begin a full-time job in my field.

The enclosed resumé gives a brief outline of my work and
education. In all the computer-related classes I have
taken, I have been ranked as one of the top three
students. As a first-year computer programming major, I
received the award for the outstanding first-year student
in programming. My natural inclination to enjoy work
involving organization and order and my study at John
Williams Community College should give me a good
foundation as a programmer.

I will be happy to supply any additional information about
myself. I am available any afternoon during the week and
all day Saturday for an interview; my telephone number at
home is 313-373-5057. I look forward to hearing from you.

Sincerely,

Alice M. Rydel

Alice M. Rydel

Enc

PLAN SHEET
FOR WRITING A LETTER OF APPLICATION WITH
A RESUMÉ

Layout Form (modified block, block, simplified block)

I will use:
modified block

Heading

My mailing address is:
1045 Drake Place
Ellisville, MA 01047

The date is
March 27, 1987

Inside Address

The person (or firm) to whom I am writing (including any identifying title after the name) is:
Ms. Adrian M. Mantee, Director of Personnel

The complete mailing address is:
Blackwell Manufacturing Company
Plattsburg, IN 47401

Salutation

I will use:
Dear Ms. Mantee:

Body

Job applied for is:
supervisor in quality control

I learned about it:
March issue What's New in Manufacturing

The reason I want the job is:
Desire to return to Indiana. Parents ill.

From the resumé I want to emphasize:

Education:
formal education in mechanical technology

Experience:
Presently employed as assistant supervisor of quality control at Barron Enterprises, Engineering Division. Have other experience in mechanical technology.

Personal data:
Have always liked working with machinery.
Current supervisor knows of this application.

Other:
none

I would like an interview on:
at convenience of Mr. Mantee, with a week's notice

I can be reached by:
mailing address given in heading

Complimentary Close

I will use:
Sincerely,

Signature (including, if applicable, the name of the firm and identifying title)

I will use:
Thomas D. Davis

Special Parts

This letter requires the following special parts:
Enclosure: Resumé

1045 Drake Place
Ellisville, MA 01047
March 27, 1987

Ms. Adrian M. Mantee
Director of Personnel
Blackwell Manufacturing Company
Plattsburg, IN 47401

Dear Ms. Mantee:

In the March issue of <u>What's New in Manufacturing</u>, I read
your advertisement seeking supervisory personnel for your
multimillion-dollar plant expansion program. Please
consider this letter as my application for a supervisory
position in Quality Control.

Presently I am Assistant Supervisor of Quality Control at
Barron Enterprises, Engineering Division. My work here is
pleasant and I enjoy it, but my wife and I want to move
back to Indiana because of the failing health of our
parents, who live there.

The enclosed resumé gives a brief outline of my training
and experience. From my earliest job while a sophomore in
high school, I have liked working with machines of any
kind. This keen interest, developed by my formal education
and experience in various aspects of mechanical
technology, has given me a background that should be
of immense value to a supervisor.

I will be happy to supply any additional information
concerning my background or present employment. My
immediate supervisor here at Barron Enterprises is aware
that I am looking for a supervisory position with a
company in the Midwest and why I am doing so. I would like
to have an interview with you, and, although distance is
an important factor, I would be glad to fly out at your
convenience, if I had at least a week's notice.

Sincerely,

Thomas D. Davis

Thomas D. Davis

Enclosure: Resumé

Examples of Resumés. Following are three examples of resumés. The first two examples illustrate student-written traditional resumés. The third example, a skills resumé written by an experienced technician, includes a filled-in Plan Sheet.

APPLICATION FOR POSITION OF COMPUTER PROGRAMMER

ALICE M. RYDEL

3621 Bailey Drive   Spring 1987
Big Rapids, MI 39207
(579) 456-2156

Career Goal

To gain experience as a computer programmer and eventually
become head programmer for a large insurance firm.

Education

1985–1987: Will receive AAS degree in Computer
Programming, John Williams Community College, Big Rapids,
Michigan. Related Courses: RPG Programming, Systems
Analysis and Design, Principles of Management, Accounting.

1981–1985: Murrah High School, Jackson, Michigan. Took
college preparatory courses and basic business courses.

Experience

1985–present: Part-time assistant to Tom Lewis, Director
of Data Processing, John Williams Community College.

1985–present: Part-time (summers and holidays) night
supervisor in Data Processing Department, Midwestern Bell
Telephone Company, Big Rapids, Michigan.

Personal Data

Age: 22. Weight: 120 pounds. Height: 5' 6". Leisure
activities: sports, reading.

Awards

Outstanding First-Year Student in Programming.

References

Mr. Tom Lewis, Director of Data Processing, John Williams
Community College, Big Rapids, Michigan 39207. Telephone:
(579) 456-9596

Ms. Lauren Watson, Supervisor of Data Processing
Department, Midwestern Bell Telephone Company, Big Rapids,
Michigan 39207. Telephone: (579) 467-3271.

Dr. Edith Rogers, Professor of English, John Williams
Community College, Big Rapids, Michigan 39207. Telephone:
(579) 456-9512

JAMES E. BROWN, JR.

206 Davis Drive
Jackson, Mississippi 39209
601-948-7660

MARITAL STATUS: Single   HEIGHT: 6'1"
AGE: 27   WEIGHT: 190 pounds

JOB OBJECTIVE

To begin work in the technical department of a company
dealing with electronics with the purpose of qualifying
for full management responsibilities. No geographic
limitations

EDUCATION
Louisiana State University

Class: May 1982   Minor: Speech
Degree: B.S.   G.P.A. 3.40 Overall
Major: Industrial   3.31 in major
        Management   field of study

HONORS

Dean's List, Beta Gamma Sigma National Honor Society
in Business Administration, Associated Students Service
and Leadership Award, Member of Who's Who Among American
Colleges and Universities

ACTIVITIES

Station Manager and Head Engineer of college radio station
KERS-FM, member of the Society for the Advancement of
Management, Vice-President of the Young Republicans

EXPERIENCE

United States Air Force, Electronics-Communications, 6/74
to 9/78. Duties and Responsibilities: Shift Chief Long-
Haul Transmitter Site (supervised 3 persons in operation
of 52 transmitters and 2 microwave systems). Team Chief
Group Electronics Engineering Installation Agency
(supervised 3 persons on installing weather and
communications equipment). Tech-Writer (wrote detailed
maintenance procedures for electronic equipment manuals).
Instructor in Electronic Fundamentals (continuous 3-month
classes of 10 persons each)

Resumé   James E. Brown, Jr.   page 2

Summer and Part Time: Manager, Campus Apartments; Disc
Jockey of KXOA and KXRQ; Laboratory Assistant for Radio—TV
and Speech Department, Stage Technician

Special Qualifications: Federal Communications Commission
First Class Radio Telephone License. Top Secret Clearance
for Defense Work

<u>PERSONAL INTERESTS</u>

Interested in water skiing, computer games, jazz, and
building hi—fi equipment

<u>REFERENCES</u>

References available upon request

PLAN SHEET
FOR WRITING A RESUMÉ

Type of Resumé (traditional or skills) I will use: *skills*

Heading

Name: *Thomas D. Davis*   Telephone: *521-363-2371*
Address: *1045 Drake Place*
          Ellisville, Massachusetts 01047

The date is: *omit*

Job Sought *Supervisor in Quality Control*

Career Goal or Objective *omit*

Education

College   Location   Date(s) of Attendance
Cain Community College   *Cain, IN*   *1965–1968*

Degree (or degree sought):   Major Field:
AAS degree   *mechanical technology*

Subjects related to prospective job: *omit*

Special training:
Massachusetts Technical Institute
   1980 Quality Control with Computers (45 clock hours)
   1978 New Materials in Industry (45 clock hours)
   1976 Production Planning and Problems (45 clock hours)

Activities: *omit*

High School   Location   Date of Graduation
omit

Areas of emphasis or of particular interest: *omit*

Awards or special recognition and/or activities: *omit*

Work Experience (beginning with the most recent or present employment)

Date of Employment	Specific Work	Name and Address of Firm	Full Name of Supervisor
1972–present	*Asst Supervisor and Processing Foreman*	*Barron Enterprises Engineering Division Ellisville, MA 01047*	*Omit*
1967–1972	*Quality Control, Parts Inspector, Layout/Design Asst*	*Always Electric Company Pattison, IN 47312*	
1965–1967	*Assembly line*	*Bickman Manufacturing Cain, IN 47315*	
1963–1965	*Machinist*	*U.S. Navy*	

Personal Data

Age:	*born 14 June 1946*	Weight:	*omit*
Height:	*omit*	Other:	*omit*

References (give full name and complete business address)

Mr. G. Harris Carmel, Supervisor of Quality Control, Barron Enterprises, Engineering Division, Ellisville, MA 01047. (302) 571-3828

Dr. Harold C. Mantiz, Professor of Mechanical Engineering, Massachusetts Technical Institute, 3429 Elkins Avenue, Boston, MA 02107. (302) 495-3172

Mr. Alfred Leake, Chief Inspector, Always Electric Company, Pattison, IN 47321. (523) 340-5689

Skills

Skills I will list and supporting details:
Supervisory—Directed crew of 20 workers
Determined work assignments on priority
Solved shop production problems
Communications—Orally passed orders to workers
Prepared monthly reports: department, budget
Prepared daily reports: discrepancies in product conformity
Personnel—Interviewed and recommended hiring new personnel
Conducted performance evaluations
Made recommendations for raises and promotions
Budgetary—Prepared and monitored spending of half-million
Machine—Can operate common machine shop tools: lathes, milling machines, grinding machines
Can use related measuring tools and gauges

```
                  Thomas D. Davis
                  1045 Drake Place
               Ellisville, MA 01047
                  (521) 363-2371
```

Birthdate: 14 June 1946

POSITION SOUGHT: SUPERVISOR IN QUALITY CONTROL

SKILLS <u>Supervisory</u>. Directed a crew of 20 workers. Determined work assignments based on priorities. Found solutions to shop production problems.

<u>Communications</u>. Orally passed on orders to workers. Prepared monthly written reports, such as departmental report to an immediate supervisor, report on budget variances. Prepared daily written reports, such as reports on discrepancies in product conformity.

<u>Personnel</u>. Interviewed and made recommendations for hiring new personnel. Conducted performance evaluations and made recommendations for raises and promotions.

<u>Budgetary</u>. Have prepared and monitored the spending of a half-million-dollar department budget.

<u>Machine</u>. Can operate all common machine shop tools, such as lathes, milling machines, grinding machines. Can use related measuring tools and gauges.

EXPERIENCE 1972–present: Barron Enterprises, Engineering Division, Ellisville, MA 01047. Assistant Supervisor of Quality Control and Processing Supervisor

1967–1972: Always Electric Company, Pattison, IN 47312. Quality Control Checker, Parts Inspector, Layout and Design Assistant

1965–1967: Bickman Manufacturing Company, Cain, IN 47315. Assembly line worker

1963–1965: U.S. Navy. Machinist

Resumé   Thomas D. Davis   page 2

EDUCATION   1965–1968: Cain Community College, Cain, IN.
                AA degree. Major in Mechanical Technology

SPECIAL   Massachusetts Technical Institute
TRAINING   1980 Quality Control with Computers
                   (45 clock hours)
                1978 New Materials in Industry
                   (45 clock hours)
                1976 Production Planning and Problems
                   (45 clock hours)

REFERENCES   Mr. G. Harris Carmel, Supervisor of Quality
                Control, Barron Enterprises, Engineering
                Division, Ellisville, MA 01047.
                Telephone: (302) 571-3828

                Dr. Harold C. Mantiz, Professor of Mechanical
                Engineering, Massachusetts Technical
                Institute, 3429 Elkins Avenue, Boston,
                MA 02107.
                Telephone: (302) 495-3172

                Mr. Alfred Leake, Chief Inspector, Always
                Electric Company, Pattison, IN 47312.
                Telephone: (523) 340-5689

Many companies provide printed forms, such as the one on pages 308–311, for job applicants. The form is actually a very detailed data sheet. In completing the form, use a pen or typewriter, be neat, and answer every question. Some companies suggest that applicants put a dash (—), a zero (0), or NA (not applicable) after a question that does not pertain to them. Doing so shows that the applicant has read the question and not overlooked it.

The completed forms of many applicants look very similar. Therefore, it is the accompanying letter that gives you an opportunity to make the application stand out. For this letter, follow the suggestions for a letter of application given earlier in this chapter.

Following is a typical form used as an application for employment.

<table>
<tr><td colspan="2">FOR OFFICE USE ONLY</td></tr>
<tr><td>Possible Work Locations</td><td>Possible Positions</td></tr>
</table>

APPLICATION
FOR
EMPLOYMENT

(PLEASE PRINT PLAINLY)

<table>
<tr><td colspan="2">FOR OFFICE USE ONLY</td></tr>
<tr><td>Work Location _____</td><td>Rate _____</td></tr>
<tr><td>Position _____</td><td>Date _____</td></tr>
</table>

To Applicant: We deeply appreciate your interest in our organization and assure you that we are sincerely interested in your qualifications. A clear understanding of your background and work history will aid us in placing you in the position that best meets your qualifications and may assist us in possible future upgrading.

PERSONAL

Date: _____

Name _____ Social Security No. _____
　　　Last　　　　　First Initial　　　　Middle Initial

Present address _____ Telephone No. _____
　　　No.　　Street　　　City　　State　　Zip

How long have you lived at above address? _____

Previous address _____ How long did you live there? _____
　　　No.　　Street　　　City　　State　　Zip

To Applicant: READ THIS INTRODUCTION CAREFULLY BEFORE ANSWERING ANY QUESTIONS IN THIS BLOCKED-OFF AREA. The Civil Rights Act of 1964 prohibits discrimination in employment practice because of race, color, religion, sex or national origin. P.L. 90-202 prohibits discrimination on the basis of age with respect to individuals who are at least 40 but less than 65 years of age. The laws of some States also prohibit some or all of the above types of discrimination.
　　DO NOT ANSWER ANY QUESTION CONTAINED IN THIS BLOCKED-OFF AREA UNLESS THE EMPLOYER HAS CHECKED THE BOX NEXT TO THE QUESTION, thereby indicating that the requested information is needed for a bona fide occupational qualification, national security laws, or other legally permissible reasons.

☐ Are you over the age of twenty-one? _____ If no, hire is subject to verification that you are of minimum legal age.

☐ Sex: M _____ F _____　　☐ Height: _____ ft. _____ in.　　☐ Weight: _____ lbs.

☐ Marital Status: Single _____ Engaged _____ Married _____ Separated _____ Divorced _____ Widowed _____

☐ Date of Marriage _____　☐ Number of dependents including yourself _____　☐ Are you a citizen of the U.S.A.? _____

☐ What is your present Selective Service classification? _____

☐ Indicate dates you attended school:

Elementary _____ High School _____ College _____
　　　　　From　　To　　　　　From　　To　　　　　From　　To

Other (Specify type of school) _____
　　　　　　　　　　　　　　　　From　　To

☐ Have you ever been bonded? _____ If yes, on what jobs? _____

☐ Have you been convicted of a crime in the past ten years, excluding misdemeanors and summary offenses? _____ if yes, describe in full _____

Employer may list other bona fide occupational questions on line below:

☐ _____

What method of transportation will you use to get to work? _____

Position(s) applied for _____ Rate of pay expected $_____ per week

Would you work Full-Time _____ Part-Time _____ Specify days and hours if part-time _____

Were you previously employed by us? _____ If yes, when? _____

List any friends or relatives working for us _____
　　　　　　　　　　　　　　　　　　　　Name(s)

If your application is considered favorably, on what date will you be available for work? _____ 19____

Are there any other experiences, skills, or qualifications which you feel would especially fit you for work with the Company? _____
_____

EF 101-2
(FORM 101)

(Turn to Next Page)
Printed in U.S.A.

Do you have any physical defects which preclude you from performing certain kinds of work? _____ If yes, describe such defects and specific work limitations. _____

Have you had a major illness in the past 5 years? _____ If yes, describe _____

Have you received compensation for injuries? _____ If yes, describe _____

RECORD OF EDUCATION

School	Name and Address of School	Course of Study	Check Last Year Completed				Did You Graduate?	List Diploma or Degree
Elementary		╳	5	6	7	8	☐ Yes ☐ No	╳
High			1	2	3	4	☐ Yes ☐ No	
College			1	2	3	4	☐ Yes ☐ No	
Other (Specify)			1	2	3	4	☐ Yes ☐ No	

MILITARY SERVICE RECORD

Were you in U.S. Armed Forces?   Yes _____ No _____ If yes, what Branch? _____

Dates of duty: From _____ To _____ Rank at discharge _____
　　　　　　　　　Month　Day　Year　　Month　Day　Year

List duties in the Service including special training _____

Have you taken any training under the G.I. Bill of Rights? _____ If yes, what training did you take? _____

PERSONAL REFERENCES (Not Former Employers or Relatives)

Name and Occupation	Address	Phone Number

—2—

List below all present and past employment, beginning with your most recent

I

Name and Address of Company and Type of Business	From		To		Describe in detail the work you did	Weekly Starting Salary	Weekly Last Salary	Reason for Leaving	Name of Supervisor
	Mo.	Yr.	Mo.	Yr.					

II

Name and Address of Company and Type of Business	From		To		Describe in detail the work you did	Weekly Starting Salary	Weekly Last Salary	Reason for Leaving	Name of Supervisor
	Mo.	Yr.	Mo.	Yr.					

III

Name and Address of Company and Type of Business	From		To		Describe in detail the work you did	Weekly Starting Salary	Weekly Last Salary	Reason for Leaving	Name of Supervisor
	Mo.	Yr.	Mo.	Yr.					

IV

Name and Address of Company and Type of Business	From		To		Describe in detail the work you did	Weekly Starting Salary	Weekly Last Salary	Reason for Leaving	Name of Supervisor
	Mo.	Yr.	Mo.	Yr.					

V

Name and Address of Company and Type of Business	From		To		Describe in detail the work you did	Weekly Starting Salary	Weekly Last Salary	Reason for Leaving	Name of Supervisor
	Mo.	Yr.	Mo.	Yr.					

May we contact the employers listed above? _____ If not, indicate by No. which one(s) you do not wish us to contact _____

The facts set forth above in my application for employment are true and complete. I understand that if employed, false statements on this application shall be considered sufficient cause for dismissal. You are hereby authorized to make any investigation of my personal history and financial and credit record through any investigative or credit agencies or bureaus of your choice.

In making this application for employment I also understand that an investigative consumer report may be made whereby information is obtained through personal interviews with my neighbors, friends, or others with whom I am acquainted. This inquiry includes information as to my character, general reputation, personal characteristics and mode of living. I understand that I have the right to make a written request within a reasonable period of time to receive additional, detailed information about the nature and scope of this investigative consumer report.

—3—   Signature of Applicant

APPLICANT—Do not write on this page
FOR INTERVIEWER'S USE

INTERVIEWER	DATE	COMMENTS

FOR TEST ADMINISTRATOR'S USE

TESTS ADMINISTERED	DATE	RAW SCORE	RATING	COMMENTS AND INTERPRETATION

REFERENCE CHECK

*Position Number	RESULTS OF REFERENCE CHECK	*Position Number	RESULTS OF REFERENCE CHECK
I		IV	
II		V	
III			

*See Page 3

This *Application for Employment* is prepared for general usage throughout the United States. In the opinion of our legal counsel this form meets all Federal and State fair employment practice laws and complies with the Fair Credit Reporting Act. Furthermore, our attorneys have undertaken a monitoring service to advise us of any revisions to be made as a result of any future changes in the laws. However, V. W. Eimicke Associates, Inc. assumes no responsibility for the inclusion in said *Application for Employment* of any questions which may violate local and/or State and/or Federal laws.

—4—

APPLICATION FOLLOW-UP LETTERS

Frequently, after writing and mailing a letter of application (pages 293, 297), filling out and submitting a job application form (pages 308–311), or completing an interview (pages 516–522), you may need to write a follow-up letter. The content of the follow-up letter is determined by events occurring after the application or interview.

The follow-up letter may be a request for a response from the prospective employer. This letter may be written if you were promised a response by a certain date but have not received a response by that date. This letter may also be written if you have sent an application for a job you hope will be available.

A letter requesting a response may resemble the sample letter on page 317. The Plan Sheet used for planning and organizing the letter is also included.

The letter may simply thank the interviewer, remind the person of your qualifications, and indicate a desire for a positive response. The body of such a letter appears below:

> The interview with you on Wednesday, June 10, was indeed a pleasant, informative experience. Thank you for helping me feel at ease.
>
> After hearing the details about the World Bank's training program and opportunities available to persons who complete the program, I am eager to be an employee of World Bank. I believe my experience working with People's Bank and my associate degree in Banking and Finance Technology provide a sound background for me as a trainee.
>
> I look forward to hearing from you that I have been accepted as a trainee in the World Bank's training program.

See also pages 317, 523.

The follow-up letter may reaffirm the acceptance of a job and your appreciation for the job. The letter may also ask questions that you have thought of since getting the job, mention the date you will begin work, and make a statement about looking forward to working with the company. The body of such a letter may include the following:

> Thank you for your offer to hire me as consulting engineer for Wanner, Clare, and Layshock, Inc. I eagerly accept the offer. The confidence you expressed in my abilities to help the firm improve workers' safety certainly motivates me to be as effective as possible.
>
> I look forward to beginning work on Tuesday, July 10, ready to demonstrate that I deserve the confidence expressed in me.

A follow-up letter may be needed if you decide to refuse a job offer. The body of such letter may be written as follows:

> Thank you for offering me a place in the World Bank's training program. The opportunities available upon completion are enticing.
>
> Since my interview with you, the president of People's Bank, where I have worked part-time while completing my associate degree in banking and finance, has offered me permanent employment. The bank will pay for my continued schooling, allow me to work during the summers, and guarantee me full-time employment upon completing an advanced degree.

By accepting this offer from People's Bank, I can live near my aging parents and help care for them.

I appreciate your interest in me and wish continued success for World Bank.

The follow-up application letter is simply a courteous response to events following the application or interview.

On the following pages are a Plan Sheet, filled in by the student before writing, and a follow-up letter requesting a response from a prospective employer.

PLAN SHEET
FOR WRITING AN APPLICATION FOLLOW-UP LETTER

Layout Form (modified block, block, simplified block)

I will use:
modified block

Heading

My mailing address is:
2121 Oak Street
Fort Collins, CO 80521

The date is:
July 19, 1987

Inside Address

The person (or firm) to whom I am writing (including any identifying title after the name) is:
Mr. William Hatton, Personnel Director

The complete mailing address is:
World Bank
20 East 53rd Street
New York, NY 10022

Salutation

I will use:
Dear Mr. Hatton:

Body

The purpose of the letter is (check one):

_____X_____ to request a response from a prospective employer
_____ to thank the interviewer
_____ to reaffirm acceptance of a job
_____ to refuse a job offer

The opening statement(s) I will use is:

On June 10 interviewed by Mr. John Salman, your representative, for a place in the World Bank's training program. Mr. Salman told me that I would be notified about my application by July 1. Middle of July; have not received a response.

I will explain the purpose of the letter by:

Must make certain decisions by August 1; need to hear from you about my employment possibilities with World Bank.

The closing statement(s) I will include is:

Eager to become an employee of World Bank; would appreciate a response.

Complimentary Close

I will use:
Sincerely,

Signature (including, if applicable, the name of the firm and identifying title)

I will use:
Jayne T. Mannos

Special Parts

This letter requires the following special parts:
omit

2121 Oak Street
Fort Collins, CO 80521
July 19, 1987

Mr. William Hatton, Personnel Director
World Bank
20 East 53rd Street
New York, NY 10022

Dear Mr. Hatton:

On June 10 I was interviewed by Mr. John Salman, your
representative, for a place in the World Bank's training
program. Mr. Salman told me that I would be notified about
my application by July 1. Although it is the middle of
July, I have not received any response.

Since I must make certain decisions by August 1, could
I please hear from you about my employment possibilities
with World Bank.

I am, of course, quite eager to become an employee of
World Bank and hope that I will receive a positive
response.

Sincerely,

Jayne T. Mannos

Jayne T. Mannos

General Principles in Writing Memorandums and Letters

1. *Memorandums are typically used for written communication between persons in the same company.* Memorandums are used to convey or confirm information.

2. *A memorandum has only two regular parts: the heading and the body.* The heading typically includes the guide words TO, FROM, SUBJECT, and DATE plus the information. The body of the memorandum is the message. The writer usually initials the memo immediately following the typed "From" name; some writers prefer to give their signature at the end of the memo.

3. *An effective business communication is written on good-quality stationery, is neat and pleasing to the eye, and follows a standardized layout form.*

4. *The six regular parts of a business letter are the heading, inside address, salutation, body, complimentary close, and signature.* The parts follow a standard sequence and arrangement.

5. *A letter may require a special part.* Among these are an attention line, a subject line, an identification line, an enclosure line, a copy line, a personal line, and a mailing line.

6. *The envelope has two regular parts: the outside address and the return address.* A special part, such as an attention line or a personal line, may also be needed.

7. *The content of an effective letter is well organized, has the "you" emphasis, stresses a positive approach, uses natural wording, and is concise.*

8. *Among the most common types of letters written by businesses and individuals are letters of inquiry, order, sales, claim and adjustment, collection, job application, and job application follow-up.* Each type of letter requires special attention as to purpose, inclusion of pertinent information, and consideration of who will be reading the letter and why.

9. *Job application materials present the qualifications of the applicant, including education, work experience, personal information, and references.* This information may be presented as a resumé or on a printed job application form.

Application 1   Writing Memorandums and Letters

Make a collection of at least ten memorandums and letters of various types and bring them to class. For each communication, answer these questions:

1. What layout form is used?
2. Is a letterhead used? What information is given in the letterhead?
3. Is the communication neat and pleasing in appearance? Explain your answer.
4. What special parts of a memorandum or letter are used?
5. What is the purpose of the communication?
6. Does the communication have the "you" emphasis? Explain your answer.

Application 2 Writing Memorandums and Letters

Select one of the communications you collected for the preceding application. Evaluate the item according to the General Principles in Writing Memorandums and Letters, above, and any special instructions for this type of communication. Hand in both the communication and the evaluation.

Application 3 Writing Memorandums and Letters

A. Assume that you are the president of an organization. The date and the location of the next regular meeting have been changed. You need to send this information to the members of the organization.

or

B. Assume that you are an employee in a company. You have an idea for improving efficiency that should lead to a larger margin of profit for the company. Present this idea in writing to your immediate supervisor. State the idea clearly and precisely, and give substantiating data.

a. Fill in the Plan Sheet for Writing a Memorandum on page 325.
b. Write a preliminary draft of the memorandum.
c. Revise. See inside back cover.
d. Write the final draft of the memorandum.

Application 4 Writing Memorandums and Letters

Examine the content and form of the following body of a letter addressed to *Popular Science Digest*. Point out every item that keeps the paragraph from being clear.

In regard to your article a while back on how to make a home fire alarm system in *Popular Science Digest*, which was very interesting. I would like to obtain more information. Would also like to know the names of people who have had good results with same. Give me where they live, too.

Application 5 Writing Memorandums and Letters

Rewrite the body of the letter in Application 4.

Application 6   Writing Memorandums and Letters

A. Prepare a letter to a person such as a former employer or a former teacher asking permission to use the person's name as a reference in a job application.

or

B. Write a letter to the appropriate official in your college requesting permission to take your final examinations a week earlier than scheduled. Be sure to state your reason or reasons clearly and effectively.

a. Fill in the Plan Sheet for Writing a Letter of Inquiry on pages 327–328.
b. Write a preliminary draft of the letter.
c. Revise. See inside back cover.
d. Write the final draft of the letter.

Application 7   Writing Memorandums and Letters

A. Prepare an order letter for an item you saw advertised in a newspaper or magazine.

or

B. Assume that you are the instructor in one of your lab courses and you have been given the responsibility of ordering several new pieces of equipment.

a. Fill in the Plan Sheet for Writing an Order Letter on pages 329–330.
b. Write a preliminary draft of the letter.
c. Revise. See inside back cover.
d. Write the final draft of the letter.

Application 8   Writing Memorandums and Letters

Find a printed order form and make out an order on it. It is usually wise to write out the information on a sheet of paper and then transfer it to the order form.

Application 9   Writing Memorandums and Letters

A. Assume that you work part-time as a salesclerk in a clothing store and that you are paid on a commission basis. Write a sales letter to be sent to a number of people you know.

or

B. Write a sales letter, to be distributed to students on campus, telling about a special sale in the campus bookstore.

a. Fill in the Plan Sheet for Writing a Sales Letter on pages 331–332.
b. Write a preliminary draft of the letter.
c. Revise. See inside back cover.
d. Write the final draft of the letter.

Application 10 Writing Memorandums and Letters

A. Assume that you ordered an item, such as a pair of shoes, a set of wheel covers, a ring, or a bowling ball, and the wrong size was sent to you. Write a claim letter requesting proper adjustment.

or

B. Write a claim letter requesting an adjustment on a piece of equipment, a tool, an appliance, or a similar item that is not giving you satisfactory service. You have owned the item two months and it has a one-year warranty.

a. Fill in the Plan Sheet for Writing a Claim or an Adjustment Letter on pages 333–334.
b. Write a preliminary draft of the letter.
c. Revise. See inside back cover.
d. Write the final draft of the letter.

Application 11 Writing Memorandums and Letters

Write an adjustment letter in response to Application 10.

a. Fill in the Plan Sheet for Writing a Claim or an Adjustment Letter on pages 335–336.
b. Write a preliminary draft of the letter.
c. Revise. See inside back cover.
d. Write the final draft of the letter.

Application 12 Writing Memorandums and Letters

As credit manager of the local college bookstore, write a series of collection letters to be mailed to students with accounts delinquent for varying periods of time.

a. Fill in the Plan Sheet for Writing Collection Letters on pages 337–338.
b. Write a preliminary draft of the letter.
c. Revise. See inside back cover.
d. Write the final draft of the letter.

Application 13 Writing Memorandums and Letters

Prepare a traditional resumé or a skills resumé or both, as directed by your instructor.

a. Fill in the Plan Sheet for Writing a Resumé on pages 339–340.
b. Write a preliminary draft of the resumé.
c. Revise. See inside back cover.
d. Write the final draft of the resumé.

Application 14 Writing Memorandums and Letters

Write a letter of application answering an advertisement in the "Help Wanted" section of a newspaper. Include with your letter one of the resumés you prepared in the preceding application. Remember that although the resumé is a full listing, the accompanying letter stresses information relevant to the specific job.

a. Fill in the Plan Sheet for Writing a Letter of Application with a Resumé on pages 341–342.
b. Write a preliminary draft of the letter.
c. Revise. See inside back cover.
d. Write the final draft of the letter.

Application 15 Writing Memorandums and Letters

Secure a job application form. (The form used by the U.S. government is especially thorough.) Fill in the form neatly and completely. It is usually wise to write out the information on a separate sheet of paper and then transfer it to the application form.

Application 16 Writing Memorandums and Letters

A. Write an application follow-up letter in which you accept the job applied for in Application 14.

or

B. Write an application follow-up letter in which you refuse the job offered as a result of the Application 14 letter.

a. Fill in the Plan Sheet for Writing an Application Follow-up Letter on pages 343–344.
b. Write a preliminary draft of the letter.
c. Revise. See inside back cover.
d. Write the final draft of the letter.

USING PART II: SELECTED READINGS

Application 17 Writing Memorandums and Letters

Read the article "Clear Writing Means Clear Thinking Means . . ." by Marvin H. Swift, pages 616–621.

a. Under "Suggestions for Response and Reaction," page 621, answer Number 2.
b. Under "Suggestions for Response and Reaction," page 621, write Number 3.

Application 18 Writing Memorandums and Letters

Read "Starch-Based Blown Films," by Felix Otey and others, pages 624–631. Under "Suggestions for Response and Reaction," page 631, write Number 4.

PLAN SHEET
FOR WRITING A LETTER OF INQUIRY

Layout Form (modified block, block, simplified block)

I will use:

Heading

My mailing address is:

The date is:

Inside Address

The person (or firm) to whom I am writing (including any identifying title after the name) is:

The complete mailing address is:

Salutation

I will use:

Body

Through this letter I am seeking:

The reason for my inquiry is:

To make my inquiry clear and specific, I need to include this information:

Complimentary Close

I will use:

Signature (including, if applicable, the name of the firm and identifying title)

I will use:

Special Parts

This letter requires the following special parts:

PLAN SHEET
FOR WRITING AN ORDER LETTER

Layout Form (modified block, block, simplified block)

I will use:

Heading

My mailing address is:

The date is:

Inside Address

The person (or firm) to whom I am writing (including any identifying title after the name) is:

The complete mailing address is:

Salutation

I will use:

Body

The items I am ordering and their identification (model number, catalog number, size, color, weight, finish, etc.) are:

Item	Quantity	Identification	Cost

The manner of payment (check, money order, credit card, COD, charge, etc.) is:

The method of shipping (parcel post, truck freight, railway freight, air express, etc.) is:

Other information or instructions:

Complimentary Close

I will use:

Signature (including, if applicable, the name of the firm and identifying title)

I will use:

Special Parts

This letter requires the following special parts:

PLAN SHEET
FOR WRITING A SALES LETTER

Layout Form (modified block, block, simplified block)

I will use:

Heading

My mailing address is:

The date is:

Inside Address

The person (or firm) to whom I am writing (including any identifying title after the name) is:

The complete mailing address is:

Salutation

I will use:

Body

To get the reader's attention and arouse interest, I will:

The product (or service) I am selling is:

The product can be described in this way:

The product will benefit or profit the reader in these ways:

The plan that I will present whereby the reader can immediately begin to use the product is:

Complimentary Close

I will use:

Signature (including, if applicable, the name of the firm and identifying title)

I will use:

Special Parts

This letter requires the following special parts:

PLAN SHEET
FOR WRITING A CLAIM OR AN
ADJUSTMENT LETTER

Layout Form (modified block, block, simplified block)

I will use:

Heading

My mailing address is:

The date is:

Inside Address

The person (or firm) to whom I am writing (including any identifying title after the name) is:

The complete mailing address is:

Salutation

I will use:

Body

Identification of the transaction (items purchased, date, invoice number, model number, style, size, etc.):

Statement of the problem or complaint:

Desired action or action that will be taken:

Complimentary Close

I will use:

Signature (including, if applicable, the name of the firm and identifying title)

I will use:

Special Parts

This letter requires the following special parts:

PLAN SHEET
FOR WRITING A COLLECTION LETTER

Layout Form (modified block, block, simplified block)

I will use:

Heading

My mailing address is:

The date is:

Inside Address

The person (or firm) to whom I am writing (including any identifying title after the name) is:

The complete mailing address is:

Salutation

I will use:

Body

Appeal to customer:

Reference to balance due and payment:

Problem(s) preventing payment:

Action the company owed has taken:

Action the company owed will take:

Complimentary Close

I will use:

Signature (including, if applicable, the name of the firm and identifying title)

I will use:

Special Parts

This letter requires the following special parts:

PLAN SHEET
FOR WRITING A RESUMÉ

Type of Resumé (traditional or skills) I will use:

Heading

Name:
Address: Telephone:

The date is:

Job Sought

Career Goal or Objective

Education

College   Location   Date of Graduation

Degree (or degree sought)   Major Field

Subjects related to prospective job:

Special training:

Activities:

High School   Location   Date of Graduation

Areas of emphasis or of particular interest:

Awards or special recognition and/or activities:

Work Experience (beginning with the most recent or present employment)

Date of Employment	Specific Work	Name and Address of Firm	Full Name of Supervisor

Personal Data

Age: Weight:
Height: Other:

References (give full name and complete business address)

Skills

Skills I will list and supporting details:

PLAN SHEET
FOR WRITING A LETTER OF APPLICATION WITH A RESUMÉ

Layout Form (modified block, block, simplified block)

I will use:

Heading

My mailing address is:

The date is:

Inside Address

The person (or firm) to whom I am writing (including any identifying title after the name) is:

The complete mailing address is:

Salutation

I will use:

Body

Job applied for is:

I learned about it:

The reason I want the job is:

From the resumé I want to emphasize:

Education:

Experience:

Personal data:

Other:

I would like an interview on:

I can be reached by:

Complimentary Close

I will use:

Signature (including, if applicable, the name of the firm and identifying title)

I will use:

Special Parts

This letter requires the following special parts:

PLAN SHEET
FOR WRITING AN APPLICATION
FOLLOW-UP LETTER

Layout Form (modified block, block, simplified block)

I will use:

Heading

My mailing address is:

The date is:

Inside Address

The person (or firm) to whom I am writing (including any identifying title after the name) is:

The complete mailing address is:

Salutation

I will use:

Body

The purpose of the letter is (check one):

_____ to request a response from a prospective employer
_____ to thank the interviewer
_____ to reaffirm acceptance of a job
_____ to refuse a job offer

The opening statement(s) I will use is:

I will explain the purpose of the letter by:

The closing statement(s) I will include is:

Complimentary Close

I will use:

Signature (including, if applicable, the name of the firm and identifying title)

I will use:

Special Parts

This letter requires the following special parts:

Chapter 8
Reports: Conveying Needed Information

Objectives

Upon completing this chapter, the student should be able to:

- Write a definition of reports
- Explain the difference between school reports and professional reports
- Select and use appropriate formats for presenting reports
- Use headings in reports
- Use effective layout and design in reports
- List and identify common types of reports
- Give a reading report
- Give a periodic report
- Give a progress report
- Give an investigative report
- Give an observation report
- Give a feasibility report
- Give a proposal

Introduction

The word *report* covers numerous communications that fulfill many purposes. You may be concerned about your monthly financial report (a bank statement) or the grade report that you will receive at the end of the term. Or you may be busy polling students about dormitory hours to support your committee's recommendations to the housing council. You may face a due date for a supplementary reading report, or a laboratory report, or a project report.

In the business, industrial, and governmental worlds, too, reports are a vital part of communication. A memorandum to the billing department, a weekly production report to the supervisor, a requisition for supplies, a letter to the home office describing the status of bids for a construction project, a performance report on the new computer, a sales report from the housewares department, a report on the availability of land for a housing project, a report to a customer on an estimate for automobile repairs—these are but a few examples of the kinds of reports and the functions they serve.

Scope of Report Writing

Although report writing is becoming a firmly established part of the curriculum for students who plan to enter business, industry, and government, difficulties arise in deciding on the specifics to be taught in the classroom. For one thing, there is no uniformity in report classifications. Depending upon the business, the industry, the particular branch of government, or the textbook, reports may be classified on one or more bases, such as subject matter, purpose, function, length, frequency of compilation, type of format, degree of formality, or method by which the information is gathered. Similarly, there is a lack of uniformity in terminology.

These difficulties underscore the aliveness, the contemporary pace, the elasticity of report writing. Only when report writing has become a relic—serving no practical usefulness—will actual reports fall into well-defined categories.

It is unrealistic, therefore, to draw sharp boundary lines between types of reports or to try to cover all the situations and problems involved in report writing. However, it is quite realistic and practical for you to:

☐ Become acquainted with the general nature of report writing.

☐ Develop self-confidence by learning basic principles of report writing. Thus, when you are given the responsibility for writing a report, you can analyze the need and then fulfill your assignment efficiently.

☐ Study and practice writing several common types of reports that you are most likely to encounter as a student and as a future employee.

Definition of "Report"

A very basic definition of a report might be this: A report is technical data, collected and analyzed, presented in an organized form. Another definition: A report is an objective, organized presentation of factual information that answers a request or supplies needed data. The report usually serves an immediate, *practical* purpose; it is the basis on which decisions are made. Generally, the report is requested or authorized by one person (such as a teacher or employer) and is prepared for a particular, limited audience.

Reports may be simple or complex; they may be long or short; they may be formal or informal; they may be oral or written. Characteristics such as these are determined mainly by the purpose of the report and the intended audience.

QUALITIES OF REPORT CONTENT

Reports convey exact, useful information. That information, or content, should be presented with accuracy, clarity, conciseness, and objectivity.

Accuracy. A report must be accurate. If the information presented is factual, it should be verified by tests, research, documentary authority, or other valid sources. Information that is opinion or probability should be distinguished as such and accompanied by supporting evidence. Dishonesty and carelessness are inexcusable.

Clarity. A report must be clear. If a report is to serve its purpose, the information must be clear and understandable to the reader. The reader should not have to ask: *What does this mean?* or *What is the writer trying to say?* The writer helps to ensure clarity by using exact, specific words in easily readable sentence patterns, by following conventional usage in such mechanical matters as punctuation and grammar, and by organizing the material logically. (See Part III, Handbook, particularly Chapter 1, The Sentence; Chapter 3, Grammatical Usage; and Chapter 4, Mechanics.)

Conciseness. A report must be concise. Conciseness is "saying much in a few words." Unnecessary wordiness is eliminated, and yet complete information is transmitted. Busy executives appreciate concise, timesaving reports that do not compel them to wade through bogs of words to get to the essence of the matter. Note the example that follows.

WORDY: After all is said and done, it is my honest opinion that the company and all its employees will be better satisfied if the new plan for sick leave is adopted and put into practice.

CONCISE: The company should adopt the new sick leave plan.

Revise a report until it contains no more words than those needed for accuracy, clarity, and correctness of expression. For further discussion of conciseness, see pages 11–12.

Objectivity. A report must be objective. Objectivity demands that logic rather than emotion determine both the content of the report and its presentation. The content should be impersonal, with no indication of the personal feelings and sentiments of the writer. For instance, a report on the comparison of new car warranties for six makes of automobiles ordinarily should not reveal the writer's preference among the six or the make of automobile he or she drives. The report must be organized logically and its appeal aimed at understanding.

Essential to objectivity is the use of the denotative meaning of words—the meaning that is the same, insofar as possible, to everyone. Denotative meanings of words are found in a dictionary; they are exact and impersonal. Such meanings contrast with the connotative meanings, which permit associated, emotive, or figurative overtones. The distinction between the single denotative meaning and the multiple connotative meanings of a word can be illustrated by examples:

WORD: war
DENOTATIVE MEANING: open, armed conflict
CONNOTATIVE MEANING: death, injustice, Vietnam, freedom, cruelty, necessary evil, God vs. Satan, draft, soldiers, high taxes, destruction, bombing, orphans

WORD: work
DENOTATIVE MEANINGS: employment, job
CONNOTATIVE MEANINGS: paying bills, happiness, curse of Adam, accomplishment, 9 to 5, satisfaction, adulthood, alarm clock, fighting the traffic, income, new car, sweat, sitting at a desk, nursing

For further discussion of denotation and connotation, see page 9.

School Reports and Professional Reports

Reports can be classified according to the function they serve: school reports, which are a testing device for students; and professional reports, which are used in business, industry, and government.

Both school reports and professional reports are important to you—but in different ways and for different reasons. School reports, of course, are of more immediate concern simply because you are a student and because making reports is a widely used learning technique. In the school report it is you the student who is important; emphasis is on you as a learner and as a developer of potential skills. The report shows completion of a unit of study; it reflects the understanding, thoroughness, and intelligence with which you have carried out an assignment. The school report also gives you practice in presenting the kinds of reports that your future job may require. Such practice can be invaluable in helping you prepare to meet on-the-job demands.

Professional reports—the reports used in business, industry, and government—serve a different function from that of school reports. In professional reports, emphasis is on the information that the report contains and on serving the needs of the recipient. The crucial question is this: How well does the report satisfy the needs of the person to whom the report is sent? On the other hand, the instructor who assigns the school report ordinarily neither needs nor will use the information; the instructor is interested in the report only insofar as it reflects the educational progress of the writer.

In both school and professional reports, desirable qualities are accuracy, clarity, conciseness, and objectivity. But you should understand that some of the information given in a school report is not needed in a professional report. For example, in a student investigative report, your instructor may require information on conventional theory and procedure (to be sure that you understand and can explain them). In business, industry, and government, such information ordinarily would be unnecessary because it is taken for granted that you are knowledgeable about recognized theories and accepted procedures.

Format of the Report

Most reports (even those presented orally) are put in writing to record the information for future reference and to ensure an accurate, efficient means of transmitting the report when it is to go to several people in different locations. A report may be given in various formats: on a printed form, as a memorandum, as a letter, in a conventional report format, or as an oral report. The format may be prescribed by the person or agency requesting the report; it may be suggested by the nature of the report; or it may be left entirely to the discretion of the writer.

PRINTED FORM

Printed forms are used for many routine reports, such as sales, purchase requests, production counts, general physical examinations, census information, delivery reports. Printed forms call for information to be reported in a prescribed, uniform manner, for the headings remain the same and the responses—usually numbers or words and phrases—are expected to be of a certain length.

Printed forms are especially timesaving for both the writer and the reader. The writer need not be concerned with structure and organization; the reader knows where specific information is given and need not worry about omission of essential items. However, printed forms lack flexibility: they can deal with only a limited number of situations. Further, they lack a personal touch that provides an opportunity for the writer to express his or her individuality.

In making a report that uses a printed form, the primary considerations are accuracy, legibility, and conciseness. (See the report on page 365.)

MEMORANDUM

Memorandums and report letters are used in similar circumstances, that is, if the report is short and contains no visual materials. Unlike a report letter, however, the memorandum is used primarily within a firm. (For a detailed discussion of memorandums, see pages 247–253.)

Reports in memorandum formats are illustrated on pages 369, 371, and 405–406.

LETTER

As with the memorandum, the letter is often used for a short report (not more than several pages) that does not include visual materials. The letter is almost always directed to someone *outside* the firm.

The report letter should be as carefully planned and organized as any other piece of writing and should observe basic principles of letter writing. (See Chapter 7, Memorandums and Letters.) The report letter follows conventional letter writing practices concerning heading, inside address, body, salutation, and signature; however, the conventional complimentary close of a report letter is "Respectfully submitted." A subject line is usually included. The report letter often is longer than other business letters and may have internal headings if needed for easier readability. The degree of formality varies, depending on the intended reader and purpose.

Although report letters are widely used, some firms discourage their use to avoid the possible difficulty occasioned if they are filed with ordinary correspondence.

A report given in a letter format is illustrated on pages 391–392.

CONVENTIONAL REPORT FORMAT

Conventional report formats include both *nonformal* and *formal* reports. Nonformal and formal, vague though often-used terms, refer in actual practice to report length and to the degree to which the report is "dressed up."

Nonformal Reports. The nonformal, or informal, report usually is only a few pages in length, is designed for circulation within an organization or for a named reader, and includes only the essential sections of the report proper:

Introduction (or purpose)
Body (procedure, discussion of results)
Conclusions and recommendations

The preliminary matter and end matter of the formal report are omitted. Examples of nonformal reports appear on pages 361–363, 375–377, 381, 382–386, 401–403, and 407–410.

Formal Reports. The formal report has a stylized format evolving from the nature of the report and the needs of the reader. Often the formal report is long (eight to ten pages, or more), is designed for circulation outside an organization, and will not be read in its entirety by each person who examines it.

Parts. Conventional parts of the formal report have developed to improve reporting efficiency and to increase readability. These parts, in their layout and organization, are simply tools or devices for aiding the communication process; therefore, the writer should combine, omit, or vary them to accommodate the purpose and the intended audience of the report. The conventional parts may be listed as follows:

Preliminary matter:	Transmittal memorandum or letter
	Title page
	Table of contents
	Lists of tables and figures
	Summary or abstract
Report proper:	Introduction
	Body (procedure, discussion of results)
	Conclusions and recommendations
End matter:	Appendix
	Bibliography

1. *Preliminary Matter.* This includes the transmittal memorandum or letter, title page, table of contents, list of illustrations, and summary or abstract.

The **transmittal memorandum** (if the report is delivered within a company) **or transmittal letter** (if the report is mailed) affords the writer an opportunity to make needed comments on the report; for instance, identification of the report, reason for the report, how and when the report was requested, problems associated with the report, or reasons for emphasizing certain items. The length of this covering memorandum or letter varies, depending on the circumstances. The communication may be as simple as: "Enclosed is the report on customer parking facilities, which you asked me on July 5 to investigate"; or it may be several pages in length. (For further discussion, see Chapter 7, Memorandums and Letters.)

The **title page** contains a listing of the title of the report (as exact, specific, and complete as possible), the name of the person or organization for whom the report is made, the name of the person making the report, and the date. The arrangement of these items on the page should be pleasing to the eye.

The **table of contents,** which functions as an outline, is a valuable part of any report of more than a few pages. It shows the reader at a glance the scope of the report, the specific points (headings) that are covered, the organization of the report, and page numbers. Indentation is used to indicate the interrelations of subdivisions and main divisions. Wording of headings in the contents should be exactly as it is in the report.

If needed, separate **lists of tables and figures** follow the table of contents. (Tables are referred to as tables; graphs, charts, drawings, maps, and other illustrations are usually referred to as figures.) The lists are helpful if more than five or six tables and figures are used, or if they form a significant part of the body of the report. All captions are typed exactly as they appear in the text, with the beginning page for each table or figure. The usual practice is to capitalize only the first letter of the words *table* and *figure* and to use arabic numerals (1, 2, 3, etc.) for numbering. Quite acceptable, however, are the decimal system of numbering, and the use of all capitals for the word *table* and roman numerals: TABLE IV.

The **summary or abstract** (which may precede the table of contents) appears on a separate page. It is very important, especially when the report is to be passed upward through a chain of command. It gives the content of the report in a highly condensed form; it is a brief, factual pulling together of the essential, or central, points of the report. Included are an explanation of the nature of the problem; the procedure used in studying the problem; and results, conclusions, and recommendations. The summary, in length no more than 10 percent of the whole, should represent the *entire* report. (For further discussion, see Informative Summary, pages 217–222.)

2. *Report Proper.* The report proper consists of three or four parts: introduction, body, and conclusions and recommendations (conclusions and recommendations may be treated separately or together).

The **introduction,** depending on the nature of the report, may be simply an introductory paragraph or an expanded formal introduction with a separate heading. A simple introductory paragraph gives an overview of the subject. In addition, the introduction may indicate the general plan and organization of the report (especially if no table of contents is provided), give a summary of the report (if no formal summary or abstract is included), provide background information, and explain the reason for the report. The formal introduction, usually entitled "Introduction," typically deals with aspects of reporting, not of subject matter: name of the person or group authorizing the report, function the report will serve, purpose of the investigation, nature of the problem, significance of the problem, scope of the report, historical background (previous study of the problem), plan or organization of the report, definition or classification of terms, and methods and materials used in the investigation.

The **body** is the major section of the report; it presents the information. The body includes an explanation of the theory on which the investigative approach is based, a step-by-step account of the procedure, a description of materials and equipment, the results of the investigation, and an analysis of the results.

The **conclusions and recommendations** complete the report proper—unless the report is purely informational, in which case the report may conclude with a summary of the main points. Ordinarily, however, a formal technical report is investigative in nature. Thus the conclusions (deductions or convictions resulting from investigation) and recommendations (suggested future activity) are a very significant part of the report. In fact, in some reports they may be *the* most significant part and as such should precede rather than follow the body of the report. The basis on which the conclusions were reached should be fully explained, and conclusions should clearly derive from evidence given in the report. The conclusions, stated positively and specifically, usually are listed numerically in the order of importance. The recommendations parallel the conclusions. Depending on the nature of the report, the conclusions and recommendations may be together in one section or they may be listed separately under individual headings.

3. *End Matter.* An **appendix** and a **bibliography** may or may not be needed, depending on the nature of the report. An appendix should include support-

ing data or technical materials (tables, charts, graphs, questionnaires, etc.) in which most readers are not primarily interested. It includes information that supplements the text but if given in the text would interrupt continuity of thought. A bibliography lists the references used in writing the report, including both published and unpublished material. Unlike literary reports, the items in scientific and technical reports may be numbered and listed sequentially, that is, in the order in which they are first mentioned in the report. (For a discussion of bibliographical forms, see pages 473–474.)

Organization. The various parts of the formal technical report may be combined or rearranged, depending on the needs of the writer and reader and on the nature of the report. The report is divided into sections, each with a heading. The headings are listed together to form the table of contents.

Examples of formal reports appear on pages 393–399 and 475–490.

ORAL REPORT

As a student you may be assigned a report, such as a book report, to give as an oral presentation only. However, because of the nature and purpose of reports, most reports will first exist in written form, which allows them to be filed for future reference.

Often reports may be presented in both written and oral forms, the oral presentation emphasizing the major aspects of the report. For example, the treasurer of a small organization might give the latest treasury report orally; the report might include the amount of money in the treasury at the beginning of the previous accounting period, additional income, amount of expenditures, and the total remaining for the next period. The written report, prepared as a record, would probably include these same figures, as well as an itemized listing of sources and amounts of money received and creditors and amounts of money paid out. This written report might be handwritten or typed. However, large corporations send out printed reports, often lengthy and sometimes elaborate, showing profits and/or losses, dividends paid to stockholders, and decisions regarding future payments of dividends. At the stockholders' convention significant aspects of the report may be presented orally by one or more members of the corporation.

Suggestions for giving an oral presentation are made in Chapter 10, Oral Communication.

Visuals

Frequently visuals make information clearer, more easily understood, and more interesting. If a report, therefore, can be made more meaningful through visual materials, decide what kinds can be used more effectively, and use them.

For a detailed discussion of visuals, see Chapter 11, Visuals.

Headings

Headings are important in a report. They are an integral part of the layout and design of the report and contribute to readability and comprehension of the content.

□ Headings give the reader a visual impression of major and minor topics and their relation to one another.
□ Headings reflect the organization of the report.
□ Headings remind the reader of movement from one point to another.
□ Headings help the reader retrieve specific data or sections of the report easily.
□ Headings make the page more inviting to read by dividing what otherwise would be a solid page of unbroken print.

To be effective, headings must be visually obvious; they must clearly reflect the distinction between major sections and minor supporting sections.

For suggested systems of headings for reports, see pages 7–8. Study the use of headings in the reports on pages 371, 375–377, 381, 382–386, 391–392, 393–399, 401–403, 405–406, and 407–410. For a discussion of headings in an outline, see pages 128–129.

Layout and Design

Critical to the effectiveness of a report are its layout and design. Layout pertains to how the material is laid out, or placed, on the page. There must be ample white space so that the text will be visually receptive. Just as important, material to be emphasized should be indicated through such techniques as uppercase (capital) letters, underlining, color, boxes, and the like. (For detailed discussion, see pages 4–5, 24, and 25.)

Design has to do with the principles of visual composition. Careful attention should be given to such matters as report format, choice of visuals, spacing, and kinds and sizes of typefaces. (For further discussion, see pages 5–7, 24, and 25.)

Common Types of Reports

Among the common types of reports that you as a student and as a future technician may encounter are the reading report, the periodic report, the progress report, the investigative report, the observation report, the feasibility report, the proposal, and the research report. The following discussion examines each of these types of reports (except the research report, which is discussed separately in the next chapter) as to purpose, uses, main parts, and organization. Both student and professional reports illustrating the different types are given in several formats (on a printed form, as a memorandum, as a letter, and in the various conventional report formats).

READING REPORT

Purpose. A reading report on an article or book describes the general nature of the work, summarizes, analyzes, and evaluates it.

The reading report is one form of the evaluative summary (see Chapter 6, The Summary, especially pages 222–224).

Uses. A reading report indicates an individual's comprehension, appreciation, and reaction to a work. In looking at the reading report, the intended audience looks for the answers to several questions: How thoroughly has the individual read the original work? How well does he or she understand it? Can he or she intelligently discuss its significant aspects? How well does he or she appreciate its literary qualities? How valid is his or her evaluation of the work?

Main Parts. Whether the requested report is on a work of imaginative literature (novels, plays, short stories, poems) or on an informative work of nonfiction is immaterial. The report usually has four main parts: (1) identification of author and work, (2) summary, (3) analysis, (4) evaluation.

1. *Identification of Author and Work.* In no more than two or three sentences, give such information as the author's name and any helpful comments about him/her (such as relationship to a particular period or school of thought); publishing information (for a book, publishing company and date of publication; for an article, title of periodical, volume, date, and page numbers); audience; and the work in literary and historical perspective.

   Note: The publishing (bibliographical) information may also be given as a separate entry preceding or following the reading report. See Chapter 6, The Summary, especially page 225, and Chapter 9, The Research Report, especially pages 473–474.

2. *Summary.* Keep the summary brief—usually no more than one-fifth of the entire report. For a nonfiction work, give the subject, scope (extent of coverage), theme or central idea, and method of presenting the material. For a work of fiction, give the setting (time and place) and atmosphere, characters, plot, and theme or central idea.

   If the work has divisions, such as parts, chapters, or stanzas, take them into consideration. Use short, strategic quotations, if needed, to help convey the flavor or tone of the work.

3. *Analysis.* Analysis, the process of dividing a subject into parts in order to understand the whole better, forms the largest section of the report. Show that you have read the work thoughtfully and carefully and can discuss it intelligently. The intended audience will seek answers to: Does this report reflect an alert reader? What insights have been gained from reading the original work?

   Analysis of a work should contain discussions on:

   Meaning—What particular ideas or group of related ideas does the author emphasize? What is the relationship of these ideas to the work as a whole; that is, how do they relate to the organization and style of the work? How sound are these ideas?

   Structure—What is the form or organization of the work? Why is the organization as it is? Is it satisfactory, or could it have been improved?

   Style—What particular stylistic devices (such as dialogue, irony, understatement) are used? How would you describe the vocabulary and word choice? What are the main characteristics of the sentence patterns? Is the style straightforward and literal, or figurative? What adjective best describes the style (e.g., lively, dull, wordy, scholarly, simple)? Is the style appropriate to the content? Or should the author have used different levels of language, vocabulary, sentence patterns, and the like?

4. *Evaluation*. This part usually comprises about one-fifth of the report and gives your reasoned reactions to the work. Did the work succeed in fulfilling its purpose? Was the work objective? Biased? How did the work affect you? Would you recommend it to others to read? Did you enjoy or like the work? Why or why not?

Give your honest opinions regarding the work; at the same time, however, be fair in placing the blame: Was the failure in the work itself or was it in you? As you become a more disciplined reader, you will find that you also become more skillful in evaluating.

Organization. A reading report, or any other piece of writing, should be coherent and permit the reader to go smoothly from one part to another. Begin the report with a brief introductory paragraph that helps the reader to understand the organization of your report and the aspects of the subject that are emphasized. This introductory paragraph may be combined with the first main part: identification of author and work. Then give the other main parts: summary, analysis, evaluation. Use transitions between sentences and paragraphs to give continuity.

The following reading report (with Plan Sheet) responds to an assignment in a career exploration class.

PLAN SHEET
FOR GIVING A READING REPORT

Analysis of Situation Requiring a Report

What is the subject of the report?
article, "Does a Degree Tell Us What a Student Has Learned?"

For whom is the report intended?
instructor

How will the report be used?
to evaluate a class assignment

In what format will the report be given?
conventional report format

When is the report due?
Dec. 3, 1988

Identification of Author and Work

John Ashcroft, "Does a Degree Tell Us What a Student Has Learned?" <u>Phi Delta Kappan</u> (Nov. 1986), pp. 225–227

Summary

The article contains information on the use of knowledge, the present concept of learning, the Task Force's ideas of education reformations, and preparing students for life and work. The college must respond to the individual needs of the student in relation to society. Assessment programs need to be devised to assure adequate preparation for students before they enter the business world. The universities and colleges must take on the task of supplying the students, community, and state councils with school progressments.

Analysis

Author uses a conservative style; is very direct and precise in getting across his ideas to the reader. Style—straightforward (informative for intended audience). Seems to have researched and prepared material, to be interested in the well-being of students, and gives suggestions for reformation in education.

Evaluation

Author is successful in expressing his concerns and views. However, some redundant material could be left out. The article is informative and educational for the person interested in improving education and preparing students for life and work. I recommend the article.

Types and Subject Matter of Visuals to Be Included in the Report

omit

Reading Report:

"Does a Degree Tell Us What a Student Has Learned?"

Kim McKay
ENG 1113 BT
December 3, 1986

Reading Report:

"Does a Degree Tell Us What a Student Has Learned?"

John Ashcroft, "Does a Degree Tell Us What a Student Has
Learned?" Phi Delta Kappan (November 1986), 225–227.

   The article consists of certain points and
recommendations pertaining to the assessment of
institutional missions. The organization of the article
helps the reader follow with ease the main idea the author
is trying to convey, taking into consideration the
personal welfare and future of the individual student,
college, and community. Much emphasis is placed on the use
of the knowledge obtained from a university in a career.
These ideas are presented by factual statements obtained
from research. The need for future reformations is also
suggested and certain ideas are portrayed.
   Ashcroft stresses the importance of "especially
developing abilities to use knowledge." The idea of
learning is assumed to take place as long as students take
classes, accumulate hours, and strive toward obtaining a
degree. Students should be learning during the time they
spend at college. The students should acquire the
knowledge of how to organize and develop their education
for career use.
   Most of Ashcroft's ideas are in conjunction with those
of the Task Force on College Quality. They suggest the need
for certain educational improvements, but also emphasize
that the universities and colleges themselves need to devise
"systematic programs to evaluate the quality of student
learning, academic programs, and curricula." Another plus
for assessment programs is the ability of the institutions
to meet the demand for information provided for parents,
students, taxpayers, and state policymakers. The
information would supply the person in concern with
accurate information on students, programs, and
institutional performance.

The Task Force calls for the enforcing of these standards on colleges and universities for the main purpose of collecting information on student outcomes. As colleges and universities identify their missions and strive for quality, the Task Force "encourages state policymakers to provide incentives for institutions to develop and implement assessment programs that will positively affect teaching and learning."

Ashcroft concludes that "these resources are certainly fundamental to accomplishing institutional missions; however, they do not by themselves guarantee a high-quality, competitive education." Institutions need to be judged by the level of intelligence the students obtain from the institution for the preparation of life and work.

Ashcroft uses a conservative style in presenting his concerns and views of university missions and the role they play in the future of the student and community. He is very direct and precise in getting across his ideas to the reader. He writes as one who has thoroughly researched his material and evidently has a genuine interest and concern in the education of students and their future places in society.

Ashcroft is very straightforward in presenting his concerns and views of the student/college relationship and its place in the community. There is, however, redundant material of certain ideas pertaining to the reformation of university missions that could have been avoided. The article is based on factual knowledge from research and the author's personal knowledge. I recommend the reading of this article to all persons interested in the improvement of higher learning and the preparation of students for life and work.

I found the article very informative and educational. After reading the article, I have a new understanding and approach which should be taken when obtaining an education.

PERIODIC REPORT

Purpose. A periodic report gives information at stated intervals, such as daily, weekly, monthly, or annually, or upon completion of a recurring action.

Uses. A daily report of absentees, weekly payroll reports, income tax returns, inventories, a sales report upon coverage of a particular geographical area, budget requests, semester grade reports, monthly bank statements—these are but a few examples of the specific uses that periodic reports serve in practically all phases of business activity.

Main Parts. In its simpler, more common forms, the periodic report primarily gives specific measurable quantities, brief responses, and explanations. Data usually are given on a printed form, although the report form may vary from a memorandum or a letter to a full-scale, formal report, such as a large corporation's annual report to its stockholders. The periodic report, whatever its form, specifies the period of time covered, the subject dealt with, and the pertinent data concerning the subject.

Organization. The organization of a periodic report usually presents little difficulty because of the nature of the material. Typically, the information is presented by categories or chronologically (time order in which events occurred). In a firm, the special-purpose periodic report tends to settle into a uniform pattern since it covers the same or similar items each time (thus it permits the use of timesaving printed forms).

An example of a periodic report on a printed form is given on the next page. Then follows a periodic report in the form of a memorandum (with Plan Sheet), written by a student employed on a part-time basis in a bank records center.

MISSING PERSON AND RUNAWAY REPORT

LAST   FIRST   MIDDLE

NAME __DOE   JANE__   SEX & RACE __F/W__

ADDRESS __1689 St. Anthony Street__   CITY __Waterford, Mass.__

AGE __16__   DATE OF BIRTH __June 23, 1971__   HAIR __Brown__   EYES __Brown__

HEIGHT __5'2"__   WEIGHT __110__ BUILD __Medium__   COMPLEXION __Fair—Freckled__

TATTOO MARKS __None__   SCARS __Appendectomy__

NATIONALITY __American__   DEFORMITIES __None__

EDUCATION __10th grade__   OCCUPATION __Schoolgirl__

DRESS __Polka dot—blue__ SUIT ____ HAT __No__ SHOES __White Keds__

SKIRT _____   PANTS ____ COAT ____ HOSE _____

BLOUSE _____   SHIRT ____ JACKET ____ PURSE __White shoulder bag__

SWEATER __Navy blue cardigan__   OTHERS _____

MISSING SINCE __March 26, 1987__

PROBABLE DESTINATION __Memphis, Tenn.__

TRAVELING __Hitchhiking__ MAKE ____ MODEL ____ COLOR ____ TAG NO ____

TRAVELING WITH __Alone__

PUBLICITY DESIRED, YES ( )   NO (X)

REPORTED BY __Mrs. J. Doe__   RELATIONSHIP __Mother__

ADDRESS __1689 St. Anthony Street__ PHONE __552—3665__

REPORTED TO __Waterford PD__   DATE __3/29/87__   TIME __1:10 PM__

GIVEN TO RADIO __Yes__ DATE __3/29/87__ TIME __1:24 PM__ BY OFFICER __Catchem__

GIVEN TO MHP, YES (X) NO ( ) COMPUTER __Yes__ DATE __3/29/87__ TIME __1:36 PM__

WHERE LOCATED _____

BY WHOM _____   DATE _____

REMARKS:

PLAN SHEET
FOR GIVING A PERIODIC REPORT

Analysis of Situation Requiring a Report

What is the subject to be reported?
COM operator's night work

For whom is the report intended?
night supervisor

How will the report be used?
to keep a record of each operator's weekly activities

In what format will the report be given?
memorandum

When is the report due?
March 25, 1988

Type of Periodic Report (daily, weekly, monthly, etc.)

weekly

Dates covered:
March 20–24 (Sunday through Thursday nights)

Information to Be Reported

Work Date/Time:	Work Completed:
Sunday, March 20 *9:30 p.m.–8:00 a.m.*	*131 fiche; 20,224 frames; reruns: 1 fiche, 8 frames* *Total: 132 fiche; 20,232 frames* *Duplicates: 696; 103 reruns* *Total duplicates: 799*
Monday, March 21 *11:30 p.m.–8:00 a.m.*	*109 fiche; 28,320 frames; reruns: 1 fiche, 178 frames* *Total: 110 fiche; 28,498 frames* *Duplicates: 920; 67 reruns* *Total duplicates: 987*
Tuesday, March 22 *11:30 p.m.–8:00 a.m.*	*73 fiche; 12,378 frames* *Duplicates: 322; 83 reruns* *Total duplicates: 405*

Work Date/Time:
Wednesday, March 23
11:30 p.m.–8:00 a.m.

Work Completed:
96 fiche; 14,956 frames; reruns: 4 fiche, 200 frames
Total: 100 fiche; 15,156 frames
Duplicates: 445; 150 reruns
Total duplicates: 595

Thursday, March 24
11:30 p.m.–8:00 a.m.

187 fiche; 43,409 frames; reruns; 19 fiche, 3001 frames
Total: 206 fiche; 46,410 frames
Duplicates: 944; 175 reruns
Total duplicates: 1119

Types and Subject Matter of Visuals to Be Included in the Report

omit

Sources of Information

daily work log

```
To: Rich Melmein, Night Supervisor

From: Bob Kersh, COM Operator

Date: March 25, 1988

Subject: Weekly report on COM operator's work, March 20—24

Work Date/Time          Work Completed

Sunday, March 20        131 fiche—20,224 frames
9:30 p.m.—8:00 a.m.     Reruns: 1 fiche—8 frames
                        Total: 132 fiche—20,232 frames
                        696 duplicates
                        Reruns: 103
                        Total duplicates: 799

Monday, March 21        109 fiche—28,320 frames
11:30 p.m.—8:00 a.m.    Reruns: 1 fiche—178 frames
                        Total: 110 fiche—28,498 frames
                        920 duplicates
                        Reruns: 67
                        Total duplicates: 987

Tuesday, March 22       73 fiche—12,378 frames
11:30 p.m.—8:00 a.m.    Reruns: None
                        322 duplicates
                        Reruns: 83
                        Total duplicates: 405

Wednesday, March 23     96 fiche—14,956 frames
11:30 p.m.—8:00 a.m.    Reruns: 4 fiche—200 frames
                        Total: 100 fiche—15,156 frames
                        445 duplicates
                        Reruns: 150
                        Total duplicates: 595

Thursday, March 24      187 fiche—43,409 frames
11:30 p.m.—8:00 a.m.    Reruns: 19 fiche—3001 frames
                        Total: 206 fiche—46,410 frames
                        944 duplicates
                        Reruns: 175
                        Total duplicates: 1119
```

PROGRESS REPORT

Purpose. The progress report gives information concerning the status of a project currently under way.

Uses. Students and employees use a progress report to describe investigations to date, either at the completion of each stage or as requested by a supervisor. For the student, the progress report can signal the teacher that assistance or direction is needed. In industry and business, the progress report keeps supervisory personnel informed so that timely decisions can be made accordingly.

Main Parts. The progress report (1) describes briefly previous work on the project, (2) discusses in detail the specific aspects that are currently being dealt with, and (3) often states plans for the future. Unexpected developments or problems encountered in the investigation are collected in one section or at the points in other parts of the report where they logically arise.

Organization. The three parts of the report (previous work, current work, future plans) form a natural, sequential order for presenting the information. For easier readability, a heading may be used for each part.

If several progress reports are to be made on the same project, they should all be organized similarly.

Two examples of a progress report follow. The first describes the status of a construction project, in memorandum format. The second report (with Plan Sheet) was written by a student to assess progress toward his educational goals.

HARMONY CONSTRUCTION CO.

781 WEST AMES STREET   P.O. BOX 4912
OMEGA, OHIO 42761
PH. (513) 812 - 4000

Project: Vo-Tech Building—Nichols Technical Institute
Date: November 7, 1986
Time: 1:30 p.m.
Location: Job Site

Attendance:

        Ken McCollum
        T. E. Weatherman
        Kelly Flaherty
        Gale Singleton
        Jean Malley
        Bob West
        F. Ray Sullivan
        Paul Gatenby IV
        Ellis Mason
        Joseph Haley
        Carl Bailey
        Jerry Callom
        Brad Hilton

1. Progress to Date:

   Exterior framing and sheathing 90% complete. 1st floor
   interior partitions 85% complete. Electrical work on
   schedule. Mechanical duct work two weeks behind
   schedule. Plumber still has not gotten tank for solar
   heat system. Masonry work 90% complete (waiting on
   plumber). Stucco work on schedule. Painting started.

2. Schedule for Next Month:

   Start exterior work 11-20. Complete exterior framing
   and sheeting 11-21. Start aluminum store front and
   glass 11-17. Complete interior partition framing 11-28.
   Complete elevator installation 11-28. Complete flashing
   and sheet metal 11-28.

3. General Discussion:

   Housekeeping: All parties present were informed of the
   General Contractor's intentions to keep the job site
   cleaned up, and that it was each subcontractor's
   responsibility to clean up after the workers.

   Safety: The necessity of wearing hard hats.

   Job Progress: Each sub present commented on progress or
   the lack of it.

GENERAL CONTRACTING

PLAN SHEET
FOR GIVING A PROGRESS REPORT

Analysis of Situation Requiring a Report

What is the subject to be reported?
work as a student at HCC

For whom is the report intended?
major adviser

How will the report be used?
to evaluate progress toward educational goals

In what format will the report be given?
conventional report format

When is the report due?
May 2, 1988

Purpose of the Project

to acquire an Associate in Applied Science degree in electronics engineering technology from HCC

Work Completed; Comments

Semester I, 1987–1988

ENG	*1113*	*Eng Comp I (technical)*
TEL	*1356*	*Electricity for Electronics*
TDR	*1553*	*Fund of Dft*
TRS	*1613*	*Tech Math I*
HPR	*1111*	*PE*

GPA 3.0 I really enjoyed the electronics course. All the courses were harder than I expected.

Work in Progress; Comments

Semester II, 1987–1988

ENG   1123   *Eng Comp II (technical)*
TEL   1376   *Electronic Devices and Circuits*
TRS   1623   *Tech Math II*
TRS   1813   *Tech Applied Phy II*
HPR   1121   *PE*

"B" average so far. Total work to be completed by the end of spring semester, 32 semester hours. Possible 98 quality points.

Work to Be Completed; Comments

Semesters I and II, 1988–1989

TRS 1223 Ind Psy	*3*
Electronics	*24*
Social Studies	*3*
Physics	*3*

I will need help from my faculty adviser in choosing most of the specific courses. At end of year should have completed 65 semester hours and 175–200 quality points.

Types and Subject Matter of Visuals to Be Included in the Report

Table showing work completed 1st sem., 1987–1988
Table showing work completed 2nd sem., 1987–1988
Table showing work to be completed in 1988–1989

Sources of Information

college catalog, my grade reports for each semester

My Work as a Student at Haws Community College—
Purpose, Progress, and Projections

Robert Lewis, May 2, 1988

In August of 1987 I enrolled in Haws Community College
to attain skills in electronics engineering technology and
to acquire an Associate in Applied Science degree in that
field. The program requires a minimum of 65 semester hours
and 130 quality points for completion. Now in the second
semester of the four-semester program, I am finding my
courses very satisfying. The courses are harder than I
expected, but I feel that I am learning valuable
information and that I am accomplishing my goal.

1987–1988

Semester I. During the first semester of the 1987–1988
academic year I completed these courses with the following
hours, grades, and quality points.

COURSES		HOURS	GRADE	QUALITY POINTS
ENG 1113	English Composition I (Technical)	3	C	6
TEL 1356	Electricity for Electronics	6	A	24
TDR 1553	Fundamentals of Drafting	3	C	6
TRS 1613	Technical Mathematics I	3	B	9
HPR 1111	Physical Education	1	A	4
	Total	16		49

These were the required courses for the first semester
of the two-year course in electronics engineering
technology. My quality point average for the semester was
3.0 on a 4.0 scale. I am proud of this average and I had
to work hard for it, especially for the A in Electricity
for Electronics. Perhaps the main reason that I
experienced such success in this course was that I really
liked it (it was the only actual electronics course I took
this semester). Too, it met every day; thus I had to study

every day. The course that I found most difficult was Fundamentals of Drafting; the course that I found most challenging was freshman English.

<u>Semester II</u>. During the second semester of the 1987–1988 academic year I am currently enrolled in the following courses.

COURSES	HOURS
ENG 1123 English Composition II (Technical)	3
TEL 1376 Electronic Devices and Circuits	6
TRS 1623 Technical Mathematics II	3
TRS 1813 Technical Applied Physics I	3
HPR 1121 Physical Education	1
	16

These are all required courses. Thus far in the second semester I have maintained a "B" average, and I am still experiencing the most success with the electronics–related courses.

At the end of this school year, I should have completed 32 semester hours and acquired a possible 98 quality points.

<u>1988–1989</u>

<u>Semesters I and II</u>. During the 1988–1989 academic year I plan to take the additional courses and hours required for graduation in electronics engineering technology. These include one specified course and my choice of courses in several areas.

COURSES	HOURS
TRS 1223 Industrial Psychology	3
Technical Electronics	24
Social Studies	3
Physics	3
	33

Except for Industrial Psychology, I will need help from a faculty adviser in choosing the specific courses. At

this point, I think I want to take Digital Fundamentals
(6 hours) and Linear Integrated Circuits (6 hours). I am
considering taking Digital Integrated Circuits (6 hours),
Introduction to Microprocessors (6 hours), and Programming
Fundamentals (3 hours) or Robotics (3 hours).

Upon completion of the sophomore year, I plan to have
the required 65 semester hours of work plus many more than
the 130 minimum quality points required.

The End Result
It is my aim to graduate from Haws Community College
with an Associate in Applied Science degree in electronics
engineering technology. Upon completion of this program of
study, I expect to be qualified to enter the field of high
technology as a microelectronics technician.

INVESTIGATIVE REPORT

Purpose. An investigative report presents results of experimentation, re-search, or testing. The investigation may be connected with any field of study—business, chemistry, physics, data processing, home economics, fire science, electronics, nursing, and so on.

Uses. This is a common kind of report for both students and on-the-job workers. It may simply report the results of tests, the purpose for the tests being implicit in their nature (such as a blood-type test). Some types of investigative reports, however, are very involved: They not only give test results but also the results are applied to specific problems or situations. Recommendations may be included as well.

Main Parts. For most investigative reports, a basic, fairly well-established form may be used. Its typical parts are as follows:

Title page

Object (also called Purpose)

Theory, Hypothesis

Method (also called Procedure)

Results

Discussion of Results (also called Comments)

Conclusions

Appendix

Original Data

Organization. The various divisions may be combined or rearranged to suit the needs of the writer and the reader. For assignments, use the division headings listed above, unless another form of presentation has been specified.

Study the two examples of student-written investigative reports that follow. The first report (with Plan Sheet) is concerned with testing two body fillers in an auto body and fender repair shop. The next report concerns an electronic experiment (this report is reprinted courtesy of the Mississippi State Department of Education).

PLAN SHEET
FOR GIVING AN INVESTIGATIVE REPORT

Analysis of Situation Requiring a Report

What is the subject to be reported?
testing of two types of body filler

For whom is the report intended?
shop instructor, Mr. McPhillip

How will the report be used?
to determine whether the shop should use the less expensive body filler

In what format will the report be given?
conventional report format

When is the report due?
July 15, 1988

Object or Purpose of the Study

to compare two types of body filler to see if the $3.21 filler is as good as the $4.55 filler

Theory, Hypothesis, Assumption

The cheaper body filler may be just as good as the body filler we use now.

Method or Procedure

For two weeks the cheaper body filler was used under the same conditions that the more expensive body filler was used.

Results; Observations; Comments

The cheaper body filler cuts more easily, is more flexible, and has fewer pinholes than the other filler.

Conclusions; Recommendations

We should start using the cheaper filler. It's better and it can save the shop about $188.00 per month. The shop uses about 140 cans of body filler a month.

Types and Subject Matter of Visuals to Be Included in the Report

omit

AUTO BODY AND FENDER REPAIR

Report written by Warren Daniels
for Instructor Herbert C. McPhillip

July 15, 1988

OBJECT. To compare Type I body filler that costs $4.55 per can with Type II body filler that costs $3.21 per can.

THEORY. The less expensive Type II body filler may be just as good as the more expensive Type I body filler, which is used in the shop. A representative of the company that manufactures Type I claims that Type II has more talcum and is therefore less desirable.

METHOD. For two weeks, Type II filler, rather than Type I, was used in the shop. It was tried on several jobs of varying difficulty. Each time it was mixed in the same way that Type I had been mixed. The same kind of primer was also used to cover the filler.

CONCLUSIONS: Type II body filler cuts more easily, is more flexible, and has fewer pinholes than Type I. Type II is not only as good as Type I, it is better. Since the shop uses about 140 cans of body filler per month, the shop will save approximately $188.00 per month.

$$
\begin{array}{rcl}
140 \times \$4.55 & = & \$637.00 \\
140 \times \$3.21 & = & \underline{449.40} \\
& & \$187.60
\end{array}
$$

RECOMMENDATIONS.   1. Only Type II body filler should be used in the shop.
2. The money that can be saved by using the less expensive body filler should be placed in a fund to purchase new shop equipment.

INVESTIGATIVE REPORT: EMITTER FOLLOWER

Elicia Longe

TEL 1326–02

Electronic Circuits I

Instructor: H. J. Johnson

February 8, 1988

Purpose: The purpose of this experiment was to design, construct, and test an emitter follower circuit which might be used as an impedance matching device. An approximation technique was employed in formulating the design characteristics.

Procedure: The circuit utilized is shown in Figure 1. For construction, standard 10% resistor values, nearest to those calculated, were used. Measurements were made with a digital multimeter and recorded. An audio signal of 1 kHz was fed to the input of the emitter follower and the output waveforms were observed on an oscilloscope. The input signal level was adjusted to provide maximum undistorted output. Input and output signal levels were measured and recorded. The input audio frequency was lowered until the output voltage across R_L dropped to .707 of the 1 kHz level. The resultant reading was recorded as the lower 3 dB frequency limit. The input audio frequency was then increased to a point above 1 kHz where a .707 reading was obtained, and this reading was recorded as the

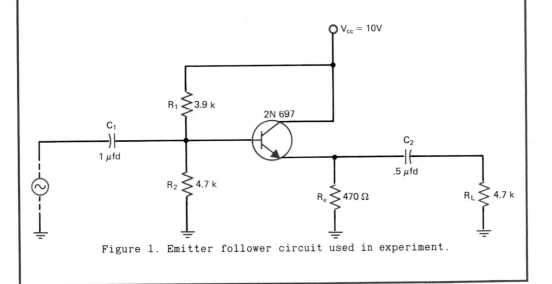

Figure 1. Emitter follower circuit used in experiment.

upper 3 dB frequency limit. The audio signal generator was
then reset to 1 kHz. The multimeter was placed in series
with the audio signal generator in the input circuit, and
the signal current was measured and recorded.
    See Table 1.

<u>Design</u>

1. $V_{ce} = .5 \times 10 = 5V$

2. $R_e = Z_{out} = 500\Omega$ (470Ω standard value)

3. $I_e = \dfrac{10 - 5}{470} = \dfrac{5}{470} = 10.6mA$

4. $V_b = 5 + .7 = 5.7V$

5. Stability Factor (S) $\simeq 10 = \dfrac{R_z}{470\Omega}$

6. $I_2 \simeq \dfrac{5.7V}{4.7k\Omega} = 1.21mA$ (where $I_1 \simeq I_2$)

7. $R_1 \simeq \dfrac{10V - 5.7V}{1.23mA} = \dfrac{4.3V}{1.23mV} = 3.5k\Omega$

8. $A_v \simeq \alpha = \dfrac{80}{81} = .987$

9. $R_b = \dfrac{3.5k(4.7k)}{3.5k + 4.7k} = \dfrac{16.45M}{8.2k} = 2.24k\Omega$

10. $C = \dfrac{1}{6.28(100)(4.7k\Omega)} = .71 \ \mu fd$

11. $C_2 = \dfrac{1}{6.28(100)(4.7k\Omega)} = .338\mu fd$

Table 1 Measurements

Bias Conditions Pertaining to All Measurements V_{be} = .62V V_{ce} = 4.9V I_e = 10.9mA			
Voltage Gain	Maximum undistorted output at 1khz = 8.1V p–p = 2.86V rms	Maximum input–8.65V p–p 3.05V rms	$A_v \simeq \dfrac{8.1}{8.6}$ $A_v \simeq .942$
3 dB Roll–off Frequencies	Voltage at roll–off points = .707 of voltage at 1kHz point = .707 × 8.1 = 5.72V	Lower roll–off freq. = 84 hZ	Upper roll–off freq. = 230 kHz
Signal Input Current and Impedance	Signal current = 1.58 mA	$Z_{in} = \dfrac{V_{in}}{I_{in}}$ $Z_{in} = \dfrac{3.05V_{rms}}{1.58\ mA}$	$Z_{in} = 1.93k\Omega$

Observations

1. Bias Considerations

	Design	Achieved	% Error
Vce	5.00V	4.90V	2.00%
Vbe	0.70V	0.62V	11.40%
Ie	10.60mA	10.90mA	2.83%

Comparison of the achieved Vce, Vbe, and Ie shows the achieved conditions were very close to the design values.

2. Gain Characteristics

Theoretical $A_v \simeq \alpha = .988$
Achieved $A_v = .942$ or $.936$
% Error $= 4.56\%$

3. Frequency Response
The achieved low 3dB roll-off point was 84Hz instead
of the design 100 Hz. This resulted from using coupling
capacitors which had values larger than the design values.
This lowered the low frequency cutoff point.

4. Input Impedance
The theoretical and calculated input impedance differed
by 13.8%.

5. Transistor Beta
A transistor with a lower Beta will have little effect
on dc bias conditions because the circuit is Beta
independent. If the Beta falls off too much, the voltage
gain could possibly suffer.

6. Stability
Increasing the input to output impedance ratio would
decrease the stability of the circuit.

Purpose. The observation report gives the results of a visit to a particular location or site. Major sources of information for this report are personal observation, experience, and knowledgeable people. (See Chapter 10, Oral Communication, pages 522–523, for how to conduct an informational interview.)

Uses. Observation reports are used in many ways. For example, they are important in estimating the value of real estate or the cost of repairing a house; establishing insurance claims for damage from a tornado or blizzard; improving production methods in a department or a firm; choosing a desirable site for a building, a highway, a lake; or serving as an educational experience for a prospective employee or interested client. The observation report gives an accurate, objective explanation and analysis of a situation so that appropriate action can be taken.

Main Parts. Since the observation report has a variety of uses and includes various kinds of information, it has no established divisions or format. The report may include such parts as a review of background information, an account of the investigation, an analysis and commentary, and conclusions and recommendations. For a student (or any other interested person), an observation report on a visit to a company might include a description and explanation of its physical layout; the personnel, materials, and equipment involved; the individual activities that comprise the major function of the company; and comments.

Organization. At the beginning of the report, state the purpose of the report, the specific site or facility or division observed, and the aspects of the subject to be presented. Then give the results of the investigation, followed by conclusions and recommendations. If the report is more than a paragraph or two, use headings. Include sketches, diagrams, charts, and other visual materials when they make explanation and description simpler or clearer.

Two examples of observation reports follow. The first (with Plan Sheet) is a report in the form of a letter on an inspection of a communication site. The second report—tending to be formal with the preliminary matter of a transmittal letter, title page, and summary—concerns the field inspection of a pond dam.

PLAN SHEET
FOR GIVING AN OBSERVATION REPORT

Analysis of Situation Requiring a Report

What is the subject to be reported?
inspection of communication site at Pensacola, FL

For whom is the report intended?
Cole-Meyers Electric Company

How will the report be used?
to determine efficiency of present operation

In what format will the report be given?
letter

When is the report due?
Nov. 28, 1988

Object or Purpose of the Study

to inspect power supply, equipment room, and control room of communication site at Pensacola; inspection requested by letter, Oct. 31, 1988

Theory, Hypothesis, Assumption

omit

Method or Procedure

on-site, personal investigation; standard tests and checks made on equipment

Results; Observations; Comments

Transformers operating at peak operating condition. Automatic switchover equipment working properly. Converter equipment output has less than 3% distortion. No major outage since Oct. 2, 1983. Multiplex equipment is adequate; one branch operating on auxiliary power with no backup. This to be corrected by end of November. Control room well supervised and efficiently operated.

Conclusions; Recommendations

Communication site at Pensacola is one of the best in the Southeast. Personnel are satisfied. Sum of average outages: 30 min. per week.

Types and Subject Matter of Visuals to Be Included in the Report

omit

Electric Troubleshooters, Inc.

810 Fourth Avenue
Waterbury, Connecticut 01836
P.O. Box 817
203-902-8165

November 28, 1988

Cole—Meyers Electric Company
1521 East Second Street
Great Neck, NY 10120

Attention: Mr. R. M. Rothchild

Cole—Meyers Electric Company:

In accordance with your letter of instructions of October
31, 1988, I have made a personal inspection of the
communication site at Pensacola, Florida, and submit the
following report.

Power Supply

<u>Transformers</u>. The transformers are in peak operating
condition. According to the log, two transformers are
operating at the same time. One supplies the power, and
the other is a standby.

<u>Automatic Switchover</u>. The automatic switchover equipment
is working properly. Standard tests were performed, and
not once did the signal lose phase from one transformer
to another.

Equipment Room

<u>Converters</u>. The converter equipment has an output with
less than 3% distortion. A routine check for distortion is
made every hour and recorded according to regulations.
Very little outage is recorded, and no major outage has
been recorded since the cable cut of October 2, 1983.

<u>Multiplex</u>. The multiplex equipment is adequate with one
exception. The second branch is operating on auxiliary
power with no backup. A power supply has been ordered and
should arrive November 28.

Cole—Meyers Electric Company
November 28, 1988
page 2

Control Room
The control room is supervised, and everyone is well
trained in control facilities. The outage record and the
logs are kept accurately and up to date.

Conclusion
The site at Pensacola is one of the best in the Southeast.
The personnel are satisfied and no complaints were filed.
The average outages add up to about 30 minutes a week,
which is the best record in the Southeast.

Very truly yours,

J. M. Black

J. M. Black

Formal Observation Report
Letter of Transmittal

393

Reports: Conveying
Needed
Information

1135 Combs Street
Jackson, MS 39204
6 April 1988

Mr. Harry F. Downing
4261 Marshall Road
Jackson, MS 39212

Dear Mr. Downing:

Attached is the field inspection report of the pond dam on your property. The dam is considered to be in stable physical condition although some minor seepage and erosion were discovered. Recommendations for correcting these are included in the report.

It has been a pleasure to work with you on this project.

Sincerely yours,

Paul Kennedy

Paul Kennedy

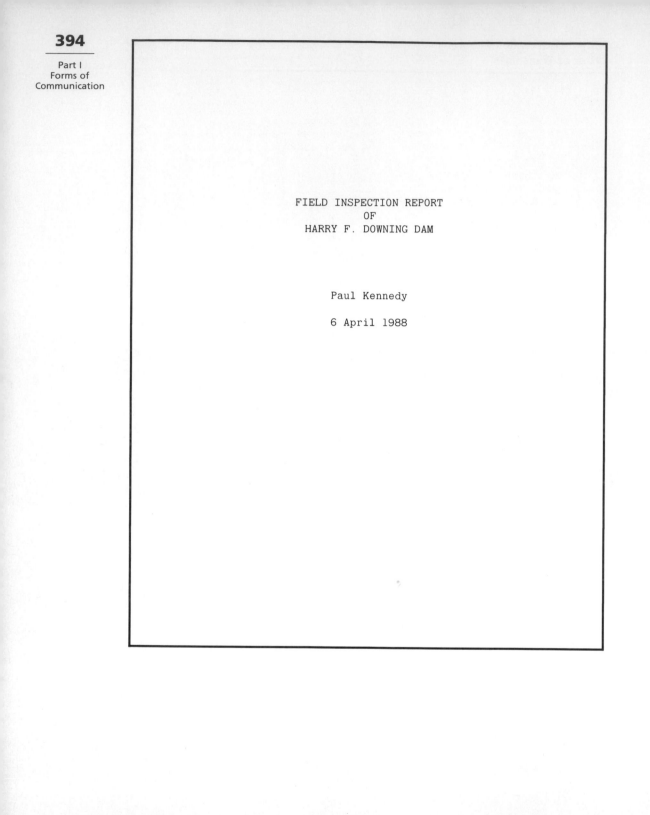

```
        FIELD INSPECTION REPORT
                  OF
          HARRY F. DOWNING DAM



             Paul Kennedy

             6 April 1988
```

SUMMARY

On 2 April 1988 an unofficial inspection of the dam of
the Harry F. Downing Pond was conducted by Paul Kennedy.
This dam is considered to be in stable physical
condition although some minor seepage and erosion were
discovered. This conclusion was based on visual
observations made on the date of the inspection.

FIELD INSPECTION REPORT

HARRY F. DOWNING POND
HINDS COUNTY, MISSISSIPPI
PEARL RIVER BASIN
CANY CREEK TRIBUTARY

PURPOSE

The purpose of this inspection was to evaluate the structural integrity of the dam of the Harry F. Downing Pond, which is identified as MS 1769 by the National Dam Inventory of 1973.

DESCRIPTION OF PROJECT

Location. Downing Pond is located two miles SE of Forest Hill School, Jackson, Mississippi, in Section 23, Township 6, Range 1 East (see Figure 1).

Hazard Classification. The National Dam Inventory lists the location of Downing Pond as a Category 3 (low-risk) classification. Personal observation of areas downstream confirm this classification since only a few acres of farmland would be inundated in the case of a sudden total failure of the dam.

Description of Dam and Appurtenances. The dam is an earth fill embankment approximately 200 feet in length with a crown width of 6 to 10 feet. The height of the dam is estimated to be 16 feet with the crest at Elevation 320.0 M.S.L. (elevations taken from quadrangle maps). Maximum capacity is 47 acre feet. The only discharge outlet for the pond is an uncontrolled overflow spillway ditch in the right (east) abutment. The spillway has an entrance crest elevation of 317.0 M.S.L. and extends approximately 150 feet downstream before reaching Cany Creek. The total intake drainage area for the pond is 30 acres of gently rolling hills.

Design and Construction History. No design information has been located. City records indicate that the dam was constructed in 1940 to make a pond for recreational purposes.

FINDINGS OF VISUAL INSPECTION

Dam. Apparently the dam was constructed with a 1V or 2H slope. This steep downstream slope is covered with dense vegetation, which includes weeds, brush, and several large trees. These trees range from 15 to 20 feet in height (see Photos 1 and 2). These trees have not likely affected the dam at the present time, but decaying root systems may eventually provide seepage paths.

A normal amount of underseepage was observed about halfway along the toe of the dam (see Photo 3). This seepage was not flowing at the time of the inspection but should be watched closely during high-water periods. The upstream face of the dam has several spots of erosion near the water's edge due to the lack of sod growth. Apparently topsoil was not placed after construction.

Overflow Spillway. The uncontrolled spillway shows no signs of erosion and is adequately covered with sod growth (see Photo 4).

RECOMMENDATIONS

It is recommended that the owner:
1. Periodically inspect the dam (at least once a year).
2. Prevent the growth of future trees on the downstream slope.
3. Install a gage and observe the flow of underseepage as compared to pool levels.
4. Fill areas of erosion and place topsoil and sod to prevent future erosion.

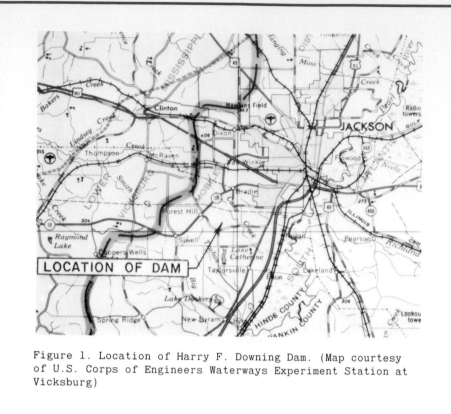

Figure 1. Location of Harry F. Downing Dam. (Map courtesy
of U.S. Corps of Engineers Waterways Experiment Station at
Vicksburg)

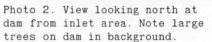

Photo 1. View looking east from left (west) abutment. Note large trees on dam at left.

Photo 2. View looking north at dam from inlet area. Note large trees on dam in background.

Photo 3. View looking west along toe of dam. Note seepage at left.

Photo 4. View looking upstream of uncontrolled overflow spillway ditch. Note adequate sod growth.

FEASIBILITY REPORT

Purpose. The feasibility report is a systematic analysis of what is possible and practicable, of what is capable of being accomplished or brought about. The feasibility report offers solutions to the problem of "Should we do this?" or "Which of these choices should I select?" or "Which course of action should the company take?" Analysis for a feasibility report is very similar to analysis in an effect-cause or comparison-contrast situation (see Chapter 5, Analysis Through Effect-Cause and Comparison-Contrast, particularly "Problem Solving," pages 181–183).

Uses. Feasibility reports are frequently used in business, industry, and government. Their data serve as the basis for significant decisions. Should our company move its central headquarters from this geographical area to another part of the country? Should we rent or buy? What is the best location in this city for my fast-food franchise? Which investment company can meet the needs of my client? Should I spend the estimated $2000 in repairs/overhaul of my four-year-old car or should I buy a new one? Which offer should we accept? The feasibility report gives a complete, accurate analysis of the possibilities and presents recommendations.

Main Parts. Typically the main parts of the feasibility report are these:

Summary

Conclusions

Recommendations

Introduction: background, purpose, definition, scope

Discussion

The main parts and thus the headings vary, depending on the subject, audience, and particular considerations.

Organization. The summary, conclusions, and recommendations may be combined. They usually come first in a feasibility study because they—particularly the recommendations—are the focus of the report. Then typically follows a section that gives background information, explains the purpose of the study, defines and describes the subject, and explains the scope of the study. The lengthiest section of the report is the discussion; here full data are given with analysis and commentary on each possible solution or option, and substantiation of all recommendations. The discussion may be organized by the several comparison-contrast patterns (point by point, subject by subject, similarities/differences) however, the most usual pattern is point by point, that is, criterion by criterion (a criterion is a standard by which an item is judged). The criteria, for instance, for selecting a new car might be cost, gasoline mileage, standard equipment, and warranty coverage.

For discussion of comparison-contrast organizational patterns, see pages 193–195 and 672–673.

An example of a feasibility report follows. In the study of available unfurnished apartments, note how the data are concisely presented in a table.

COST FEASIBILITY REPORT OF AVAILABLE UNFURNISHED
APARTMENTS FOR THE LINDA BRIDGERS FAMILY
6 FEBRUARY 1987

<u>Purpose</u>: The purpose of this feasibility study was to
locate and rent an unfurnished apartment whose location
would cut down on my daily driving mileage to Harper
Community College, Rayfold, Minnesota, from my present
location in Brandon, Minnesota (56 miles round-trip). The
criteria that I established for the unfurnished apartment
are the following:

- 2 bedrooms
- $1\frac{1}{2}$ baths
- washer and dryer connections
- children and pets permitted
- $350.00 or less per month
- $200.00 or less deposit (preferably no deposit)
- 1-year lease (minimum) honored

<u>Method</u>: From the classified ads, section c, of the 1
February 1987 <u>The Clarion-Ledger/Jackson Daily News</u>, I
found several listings for unfurnished apartments. I then
telephoned six of these that might possibly fit my
criteria.

<u>Data</u>: One of the six listings I called is located in north
Jackson and has all but two of my criteria, those being
only 1 bath instead of $1\frac{1}{2}$ and they honor no lease. Three
of the listings are in northeast Jackson; all three rent
for less than $350 each, but they do not allow children or
pets. The listing I called in south Jackson has only the
2-bedroom item from my list of criteria. The sixth listing
I called is for an apartment located in Clinton,
Minnesota; it fits all of my criteria except for the
number of baths. The data for all six apartments are given
in Table 1.

<u>Conclusion</u>: Even though the apartment located in north
Jackson would be better for us financially, it does not
honor a lease. The three apartments in northeast Jackson

and the one in south Jackson are out because of the
children and pet restrictions. We have chosen the
apartment located in Clinton, Minnesota. This apartment
fits all of my criteria except one, and this location
requires only 16 miles round-trip to Harper (significantly
closer than the present Brandon apartment).

TABLE 1 UNFURNISHED RENTAL APARTMENTS

Location	Number of Bedrooms	Number of Baths	Washer and Dryer Connections	Children	Pets	Rent Monthly	Deposit	Lease
North Jackson	2	1	YES	YES	YES	$250.00	$125.00	NO
Northeast Jackson A	2	1½	YES	NO	NO	$250.00	$200.00	1 yr.
Northeast Jackson B	2	2	YES	NO	NO	$250.00	$150.00	6 mo. to 1 yr.
Northeast Jackson C	2	1½	YES	NO	NO	$320.00	$150.00	6 mo.
South Jackson	2	1	NO	NO	NO	$400.00	$250.00	NO
Clinton	2	1	YES	YES	YES	$350.00	$200.00	1 yr.

PROPOSAL

Purpose. A proposal is a well-thought-out plan. The plan may be to start something new, or the plan may be to change the way things are now being done. The idea of the proposal, a plan for action, might come from an employee who perceives a need and on his or her own thinks up a way to do something or to do something in a different way; or the idea of the proposal may come from an employer or other person of responsibility, who requests that a proposal be submitted. A proposal is a form of persuasion (see "To Persuade," pages 507–509, in Chapter 10, Oral Communication). The purpose of a proposal is to persuade the reader to accept or support an idea, to buy a product or service, or to change a situation (see "Problem Solving," pages 181–183, in Chapter 5, Analysis Through Effect-Cause and Comparison-Contrast).

Uses. Proposals serve many uses. Proposals may include such divergent uses as a letter from a student government association to the president of the college setting forth proposed changes in the class attendance/withdrawal policy; a memo from a sales associate in a clothing store to the department manager detailing the need for additional salespersons; a multivolume document from an arms manufacturer to the Department of Defense bidding on a contract; a proposal from a college professor to a foundation for a grant of money to research visual hindrances in preschool children.

Main Parts. Since the proposal has a variety of uses, since it may be unsolicited or solicited, since it may be nonformal or formal, and since it may be written by one person or a group, it has no set format or divisions. Depending on the purpose of the proposal and the audience, however, you might include these parts:

Purpose
Problem
Proposed solution or plan, including methods or procedure
Qualifications of the writer (if needed)
Conclusion

Organization. At the outset, give an overview of the situation with a statement of the problem; describe, define, explain the problem in detail. Then present the thrust of the proposal—the proposed solution to the problem or situation. The proposed solution includes what will be done, by whom, how, when, with what cost, and with what expected results.

Two examples of proposals follow. The first, in memorandum format, is from a student to a teacher concerning the choice of topic for a major report. The second example proposes changes of a deli-grocery into a restaurant.

TO: Dr. Nell Ann Pickett
FROM: Janet Thomas, ENG 1123-AW
DATE: April 20, 1988
SUBJECT: Topic for library research report

Statement of Problem

For my report I would like to explore Alzheimer's disease from the standpoint of the person who provides care to a family patient at home. The caretaker is usually a spouse, a son or a daughter, or some other close relative.

While there seems to be ample material about what is known and not known about Alzheimer's disease itself, I have not yet found any articles dealing only with how family members cope with caring for a patient at home.

Reason for Choice of Topic

As a nursing major who plans to specialize in geriatrics, I have a keen interest in dysfunctions and diseases of the elderly. Although Alzheimer's disease is not restricted to the elderly, it is certainly a debilitating illness for many older people.

In addition to my professional interest, I have a personal interest in the disease and especially in the role of the caretaker of the diseased person. My grandfather has Alzheimer's disease; for three years I have observed my grandmother, who, with the help of my father and two aunts, has taken care of my grandfather. I would like to research this topic so that I can better understand the emotional and physical stress and frustration that my grandmother and other family members have experienced as they have tried to care for my grandfather.

Procedure

I will research this topic in the college library and in the library at the Nursing/Allied Health Center. Also I will structure a list of questions and then interview these people (and perhaps others): a home health nurse who weekly visits several homes with Alzheimer's disease

patients, several nonrelative caretakers of Alzheimer's
disease patients, my grandfather's doctor, the director of
the county mental health association, an appropriate
person at the state health department, and a social worker
at the regional hospital.

Possible Problems
    There should be no problem in finding numerous articles
and other information on Alzheimer's disease itself. But
sorting out the information about the caretaker and the
changes in her/his life-style and attitude toward life
will probably take a lot of time.
    At this time, I foresee no problems in meeting the
deadline of May 11 for the report or in meeting the
interim deadlines for working outline, bibliography cards,
note cards, and preliminary draft.

Need for Consultation with Instructor
    I would like for you to look over my interview
questions when I have them made out. I'm thinking about
taking a tape recorder to record each interview, but I'm
not sure how that would work out (especially with elderly
caretakers).

Proposal for Remodeling Chamoun's Restaurant
Clarksdale, Missouri
Mary Louise Chamoun
November 4, 1988

Purpose
    The purpose of this proposal is to present remodeling
and expansion ideas for Chamoun's Restaurant, in
Clarksdale, Missouri.

Present Conditions
    Chamoun's is a family-owned and operated grocery and
delicatessen. The business was established in 1967 by
Chabik and Louise Chamoun as a grocery and has since
become a deli serving a variety of Lebanese, American, and
Italian specialties. With the many requests for evening
dining, the question of remodeling and expanding has
become dominant in family conversation.
    Chamoun's is presently open from 8:00 a.m. to 7:00 p.m.
Monday through Saturday. The store operates a small deli
containing seven small tables and a carry-out service with
the small grocery business. Figure 1 shows the present
floor plan.

Remodeling Recommendations
    An informal estimated cost of remodeling and expanding
the restaurant is $15,000 to $20,000. The restaurant hours
would be 11:00 a.m. to 10:00 p.m. Monday through Saturday
and 11:00 a.m. to 2:00 p.m. on Sundays.
    My recommendations for expansion and remodeling (see
Figure 2) are as follows:

1. Add a 20' × 20' dining room area to accommodate
   65–85 customers.
2. Enclose and furnish with industrial fittings and
   appliances the partially exposed grill/kitchen
   area.
3. Remove the gasoline pumps and replace the porch
   area with an enclosed glass waiting area.
4. Add a parking lot to accommodate 40–55 cars.

   5. Remodel the present lunch dining area for the
      large lunch crowd.
   6. Brick the outside of the restaurant, and
      landscape the grounds.
   7. Decorate the interior with Middle-East
      furnishings, to accentuate the name of the
      restaurant.
   8. Install an intercom stereo system for customer
      dining pleasure.

   The business is well established and the food is top
quality and well known in Clarksdale and the surrounding
area. With the demand in Clarksdale for good restaurants,
it would be quite profitable to remodel and to offer
evening dining.

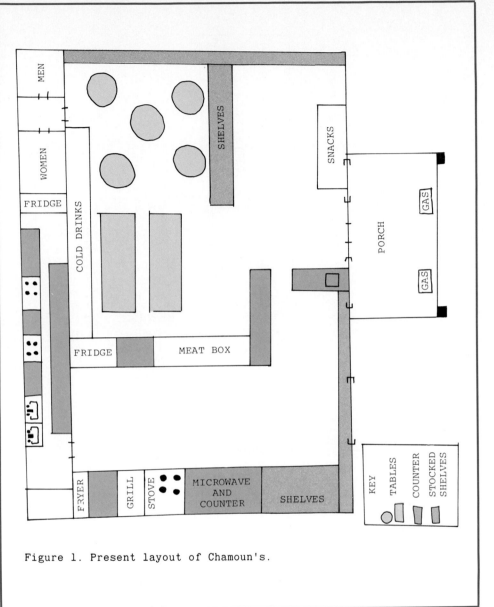

Figure 1. Present layout of Chamoun's.

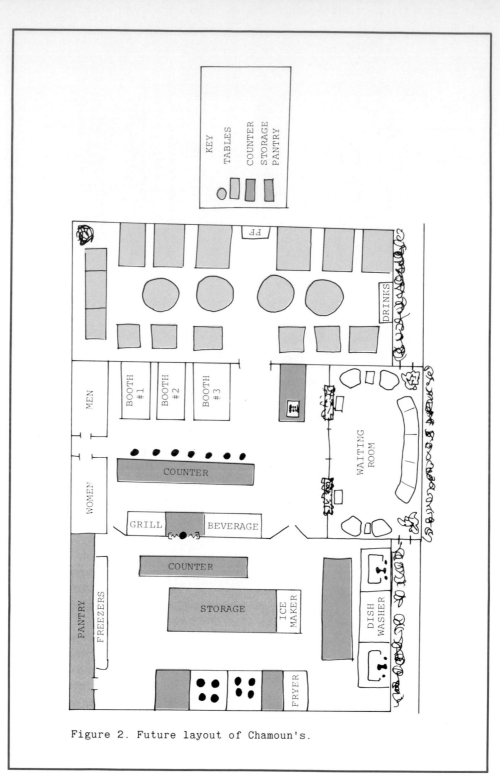

Figure 2. Future layout of Chamoun's.

The research report, prepared after a careful investigation of all available resources, is similar to what some students refer to as the research paper or term paper. The research report requires note-taking, documentation, and the like. For a full discussion, see Chapter 9, The Research Report.

General Principles in Giving a Report

1. *A report must be accurate, clear, concise, and objective.*
2. *Reports may be classified according to the function they serve.* They may be school reports, which are a testing device for students. They may be professional reports, which are used in business, industry, and government.
3. *Reports may be given in various formats.* These formats include printed forms, memorandums, letters, conventional report formats (nonformal or formal), and oral presentations.
4. *Reports frequently include visuals.* Visuals may make information clearer, more easily understood, and more interesting.
5. *Reports usually include headings.* Headings reflect the major points of the report and their supporting points.
6. *Layout and design are integral aspects of effective reports.*
7. *Common types of reports include the reading report, the periodic report, the progress report, the investigative report, the observation report, the feasibility report, the proposal, and the research report.*

Procedure for Giving a Report

1. *Determine the audience and the purpose of the report.* It is essential to know who will be reading or listening to the report and why. Plan the report accordingly.
2. *Review the common types of reports.* Think about whether the report you are preparing might be classified as a reading report, a periodic report, a progress report, an investigative report, an observation report, a feasibility report, a proposal, or a research report, or a combination of two or more of these types.
3. *Investigate and research the subject, being careful to take accurate notes and list sources used.* Make use of all available sources—print and nonprint materials, knowledgeable people, experiments, observation.
4. *Select an appropriate format for the report.* Decide whether the report will be oral or written or both. Decide whether to use a printed form, a memorandum, a letter, or a conventional report format.
5. *Organize the report, using headings.* Arrange the material in a logical order, as determined by audience, purpose, and subject matter.
6. *Plan and prepare visuals.* Decide which information should be given in visuals and what type of visuals to use. Then carefully prepare the visuals and place them in the report where they are most effective.

7. *Plan the layout and design of the report.* Think through how the text and the visuals can be most effectively placed on the page, and think through how you want each page and the report as a whole to appeal visually to the reader.

8. *Remember that a report should be accurate, clear, concise, and objective.*

Application 1   Giving Reports

In a standard desk dictionary, look up the word *report;* list all the meanings that are given. Indicate the meanings that you think are related to report writing.

Application 2   Giving Reports

Interview at least four people who are employed in business, industry, or government. Find out the kinds of writing they must do. Determine which of these kinds of writing could be classified as report writing. Present your findings in a one-page report.

Application 3   Giving Reports

Make a collection of at least five reports. Using the criteria of accuracy, clarity, conciseness, and objectivity, analyze the reports. Present your analysis in a one-page report.

Application 4   Giving Reports

A. Make a reading report on a chapter in one of your textbooks.

or

B. Find an article in a current periodical related to your major field of study. Make a reading report on the article.

a. Fill in the Plan Sheet on pages 417–418.
b. Write a preliminary draft.
c. Revise. See inside back cover.
d. Write the final draft.

Application 5   Giving Reports

A. Write a daily periodic report on your activities in office, field, shop, or laboratory. Then expand this report to a weekly periodic report on your activities in office, field, shop, or laboratory. Write the report in the form of a memo-randum.

or

B. Assume that you have an annual scholarship of $1800 from a company related to your major field of study. Every three months you must report how you have spent any or all of the money and how you intend to spend any remaining portion. Write the report in the form of a letter.

  a. Fill in the Plan Sheet on pages 419–420.
  b. Write a preliminary draft.
  c. Revise. See inside back cover.
  d. Write the final draft.

Application 6   Giving Reports

A. Write a report showing your progress toward completing a degree or receiving a certificate in your major field.

or

B. Write a weekly progress report on your work in a specific course or project (math, science, research, specific technical subject). Using the same course or project, write a monthly progress report on your work.

  a. Fill in the Plan Sheet on pages 421–422.
  b. Write a preliminary draft.
  c. Revise. See inside back cover.
  d. Write the final draft.

Application 7   Giving Reports

A. Write an investigative report on an experiment or test.

or

B. Assume that you have been experimenting with three different brands of the same product or piece of equipment in an effort to decide which brand name you should select for your home, shop, or lab. Write an investigative report to be sent to all three manufacturers.

  a. Fill in the Plan Sheet on pages 423–424.
  b. Write a preliminary draft.
  c. Revise. See inside back cover.
  d. Write the final draft.

Application 8 Giving Reports

A. Visit a business or industry related to your major field of study, a governmental office or department, or an office or division of your college, to survey its general operation. Write an observation report on your visit.

or

B. Write an observation report including conclusions and recommendations on one of the topics below or on a similar topic of your own choosing.

1. Parking problems on campus
2. On-the-job training for laboratory technicians or persons in other health-related occupations
3. Condition of a structure (building, dam, fire tower, etc.) or piece of equipment
4. Efficiency of the registration process on campus
5. Conditions at a local jail, prison, hospital, rehabilitation facility, mental institution, or other public institution
6. The present work situation of last year's graduates in your major field of study
7. Employment opportunities in your major field of study
8. Services by an organization such as the Better Business Bureau or the Chamber of Commerce

a. Fill in the Plan Sheet on pages 425–426.
b. Write a preliminary draft.
c. Revise. See inside back cover.
d. Write the final draft.

Application 9 Giving Reports

Write a feasibility report including conclusions and recommendations on one of the topics below or on a similar topic of your own choosing.

1. Purchase or renting of an item, such as a car, computer, or apartment
2. Selection of a site for a business or an industry

a. Fill in the Plan Sheet on pages 427–428.
b. Write a preliminary draft.
c. Revise. See inside back cover.
d. Write the final draft.

Application 10 Giving Reports

Write a proposal on one of the topics below or on a similar topic of your choosing.

1. Need for purchasing new equipment for a laboratory or shop

2. Rearrangement of equipment in a laboratory or shop
3. A color scheme and furnishings for lounges in a college union building
4. A landscaping plan, a type of air conditioning system, an interior decorating plan (or some other area related to your major field of study) for a housing project
5. A system of emergency exit patterns and directions for several campus buildings

a. Fill in the Plan Sheet on pages 429–430.
b. Write a preliminary draft.
c. Revise. See inside back cover.
d. Write the final draft.

Application 11   Giving Reports

As directed by your instructor, adapt the reports you prepared in the applications above for oral presentation. Ask your classmates to evaluate your speech by filling in an Evaluation of Oral Presentations from pages 527–535.

USING PART II: SELECTED READINGS

Application 12   Giving Reports

Read "Clear Writing Means Clear Thinking Means . . . ," by Marvin H. Swift, pages 616–620. What points in this article are important in writing reports?

Application 13   Giving Reports

Read "The World of Crumbling Plastics," by Stephen Budiansky, pages 622–623, and "Starch-Based Blown Films," by Felix H. Otey and others, pages 624–630. Compare the two reports from the standpoint of content, purpose, and audience.

PLAN SHEET
FOR GIVING A READING REPORT

Analysis of Situation Requiring a Report

What is the subject of the report?

For whom is the report intended?

How will the report be used?

In what format will the report be given?

When is the report due?

Identification of Author and Work

Summary

Analysis

Evaluation

Types and Subject Matter of Visuals to Be Included in the Report

PLAN SHEET
FOR GIVING A PERIODIC REPORT

Analysis of Situation Requiring a Report

What is the subject to be reported?

For whom is the report intended?

How will the report be used?

In what format will the report be given?

When is the report due?

Type of Periodic Report (daily, weekly, monthly, etc.)

Dates covered:

Information to Be Reported

Work Date/Time: Work Completed:

Types and Subject Matter of Visuals to Be Included in the Report

Sources of Information

PLAN SHEET
FOR GIVING A PROGRESS REPORT

Analysis of Situation Requiring a Report

What is the subject to be reported?

For whom is the report intended?

How will the report be used?

In what format will the report be given?

When is the report due?

Purpose of the Project

Work Completed; Comments

Work to Be Completed; Comments

Types and Subject Matter of Visuals to Be Included in the Report

Sources of Information

PLAN SHEET
FOR GIVING AN INVESTIGATIVE REPORT

Analysis of Situation Requiring a Report

What is the subject to be reported?

For whom is the report intended?

How will the report be used?

In what format will the report be given?

When is the report due?

Object or Purpose of the Study

Theory, Hypothesis, Assumption

Method or Procedure

Results; Observations; Comments

Conclusions; Recommendations

Types and Subject Matter of Visuals to Be Included in the Report

PLAN SHEET
FOR GIVING A FEASIBILITY REPORT

Analysis of Situation Requiring a Report

What is the subject to be reported?

For whom is the report intended?

How will the report be used?

In what format will the report be given?

When is the report due?

Background, Purpose, Definition, Scope of the Study

Method or Procedure

Results; Observations; Comments

Conclusions; Recommendations

Types and Subject Matter of Visuals to Be Included in the Report

PLAN SHEET
FOR GIVING A PROPOSAL

Analysis of Situation Requiring a Proposal

What is the subject of the proposal?

For whom is the proposal intended?

How will the proposal be used?

In what format will the proposal be given?

When is the proposal due?

Purpose of the Proposal

Problem

Proposed Solution or Plan, Including Method or Procedure

Conclusion

Other Information

Types and Subject Matter of Visuals to Be Included in the Proposal

Chapter 9

The Research Report: Becoming Acquainted with Resource Materials

Objectives

Upon completing this chapter, the student should be able to:

- Select an appropriate subject for a research report
- State the general problem being investigated in a report
- Divide the general problem into specific problems or questions (preliminary outline)
- Make a working bibliography
- Use the public (card) catalog
- Use periodical indexes
- Locate source materials for a research report
- Write a direct quotation note card
- Write a paraphrase note card
- Write a summary note card
- Formulate the central idea of a research report
- Make an outline for a research report
- Document a research report
- Write a complete research report

Introduction

Some of the writing and speaking you will do as a student—term papers, book reviews, reports on various topics, speeches—will require research. Also, assignments you may be given as an employee in business or industry—feasibility reports, proposals, process explanations, or even a speech to a local civic club—may require research. The information gathered through research can be organized into a final presentation to accomplish a stated purpose for a specific audience.

To research a topic thoroughly, you need to know the possible sources of information and how to use these sources effectively.

This chapter provides a general guide to library research, one type of research that can be used to prepare and write a research report. It does not attempt, however, to deal with the more intricate points in library research or with the methods used in pure research or in scientific investigation, two other types of research. See pages 624–630 for an example of a scientific report. It does not attempt to present every acceptable method and form for writing a research report. It does attempt to present in a logical, step-by-step sequence a procedure for researching and writing a successful report.

The first part of the chapter explains basic library research, and the second part outlines and explains the procedure for writing a research report.

Locating Materials

Materials about a subject are usually located by a systematic, organized search of available sources. Therefore, to begin research, you must first have some idea of possible sources of information and know how to make a systematic search of the sources.

Sources of Information

Possible sources of information can be classified as:

1. Personal observation or experience
2. Personal interviews
3. Free or inexpensive materials
4. Library materials

An investigation of these general sources will help you to compile a list of specific sources—books, periodicals, people, agencies, companies—that can supply needed information about a subject or topic.

PERSONAL OBSERVATION OR EXPERIENCE

Personal observation or experience can make writing more realistic and vivid, and sometimes this observation or experience is essential. If writing, for instance, about the need for more up-to-date laboratory equipment in local technical education centers, personal inspections would be quite helpful.

Do not be misled, however, by surface appearances. Remember that the same conditions and facts may be interpreted in an entirely different way by another person. For this reason, do not depend completely on personal observations or personal experiences for information. Every statement given as fact in a research presentation must unquestionably be based on validated information.

PERSONAL INTERVIEWS

Interviewing persons who are knowledgeable about a topic lends human interest to the research project. Talking with such people may also prevent chasing up blind alleys and may supply information unobtainable elsewhere.

Since methods of getting facts through interviews require almost as much thought as the methods involved in library work, interviewing should be carried out systematically. The first thing to remember is courtesy. In seeking an appointment for an interview, request cooperation tactfully. Second, give thoughtful preparation to the interview. Be businesslike in manner and prepare your questions in advance. Finally, keep careful notes of the interview. In quoting the person interviewed, be sure to use his or her exact words and intended meaning. And remember, although the personal interview is a valuable source of information, it offers only one person's opinions. (See Chapter 10, Oral Communication, pages 522–523, for a discussion of how to conduct an informational interview.)

FREE OR INEXPENSIVE MATERIALS

Many agencies distribute free or inexpensive pamphlets, documents, and reports that contain much valuable information. The U.S. Government Printing Office, the various departments of the United States Government, state and local agencies, industries, insurance companies, labor unions, and professional organizations are just a few of the sources that may supply excellent material on a subject.

Not at all unusual is the case of a student writing about Teflon who requested information from the DuPont Company. In her letter she stated that as a student in mechanical technology she was writing a library research paper about industrial uses of Teflon. Within a few days the student received a packet of extremely helpful materials. Some of the material was so new that it was months before it began to appear in periodicals. Some of the material consisted of reprints of magazine articles that the student did not have access to. And some of the individual pieces of material she received would never appear in magazine or book form.

In seeking material from various agencies, make the request specific. For instance, a student writing a research report about aluminum as a structural building material might be disappointed in the response to a request to Alcoa for "information about aluminum." Such a request is difficult if not impossible for the company to fill because the particular need of the writer is not specified. A request for "information about aluminum as a structural building material," however, specifies the subject and thus encourages a more satisfactory response.

In requesting materials, consider the time element. Do not assume that all materials will arrive immediately or that all or even most of the information for a research report will come from these requested materials. Depending too much on material to arrive before the due date of the report could be disastrous.

LIBRARY MATERIALS

Much research is carried out in a library, whether a school library, a public library, or a specialized library such as a medical library, a business library, or a law library. The library contains a wealth of information in printed materials—books, pamphlets, newspapers, and other periodicals—and in audiovisual materials—tapes, films, microforms, recordings. But this information is useless to the researcher until it is found and used to accomplish a purpose.

To simplify finding specific materials within the total library collection, which may be thousands or even millions of items, libraries use a system of classifying, cataloging, and indexing information. With a knowledge of the system and a little curiosity, the researcher is well on the way to finding needed information.

Systems of Classifying Books. Libraries classify books according to one of two systems: the Dewey Decimal Classification or the Library of Congress Classification.

The older system, the one that most smaller libraries follow, is the Dewey Decimal Classification, which uses numbers to divide all books into ten basic groups:

000	General works
100	Philosophy and related disciplines
200	Religion
300	Social sciences
400	Language
500	Pure sciences
600	Technology (Applied sciences)
700	The arts

800   Literature (belles-lettres)

900   General geography, biography, history

These basic groups are divided and then subdivided numerous times.

In addition to the books in the first group, 000 General works, the books that the technical student will be using most often are those in the 500s, 600s, and 700s. The major divisions of these groups are as follows:

500   Pure sciences
    510   Mathematics
    520   Astronomy and allied sciences
    530   Physics
    540   Chemistry and allied sciences
    550   Sciences of earth and other worlds
    560   Paleontology, paleozoology
    570   Life sciences
    580   Botanical sciences
    590   Zoological sciences
600   Technology (Applied sciences)
    610   Medical sciences, medicine
    620   Engineering and allied operations
    630   Agriculture and related technologies
    640   Home economics and family living
    650   Management and auxiliary services
    660   Chemical and related technologies
    670   Manufacturers
    680   Manufacture of products for specific uses
    690   Buildings
700   The arts
    710   Civic and landscape art
    720   Architecture
    730   Plastic arts and sculpture
    740   Drawing, decorative and minor arts
    750   Painting and paintings
    760   Graphic arts, print making and prints
    770   Photography and photographs
    780   Music
    790   Recreational performing arts

The Library of Congress Classification, used by many large libraries and all government libraries, designates books by letter-number combinations. The 20 main classes are these:

A   General works
B   Philosophy, psychology, religion
C   History, auxiliary sciences
D   History and topography (except America)

E and F   American history

      G   Geography, anthropology

      H   Social sciences

      J   Political science

      K   Law

      L   Education

      M   Music

      N   Fine arts

      P   Languages and literature

      Q   Science

      R   Medicine

      S   Agriculture, plant and animal industry

      T   Technology

      U   Military science

      V   Naval science

      Z   Bibliography, library science

The following example shows how the book *Lasers and Their Applications* would be classified using the Library of Congress and the Dewey Decimal designations.

Library of Congress: TK7872.L3S7
Dewey Decimal: 621.329

Public (Card) Catalog. The single most useful item in the library is the public catalog, formerly called the card catalog. The public catalog, once available only in the form of 3- by 5-inch cards, is increasingly being made available in such electronic forms as computer-output-microform (COM) catalogs and/or on-line computer systems.

Microfiche, usually referred to as fiche, is a sheet of microfilm typically 4 by 6 inches and capable of accommodating and preserving a considerable number of pages in reduced form. One 4- by 6-inch fiche can, on the average, accommodate the information from 1600 3- by 5-inch cards in the card catalog. The COM catalog is produced in this way: The information contained in the public catalog is input and stored in a computer to which new acquisitions are routinely added. Periodically, a computer tape of the entire catalog is retrieved and reproduced on either microfiche or microfilm. An on-line computer system makes available quickly and accurately millions of items of information stored in a central computer. A terminal links the user to the computer.

The entries in the public (card) catalog are arranged alphabetically according to the first important word in the heading (first line) on the entry. The heading of each entry is determined by the type of entry: subject entry, author entry, or title entry. Only the heading for each entry differs; the other information is the same.

A typical entry in the public (card) catalog contains the following information (see the corresponding colored numbers on the last sample entry on page 439):

1. Call number. (This is the designation used for classifying and shelving the book.)
2. Heading. (This is a subject entry; therefore the subject, typed in red or in all capital letters if on a card, is on the first line.)
3. Author's name, usually followed by year of birth and year of death. (A dash—indicates that the author was still living at the time the card was printed.)
4. Complete title of book
5. City of publication. (If New York were not a well-known city, the state would also have been given.)
6. Publishing company
7. Date of publication
8. Number of pages in the book. (Roman numeral indicates pages of introductory material.)
9. Height in centimeters (a centimeter is 0.4 inch)
10. International Standard Book Number
11. Cost of the book
12–13. Other headings by which the book is cataloged
14. Library of Congress catalog card number

As you begin looking in the public (card) catalog for books containing information on your subject, you may not know any authors or titles to consult. Therefore, look for subject entries. Suppose you were writing a paper entitled "Medical Applications of the Laser." If you do not know any title or author concerning the subject, first look in the "L" section for "Laser" and "Lasers." However, do not stop after looking under these two subject headings; look also under other related subject headings. Some of your most important information might be cataloged under such subjects as "Medicine" or "Surgery" or "Physics."

Sample Author, Title, and Subject Entries

Author Entry

```
309.262
C93t    Currie, Lauchlin Bernard.
            Taming the megalopolis: a design
        for urban growth/by Lauchlin Currie.
        --2nd ed.--Oxford; New York: Pergamon
        Press, c1983.
            ix, 127 p.; 26 cm.
```

Title Entry

439

The Research
Report: Becoming
Acquainted with
Resource Materials

```
309.262      Taming the megalopolis.
C93t      Currie, Lauchlin Bernard.
             Taming the megalopolis: a design
          for urban growth/by Lauchlin Currie.
          --2nd ed.--Oxford; New York: Pergamon
          Press, c1983.
             ix, 127 p.; 26 cm.
```

Subject Entry
(Numbers in color added for explanation; see corresponding list of items on
page 438.)

```
 1309.262   2Cities and towns-Planning-1945-
    C93t    3Currie, Lauchlin Bernard.
             4Taming the megalopolis: a design
          for urban growing/by Lauchlin Currie.
          --2nd ed.--Oxford; 5New York 6Pergamon
          Press,
         7c1983    8        9
             ix, 127 p.; 26 cm.
                                    11
            10ISBN 0-08-020980-7: $14.10

         12-131. Cities and towns--Planning--1945-
          I. Title
                             1483-6848
                             MARC
```

```
WORD PROCESSING.
  Lindsell, Sheryl L.
    Proofreading and editing for word
processors / Sheryl L. Lindsell. New York :
Arco Pub., c1985.
    vi. 120 p. : ill. ; 27 cm.
mll235621
    0668060883 (hard). -- 0668060921
    (pbk.)
    1. Word-processing. 2. Editing. 3.
Proofreading I. Title.
652.5 L64p RAY
```

Note that the call number, rather than appearing in the upper left-hand corner, is given at the end of the entry (RAY indicates the location in the library system that has one or more copies of the book).

Library of Congress Subject Headings. Particularly useful in looking up subject entries is the *Library of Congress Subject Headings* (and its supplements). This book gives subject descriptors used by the Library of Congress, and subsequently by most other libraries. An excerpt indicating some of the subject descriptors for "Laser" and "Lasers" is given below and on the next page.

EXCERPT FROM *LIBRARY OF CONGRESS SUBJECT HEADINGS*

• • •

Laser beams
  sa Laser fusion
        Laser-plasma interactions
        Laser plasmas
        Laser pulses, Ultrashort
  x   Beams, Laser
        Laser radiation
  —Atmospheric effects *(QC976.L)*
        x Atmospheric effects on laser
             beams
        xx Meteorological optics
  —Diffraction *(QC446.2)*
  —Scattering
        sa Laser speckle

• • •

Laser coagulation
  sa Lasers in surgery
  x   Laser photocoagulation
  xx Lasers in surgery
        Light coagulation

• • •

Lasers *(Indirect)*
  sa Astronautics—Optical
             communication systems
        Atmosphere—Laser observations
        Carbon monoxide lasers
        Cartography—Laser use in
        Chemical lasers
        Dye lasers
        Far infrared lasers
        Fingerprints—Laser use in
        Free electron lasers
        Gamma ray lasers
        Gas lasers
        Injection lasers
        Laser industry
        Laser interferometer
        Laser recording
        Laser transitions
        Lasers in controlled fusion
        Lasers in plasma diagnostics
        Lasers in plasma research

Lasers   *(Indirect)*
    Lunar laser ranging
    Metallurgy—Laser use in
    Mode-locked lasers
    Molecular gas lasers
    Negative temperature
    Neodymium glass lasers
    Nonlinear optics
    Nuclear-pumped lasers
    Plasma lasers
    Precipitation (Meteorology)—
      Laser observations
    Printing, Practical—Laser use in
    Rare earth lasers
    Ruby lasers
    Semiconductor industry—Laser
      use in
    Semiconductor lasers
    Solid-state lasers
    headings beginning with the word
      Laser
  x Light amplification by stimulated
      emission of radiation
    Masers, Optical
    Optical masers
 xx Astronautics—Optical
      communication systems
    Infrared sources
    Light
    Light amplifiers
    Light sources
    Nonlinear optics
    Optical pumping
    Optroelectronic devices
    Photoelectronic devices
    Photons
—Diagnostic use
  xx Diagnosis
      Lasers in medicine
—Effect of radiation on
  x Lasers, Effect of radiation on
  xx Radiation

—Military applications   *(UG486)*
  xx Military engineering
      Weapons systems
—Mirrors
  x Laser mirrors
  xx Mirrors
—Resonators
  x Laser resonators
  xx Resonators
—Windows
  x Laser windows
      Windows, Laser

• • •

Lasers in controlled fusion
  xx Controlled fusion
      Lasers
Lasers in isotope separation
  xx Isotope separation
Lasers in medicine   *(R857.L37)*
  sa Holography in medicine
      Lasers—Diagnostic use
      Lasers in surgery
  xx Medical instruments and
      apparatus
Lasers in mining   *(TN292)*
  xx Mining engineering
Lasers in physics
  xx Physics
Lasers in plasma diagnostics
  xx Lasers
      Plasma diagnostics
Lasers in plasma research
  xx Lasers
      Plasma (Ionized gases)—Research
Lasers in surgery *(RD73.L3)*
  sa Laser coagulation
  xx Laser coagulation
      Lasers in medicine
      Surgical instruments and
      apparatus

• • •

SYMBOLS

    sa (see also)   indicates a reference to a related or subordinate topic

    x (see from)   indicates a reference from a term not itself used as a heading

  xx (see also from)   indicates a related heading from which a *sa* reference is made

As you begin looking in the public (card) catalog for sources of information, first check to see if the alphabetizing is letter by letter or word by word. Most dictionaries alphabetize letter by letter, but encyclopedias and indexes may use either alphabetical order or some other order such as chronological, tabular, regional. Knowing this information is essential; otherwise you might incorrectly assume the library has no materials on a subject. The following example illustrates these two methods of alphabetizing:

Letter by Letter	Word by Word
art ballad	art ballad
art epic	art epic
article	art lyric
art lyric	art theatre
art theatre	Arthurian legend
Arthurian legend	article
artificial comedy	artificial comedy
artificiality	artificiality

INDEXES

Indexes are to periodicals (magazines, newspapers) what the public (card) catalog is to books. By consulting indexes to periodicals, specific sources of information on a subject can be found without looking through hundreds of thousands of magazines.

Much of any needed information on a technical or business subject is likely to come from magazines. For one thing, there is a great deal of information in magazines that is never published in book form. Furthermore, because the writing and publishing of a book usually takes at least a year, the information in magazines is likely to be much more current.

At the beginning of any index, there are directions for use, a key to abbreviations, and a list of periodicals indexed.

Magazine Index. The *Magazine Index* is on microfilm. Because of its coverage, organization, and ease of use, the *Magazine Index* makes magazine research simple. It is a "continuous" five-year index of all significant items in some 400 of the most popular magazines in America. It includes all articles indexed in the *Readers' Guide to Periodical Literature* (discussed later in this chapter) and more. The *Magazine Index* indexes articles, short stories, poems, biographical pieces, recipes, reviews (books, movies, records, restaurants, concerts, etc.), product evaluations, editorials, and the like. Coverage began in 1976.

The *Magazine Index* provides a five-year "rolling" cumulation; that is, as a new month is added, the corresponding month five years previous is deleted from the microfilm. Cumulative microfiche of deletions are provided periodically, so the subscriber will have a complete listing.

Given on the next page are printed excerpts from the *Magazine Index* for the heading WORD PROCESSING for the period 1 June 1982–1 November 1986.* The some 600 entries in the full listing were categorized under 56 subheadings.

---

* *Magazine Index.* Copyright © 1986 by Information Access Company. Material reproduced by permission of the publisher.

WORD PROCESSING

The write stuff. (advertisement) Savvy
v6-March '85-pA56(1)

—ACHIEVEMENTS AND AWARDS

1982 - WP executive: pioneering on an
international scale. (Mary Jo Greil) il
Modern Office Procedures v27-June
'82-p62(2)   19J2271

—AIDS AND DEVICES

G-Spell: $9.95 spelling checker.
(computer program) (evaluation) il
PC Magazine v5-Oct 14 '86-p59(1)

Text entry without using the
keyboard. (evaluation) il Personal
Computing v10-July '86-p134(1)

• • •

—ANALYSIS

New ways to outline your ideas. (idea
processors can assist organization) il
Personal Computing v10-Sept
'86-p79 (5)

The vagaries of WYSIWYG. (word
processing) (column) by Peter
Norton PC v4-Sept 3
'85-p83(3)   28K2879

Getting started with word processing
by M. David Stone il Science
Digest v93-June '85-p59(2)   32A6506

• • •

—APPRECIATION

Learning to tough type. (editorial) by
William F. Buckley Jr. il Computers
& Electronics v22-Aug
'84-p24(3)   23K1696

The right wording. (word processors)
(column) by Ken Uston il Penthouse
v15-Sept '83-p48(1)

—AUTOMATION

IBM lowers prices on Displaywriter
and several other office products.
Office Administration and
Automation v44-Aug
'83-p17(1)   19B0713

• • •

—CASES

Convergent, Savin reach accord.
Office Administration and
Automation v44-March
'83-p13(1)   16M5834

—COMPUTER PROGRAMS

First Choice integrates ease, economy,
options. (integrated software)
(evaluation) il PC Magazine v5-Oct
14 '86-p33(3)

• • •

—DO-IT-YOURSELF

Word processing: 7 commandments.
Writer's Digest v64-Feb '84-p53(2)

—ECONOMIC ASPECTS

Commit or hold off? The short- vs. the
long-term tug of war. (office automation)
il Administrative Management v43-Sept
'82-p24(4)   12M3880

• • •

—PHYSIOLOGICAL ASPECTS

Precautions must be taken to eliminate
eyesight health hazards for operators
of video display terminals. (column)
Administrative Management v43-Oct
'82-p74(1)   09B1567

—PLANNING

Administrative managers tell 'my
biggest challenge.' il Administrative
Management v43-Oct
'82-p30(3)   09B1525

• • •

—SECURITY MEASURES

Theft of computer time (or, is your
operator secretly writing romances?)
il Office Administration and
Automation v44-Nov
'83-p41(3)   19K4395

—SERVICES

High-tech temps; sophisticated skills
have earned temporary workers a
permanent spot in the automated
office. il PC Week v2-Aug 20
'85-p45(4)   30C6376

Typing out a $1 million idea. (Money
Makers) (starting a word processing
company) (column) by Leslie
Laurence il Money v14-May
'85-p27(2)   27K5730

• • •

The last entry in the *Magazine Index* excerpts shows that "Typing Out a $1 Million Idea" is in the column Money Makers with the subtitle "Starting a Word Processing Company." By Leslie Laurence, the illustrated (il) article appears in *Money*, volume 14, the May 1985 issue, on page 27. The article, for copying, is two pages and in the "Magazine Collection" (see following paragraph) is number 27K5730.

Magazine Index, for all practical purposes, is one of the first resources to which a student researcher should turn.

☐ The headings, subheadings, and cross references can be quite valuable in suggesting subject areas and topics for a library research report. In particular, the subheadings with subsequent entries indicate restricted, usable possibilities for report topics and titles.

☐ Headings in *Magazine Index* conform to the *Library of Congress Subject Headings*. Magazine search can easily be coordinated with search for material in the public (card) catalog.

☐ The chronological listing of entries under each subheading reflects the order of most recent information first.

☐ "Hot Topics" aids in quickly locating reference to popular issues. A spin-off of *Magazine Index*, "Hot Topics" is a quarterly printout (in a loose-leaf binder) of current magazine articles on such topics as abortion, child abuse, computer crime, drug testing, and terrorism.

☐ The "Magazine Collection" gives ready reference to selected articles. The "Magazine Collection" consists of sets of 16 mm microfilm cartridges containing the complete texts of articles from many periodicals indexed in the *Magazine Index*. The "Magazine Collection" dates back to 1980 and is updated twice a month.

Readers' Guide to Periodical Literature. The *Readers' Guide* is a general index of over 180 leading popular magazines published from 1900 to the present. There are two issues a month except for January, February, May, July, August, and November, when there is only one issue a month. The second issue of each month includes the material from the first issue of that month. There are cumulative issues every three months, and a bound cumulation each year. This cumulation saves the researcher from having to look in so many different issues and keeps the index up to date.

The main body of the *Reader's Guide* is a listing, by subject and by author, of periodical articles. Each entry gives the title of the article, the author (if known), the name of the magazine, the volume, the page numbers, and additional notations for such items as bibliography, illustration, or portrait. Following the main body of the index is an author listing of citations to book reviews.

The excerpt on the next page from the August 1986 issue of the *Readers' Guide to Periodical Literature* is the complete listing for the subject "Word processors and processing."*

The first entry gives this information: title of article—"Can We Talk? [R. Kurzweil's VoiceWriter]" (annotation inserted in brackets by the editors of *Readers'*

* *Readers' Guide to Periodical Literature*. Copyright © 1986 by the H. W. Wilson Company. Material reproduced by permission of the publisher.

Word processors and processing

Can we talk? [R. Kurzweil's VoiceWriter] G. M. Henry. il por *Time* 127:54 Ap 28 '86

A lower-tech solution for home word processing. G. Lewis. *Bus. Week* p72 Je 9 '86

The right tool for the writing job. C. J. Mullins. *Work Woman* 11:80+ Ap '86

Word machines—a new touch to your typing. W. J. Hawkins. il *Pop Sci* 228:92-4+ My '86

Poetry use

Poetry processing: computers and the mundane tasks of prosody. K. E. Eble. *Change* 18:62 My/Je '86

Programming

Lettrix, A. R. Miller, il *Byte* 11:299–300+ My '86

Low-cost spreadsheet with word processing [Farsight] M. Antonoff. *Pers Comput* 10:178 My '86

Making the numbers look good [spreadsheet/word processor combinations] M. S. Downing. *Pers Comput* 10:225 Je '86

Producing efficient forms. M. Liskin. il *Pers Comput* 10:39+ My '86

Scientific word processor [Tech/Word] *Byte* 11:44 My '86

Stepping up to a laser. L. Wood. il *Pers Comput* 10:87–91+ Ap '86

Two top word processors [Microsoft Word and WordPerfect] il *Pers Comput* 10:31+ Je '86

Upgrade fever: WordStar 2000 Release 2, dBase III Plus, Volkswriter 3, and KeepTrack Plus. E. Shapiro. *Byte* 11:329–30+ Je '86

Word Finder—an electronic thesaurus. *Radio-Electron* 57 Computer Digest:5 May '86

Scientific use

Scientific word processor [Tech/Word] *Byte* 11:44 My '86

Study and teaching

People are still the best computer components around [M. C. Holmes trains Manpower's staff] E. Alvarez. il por *Work Woman* 11:38 My '86

Guide to identify the particular word processor); author—G. M. Henry; illustration(s) and portrait(s); magazine—*Time*; volume 127; page 54; date—April 28, 1986.

With cumulative quarterly and yearly issues, the *Readers' Guide* is an easy-to-use, up-to-date source of articles in general interest magazines such as *Nation s Business, Popular Mechanics, Reader's Digest, Newsweek, Time, Health, Today's Education, Architectural Record, Flying,* and *Car and Driver.*

Readers' Guide Abstracts. The *Readers' Guide Abstracts* contains on typeset microfiche abstracting of all the articles and product reviews contained in *Readers' Guide* as well as all of the cross-references. Each abstract is a concise summary which presents the major points, facts, and opinions of the original article. Updated and cumulated every six weeks, each cumulation of *Readers' Guide Abstracts* contains more than one year of retrospective coverage. The first cumulation was issued in the fall of 1985 and contained retrospective coverage to September 1984. The cumulation cycle of *Readers' Guide Abstracts* is coordinated with the publication cycle of *Readers' Guide.*

Applied Science & Technology Index. The *Applied Science & Technology Index* indexes about 400 English language periodicals in the fields of aeronautics and space service, atmospheric sciences, chemistry, computer technology and applications, construction industry, energy resources and research engineering, engineering, fire and fire prevention, food and food industry, geology, machinery, mathematics, metallurgy, mineralogy, oceanography, petroleum and gas, physics, plastics, textile industry and fabrics, transportation, and other industrial and mechanical arts. The main body of the *Index* lists subject entries to periodical articles. Additionally, there is an author listing of citations to book reviews.

First published in 1958 (until 1957 it was part of the *Industrial Arts Index*), the *Applied Science & Technology Index* is issued monthly except July, and has quarterly and annual cumulations.

The arrangement of entries in the *Applied Science & Technology Index* is similar to that in the *Readers' Guide to Periodical Literature*.

Below are excerpts of entries under "Text editors" from the September 1986 issue of the *Applied Science & Technology Index*.* (The entry "Word processors" was followed by "*See* Text editors [Computer programs].")

Text editors (Computer programs)

A language for music printing. J. S. Gourlay. bibl diags *Commun ACM* 29:388–401 My '86

Lettrix [print processor] A. R. Miller, *Byte* 11:299–300+ My '86; Discussion. 11:319 Jl '86

A note on undetected typing errors. J. L. Peterson, bibl *Commun ACM* 29:633–7 Jl '86

Now its easy to process Chinese characters. T. Naegele, *Electronics* 54:40+ Jl 24 '86 pfs: WRITE—simplified word processor. *Radio-Electron* 57:CD5 Jl '86

Uniforms: an automatic forms facility. M. E. C. Hull and T. P. Wilson. bibl diags *Data Process* 28:258–64 Je '86

Will Britain's Amstrad make it big in the U.S.? D. Boothroyd, il *Electronics* 59:53+ Je 23 '86

Word-processing software, not a CAE package, is an engineer's most valued assistant. D. Asbrand. il *EDN* 31:203–4 F 6 '86

XyWrite III. J. Brown. *Pollut Eng* 18:42 Ap '86

The last entry in the excerpt indicates that the article "XyWrite III," by J. Brown, appears in *Pollution Engineering*, volume 18, page 42, in the April 1986 issue.

Business Periodicals Index. This is an index to over 300 English language periodicals in the fields of accounting, advertising and marketing, banking, building and buildings, chemical industry, communications, computer technology and applications, drug and cosmetic industries, economics, electronics, finance and investments, industrial relations, insurance, international business, management and personnel administration, occupational health and safety, paper and pulp industries, petroleum and gas industries, printing and publishing, public relations, public utilities, real estate, regulation of industry, transportation, and other specific businesses, industries, and trades.

The main body of the *Index* lists subject entries to business periodical articles. Additionally, there is an author listing of citations to book reviews.

The *Business Periodicals Index* is published monthly except August and has quarterly and annual cumulations. It is very similar in format and arrangement to the *Readers' Guide* and the *Applied Science & Technology Index*.

The general subject headings are divided into subclassifications; thus, locating articles on a particular topic is easy. These subclassifications of a general subject may be useful to the student in another way: They may serve as a guide to a suitable subject for a library research report.

Below are excerpts from the July 1986 issue of the *Business Periodicals Index*.*

Word processing
> *See also*
> Dictation
> Text processing (Computer science)

6th annual word processing update [special report] graphs tab *Manage World* 15:46–9 Ap-My '86

• • •

Costs
Word processors: what do they really cost? R. Fetzer. *Office* 103:71 F '86

Employees
The effect of technology on solving human problems. J. Joner. *Office* 102:120 N '85

Technical temps—a growing trend. M. J. Paznik, graph *Adm Manage* 47:25–9 F '86

Management
Handling the hurdles of word processing supervision. P. C. Lukanen. *Office* 102:104–5+ N '85

Supplies
> *See also*
> Word processing equipment

Word processing equipment
> *See also*
> Electronic typewriters
> Work stations (Office automation)

Is your word processing equipment now old hat? P. C. Lukanen. *Office* 103:80+ F '86

• • •

Prices
Re-enter the electronic typewriter [and personal computer software] S. M. Unley. *Manage World* 15:46–7 Ap-My '86

Printers
See Printers (Data processing systems)

Programs
See Word processing programs

Rating
The Amstrad PCW 8256 [Z80-based personal computer and word processor] D. Pountain. il *Byte* 11:333–4+ Mr '86

Security measures
Safeguarding your WP system? It can be done. L. H. Grosman. *Office* 102:156+ N '85

Specifications
Buyers' guide to word processors [table] *Office* 102:137–8 N '85

Word processing equipment industry
> *See also*
> Data General Corp.
> Harris Corp.
> NCR Corp.
> Protype Corporation
> Wang Laboratories, Inc.

Suits and claims
CPT files suit against Info. Storage Int'l. J. Schroeder. *Electron News* 32:28 Mr 24 '86

Word processing programs
DEC's WPS-PLUS/PC integrates IBM PC with VAX systems. diag *Adm Manage* 46:63 D '85

Homogenize your word processors [compatibility] M. J. Paznik. *Adm Manage* 47:14 F '86

• • •

Rating
Braille-Edit. H. Brugsch. *Byte* 11:251-2+ Mr '86

Low-cost micro WP packages top test. *Comput Decis* 18:24 Ja 28 '86

Samna Word III. B. Hodkinson. tab *Mod Off Technol* 31:68+ F '86

Word processors *See* Word Processing equipment

Other Indexes to Professional Journals. In addition to the *Magazine Index, Readers' Guide to Periodical Literature, Applied Science & Technology Index,* and *Business Periodicals Index,* the following indexes to periodicals may be useful:

Applied Arts Index (since 1913; formerly *Industrial Arts Index*)

Biological and Agricultural Index (since 1916; formerly *Agricultural Index*)

Cumulative Index to Nursing and Allied Health Literature (since 1956; formerly *Cumulative Index to Nursing Literature*)

Engineering Index (since 1884)

General Science Index (since 1980)

Hospital Literature Index (since 1945)

Public Affairs Information Service Bulletin (PAIS) (since 1915)

Social Sciences Index (since 1974; formerly part of *Social Sciences and Humanities Index*, 1965–1974, and of *International Index*, 1907–1965)

On microfilm: *Business Index* (since 1979). Has a "Business Collection" similar to the "Magazine Collection" for use with the *Magazine Index* (see pages 442–444)

Indexes to Newspapers. Newspaper indexes include:

Christian Science Monitor Index (since 1960, monthly)

The London Times Index (1906–1977, bimonthly and quarterly; since 1977, monthly)

The New York Times Index (since 1851, semimonthly, annual cumulation). Also on-line

Wall Street Journal Index (since 1950, monthly, annual volume). Also on-line

On microfilm: *The National Newspaper Index,* covering *The New York Times, Christian Science Monitor, Wall Street Journal, Los Angeles Times,* and *Washington Post* (since 1 January 1979, monthly, "Rolling" cumulation index similar to *Magazine Index*). Also on-line

The New York Times Index is especially useful. The *Times* covers all major news events, both national and international. Because of its wide scope and relative completeness, the *Times Index* provides a wealth of information. It is frequently used even without viewing the individual paper of the date cited. A brief abstract of the news story is included with each entry. Thus, someone seeking a single fact, such as the date of an event or the name of a person, may often find all that is needed in the *Index*. In addition, since all material is dated, the *Times Index* serves as an entry into other, unindexed newspapers and magazines. *The New York Times Index* is arranged in dictionary form with cross-references to names and related topics.

Indexes to Government Publications. The U.S. government prints all kinds of books, reports, pamphlets, and periodicals for audiences from the least to the most knowledgeable. The best-known index to government documents is the *Monthly Catalog of United States Government Publications,* published by the U.S. Superintendent of Documents. Each monthly issue lists the documents published that month.

U.S. Government Books is a selective list of government publications of interest to the general public.

Also available from the U.S. government are some 250 subject bibliographies of government publications. These are listed in the *Subject Bibliographies Index;* subjects of bibliographies are as varied as "Accidents and Accident Prevention," "United States Army in World War II," and "Zoology."

A guide to reports from research is *Government Reports Announcements* (GRA) and its accompanying index *Government Reports Index,* both published twice a month.

Government Reference Books is an annotated guide that provides access to important reference works issued by agencies of the United States government. The guide is published every two years; the entries are arranged by subject.

Essay Index. It is often difficult to locate essays and miscellaneous articles in books. The *Essay and General Literature Index* is an index by author, subject, and some titles of essays and articles published in books since 1900. The index is kept up to date by supplements. Subjects covered include social and political sciences, economics, law, education, science, history, the various arts, and literature.

OTHER "HELPS" IN LIBRARY RESEARCH

In addition to the public (card) catalog and the indexes, other tools or sources of information may be valuable in research.

Computer-Aided Search. In a computer search, various databases (databases are electronic file cabinets) produced by the government, by nonprofit organizations, or by private companies are accessed through a computer terminal.

Some indexes are available in printed form and in a database, in which case the information is the same but the format (or appearance) is different. Also, the database may contain additional access points because more subject approaches may be used and key words in the titles may be searched.

Using the computer terminal to access the database is a two- or three-step procedure. First, contact is established with the computer containing the database via telephone lines. Some computers contain only one database; some contain many. The database needed must be accessed. Then, using carefully formulated search logic, the computer searches the database for the desired materials.

Search strategies should be formed, using one or more key words before going "on-line" (on-line is the time actually spent interacting with the computer). A thesaurus or a subject heading list, if available for the index, should be used. These have "see" and "see also" terms to help in selecting the proper subject headings to use.

During the computer search, the search strategy may be modified if the results are not satisfactory. Thus, if no items are located, another term may be used. If too many citations are found, the search may be limited. Limitations may include time, place, or more specific descriptors.

A computer search is fast (particularly for searches covering several years), current (some information is available on-line weeks or months before the printed version), accurate, thorough (more in-depth coverage since there is more access to the information), and convenient. Printouts of citations can also save time and energy.

Computer searches require the assistance of a librarian and during busy times may need to be scheduled in advance. In addition, computer searches may be expensive, although the cost of using databases varies greatly. The user is charged for on-line time (cost per minute of time the user is "connected" to the database) and for each citation printed. Some users have the citations printed off-line and mailed to cut costs, but it may take two days to a week to receive them.

Most thorough searches for information require several indexes. Computer searches may be considered to complement manual searches when many issues of a printed index must be consulted, when numerous sources are needed, and when currency is important.

Traditionally, database searches have been used for locating citations for journal and newspaper articles on a subject. Now with entire works—periodical articles, books, and even encyclopedias on-line—databases may be used for locating detailed financial data and directory listings on companies, statistics, biographical information, news-wire stories, information on colleges and universities, and quotations. The information itself is becoming on-line, not just the citations to it.

NewsBank. NewsBank is a microfiche newspaper reference service covering the social, health, legal, political, international, economic, and scientific fields. Articles are collected from more than 300 leading newspapers all over the United States and indexed by subject. The printed indexes are published monthly and are cumulated quarterly in March, June, and September, with an annual cumulation in December. The articles, reproduced on microfiche, are provided monthly with the index. NewsBank is a particularly useful help because of the on-the-spot availability of all indexed articles.

Given below is an excerpt from the August 1986 Index with an explanation of how to use the Index.*

COMPUTERS AND COMPUTER INDUSTRY

See also **Electronic Equipment and Supplies;** Semiconductor Industry; **Teaching Materials and Equipment**—computers

applications
　　Articles are indexed under the subject of the application

copyrights and patents
　lawsuits and court orders
　　Texas—BUS 90:E12

employee benefits
　early retirement offered
　　Wang Laboratories—EMP 38:F7

employment market
　layoffs
　　Convergent Technologies—EMP 38:F8

financial problems
　. . .
　　Floating Point Systems—BUS 90:E13-F1

mergers and acquisitions
　management
　　Burroughs/Sperry—BUS 90:F2

new products
　. . .
　　IBM—BUS 90:F3

personal computers
　competition—BUS 90:F4-5

plant closings
　impact on employees
　　Data General—EMP 38:F10

profiles
　. . .
　　Commodore International—BUS 90:F6-7

research and development
　artificial intelligence
　　IBM—BUS 90:F8-9

sales
　　See also **Retail Stores and Trades**—computer stores

How to use the
NewsBank Index (Sample Entry)

AIR POLLUTION←1   1986
    automobiles ←———2 4 5 6   ↑
    smoking vehicles ←—┘ ↓ ↓ ↓   7
    3——→Colorado: Denver—ENV 79:B3

1 - Major Subject Heading   5 - Microfiche Card Number
2 - Sub-Headings   6 - Microfiche Grid Coordinates
3 - Geographic Location   7 - Microfiche Year (at top of index page)
4 - Microfiche Category

The *NewsBank Electronic Index* uses CD-ROM (Compact Disk-Read Only Memory) and the personal computer to quickly and easily access nearly five years of newspaper articles in NewsBank, *Names in the News*, and *Review of the Arts*. Thus, the more than 120,000 newspaper articles per year on the microfiche for these three NewsBank products are indexed in the *NewsBank Electronic Index*, which began in 1986 and is updated monthly.

Business NewsBank, another NewsBank product, is similar to NewsBank. It provides a printed index to business articles from over 300 newspapers in the United States. The articles—which pertain to companies, industries, products, and personnel—are reproduced on microfiche and accompany the index each month. A separate section of the index gives geographic access to the information.

Social Issues Resources Series. A valuable help in library research is the Social Issues Resources Series (SIRS). SIRS is a selection of articles reprinted from newspapers, magazines, journals, and government publications. The articles are organized into volumes with each volume, in a loose-leaf notebook format, dealing with a different social issue. Among the issues addressed (and thus titles for the volumes) are Aging, Alcohol, Communication, Consumerism, Corrections, Defense, Energy, Family, Food, Health, Human Rights, Mental Health, Pollution, Privacy, Sports, Technology, and Women. Each volume contains a minimum of 20 articles and is updated with an annual supplement of another 20 articles. (Each volume eventually contains 100 articles.) The articles are selected to represent various reading levels, differing points of view, and many aspects of the issue.

SIRS Science is a series of five volumes covering Earth Science, Physical Science, Life Science, Medical Science, and Applied Science. Similar to the regular SIRS volumes, these also contain articles selected from various sources and held in loose-leaf notebooks. Each of the volumes contains 70 indexed articles in chronological order. The volumes are annual.

Audiovisual Materials. Most colleges have an audiovisual department, a media or learning resources center, or a library department that supervises audiovisual materials. Check with the librarian or person in charge concerning the cataloging of such materials as audiotapes, videotapes, phonograph recordings, films, filmstrips, and slides. These items may be alphabetized in the public (card) catalog or in a separate catalog. They may contain valuable information on a research topic.

Periodicals Holdings List. The periodicals holdings list is a catalog of all the periodicals in a library. The alphabetical list gives the name of each magazine and newspaper and the dates of the available issues. The list may also indicate whether the issues are bound or unbound and where in the library the periodicals are located.

The periodicals holdings list may be in various forms, such as a drawer of 3- by 5-inch cards, a visible index file, a typed sheet, a microform, an automated or available on-line printout. Some libraries combine the periodicals holdings list with the public (card) catalog.

It is essential to know where the periodicals holdings list in a library is located. When in the various periodical indexes you find titles of articles that seem usable, you need to know if the library has the specified magazines. The periodicals holdings list will give this information; you can save time by checking the list yourself.

Reference Works. Libraries are filled with all kinds of reference books. Common among these are encyclopedias, dictionaries, books of statistics, almanacs, bibliographies, handbooks, and yearbooks.

Encyclopedias. Encyclopedias may be general encyclopedias or specialized encyclopedias. General encyclopedias may be a good starting point for locating information about a subject and for gaining an overall view of it. *Americana, Britannica, Colliers,* and *World Book* are well-known general encyclopedias that provide articles written for the general reader.

A valuable specialized encyclopedia is the *McGraw-Hill Encyclopedia of Science and Technology,* an international reference work in 20 volumes including an index. The articles, arranged in alphabetical order (word by word), cover pertinent information for every area of modern science and technology. Each article includes the basic concepts of a subject, a definition, background material, and multiple cross-references.

Specialized encyclopedias are numerous. Included are the *Encyclopedia of American Forest and Conservation History, International Encyclopedia of Population, Encyclopedia of Psychology, Encyclopedia of Crime and Justice, The Encyclopedia of Psychoactive Drugs, Encyclopedia of Superstitions, Encyclopedia of Textiles, Encyclopedia of Advertising, Glen G. Munn's Encyclopedia of Banking and Finance,* and *The Encyclopedia of Alcoholism.*

Dictionaries. General-use dictionaries include *The American College Dictionary, The American Heritage Dictionary of the English Language, The Random House Dictionary of the English Language,* and *Webster's New Collegiate Dictionary.* Specialized dictionaries include those of computer languages, architecture, welding, decorative arts, technical terms, the occult, U.S. military terms, Christian ethics, and economics.

Books of Statistics. Statistics are often an important part of a report because they support conclusions, show trends, and lend validity to statements. Since statistics change frequently, currency is usually essential.

Up-to-date statistics on a subject may be located in such sources as almanacs, yearbooks, and encyclopedias as well as recent newspaper and magazine articles on the topic. Statistical information on the past, such as wages in the 1930s, may be located in *Historical Statistics of the United States, Colonial Times to 1970.*

Statistics for the United States and other countries in such areas as industry, business, society, and education are found in *Statistics Sources*.

For annual updates of statistical data, see the *Statistical Abstract of the United States* (since 1878) and the *Statistical Yearbook* (since 1949), which covers yearly events for about 150 countries.

Two monthly indexes which provide access to a large area of statistical information are *American Statistical Index* (ASI) and *Statistical Reference Index* (SRI). ASI (since 1973) indexes most statistical sources published by the federal government and SRI (since 1980) indexes statistical information in U.S. sources other than those published by the federal government. Both of these complementing indexes have annual cumulations.

For statistical works held by the library, look in the public (card) catalog under the subject headings "Statistics" and "United States—Statistics." For statistical works just on your topic, also look under your topic for the subheading "Statistics," for example "Agriculture—Statistics."

Almanacs. Perhaps the best-known general almanac is *The World Almanac and Book of Facts*. It includes such diverse facts as winners of Academy Awards, accidents and deaths on railroads, civilian consumption of major food commodities per person, fuel economy in 1986 cars, National Football League champions, notable tall buildings in North American cities, and major new U.S. weapons systems.

Other almanacs include *Information Please Almanac* and the *Readers' Digest Almanac and Yearbook*.

Bibliographies. Bibliographies are guides to sources of information on specific subjects. They do not generally provide answers but rather list sources, such as periodicals, books, pamphlets, and audiovisuals, which contain the needed information.

Most bibliographies contain specific citations to information on a subject; however, others, such as the *Encyclopedia of Business Information Sources*, give types of sources available on a subject. The *Encyclopedia of Business Information Sources* lists by subject such types of sources as encyclopedias and dictionaries, handbooks and manuals, periodicals, statistical sources, and almanacs and yearbooks. This type of work is particularly useful when researching a topic about which little is known or few sources seem to be available.

For bibliographies available in the library on a specific subject, consult the public (card) catalog. Look under the subject for the subheading "Bibliography," for example, "Business—Bibliography," "Marketing—Bibliography," or "Nursing—Bibliography."

Handbooks. Handbooks contain specialized information for specific fields. The variety of handbooks is illustrated by the following examples. There are handbooks for word processor users, secretaries, electronics engineers, prospectors, and construction superintendents, as well as handbooks for the study of suicide, transistors, air conditioning systems design, food additives, simplified television service, and law of sales, to list a few.

Persons interested in information about job opportunities will find the *Occupational Outlook Handbook* quite useful. Published biennially by the U.S. Bureau of Labor Statistics, it describes about 200 occupations and for each gives the outlook, the number currently employed, the possible salary, educational requirements, possibility for advancement, and so on.

Yearbooks. Yearbooks, as the name suggests, cover events in a given year. An encyclopedia yearbook updates material covered in a basic set—this is an example of a general yearbook.

Specialized yearbooks include the *Yearbook of Higher Education, Yearbook of Agriculture, Yearbook of American and Canadian Churches, Yearbook of Drug Therapy, Yearbook of World Affairs, Yearbook of Labor Statistics,* and *Yearbook of Science and the Future*.

Note that the above examples of encyclopedias, dictionaries, books of statistics, almanacs, bibliographies, handbooks, and yearbooks illustrate the number and variety of these types of reference works. A quick glance at *Books in Print,* under these categories, further emphasizes the number and variety.

Guides to Reference Works. For a quick review of what reference materials exist, the following guides are useful. If you find in these guides materials not available in your library, ask your librarian about an interlibrary loan. But remember that interlibrary loans take time.

Periodicals. *Ulrich's International Periodicals Directory* includes an alphabetical listing (by subject and title) of in-print periodicals, both American and foreign, and lists some of the works which index the periodicals.

Books. Sheehy's *Guide to Reference Books* lists approximately 10,000 reference titles of both general reference works and those in the humanities, social sciences, history, and the pure and applied sciences. This guide and its supplement also include valuable information on how to use reference works.

The information in Sheehy's *Guide* may be updated by using *American Reference Books Annual* (ARBA), which lists by subject the reference books published each year in the United States.

The *Cumulative Book Index* lists books printed in the English language each year. *Books in Print* is a listing of books in print in America.

Vertical File. Much printed information exists in other than book and magazine forms. Pamphlets, booklets, bulletins, clippings, and other miscellaneous unbound materials are usually in a collection called the vertical file. These materials are filed or cataloged by subject. The vertical file and the *Vertical File Index,* which lists current pamphlets, may be valuable sources of information.

Planning and Writing a Research Report

Now that you are familiar with the four general sources of information—personal observation or experience, personal interviews, free or inexpensive materials from various agencies, and library materials and their use—you are ready to begin planning and writing a research report.

A research report is the written result of an organized investigation of a specific topic. It requires systematic searching out and bringing together information from various sources. It requires that the gathered information be presented in a conventional, easy-to-follow form.

Writing a research report can be profitable and rewarding. Knowing how and where to search for information and then how to present that information in a practical, understandable, and logical manner are real accomplishments.

The procedure for writing a research report centers around four major steps:

1. Selecting a subject and defining the problem
2. Finding the facts
3. Recording and organizing the facts
4. Reporting the facts

SELECTING A SUBJECT AND DEFINING THE PROBLEM

Selecting a Subject. The success of the research project depends on a wise choice of subject. If a specific subject is assigned, of course the selection problem is solved. However, if you choose the subject, consider the following guidelines.

1. *Choose a subject that fulfills a need.* Doing research should not be a chore or just another assignment—it should be an adventure that fulfills a need to know. Investigating some aspect of a future vocation, following up a question or a statement in a class, finding out more about an invention or a discovery, wanting to know the functions and uses of a particular mechanism—whatever the subject, let it be something that appeals to you because you want and need to know the information.

2. *Choose a subject that can be treated satisfactorily in the allotted time.* No one expects a college student to make an earthshaking contribution to human knowledge by presenting the result of months and years of research and study. Library research gives the student experience in finding, organizing, and reporting information on a specific subject. Therefore, the topic chosen should be sufficiently limited for adequate treatment in the few weeks allotted for the paper. The subject "Dogs," for instance, is much too broad to try to cover in a few weeks. Even "Hunting Dogs" would require far more time than is available. But "The Care and Training of Bird Dogs" is a specific topic that could reasonably be investigated in the usual time designated for a research report.

3. *Choose a subject on which there is sufficient available material.* Before deciding definitely on a subject, check with instructors and the library to be sure that the topic has sufficient accessible material. Generally, it is wise to choose another topic if the principal library you use has little or no information on the proposed topic. If the topic is highly specialized or has information in only very select journals, locate several references before deciding definitely to use the topic. Sometimes a news broadcast or a current event may suggest a subject. Again, be sure that enough material is available for an effective report.

Defining the General Problem. After the selection of the subject for the research report, the central problem or idea being investigated should be specifically stated. The formulation of this central problem, or basic question, is important because it determines the kind of information to look for. In writing a report, the idea is not simply to gather information that happens to fall under a general heading; the idea is to gather information that relates directly to the subject and that can give it form and meaning. For the subject "The Effects of Alcohol on People," for example, the central problem might be defined as the following: "How does alcohol affect people: How does it affect their bodies, behavior, relationships with other people?" This question gives direction to the researcher's reading and to the investigation as a whole.

Defining the Specific Problems. As a guide in searching for information, divide the central problem into specific problems or questions. These questions serve as a preliminary outline for more selective reading. Again, take the subject "The Effects of Alcohol on People" with the central problem, "How does alcohol affect people: How does it affect their bodies, behavior, relationship with other people?" Divide this general question into smaller ones, such as these:

1. Does alcohol kill brain cells?
2. Why do people drink?
3. Why do some people become silly and giggle a lot and others become morose and withdrawn when they drink?
4. What organs of the body does alcohol affect?
5. How does excessive drinking affect family relationships?
6. How much alcohol does it take to have an effect on the body?
7. Why can some people take or leave alcohol while others become addicted to it?

Other questions, of course, might be added to these.

As the investigation proceeds and the body of collected data increases, change or drop or add to the questions in the preliminary outline.

FINDING THE FACTS

Finding the facts for a research report requires a search of all available resource materials on the subject of the report. As discussed earlier in this chapter, the possible sources of information can be classified as: (1) personal observation, (2) personal interviews, (3) free or inexpensive materials from various agencies, and (4) library materials.

Make a thorough, systematic search of all available materials. A good place to start is a general encyclopedia. From there move to the public (card) catalog, remembering to check every possible topic you can think of that is related to the subject of the report. Then check indexes to periodicals; if the subject requires current information, check only the issues of current date. If, however, the date of publication is not a factor, check in all available issues. Remember to check in more than one index, too. You might start with the *Magazine Index* or the *Reader's Guide* and then the *Applied Science & Technology Index* or the *Biological and Agricultural Index* or one of the other specialized indexes.

A check of these basic resources gives you a good start in finding the facts. Then you can go to any other available resources.

RECORDING AND ORGANIZING THE FACTS

Recording and organizing information is a key step in writing a successful research report. This step involves evaluating resource material, compiling a working bibliography, taking notes, stating the central idea of a report and constructing a formal outline, and arranging the note cards to fit the outline.

Evaluating Resource Material. You must *evaluate* resource material, for all sources are not equally useful or reliable. To evaluate material carefully, you need to know the difference between primary and secondary sources, to separate fact from opinion, and to use the nontextual qualities of a source in evaluating it.

Primary and Secondary Sources. Primary sources are those giving firsthand accounts of an event or condition. Material written by Dr. Jonas Salk on the development of the Salk polio vaccine would be a primary source. Secondary sources are works written about firsthand accounts. A magazine article about Dr. Salk would be a secondary source. Although secondary sources are generally more readily available, use primary sources whenever possible.

Fact and Opinion. Separating fact from opinion is essential in evaluating a source. A fact is an item of information that can be proved, such as "The book has 875 pages" or "Hereford is a breed of cattle." An opinion is an interpretation of a fact, or an idea about a fact, such as "This book has 875 *exciting* pages" or "Hereford is the *most profitable* breed of cattle." The researcher must be able to distinguish facts from the writer's opinions.

Nontextual Qualities. The nontextual qualities of a source may be helpful in evaluation. The reputation of the author and of the publishing company, date of publication, table of contents, introductory material, bibliography, index—each gives an indication of the usefulness and reliability of a source. An article about the surface of the moon written in 1968 would not be as reliable as an article written by Armstrong and Aldrin after their moon landing in 1969 or one written by astronauts who visited the moon later.

Compiling a Working Bibliography. Once you have selected a subject and defined the problem to be researched, the next step is to begin research. As you locate possible specific sources, list these sources (books, periodicals, and so on) to compile a working bibliography or tentative bibliography.

Record bibliographical information for each possible source on a separate card (usually 3- by 5-inch or 4- by 6-inch cards). To save time later, record the information in the same order and form to be used for the final bibliography. You will notice the form is different from the form on a public (card) catalog entry or in an entry in an index.

The following samples illustrate an acceptable form for recording on cards bibliographical information for commonly used sources: a book, an essay in a book, a journal article, a popular magazine article, a newspaper item, a NewsBank item, a reprinted item, an encyclopedia article, a pamphlet, and a personal interview. Forms for other sources are given on pages 473–474.

Bibliography Card for a Book. A bibliography card for a book includes the following items:

1. Author's or editor's name (last name first for easy alphabetizing)
2. Title of the book (underlined)
3. Edition (if other than the first)
4. Volume number (if the work is in several volumes)
5. City of publication (followed by a colon)
6. Publishing company
7. Date of publication
8. Call number. (Including the call number of the book saves time if you need to refer to it again; you do not have to look it up again in the public [card] catalog. The lower· or upper left-hand corner is a good place for the call number.)

In addition, it may be helpful to write the following information on the bibliography card:

9. Notation as to the contents of the book
10. Library where book is located (especially if several libraries are used). Place the name of the library in the lower or upper left-hand corner, below the call number.

Sample Card for a Book

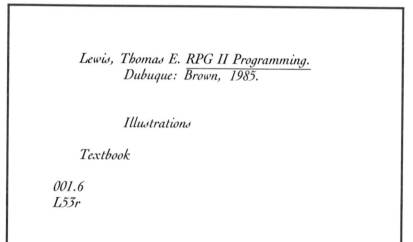

Lewis, Thomas E. RPG II Programming.
Dubuque: Brown, 1985.

Illustrations

Textbook

001.6
L53r

Bibliography Card for an Essay in a Book. If a possible source of information is an essay in a book, record the following data on a bibliography card:

1. Author of the essay (last name first)
2. Title of the essay (in quotation marks)
3. Title of the book (underlined)
4. Editor of the book, if different from the author of the essay
5. City of publication
6. Publishing company
7. Date of publication
8. Page numbers of the essay

Sample Card for an Essay in a Book

459

The Research
Report: Becoming
Acquainted with
Resource Materials

> *Bracken, Paul. "Accidental Nuclear War."*
> *Hawks, Doves, and Owls. Ed. Graham*
> *T. Allison, Albert Carnesale, and Joseph*
> *S. Nye, Jr. New York: Norton, 1985.*
> *25–53.*

Bibliography Card for an Article in a Journal. On each bibliography card for an article from a journal, record:

1. Author's name (last name first)
2. Title of the article (in quotation marks)
3. Title of the journal (underlined)
4. Volume number
5. Date of publication (in parentheses)
6. Page numbers

Sample Card for an Article in a Journal

> *Curington, William P. "Safety Regulation and*
> *Workplace Injuries." Southern Economic*
> *Journal 53 (1986): 51–72.*

Bibliography Card for an Article in a Popular Magazine. On each bibliography card for an article from a popular magazine, record:

1. Author's name (last name first), if given
2. Title of the article (in quotation marks)

3. Title of the magazine (underlined)
4. Date of publication
5. Page numbers

Sample Card for an Article in a Popular Magazine

> *Wallis, Claudia. "Viruses." <u>Time</u> 3 Nov.*
> *1986: 66 – 70, 73 – 74.*

Bibliography Card for an Item in a Newspaper. On each bibliography card for an item from a newspaper, record:

1. Author's name (last name first), if given
2. Title of the item (in quotation marks)
3. Name of the newspaper and, if applicable, the edition. (If needed for publication clarification, insert in brackets the name of the city or state, or both.)
4. Date
5. Section and page

Sample Card for an Item in a Newspaper

> *Kilpatrick, James J. "Drug Test Not*
> *'Unreasonable'." <u>The Clarion-Ledger</u>*
> *[Jackson, MS] 1 Oct. 1986, final ed.:*
> *A11.*

Bibliography Card for an Item from NewsBank. For a newspaper item appearing in NewsBank (on microfiche) it is necessary to give two sets of bibliographical data: citation of the newspaper in which the item originally appeared and citation of the NewsBank location. On each bibliography card for an item from NewsBank, record:

1. Author's name (last name first), if given
2. Title of the item (in quotation marks)
3. Name of the newspaper. (If needed for publication clarification, insert in brackets the name of the city or state, or both.)
4. Date
5. NewsBank identification, subject heading, year, fiche number and grid location

Sample Card for an Item from NewsBank

> *Wyss, Bob. "The Road to a Solid Waste Incinerator." Providence [RI] Journal 23 Feb. 1986. NewsBank, Environment, 1986, 16:G2.*

Bibliography Card for an Item Reprinted from Another Source. For an item reprinted (rpt.) from another source, it is necessary to give two sets of bibliographical data: citation of the original source of the item and citation of the immediate source of the item. (This bibliography card is somewhat similar to a bibliography card for an item from NewsBank.) For the citation of the original source, follow the standard format for citing the particular item: essay from a book, a magazine article, newspaper item, or whatever. Then identify the immediate source of the item.

For example, for a bibliography card of a magazine article reprinted in the Social Issues Resources Series, record:

1. Author's name (last name first), if given
2. Title of the article (in quotation marks)
3. Title of the magazine (underlined)
4. Date of publication
5. Page numbers
6. Social Issues Resources Series identification, volume title, volume number, article number

> *Schiffres, Manuel. "The Shadowy World of
> Computer Hackers." U.S. News &
> World Report 3 June 1985: 58–60. Rpt.
> in Social Issues Resources Series, Crime
> 3: art. 51.*

Bibliography Card for an Encyclopedia Article. For a bibliography card for an encyclopedia article, record:

1. Author's name (last name first), if given. (Encyclopedia articles are often signed with initials; it may be necessary to look in a separate listing for identification of the initials.)
2. Title of the article (in quotation marks)
3. Name of the encyclopedia (underlined)
4. Edition year. (Note that volume and page numbers are given if the reference is to material on a single page of an article covering several pages.)

Note: For other than well-known encyclopedias, give full publication information as for any other book.

Sample Card for an Encyclopedia Article

> *"Computer." McGraw-Hill Encyclopedia of
> Science and Technology. 1987 ed.*

Bibliography Card for a Pamphlet. For a bibliography card for a pamphlet or any individual printed work of less than book length, give the following:

1. Author's or editor's name, if given (last name first), or name of sponsoring organization
2. Title of pamphlet (underlined)
3. Other identifying information, such as series name and number
4. City of publication
5. Publishing company
6. Date of publication. (It might be helpful to remember that a pamphlet reference is very similar to a book reference.)

Sample Card for a Pamphlet

> United States. Dept. of the Interior. Fish and
> Wildlife Service. Dioxin Hazards to Fish,
> Wildlife, and Invertebrates: A Synoptic
> Review. Washington: GPO, 1986.

Bibliography Card for a Personal Interview. On each bibliography card for a personal interview (for a telephone interview, substitute the word *telephone*), record:

1. Interviewee's name (last name first)
2. Kind of interview and, if applicable, subject of the interview
3. Date of the interview

Sample Card for a Personal Interview

> Strong, Alicia M. Personal interview on
> employment opportunities. 2 Feb. 1986.

Taking Notes. While investigating each source in the working bibliography, save time by first consulting the table of contents and the index for the exact pages that may contain helpful information. After locating the information, scan it to get the general idea. Then, if the information is of value, read it carefully. If certain material may be specifically used in the paper, write it down on note cards.

General Directions for Taking Notes. Regardless of the shortcuts you may think you have devised for taking notes, following these directions will be the shortest shortcut:

1. Take notes in ink on either 3- by 5- or 4- by 6-inch cards. Ink is more legible than pencil, and cards are easier to handle than pieces of paper.
2. If it is difficult to decide when to take notes and when not to, stop *trying* to take notes. Read a half dozen or so of the sources, taking no notes. Study the preliminary outline. (Do not hesitate to revise the outline—it is purely for your use.) Then return to the sources with a clearer idea of what facts are needed.
3. Write one item of information on a note card. Write on only one side of the card. Cards can then be rearranged easily.
4. Always write notes from different sources on different cards.
5. Take brief notes—passages that may possibly be quoted in the paper, statistics, and other such specific data, proper names, dates, and only enough other information to jog the memory. Writing naturally or well cannot be done if the paper is based solely on notes rather than on overall knowledge and understanding. One caution: Write enough so that you can use the information after it is "cold."
6. Write a key word or phrase indicating the content of each note in the upper left corner for quick reference. This key word or phrase will be especially helpful in arranging cards according to topic; often these key words or phrases become headings in the final outline.
7. In the upper right corner, identify the source (author's name and, if necessary, title of the work) and give the specific page number from which the material comes. This information in the right corner keys the note card to the bibliography card, needed for writing notes and bibliography.
8. Remember that proper credit must be given in the paper for all borrowed material—quotations, exact figures, or information and ideas that are not widely known among educated people. Remember, too, that an idea stated in one's own words is just as much borrowed as an idea quoted in the author's exact words. Therefore, take notes carefully so that you know which notes are direct quotes and which are paraphrases or summaries.

Types of Note Cards. Notes may be direct quotations, paraphrases, or summaries.

Direct Quotation   Some notes will be direct quotations, that is, the author's exact words.

There may be passages in which the author has used particularly concise and skillful wording, and you may want to quote them; in this case, the author's exact words are written on the note card. Generally, however, take down few direct quotations—they create a tendency to rely on someone else's phrasing and organization for the paper. If it is desirable to leave out part of a quotation, use ellipsis points, that is, three spaced dots (. . .) or four spaced dots if the omission includes an end period (. . . .). Be sure to put quotation marks around all quoted material.

Sample Note Card Containing a Direct Quotation

> *Potential of miracle chip*   *"The Computer Society" p. 467*
>
> *"Far from rendering the big computer obsolete, the miracle chip has opened the way for the design of custom-made supercomputers more powerful than anything dreamed possible a few years ago."*

Paraphrase and Summary   Most of the notes for a library report will be paraphrases and summaries rather than direct quotations. A paraphrase is a restatement in different words of someone else's statement. A summary is the gist—it includes the main points—of a passage or article. In reading, get main ideas and record them in your own words. Jot down key phrases and sentences that summarize ideas clearly. Do not bother taking down isolated details and useless illustrative material. It is not necessary to write notes in complete sentences, but each note should be sensible, factual, and legible—and meaningful after two weeks.

Following are examples of the paraphrase and the summary note cards.

Sample Note Card Containing a Paraphrase

> *Effect of miracle chip on*   *"The Computer Society" p. 467*
> *computer industry*
>
> *The miracle chip makes possible more powerful computers. It will lower the cost of minicomputers and enable the small computers to perform more and more functions previously requiring large computers. Such use will result in future cost reduction.*

Potential use   "The Computer Society" pp. 468–469

Key areas for potential use of miracle chips:

1. Automobiles
2. Communications
3. Office equipment

Stating the Central Idea and Constructing a Formal Outline. At the very beginning of the research project a basic question or problem was formulated; then this general question was divided into specific questions, or aspects, as a preliminary outline. As investigation has progressed, you probably have added to, changed, and perhaps even omitted some of the original questions. Thus, the present list of questions looks quite different from the beginning list. (The preliminary outline should be revised and brought up to date after all notes are taken.)

By this stage in the research you should have clearly in mind the central idea of the report. Write this central idea (thesis statement or controlling idea) as a one-sentence summary at the top of what will be the working outline page.

From the central idea, the list of questions or divisions in the revised preliminary outline, and the key words in the upper left corner of the note cards, structure a formal outline to serve as the working outline for the first draft of the paper. Every major division in the outline should reflect a major aspect of the central idea. Taken as a whole, all of the major divisions should cover the central idea and adequately explain it.

The formal outline may be either a sentence outline or a topic outline. Remember: in a sentence outline all the headings are complete statements (no heading should be a question), and in a topic outline all the headings are phrases or single words. The sentence outline is more complete and helps to clarify the writer's thinking on each point as he or she goes along; it brings him or her one step nearer the writing of the paper. The topic outline is briefer and shows at a glance the divisions and subdivisions of the subject. Whichever type of outline is chosen, be consistent; do not mix topics and sentences. (For further treatment of outlining, see pages 128–129, 160, and 511.)

The working outline will serve as a work plan in writing the first draft of the report; it will let you know where you are and will provide a systematic organization of the information.

Arranging the Note Cards to Fit the Outline. When the working outline is completed, mark each note card with a roman numeral, letter of the alphabet, or arabic numeral to show to what section and subsection of the outline the note card corresponds. Then rearrange the note cards accordingly. If the note cards are insufficient, look up additional material; if there are irrelevant note cards, discard them. When you have enough material to cover each point in the outline adequately, you are ready to begin writing the report. Each main heading in the outline should correspond to a major section of the report and each subheading to a supporting section.

Reporting the Facts

WRITING THE FIRST DRAFT

With the working outline for a guide and the note cards for content, write the first draft of the report rapidly and freely. Concentrate on getting thoughts down on paper in logical order; take care of grammar, punctuation, and mechanical points later. In writing the first draft, it will be necessary to write short transitional passages to connect the material on the note cards. Also, it will be necessary to rephrase notes (except for direct quotations) to suit the exact thought and style of writing. Add any headings that would help to make the material clearer. (See pages 7–8 and 354–355 for a discussion of headings.) Think through the layout and design of the report (see pages 4–7, 355).

Introduction and Conclusion. In the first paragraph or in a section headed "Introduction," state the thesis, or central idea, of the report. Let the reader know what to expect, to what extent the subject will be explored, as indicated in the major headings of the outline. The introduction, designed to attract the reader's attention, serves as a contract between writer and reader. It may be a sentence, a single paragraph, several paragraphs, or even several pages. In it, the writer commits the report to a subject, a purpose, and an organization of material. (See also Chapter 8, Reports, page 353, for further discussion of the introduction.)

In the introduction, make clear the subject to be covered and the extent of the coverage. Stating the thesis or the central idea usually is sufficient to identify the subject; a sentence listing the major headings of the outline will show the extent of coverage.

Supply any information needed to help the reader understand the material to follow; this information might include a definition of the subject itself or of terms related to the subject. Explain the significance of your investigation and reporting.

After completing the body of the paper, close with a paragraph that is a summary, a climax, conclusions, or recommendations drawn from the material that has been presented.

Quotations and Paraphrases. If the exact words of another writer are used, put them in quotation marks. If the quotation is four lines or more in length, however, instead of using quotation marks, double-space the quotation and indent each line ten spaces from the left margin.

Incorporate paraphrased matter and indirect quotations into the body of the paper, and do not use quotation marks. However, document *all* borrowed information, whether quoted directly or paraphrased. (Documentation is discussed in detail later in this chapter.)

A Note About Plagiarism. Plagiarism is implying that another person's ideas or words are your own ideas or words. Sometimes plagiarism is unintentional; sometimes intentional. To avoid the problem of plagiarism, keep the following suggestions in mind.

1. Place quotation marks around any material that you yourself did not write and credit the source. Indicate material copied on note cards by using quotation marks. In the report you may wish to introduce the material by using the name of the person who wrote it or the title of the work in which it appeared.
2. Write paraphrased material in your own style and language. Simply rearranging the words in a sentence or changing a few words does not paraphrase.

There is no excuse for plagiarism. A careful handling of information should help you avoid any problem.

Documentation. Documenting information means supplying references to support assertions and to acknowledge ideas and material that you borrow. Documentation is usually done by inserting in the text a shortened parenthetical identification or a number that corresponds to a work identified fully in a footnote at the bottom of the page, in a note at the end of the report, or, most often, in a list of "References Cited" at the end of the report. Sometimes included is a bibliography listing all the sources used in writing the report, whether specifically referred to or not. (Documentation practices are discussed in detail on pages 469–474.)

Revising the Outline and First Draft. After writing the first draft of the report, study it carefully for accuracy of content, mechanical correctness, logical organization, clarity, proportion, and general effectiveness. Check to see if the intended reader has been kept in mind. The writer may understand perfectly well what is written, but the real test of writing skill is whether or not the reader will understand the report.

Make needed revisions in the outline and the first draft. Make sure that the headings in the report correspond exactly to the headings in the outline. (For a further discussion of headings, see pages 7–8 and 354–355.) Plan and prepare visuals. Think through the kinds of visuals that will be the most effective and decide where to place the visuals. (For a detailed discussion, see Chapter 11, Visuals.)

The revising process may involve making several drafts of the outline and the report. But after the hours, days, and weeks already invested in this project, a few more hours spent polishing the report are worth the effort. Careful revision may mean the difference between an excellent and a mediocre report. (See also the checklist for revising on the inside back cover.)

WRITING THE FINAL DRAFT

When the outline, the text of the paper, and the documentation are completed, carefully recopy the pages. Prepare a letter of transmittal, a title page, and an abstract (see pages 352–353). For the entire report, follow the general format directions for writing, on the inside front cover. Double-space a typewritten report. In a handwritten report, space as if it were typewritten.

Proofread the report thoroughly. Clip the pages together in this order: letter of transmittal, title page, outline, abstract, body of the report, and as applicable: notes, list of works cited, bibliography. The research report is finished at last and ready to be handed in! (For a detailed discussion of reports, see Chapter 8, Reports, pages 347–355.)

DOCUMENTATION

Documentation provides a systematic method for identifying sources or references used in a research report.

The format or style used for documentation varies from discipline to discipline and even within disciplines. Societies or organizations of persons in a specific discipline may offer style manuals to be used in documenting papers in that discipline. Examples include the American Mathematical Society's *Manual for Authors of Mathematical Papers,* the Council of Biology Editors' *CBE Style Manual,* the American Chemical Society's *Handbook for Authors of Papers in American Chemical Society Publications,* the American Institute of Physics' *Style Manual,* and the International Steering Committee of Medical Editors' "Uniform Requirements for Manuscripts Submitted to Biomedical Journals."

Then there are basic manuals that are used by various disciplines, such as the *MLA Handbook for Writers of Research Papers* (2nd ed.; New York: MLA, 1984); *The Chicago Manual of Style* (13th ed.; Chicago: U of Chicago P, 1982); *Publication Manual of the American Psychological Association* (3rd ed.; Washington: American Psychological Assn., 1983); Turabian's *Manual for Writers of Term Papers, Theses, and Dissertations* (5th ed.; Chicago: U of Chicago P, 1987); Campbell, Ballou, and Slade's *Form and Style: Theses, Reports, Term papers* (7th ed.; Boston: Houghton Mifflin, 1986); or the United States Government Printing Office *Style Manual* (Washington: GPO, 1984).

Although documentation practices in literature, history, and the arts are somewhat uniform, documentation in other areas follows various practices. Among the science disciplines, for instance, there is little uniformity of procedure.

While the *reason* for documentation is the same in all areas—citation of sources for statements presented or to acknowledge borrowed matter—the *procedures* for citing sources are somewhat different.

The following discussion is presented with these reservations and with this advice: Regarding documentation practices, consult with the instructor assigning the report, consult with instructors in the subject area of the report, and study the practices in pertinent professional journals.

Remember: Whatever format you choose, be consistent.

Four Styles of Documentation. In documenting information, the important thing is to cite the source of the information. The source can be given in the text:

According to page 35 of James Fallows' article "The Japanese Are Different from You and Me" in the September 1986 issue of *The Atlantic,* pages 35–41, Japan seems better prepared than the West for production competition due to its emphasis on the individual worker and company rather than profit.

Such a citation within the text, however, detracts from the emphasis in the sentence; and several such citations within the text are likely to annoy the reader and hinder readability.

Various styles or systems of documentation are used, but four styles seem to dominate. They will be referred to as Style A, Style B, Style C, and Style D.

Style A, recommended by the Modern Language Association, is distinguished by a textual parenthetical listing of the reference (author's last name and page number) and at the end of the report, a full citation of the reference.

EXAMPLE:

Japan seems better prepared than the West for production competition due to its emphasis on the individual worker and company rather than profit (Fallows 35).

The full citation in the list of references:

Fallows, James. "The Japanese Are Different from You and Me." *The Atlantic* Sept. 1986: 35–41.

Style B, recommended by *The Chicago Manual of Style* (Chicago) and the *Publication Manual of the American Psychological Association* (APA), is distinguished by a textual parenthetical listing (author's last name and year of publication) and at the end of the report, a full citation of the reference.

EXAMPLE:

Japan seems better prepared than the West for production competition due to its emphasis on the individual worker and company rather than profit (Fallows 1986). [APA inserts a comma between author and year.]

The full citation in the list of references:

Chicago: Fallows, James. 1986. The Japanese are different from you and me. *The Atlantic*, September, 35–41.
APA: Fallows, J. (1986, September). The Japanese are different from you and me. *The Atlantic*, pp. 35–41.

Another Example of Style B:

Text of Paper:

Field (1971) cited 11 references to indicate that the estimated retail yield favored rams over wethers. Shelton and Carpenter (1972) found that not only is the amount of fat less in rams than in wethers or ewes, but the rate of fat deposition is also much lower at heavy weights (64 kg live weight) in rams. Kemp et al. (1972) observed that fat measurements were greater for wethers than for rams in light weight groups and that these differences between sexes became greater as weight increased. Crouse et al. (1981) found ram lamb carcasses to be significantly leaner than ewe or wether lamb carcasses.

Entries from "Literature Cited" at the End of the Paper:

Crouse, J. D., J. R. Busboom, R. A. Field, and C. L. Ferrell, 1981. The effects of breed, diet, sex, location and slaughter weight on lamb growth, carcass composition and meat flavor. J. Anim. Sci. 53:376.

Crouse, J. D., R. A. Field, J. L. Chant, Jr., C. L. Ferrell, G. M. Smith, and V. L. Harrison, 1978. Effect of dietary energy intake on carcass composition and palatability of different weight carcasses from ewe and ram lambs. J. Anim. Sci. 47:1207.

Deweese, W. P., H. A. Glimp, J. D. Kemp, and D. G. Ely, 1969. Performance and carcass characteristics of rams and wethers slaughtered at different weights. Kentucky Agr. Exp. Sta. Prog. Rep. 181.

Dolezal, H. G., G. C. Smith, J. W. Savell, and Z. L. Carpenter, 1982. Comparison of subcutaneous fat thickness, marbling and quality grade for predicting palatability of beef. J. Food Sci. 47:942.

See pages 624–630 for a report using style B documentation.

STYLE C:
Sciences and
Engineering

Style C is a number system. A number is usually placed in parentheses or in brackets (sometimes the number is written as a superscript—slightly above the regular line of type) in the text. This number refers to an entry in the numbered list of references at the end of the paper.

EXAMPLE:

Text of Paper:

Of the anatomic indices that have been developed, the most widely used is the Injury Severity Score (ISS) *(1,2)*. This index is a modification of an earlier index, the Abbreviated Injury Scale (AIS) which was originally developed for motor vehicle injuries. The AIS-ISS system of grading severity has undergone considerable evolution as experience in its use has accumulated, and an updated version appeared in 1984 *(3)*.

Entries from "References" at the End of the Paper:

1. Baker, S. P., O'Neill, B., Haddon, W., and Long, W.: The injury severity score: a method for describing patients with multiple injuries and evaluating emergency care. J Trauma 14: 187–193 (1974).
2. Baker, S. P., and O'Neill, B.: The injury severity score: an update. J Trauma 16: 882–885 (1976).
3. The abbreviated injury scale—1984 revision. American Association for Automotive Medicine, Morton Grove, Ill., 1984.

The list of references at the end of the paper may be given in (a) alphabetical order or in (b) the order in which the sources are mentioned in the text of the paper (as in the above example).

References in science papers may be written in a number of ways; there is no one uniform format. Given below is a sampling of notes using various formats. (Remember, whatever format you choose for your paper, be consistent.)

EXAMPLES (each example from a different paper):

(4) Karim, Omar A.; McCammon, J. Andrew. *J. Am. Chem. Soc.,* 108:1762–1766 (1986).

17. Paganini-Hill A, Ross RK, Henderson BE: Prevalence of chronic disease and health practices in a retirement community. *J Chron Dis* 39:699–707, 1986.

9. Matlak, T., "Station to Station: Programmable Control of Assembly," *Mechanical Engineering,* April 1986, pp. 34–40.

6. Love, L. C., Principles of Metallurgy (Reston, VA: Reston Publishing Company, 1985), chap. 2, p. 40.

Style D is the traditional, humanities system of documenting through footnotes or endnotes and with or without a list of works cited or a bibliography. A superscript number in the text is keyed to a full citation at the bottom of the page or, more commonly, at the end of the report.

EXAMPLE

Today some 21,000 cases of AIDS (acquired immune deficiency syndrome) have been documented in the United States and thousands more confirmed in 47 other nations.[1]

The full citation as a footnote or endnote:

[1]Kevin M. Cahill, "AIDS—Medical Reflections," *America*, 21–28 June 1986, p. 507.

If there is a list of references, the citation is as follows:

Cahill, Kevin M. "AIDS—Medical Reflections." *America*, 21–28 June 1986, pp. 507–508.

In Style D the first reference to a source is a complete citation, as illustrated in the example above. Frequently, reference is made to a source a second, third, or more times. Any source reference after the first usually gives the author's last name and the page number(s). The reference may also include an intelligible short title if necessary to make clear which work is cited.

First reference: [2]Laurence, J. Peter, *The Peter Pyramid, or, Will We Ever Get the Point?* (New York: William Morrow, 1986), p. 43.
[3] . . . .
[4] . . . .
Second reference: [5]Peter, p. 128.

Notice that footnotes/endnotes are indented as if they were paragraphs and that the items of information are separated by commas, not periods.

Citations List and Bibliography. A citations list or bibliography typically concludes a library research report. A citations list gives author, title, and publishing information about all source materials referred to in the text. The citations list is headed "Works Cited," "Works Consulted," "References," or a similar title.

The citations list is usually in alphabetical order. If Style C documentation (a number system) is used, however, the numbered list of citations may be either alphabetical or in the order in which the sources are mentioned in the text.

A bibliography is an alphabetical listing of all source materials—whether specifically referred to or not—used in preparing and writing a research report.

An annotated bibliography includes descriptive or evaluative comments about each entry. (See Chapter 7, The Summary, especially pages 214–217.)

The citations list and bibliography observe hanging indentation, that is, in every entry each line after the first is indented.

Examples of Citations. Given below are examples of citations to various kinds of references: books, periodicals, and nonprint materials. These examples follow Style A documentation (author-page parenthetical insertion in the text and an end-of-report citations list). The sample library research paper that concludes this chapter follows Style A documentation.

Remember that although documentation styles differ, they differ only in details—details of punctuation and abbreviation and of order in the placement of information.

CITATIONS FOR BOOKS

One author	Hordeski, Michael. *Microcomputer Design.* Englewood Cliffs: Prentice, 1986.
Two authors	Smith, Douglas A., and Richard Sonnenblick. *Laser Printing, Typesetting, and Design.* Englewood Cliffs: Prentice, 1986.
Three or more authors	Smith, Garry D., Danny R. Arnold, and Bobby G. Bizzell. *Strategy and Business Policy.* Boston: Houghton, 1985.
	Harwell, Edward M., and others. *Meat Management and Operations.* New York: Lebhar Friedman, 1985.
Corporate author	National Institute for the Food Service Industry. *Management by Menu.* 3rd ed. Dubuque: Brown, 1986.
	United States. National Commission for Employment Policy. *Training for Work in the Computer Age: How Workers Who Use Computers Get Their Training.* Washington: GPO, 1986.
Editor	Conklin, John E., ed. *Criminology.* 2nd ed. New York: Macmillan, 1986.
Essay	Burnham, David. "The Rise of the Computer State." *High Technology and Human Freedom.* Ed. Lewis H. Lapham. Washington: Smithsonian Institution, 1985. 141–46.
	Jackson, Blyden. "Richard Wright in a Moment of Truth." *The Southern Liberary Journal* 3 (1971): 3–17. Rpt. in *Richard Wright, A Collection of Critical Essays.* Ed.

Richard Macksey and Frank E. Moorer. Twentieth Century Views. Englewood Cliffs: Prentice, 1984. 182–93.

Republished book	Finck, Henry. *Romantic Love and Personal Beauty*. 1891. Havertown: R. West, 1973.
Encyclopedia article	"Field Effect Transistor." *McGraw-Hill Encyclopedia of Science and Technology*. 1987 ed.
	Kornblum, Zvi C. "Photochemistry." *Encyclopedia Americana*. 1985 ed.

CITATIONS FOR ARTICLES IN PERIODICALS

Article in a journal	McKay, R. M., and I. Garnett. "Prenatal and Postnatal Influences on Growth and Fat Measurements in Swine." *Journal of Animal Science* 63 (1986): 1095–1100.
Article in a popular magazine	Wellborn, Stanley N. "Lean Days for Space Research." *U.S. News & World Report* 10 Nov. 1986: 91.
	Gallagher, Winifred. "The Looming Menace of Designer Drugs." *Discover* Aug. 1986: 24–35.
Article reprinted in another source	Back, Jonathan, and Susan S. Lang. "Holography." *Futurist* Dec. 1985: 25+. Rpt. in Social Issues Resources Series, Technology 2: art. 58.
Item in a newspaper	Burgess, John. "Japan Faces 'Graying' of Population." *The Washington Post* 12 Oct. 1986: A21.
	"Arms Control Unchained." Editorial. *The Wall Street Journal* 17 Oct. 1986: A22.

CITATIONS FOR NONPRINT MATERIALS

NewsBank (microfiche)	Quanrud, Ted. "Suicide." *Bismarck* [ND] *Tribune* 6 Apr. 1986. NewsBank, Health, 1986, 45:A13.
Filmstrip (or film or video tape)	"Chest Trauma." Filmstrip and cassette tape. Acute Respiratory Care. Medcom, 1985. 111 frames, 24 min.
Computer software	*The Listening Inventory*. Computer software. Orange Juice Software Systems, 1985. Apple IIPlus, IIe, or IIc, 48K, DOS 3.3, disk.
Personal interview (or telephone interview or lecture)	Adams, John. Personal interview. 31 Oct. 1986.

A Sample Research Report

Following is a sample research report prepared by a student concerned about the worldwide loss of forestland. Included are the letter of transmittal, the title page, the outline, a sample table of contents page, an abstract, the report, and list of works cited. Note the use of headings throughout the paper.

6801 Paxton Road
Vicksburg, MS 39180
April 30, 1987

Dr. Nell Ann Pickett
Department of English
Raymond Campus
Hinds Junior College District
Raymond, MS 39154

Dear Dr. Pickett:

   Enclosed is the report entitled "The Future of the
World's Forests: A Look at Mother Nature on the Run" as
required in English 1123—Technical Approach. It discusses
the current problem the world is facing because of rapid
loss of forest resources, some reasons it is that way, and
some possible solutions.

   All research for this paper was done at the McLendon
Library. Since I am a Forestry major, this subject is of
great interest to me. This subject is not as widely
publicized as the current energy shortage is; yet the
magnitude and importance of these two problems are equal
and I feel all of us need to better understand this
problem in order to solve it for future generations.

                           Sincerely,

                           William E. Bratcher

                           William E. Bratcher, Jr.

THE FUTURE OF THE WORLD'S FORESTS:
A LOOK AT MOTHER NATURE ON THE RUN

by
William E. Bratcher, Jr.

English 1123–40
Raymond Campus
Hinds Junior College District
Raymond, Mississippi
May 1, 1987

THE FUTURE OF THE WORLD'S FORESTS:
A LOOK AT MOTHER NATURE ON THE RUN

Purpose Statement: The purpose of this paper is to present
the current worldwide forest situation, some reasons for
its being that way, the effects of a worldwide timber
shortage, and some possible answers to the problem we
face.

   I. Geographical analysis
  II. Forest deprivants
     A. Agriculture
     B. Forest harvesting
       1. Lumber
       2. Pulpwood
       3. Fuel
     C. Urbanization
 III. Effects of deforestation
     A. Ecological effects
       1. Water supply
       2. Erosion
       3. The greenhouse effect
       4. Oxygen
       5. Wildlife
     B. Economic effects
       1. Underdeveloped countries
       2. Developed countries
  IV. Possible solutions
     A. Research
       1. Agroforestry
       2. American forestry
     B. Reforestation
     C. Land management

NOTE: A research report might include a Table of Contents in place of an outline. Following is an example for the report on forests.

TABLE OF CONTENTS

2

ABSTRACT

A pressing international problem is the ever increasing
worldwide loss of forestland due to overuse and indirect
involvement of human beings, both intended and unintended.
Third World countries at the present time face dire
situations in respect to forestland and forest resources.
The human race is depriving itself of forest resources
with its own consumption and so-called progress. Solutions
to this very serious world problem are possible through
research, reforestation, and land management.

THE FUTURE OF THE WORLD'S FORESTS:
A LOOK AT MOTHER NATURE ON THE RUN

In the time it takes for you to read this sentence, eight acres of forest will disappear (Reiss 117).

This alarming statement is a good representation of the speed by which the world's forests are being lost. The tropical forests of South America alone are being cleared at the rate of 250,000 square kilometers a year and are believed to make up a mere half of their original area (Raven 633). In a world so dependent on forests and their products, these facts cannot be overlooked, and, as years pass, the figures become more and more ominous. What does the future hold for the world's forests? That is the question that is the basis of this report.

GEOGRAPHICAL ANALYSIS

The world's forests are rapidly disappearing all over the globe, on all the continents, and in nearly every country in the world. The problems involved in deforestation, however, are far more noticeable in the world's underdeveloped countries, which are located primarily around the equator. (See Figure 1.) This fact is due not only to the population and poor economic conditions of these countries but also to the lack of technological advancements in these countries. These countries include such states as Brazil, Peru, Colombia, and Venezuela in South America; Mexico in North America; Niger, Mali, Gabon, Zaire in Africa; and India, Burma, Bangladesh, and Thailand in Asia.

The developed countries of this world, on the other hand, are not as vulnerable to the rapid deforestation of the earth because of their low population density and because of the relative economic stability they possess. Therefore, the developed countries, which occur primarily north of the equator, have established the northern hemisphere as the last great stronghold of the earth's forests.

4

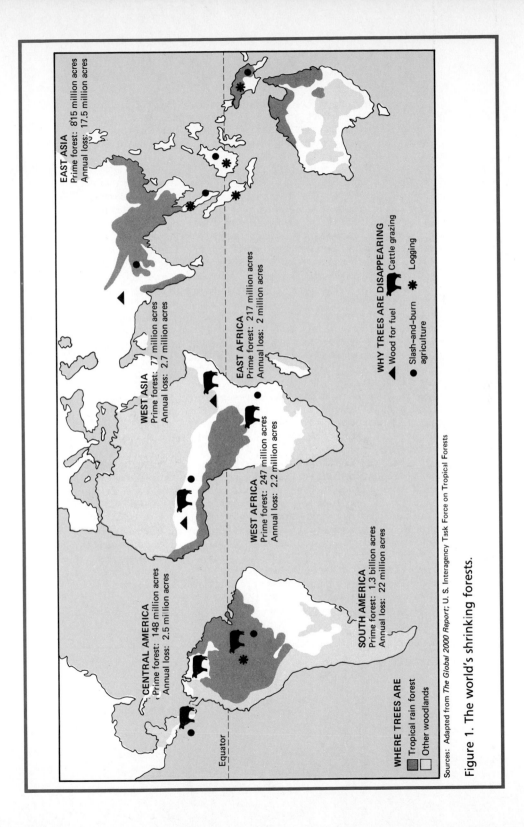

WHERE TREES ARE
▨ Tropical rain forest
☐ Other woodlands

CENTRAL AMERICA
Prime forest: 148 million acres
Annual loss: 2.5 million acres

SOUTH AMERICA
Prime forest: 1.3 billion acres
Annual loss: 22 million acres

WEST AFRICA
Prime forest: 247 million acres
Annual loss: 2.2 million acres

EAST AFRICA
Prime forest: 217 million acres
Annual loss: 2 million acres

WEST ASIA
Prime forest: 77 million acres
Annual loss: 2.7 million acres

EAST ASIA
Prime forest: 815 million acres
Annual loss: 17.5 million acres

WHY TREES ARE DISAPPEARING
▲ Wood for fuel   ⬟ Cattle grazing
● Slash-and-burn   ✳ Logging
  agriculture

Equator

Sources: Adapted from *The Global 2000 Report*; U. S. Interagency Task Force on Tropical Forests

Figure 1. The world's shrinking forests.

FOREST DEPRIVANTS

The reason behind the rapid deforestation of the earth
is our need to exploit our natural resources toward our
own good. Due to the skyrocketing rate of population
growth, consumption of forest materials is the major
threat to our forests.

Agriculture

One of the major deprivants of our forests is another
of our exploitations of the earth, agriculture. In the
rush to feed our starving world, agriculture is becoming
more and more important. Unfortunately, land needed for
crops is often cleared from existing forestland. As Erik
Eckholm states in his report on deforestation, "When soil
fertility is lost, cultivation is abandoned and the land
is often grazed. The bare soil will frequently return to
the forest, unless, as is often the case, it is first
destroyed by erosion" (14). Often, this replacement of
forest lands with crops is attempted at high altitudes
where forests thrive and crops do poorly, leaving bare,
eroded patches of earth when the crops die.

Forest Harvesting

Another form in which our forests are being destroyed
is the actual harvesting of trees for consumption. Lumber
is still the major use of forest products, the major
consumers of which are the United States and other
developed countries.

Pulpwood is another use of timber that is growing in
need of resources daily. Besides being used for paper and
pressboard, the cellulose from pulpwood can be used in the
production of cellophane, rayon, and plastic. Each Sunday
edition of The New York Times requires 153 acres of
southern pines; similarly, the paper used by McDonald's
alone for paper bags and napkins requires 315 square miles
of trees each year (The Renewable Tree). Also, paper is
the driving force behind the current "information boom,"
as the feedstock of government, education, and most
interpersonal communications (Eckholm 13).

6

Another major use of timber and forest products is
<u>fuel</u>. In a world that is rapidly losing its natural
sources of energy, more and more people are turning to
wood for fuel, an uneasy answer to a very complex problem.
Wood is the principal source of fuel for cooking and
warmth in the underdeveloped countries of Africa, Asia,
and Latin America (Maersch). Vast areas around the towns
and cities of these countries are totally denuded due to
the clearing of these forests to supply the energy-hungry
masses, and, in these poorer countries, the money is not
available for reforestation and the land is destroyed by
erosion or overrun by useless deserts.

The recent energy crunch has also shown its effects on
American forests with the increased interest shown in
heating homes with woodburning stoves. The Department of
Energy estimates that there are over 5 million woodburning
stoves in the United States and that wood provides the
United States with two-thirds as much energy as does
nuclear power. A comparison of the cost of energy obtained
from the four major fuels (Starnes 12) shows that the
economics of wood burning is sound.

COST PER MILLION Btu's

| Oil............ | $7.81 | Coal.......... | $2.38 |
| Natural Gas.... | $4.64 | Wood.......... | $2.30 |

In an ever tightening economy, these figures are
increasing the consumption of cordwood in the United
States astronomically.

Another use of wood as an energy source, which was
first developed by the Germans in World War II as an
alternative fuel source to oil, is ethanol. Ethanol is a
combustible fuel produced by breaking down the cellulose
in woody material into sugars that are fermented into
ethanol. Although ethanol is not projected to replace oil,
it is a good way to stretch our currently depleted oil
supplies to last longer than expected.

Due to new technology, it is now possible to convert

7

fiber into wood pellets by pressure. These pellets can be
used in much the same way as coal while eliminating the
bad points of coal such as mining and shoveling, not to
mention the mess involved in household heating by coal.
Trucks carrying wood pellets make deliveries to homes and
simply dump the pellets down a chute, where they are
automatically fed into a furnace (Harris 42).

Urbanization

There is yet another threat to our forests that seems
uncontrollable; that is the rapid urbanization of the
earth as the population explosion continues. Urbanization
destroys many millions of acres of forests yearly, through
clearing land for industry as well as for homes. Along
with this urbanization come the usual human tendencies to
pollute and exploit our natural resources.

EFFECTS OF DEFORESTATION

The effects of deforestation are twofold: ecological
and economic.

Ecological Effects

First and most noticeable of these effects are the
effects on the ecology. Clearcuts, areas of forests where
the land has been totally denuded, are seen all over the
world and have much the same aesthetic values as strip
mines. However, there are more ways than just the actual
loss of trees in which the loss of our forests is
affecting the earth's ecology.

The earth's water supply is one of the victims of
deforestation (Brett 10). Tree roots collect and hold
water during rains, keeping the balance of water even over
large tracts of land. Also the roots of trees hold the
soil together, preventing the formation of silt in our
waterways as water travels through its cycle.

The soil of the earth is also endangered by
deforestation. Trees are one of, if not the most,
effective preventives of erosion. The leaves of the trees
stop rainfall from hitting the ground full force and

loosening the soil for erosion. That, along with the spongy soil of the forest, which absorbs excess water, is a very good natural way of preventing erosion.

Another of the ill effects of deforestation is what scientists call the greenhouse effect. This situation is currently happening on the earth, and is growing in proportion every day. The greenhouse effect is a global warming trend due to the overabundance of carbon dioxide in the air, which is compounded by the extensive clearing and burning of tropical and temperate forests. If left unchecked, this phenomenon could melt the polar ice caps and raise the sea level by over 20 feet in 70 years (Reiss 120). The clearing and burning of the world's forests account for a full third of the carbon dioxide in the atmosphere, while removing carbon deposits from the earth (Delcourt 35). Along with the increase of carbon dioxide due to deforestation, the oxygen supply of the earth is shrinking due to the destruction of trees, which are a major source of the earth's oxygen.

Yet another victim of our greed for timber is wildlife. This problem is best shown in the deforestation of the tropics, which is expected to be complete within a few decades. The consequences of this are expected to include the extinction of upwards of one-third of the estimated three million species of plants, animals, and microorganisms in the tropics (Raven 633). In a world where extinction of species occurs repeatedly, this trend can only produce ominous circumstances that no act of ours can reverse. Once a species becomes extinct, there is no bringing it back.

These ecological disasters that appear imminent cannot be simply overlooked. They will not go away by themselves; and, like all of the problems brought up in viewing our world's deforestation, they have no simple answers.

Economic Effects

Deforestation by humans has also brought about some problems that bear direct importance on the nations of the earth, economic shortages being the most obvious and important. Economics is defined as the allocation of scarce goods and resources, which becomes more and more difficult as this scarcity increases. As Peter Raven states it: "The loss of resources worldwide is a leading factor in inflation and unemployment, but it is not often stressed by economists" (633). This callousness shown by our social leaders is only one of the many reasons the rapidly growing population of the earth is starved for natural resources, and will be even more so in the future unless the proper actions are taken.

The underdeveloped countries of the world, with their unstable political leadership and overabundance of people, are now, and have been for quite some time, feeling the hunger pains of a starving world. The question to these people is not whether or not to conserve; rather they must ask where the next day's meal or fuel for cooking and heat will come from. As is obvious, a cold and starving man is not going to consider the consequences of cutting a tree for fuel or of clearing a few acres of forest to grow his crops. As one Malaysian forestry official put it, "Our hardwoods take 90 years to grow. What bank will lend us 90-year money?" (Reiss 122). Consequently, the Third World countries of the tropics are moving rapidly toward mass starvation, social disorder, and a profound disruption of international economics (Watson and Bickers).

The developed countries of the earth are not much better off because the entire world must feel the pains eventually. The developed countries are losing the Third World as a market for their goods as well as a source of raw materials. Along with that, there will be a mass migration out of the Third World countries into the developed countries (Polsgrove 43).

Yet, on the more optimistic side, the developed countries of the world still have the technology necessary to create more of these natural resources as well as conserving them more efficiently and using them to our best advantage. Let us hope that our technology can stay one step ahead of our growing population.

10

POSSIBLE SOLUTIONS

What kind of technology exists to relieve the world of this timber crunch? It all depends on the advancement of our research technology, proper land management, and reforestation.

Research

Research in the field of forestry is currently being conducted all over the world, from the highly energy-intensive tree farms of the United States to the infant forestry program in India.

Despite the lack of research funds spent in underdeveloped countries, only 5 percent of the world total, many important advancements are being made in forestry in the Third World. First among these advancements is the advancement of a relatively new science called agroforestry.

Agroforestry is the scientific rotation of food crops and timber, thereby eliminating the destruction of millions of acres of forests, while at the same time producing much needed food. Crops are rotated with timber, using the timber growth period as rest for the soil. Agroforestry is a partial answer to the Third World countries' fuel and food problems, producing food, fuel, and forests simultaneously (Polsgrove 43).

Another form of research currently being done in the Third World countries is being done with legumes. These are small, fast-growing trees that are extremely weather resistant and can be used for food as well as for fuel and paper (Reiss 122). Two of the most successful legumes are Gmelina and Leucaena, two fast-growing tropical trees that could provide part of the answer to the Third World countries' severe timber shortage.

The American forestry situation is only marginally better. The U.S. Forest Service predicts that the volume of wood produced in the United States must double over the next 50 years to meet domestic needs (Starnes 12). This fact, along with the current inflation rate and the increasing thirst for energy (which has direct effects on

11

our timber resources), presents us with a grim outlook for the future of our forest resources.

What can the United States do to curb the tide of our growing timber shortage? The key lies, Wiegner contends, in the advancement of our forest technology. Currently, our forest technology leads the world, resulting in a greater yield per acre in the United States than anywhere else in the world. Also, the U.S. Forest Service predicts that, with proper management, the country could double, perhaps triple, the productivity of its forests within 50 years (41).

Reforestation

Another key to staying one step ahead of global deforestation is reforestation. Reforestation is necessary to stem the tide of deforestation, yet it lacks economic support in most cases; the exceptions are large industry and government, which can afford such a long-time investment. Once a seedling is planted, the economic return will not show up for at least 25 years.

Despite the ever increasing cost of reforestation, American landowners, industry, and government are seeing the need of reforestation in a growing world. In 1979, over 600 million seedlings were planted commercially in the South alone (The Renewable Tree).

Land Management

To support reforestation and advancing technology, proper land management must be practiced in order to receive the full effect of a necessary concerted effort. The responsibility for land management in the South lies primarily with the private landowner, who owns approximately 73 percent of the timberland in the South. Government land management is foremost in the West, however, where government lands make up approximately 71 percent of the forested land area.

Although the United States is leading the way in technology and development, we must share our valuable knowledge with the other countries of the world in order

to help them meet the increasing shortage of timber with
resources of their own. For just as all people are the
cause of deforestation, so should all people be the cause
of reforestation.

WORKS CITED

Brett, R. M. "A Different View of the Forest." <u>American
    Forests</u> Dec. 1982: 10, 43–44.

Delcourt, Hazel R. "The Virtue of Forests, Virgin and
    Otherwise." <u>Natural History</u> June 1981: 33–39.

Eckholm, Erik. "The Shrinking Forests." <u>Focus</u> Sept.–Oct.
    1982: 12–16.

Harris, Michael. "Farewell Forests." <u>The Progressive</u> Feb.
    1983: 42–43.

Maersch, Stephen. "Fuel That Grows on Trees." <u>Milwaukee
    Journal</u> 17 Jan. 1982. NewsBank, Consumer Affairs, 1982,
    2:C6.

Polsgrove, Carol. "The Vanishing Forests of the Third
    World." <u>The Progressive</u> Aug. 1982: 42–43.

Raven, Peter H. "Raven Warns of Rapid Tropical
    Deforestation." <u>Bioscience</u> Oct. 1982: 633.

Reiss, Spencer, "Vanishing Forests." <u>Newsweek</u> 24 Nov.
    1982: 117–22.

<u>The Renewable Tree</u>. Videotape. Educational Resources,
    1983. 60 min.

Starnes, Richard. "The Tree Eaters Are Coming." <u>Outdoor
    Life</u> Oct. 1981: 12–13.

Watson, Steve, and Chris Bickers. "Timber!" <u>Southern World</u>
    Mar./Apr. 1981: 36–38. Rpt. in Social Issues Resources
    Series, Energy 3: art. 25.

Wiegner, Kathleen K. "America's Green Gold." <u>Forbes</u> 34
    (Dec. 1982): 40–46.

14

Application 1 Writing a Research Report

Answer the following concerning your college library.

1. What are the library hours?
2. Which system of classifying books (Library of Congress or Dewey Decimal System) does the library use?
3. What is the procedure for checking out a book?
4. For how long a period of time may books be checked out?
5. Where are reserve books shelved?
6. What is the procedure for requesting a magazine?
7. Where are current magazines located?
8. If the library has audiovisual materials, where are they located?
9. Where are microfilm and microfiche materials and readers located?

Application 2 Writing a Research Report

Draw a floor plan of your college library. Indicate the location of the following: public (card) catalog; periodical indexes; periodicals holdings list; general encyclopedias; technical encyclopedias, dictionaries, and guides; biographical reference works; and works in your major field.

Application 3 Writing a Research Report

In a periodical index (such as the *Magazine Index, Readers' Guide to Periodical Literature,* or *Applied Science & Technology Index*) look up a general subject, such as hydraulics, nutrition, computers, radio receivers, metals, recreation, stock market, seawater, space research, antipollution laws, soils, glass, fibers, eye, electronic circuits, wood, internal-combustion engines, and so on. For the general subject, list at least ten subclassifications that would be sufficiently limited for a research report.

Application 4 Writing a Research Report

Choose one of the following topics (or pick a topic of your own). Formulate a possible basic question, or basic problem, to be investigated in a research report. Then divide the basic question into a list of specific questions (preliminary outline).

1. Problems in developing an electric automobile
2. Early results of the discovery of penicillin
3. Effects of marijuana on the body
4. Benjamin Franklin's minor inventions
5. Architectural problems in designing the Astrodome
6. The shooting down of Korean Airlines Flight 007
7. July 20, 1969
8. Cooking food electronically
9. The next ten years in space exploration

10. The American office—100 years ago and today
11. The laser in surgery
12. The effects of the disappearing farms on the U.S. economy
13. The United States sale of arms to Iran
14. Effects of the computer on business
15. Changes in marketing techniques

Application 5   Writing a Research Report

Explain each of the items on the following card from a public (card) catalog.

```
657
M47a3  Meigs, Walter B
          Accounting: the basis for business
       decisions [by] Walter B. Meigs, A. N.
       Mosich [and] Charles E. Johnson. 4th ed.
       New York, McGraw-Hill [1982]
          xxiv, 936p. 24cm.

          1. Accounting.  I. Mosich, A.N.,
       joint author.  II. Johnson, Charles E.,
       joint author
```

Application 6   Writing a Research Report

Turn to page 446. Look at the last three entries from the *Applied Science & Technology Index*. Explain the items in each of the entries.

Application 7   Writing a Research Report

Select a topic that is sufficiently restricted for a research report. Find and list at least two magazine articles, two books, and one encyclopedia article that contain information on the topic. Use conventional bibliographical forms. (See pages 473–474 for examples.)

Application 8   Writing a Research Report

Using the article on pages 622–623, "The World of Crumbling Plastics," do the following:

1. Make a bibliography card for the article.
2. Make a direct-quotation note card for the first sentence in paragraph 5.
3. Make a paraphrase note card of the information in the second and third sentences of paragraph 5.
4. Make a direct-quotation note card for the first sentence in paragraph 4, omitting the phrase "a chemist."
5. Make a summary note of paragraph 6.
6. Of the information on the preceding note cards, which would require notes (either footnotes or endnotes) in a paper? Why?
7. Show documentation for the summary note written in number 5 above.
8. Write a sentence about the need for a substitute for the currently used plastic sheeting (paragraphs 1, 2, and 3) in which you quote these words: "farmers spend $100 to $200 an acre . . . to dispose of the 125 million pounds of plastic mulch."

Application 9   Writing a Research Report

Arrange the following information as if it were the final bibliography for a report.

Computers and Their Applications, by Larry J. Goldstein, published in 1986 by Prentice-Hall, Inc., Englewood Cliffs, New Jersey.

Legal Aspects of Computer Use, by William J. Luddy and Stuart R. Wolk, published in 1986 by Prentice-Hall, Inc., in Englewood Cliffs, New Jersey.

Software Development Tools: A Source Book, edited by Stephen J. Andriole, published in 1986 by Petrocelli Books, Princeton, New Jersey.

"Speedy Graphics Engine Ushers in 2000-line Screens," by S. Ohr, in *Electronic Design*, the 7 August 1986 issue, vol. 34, page 38.

"A Benchmark Comparison of 32-bit Microprocessors," by T. C. Cooper and others, in *IEEE Micro*, the August 1986 issue, vol. 6, pp. 53–8.

"Protecting Programs from Pirates," by Erik Sandberg-Diment, in *The New York Times*, Section 3, page 14, columns 1–5, on 27 July 1986.

Application 10   Writing a Research Report

Write a research report as directed by your instructor. Use the schedule below as a checklist of steps in writing the report. Fill in the appropriate Plan Sheet from pages 495–502 before beginning each major step.

Schedule for Writing a Research Report

ASSIGNMENT	DATE DUE
Choice of general subject	_____
Choice of specific topic	_____
Statement of basic problem (basic question)	_____

Statement of specific questions (preliminary outline)   _____

Working bibliography   _____

Note cards   _____

Working outline   _____

Purpose statement or central idea   _____

First draft
  With correct documentation   _____

Final report, including outline and documentation
  (Your instructor may also require that bibliography
  cards, note cards, outlines, preliminary drafts, etc., be
  turned in with the final report.)   _____

Application 11   Writing a Research Report

As directed by your instructor, adapt your research report for oral presentation. Ask your classmates to evaluate your speech by filling in an Evaluation of Oral Presentations from pages 527–535.

USING PART II: SELECTED READINGS

Application 12   Writing a Research Report

Read the article "Holography," by Jonathan Back and Susan S. Lang, pages 589–594. Under "Suggestions for Response and Reaction," write Number 4.

Application 13   Writing a Research Report

Read the article "American Labor at the Crossroads," by Steven M. Bloom and David E. Bloom, pages 583–588. Under "Suggestions for Response and Reaction," write Number 5.

PLAN SHEET 1
FOR WRITING A RESEARCH REPORT:
SELECTING A SUBJECT

Technical Field

My technical field is:

Allotted Time

Today's date is:

The completed report is due on:

That means I have _____ weeks to work on this report.

Possible Subjects

Five to ten possible subjects for my research report are the following:

1.   6.

2.   7.

3.   8.

4.   9.

5.   10.

Specific Topics

From the preceding list of possible subjects, the one that interests me the most is:

I can narrow this subject to at least three topics that I could treat satisfactorily in the allotted time:

1.

2.

3.

Chosen Topic

After checking in the library and with my instructors to be sure there is sufficient available material, I have chosen the following as the topic for my research report:

Questions

I need to ask my instructor these questions:

PLAN SHEET 2
FOR WRITING A RESEARCH REPORT:
DEFINING THE PROBLEM

Subject

The topic for my research report is:

General Problem

I can state the general problem or idea that I am investigating as this question:

Specific Problems

I can divide the general problem into these specific problems or questions:

Questions

I need to ask my instructor these questions:

PLAN SHEET 3
FOR WRITING A RESEARCH REPORT:
FINDING THE FACTS

Subject

The subject for my research report is:

Sources to Be Researched

Personal Observation or Experience   In gathering information for writing, I will need to visit the following places to observe conditions:

Personal Interviews   In gathering information for writing, I may want to interview the following people:

Library Materials   The library (or libraries) I will use is (are):

A tentative bibliography of available materials includes:

Free and Inexpensive Materials   In gathering information for writing, I may write or call the following companies or agencies:

Questions

I need to ask my instructor these questions:

PLAN SHEET 4
FOR WRITING A RESEARCH REPORT:
RECORDING AND ORGANIZING THE FACTS

Subject

The subject for my research report is:

Purpose Statement or Central Idea

I can sum up the purpose or the main idea of my report in the following sentence:

Formal Outline

The formal outline is a _____ (sentence or topic) outline. The outline is as follows:

Introduction

The introductory material for the report is as follows:

Closing

The closing material for the report is as follows:

Visuals

Kinds of visuals and what they will show are as follows:

Questions

I need to ask my instructor these questions:

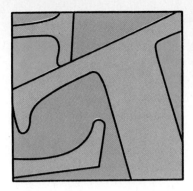

Chapter 10
Oral Communication: Saying It Clearly

503

Objectives

Upon completing this chapter, the student should be able to:

- List ways in which an oral presentation differs from a written presentation
- List and identify the modes of delivery for a formal speech
- Classify oral presentations according to general purpose
- Prepare and use visuals in oral presentations
- Prepare and deliver to a specific audience an oral presentation that meets a clearly stated purpose
- Evaluate an oral presentation
- Demonstrate a knowledge and an understanding of how to hold a job interview
- Demonstrate a knowledge and an understanding of how to hold an informational interview

Introduction

Oral expression is the most important communication skill that you need to master. In school or on the job, you probably speak thousands of times as often as you communicate in any other way. Even a moment's consideration of how people make a living will point to the importance of being able to express ideas orally.

Every time you speak, you convey something of who you are and what you think. Your vocabulary, pronunciation, grammatical usage, phrasing, and expressed ideas are aspects of speech that make an impression on your listeners and by which they then form opinions about you. Unfortunately, sometimes the impression you make can be a negative or poor one simply because you cannot orally express what you want to say. Such a situation shows a lack of self-confidence and knowledge in handling oral communication.

This chapter will acquaint you with various group and individual communication situations and will give suggestions to help you develop the self-confidence and knowledge needed for effective oral expression.

Speaking Differs from Writing

Speaking and writing have much in common because they are both forms of communication based on language. Speaking differs, however, in several important ways:

1. *Level of diction*. In speaking, a simpler vocabulary and shorter sentences of less involved structure are typically used.
2. *Amount of repetition*. More repetition is needed in speaking to emphasize and to summarize important points.
3. *Kind of transitions*. Transitions from one point to another must be more obvious in speaking. Such transitions as *first, second,* and *next* signal movement often conveyed on the printed page through paragraphing and headings.

4. *Kind and size of visuals*. Speaking lends itself to the use of exhibits and projected materials; some kinds of flat materials such as charts, drawings, and maps must be constructed on a large scale.

Study these features in the speech "To Serve The Nation: Life Is More Than a Career" on pages 632–637.

Classification as Informal and Formal

Oral communication might be broadly classified as informal and formal. The term *informal* describes nonprepared speech and *formal* describes well-planned, rehearsed speech. Most of us spend a large percentage of each day in informal communication: We talk with friends, parents, fellow workers, neighbors, other family members. Often, individuals may be asked to share views or knowledge about events, people, places, or things with a group, such as a service club, a professional organization, or a class. They may have been asked at the meeting to respond impromptu; or they may have been asked ahead of time but made little or no preparation. In both cases, their information will be shared through informal oral communication.

On the other hand, formal oral communication involves a great deal of preparation and attention to delivery. Professional and learned people are often asked to share their views and knowledge regarding their field. The deliberate, planned, carefully organized and rehearsed presentation of ideas and information for a specific purpose constitutes formal communication.

MODES OF DELIVERY FOR FORMAL PRESENTATIONS

Formal presentations may be categorized according to the speaker's mode of delivery as:

☐ Extemporaneous
☐ Memorized
☐ Read from a manuscript

In the extemporaneous mode—the most often used of the three—the speaker refers to brief notes or an outline, or simply recalls from memory the points to be made. In this way, the speaker is able to interact with the audience and convey sincerity and self-assurance. In the memorized mode, the speaker has written out the speech and committed it to memory, word for word. The memorized speech is typically lacking in spontaneity and an at-ease tone; too, the speaker has the very real possibility of forgetting what comes next. In the third mode of delivery, the speaker reads from a manuscript. While this type of delivery may be needed when exact wording is required in a structured situation, the manuscript speech has serious limitations. It is difficult for the speaker to show enthusiasm and to interact with the audience, delivery is usually stilted, and the audience may soon become inattentive.

The effective speaker considers the occasion, the audience, and the purpose of the speech in determining the mode of delivery to be used.

Classification According to General Purpose

In addition to informal and formal, another way to classify oral communication is according to its general purpose: to entertain, to persuade, or to inform. Any one of these types of communication could be presented formally or informally.

TO ENTERTAIN

Oral communication that is meant to entertain is intended to provide enjoyment for those who listen. Probably neither a technician nor a student would have many occasions to communicate solely for the purpose of entertaining except on an informal basis with friends and relatives.

TO PERSUADE

The goal of communication that is meant to persuade is to affect the listeners' beliefs or actions. As a technician you may well find yourself responsible for persuading supervisory personnel or customers or employees to change a method or a procedure, to hire additional personnel, to buy a certain piece of machinery or equipment, and so forth. Whether presenting an idea, promoting a plan, or selling a product, the same basic principles of persuasion are involved.

The art of persuasion can be summed up in two sentences: Present a need, want, or desire of the customer (buyer). Show the customer how your idea, service, or product can satisfy that need, want, or desire.

Basic Considerations in Persuasion. If you are to be able to prepare and present a persuasive speech effectively, you must be aware of several factors: needs, wants, and desires of people; and kinds of appeals.

Needs, Wants, and Desires of People. The actual needs of people in order to exist are few: food, clothing, and shelter. In addition to these physical needs, people have numerous wants and desires. Among these are the following:

- □ *Economic security*. This includes a means of livelihood and ownership of property and material things.
- □ *Recognition*. People want social and professional approval. They want to be successful.
- □ *Protection of self and loved ones*. Safety and physical well-being of self and of family and friends are important.
- □ *Aesthetic satisfaction*. Pleasant surroundings and pleasing the senses can be very satisfying.

Consideration of people's needs, wants, and desires is essential in order to effectively present an idea, plan, or product. For example, if you were to try to persuade your employer to purchase new typewriters, you probably would appeal to the employer's economic and aesthetic desires; that is, you probably would emphasize the time that could be saved with new typewriters and the improved appearance of typewritten material. Or if you were a supervisor trying to impress upon a worker the importance of following dress regulations, you would stress protection (safety) of self and the possible loss of economic security for the worker and his or her family should injury or accident occur. Or if you were to try to persuade a co-worker to take college courses in the evening, you would point out the recognition and economic security aspects—gaining recognition and approval for furthering education and skills and the possible subsequent financial rewards.

Appeals—Emotional and Rational, Direct and Indirect. Persuasion is the process of using combined emotional and rational appeals and principles. Emotional appeals are directed toward feelings, inclinations, and senses; rational appeals are directed toward reasoning, logic, or intellect. Undoubtedly, many times emotional appeals carry more weight than do rational appeals.

The most satisfying persuasion occurs when people make up their own minds or direct their own feelings toward a positive reception of the idea, plan, or product—but without being told to do so. Thus, indirect appeals, suggestions, and questions are usually much more effective than a direct statement followed by proof.

The Persuasive Presentation. The persuasive presentation involves four steps or stages: opening, need and desire intensified, supporting proof, and close. For timing in moving from one step to another, you must use your judgment by constantly analyzing conditions and audience mood.

In approaching a single listener or a small group of listeners, be sincere and cordial. A firm handshake should set a tone of friendliness. Is it natural also to exchange a few brief pleasantries (How are you?—How is business?—Beautiful weather we are having—etc.) before getting down to business. With a large group of listeners, such a personal approach is impossible. You can, however, be sincere and cordial.

Opening. In the opening, the listener's attention is aroused. Thus the opening should immediately strike the listener's interest and should present the best selling points. This may be done directly or indirectly.

DIRECT: Our new Top Quality razor blades give a closer, cleaner shave than any others on the market.
INDIRECT: Do your customers ask for razor blades that give a closer, cleaner shave?

Need and Desire Intensified. When the audience's attention is aroused, each main selling point is then developed with explanatory details.

Both emotional and logical appeals are used to show how the proposal will help satisfy one or more of these basic desires: economic security, recognition, protection of self and loved ones, and aesthetic satisfaction.

At this stage and throughout the proposal presentation, the listener may raise objections. The best way to handle these is to be a step ahead of the listener; that is, to be aware of all possible objections, prepare effective responses, and incorporate them into your main presentation.

Supporting Proof. Description and explanation must be supported by evidence or proof. Visual exhibits, demonstrations, testimony of users, examples of experience with and uses of the product, and statistics showing specifications and increased productivity are methods to prove the worth of your proposal.

Close. In closing a presentation, you usually are wise to assume the positive attitude that the audience will accept the idea or plan or will buy the product. The following suggestions reflect such an attitude: *When may we begin using this procedure? Which model do you prefer?* Reaffirmation of how the proposal will enhance the listener's business often helps to conclude the deal and to reinforce his or her satisfaction. If you detect a negative attitude, avoid a definite "no" by suggesting further consideration of the proposal or a trial use of the product and another meeting at a later date.

TO INFORM (EMPHASIS OF REMAINDER OF CHAPTER)

Of the three general purposes of oral communication—to entertain, to persuade, and to inform—the informative purpose is most frequently employed by the technician. In communicating instructions and processes, descriptions of mechanisms, definitions, analyses through classification and partition, analyses through effect and cause and through comparison and contrast, summaries, and reports (Chapters 1–6, and 8 in Part I), the speaker will have as a major goal *informing* the listeners.

Giving oral informative presentations is a very significant aspect of the technician's communication responsibilities. The next three sections of this chapter—Preparation of an Oral Presentation, Delivery, and Visual Materials—are geared to the informative speech. The basic principles discussed in these sections are, of course, applicable to any speech situation. For instance, the steps in preparing a speech are essentially the same, whether the purpose of the speech is to entertain, to persuade, or to inform.

As you study this chapter, review the two student oral presentations—"How a Common Table Mushroom Grows," pages 45–46, and "Description: The Hewlett-Packard ColorPro Plotter 7440A," pages 75–77—and the speech "To Serve the Nation: Life Is More Than a Career," pages 632–637.

Preparation of an Oral Presentation

Preparing an oral presentation includes these steps:

- ☐ Determine the specific purpose
- ☐ Analyze the type of audience
- ☐ Gather the material
- ☐ Organize the material
- ☐ Determine the mode of delivery
- ☐ Outline the speech, writing it out if necessary
- ☐ Prepare visual materials
- ☐ Rehearse

DETERMINE THE SPECIFIC PURPOSE

The general purpose of a technician's speech typically is to inform; sometimes the general purpose is to persuade, or, occasionally, to entertain. The *specific* purpose, however, must be determined if the speech is to be effective. The reason for the speech and who will use the information must be established. Data that completely, accurately, and clearly present the subject must be given, analyzed, and interpreted thoroughly and honestly. Recommendations can then be made accordingly.

ANALYZE THE TYPE OF AUDIENCE

A speech, if it is to be effective, must be designed especially for the knowledge and interest level of the intended audience or listeners. Vocabulary and style must be adapted to the particular audience. For instance, if you were to report on recent applications of the laser, your report to a group of nurses, to a group of engineers, to a college freshman class of physics students, or to a junior high science club would differ considerably. Each group represents a different level of knowledge and a different partisan interest.

GATHER THE MATERIAL

The material is gathered primarily from three sources: interviews and reading, field investigation, and laboratory research.

The extent to which one or more of these sources will be used depends on the nature of the speech. A student reading report in history, for instance, may simply call for the reading of certain material in a book. An investigation of parking facilities in a particular location may call for personal interviews plus on-site visits. Or an analysis of the hardiness of certain shrubs when exposed to sudden temperature changes may involve both field investigation and experimental observation.

ORGANIZE THE MATERIAL

To organize the material, select the main ideas; do not exceed three or four. (Remember that your audience is listening, not reading.) Arrange supporting data under each main idea. Use only the supporting data necessary to develop each main idea clearly and completely.

After the main body of material is organized, plan the introduction. Let the audience know the reason for the speech, the purpose, the sources of data, and the method or procedure for gathering the data. Then state the main ideas to be presented. The function of the introduction is to set an objective framework in which the audience will accept the information as accurate and as significant.

Plan the conclusion. It should contain a summary of the data, a summary of the significance or of the interpretation of the data, and conclusions and recommendations for action or further study.

A suggested outline for the introduction, body, and conclusion is given in the section Outline the Speech, Writing It Out if Necessary.

(The organization of a persuasive speech is treated earlier in the chapter.)

DETERMINE THE MODE OF DELIVERY

Once you have analyzed the speaking situation and gathered and organized the material, you are in position to determine the appropriate mode of delivery. Is it more appropriate to speak extemporaneously, to recite a memorized speech, or to read from a manuscript? (Of course, you may have been told which mode to use; thus the decision has already been made for you.) The memorized speech is most appropriate in such situations as competing in an oratorical contest or welcoming an important visiting dignitary. Reading from a script is most appropriate if presenting a highly technical scientific report, giving a policy speech, or the like. For most other situations, extemporaneous speaking is the most appropriate.

OUTLINE THE SPEECH, WRITING IT OUT IF NECESSARY

Outline your speech. A suggested outline form is as follows:

Introduction
I.   Reason for the speech
   A. Who asked for it?
   B. Why?
II. Purpose of the speech
III. Sources of data
IV. Method or procedure for gathering the data
V.   Statement of main ideas to be presented

Body
I.   First main idea
   A. Sub-idea
      1., 2., etc. Data
   B., etc. Sub-ideas
      1.,2., etc. Data
II., III., etc. Second, third, etc., main ideas
   A., B., etc. Sub-ideas

Conclusion
I.   Summary of the data
II. Summary of the significance or of the interpretation of the data
III. Conclusions and recommendations for action or further study

If you plan to present a memorized speech or read from a script, write out the speech. Special care should be given to manuscript form and to the construction of visuals if you are to distribute copies of the speech (copies should be distributed *after* the oral presentation, not before or during it).

PREPARE VISUAL MATERIALS

Carefully select and prepare visuals to help clarify information and to crystallize ideas. See the section that follows entitled Visual Materials.

REHEARSE

For an extemporaneous speech: From your outline, make a note card (3- by 5-inch, narrow sides up and down) of the main points that you want to make. Indicate on the card where you plan to use visuals. Rehearse the entire speech several times, using only the note card (not the full outline). Get fixed in your mind the ideas and supporting data and the order in which you want to present them. For a memorized speech: Commit to memory the exact wording of the script. As you practice the speech, put some feeling into the words; avoid a canned, artificial sound. For a speech read from a manuscript: Just because you are to read a speech doesn't mean you shouldn't practice it. Go over the speech until you know it so thoroughly that you can look at your audience almost as much as you look at the script. Number the pages so that they can be kept in order easily. Leave the pages loose (do not clip or staple them together); you can then unobtrusively slide a finished page to the back of the stack.

Some speakers find it helpful to tape record their speech once or twice while rehearsing; then they play back the recording for an objective analysis of their strengths and weaknesses.

Rehearsing your presentation several times is very important; it gives you self-confidence and it prepares you to stay within the time allotted for the speech.

Delivery

A major factor in oral communication is effective delivery, or *how* you say what you say. When giving a speech, observe the following suggestions.

WALK TO THE PODIUM WITH POISE AND SELF-CONFIDENCE

From the moment the audience first sees you, give a positive impression. Even if you are nervous, the appearance of self-confidence impresses the audience and helps you to relax.

CAPTURE THE AUDIENCE'S ATTENTION AND INTEREST

Begin your speech forcefully. Opening techniques include asking a question, stating a little-known fact, and making a startling assertion (all, of course, should pertain directly to the subject at hand).

LOOK AT THE AUDIENCE

Interact with the audience through eye-to-eye contact, but without special attention to particular individuals. You should not overdo looking at your notes, the floor, the ceiling, over the heads of your audience, or out the window.

STICK TO AN APPROPRIATE MODE OF DELIVERY

If, for instance, your speech should be extemporaneous, don't read a script to the audience.

PUT SOME ZEST IN YOUR EXPRESSION

Relax; be alive; show enthusiasm for your subject. Avoid a monotonous or "memorized" tone and robot image. Have a pleasant look on your face.

GET YOUR WORDS OUT CLEARLY AND DISTINCTLY

Make sure that each person in the audience can hear you. Follow the natural pitches and stresses of the spoken language. Speak firmly, dynamically, and sincerely. Enunciate distinctly, pronounce words correctly, use acceptable grammar, and speak on a language level appropriate for the audience and the subject matter.

ADJUST THE VOLUME AND PITCH OF YOUR VOICE

This adjusting may be necessary for emphasis of main points and because of distance between speaker and audience, size of audience, size of room, and outside noises. Be certain everyone can hear you.

VARY YOUR RATE OF SPEAKING TO ENHANCE MEANING

Don't be afraid to pause; pauses may allow time for an idea to become clear to the audience or may give emphasis to an important point.

STAND NATURALLY

Stand in an easy, natural position, with your weight distributed evenly on both feet. Bodily movements and gestures should be natural; well-timed, they contribute immeasurably to a successful presentation.

AVOID MANNERISMS

Mannerisms detract. Avoid such mannerisms as toying with a necklace or pin, jangling change, or repeatedly using an expression such as "You know" or "Uh."

SHOW VISUALS WITH NATURAL EASE

For specific suggestions, see Showing of Visuals on pages 515–516.

CLOSE—DON'T JUST STOP SPEAKING

Your speech should be a rounded whole, and the close may be indicated through voice modulation and a simple "Thank you" or "Are there any questions?"

Visual Materials

Visual materials can significantly enhance your oral presentation. Impressions are likely to be more vivid when visuals are used. In general, they are more accurate than the spoken word. Showing rather than telling an audience something is often clearer and more efficient. And showing *and* telling may be more successful than either method by itself. For instance, a graph, a diagram, or a demonstration may present ideas and information more quickly and simply than can words alone.

In brief, visual materials are helpful in several ways. They can convey information, supplement verbal information, minimize verbal explanation, and add interest.

See Chapter 11, Visuals, for a discussion of all types of visual materials for both oral and written communication.

GENERAL TYPES OF VISUALS IN ORAL PRESENTATIONS

Visuals for use with oral presentations can be grouped into three types: flat materials, exhibits, and projected materials. A brief survey of these can help you determine which visuals are most appropriate for your needs.

Flat Materials. Included in flat materials are two-dimensional materials such as the chalkboard, bulletin board, flannel board, magnetic boards, handout sheets, pictures, posters, cartoons, charts, maps, and scale drawings.

Although these are usually prepared in advance and revealed at the appropriate time (as in the picture, the chart, and the poster used in the oral presentation of the description of the ColorPro plotter, pages 75–77), sometimes they are created spontaneously during the presentation (as in the outlining of steps on the chalkboard in the oral presentation on how a mushroom grows, pages 45–46). A chalkboard or easel and paper (a pad of newsprint is excellent) serves beautifully. Actually, the visuals should be created in advance and reproduced from memory or notes during the presentation.

In using printed handout material, careful attention should be given to its time and manner of distribution. The main thing that the speaker should guard against is competing with his or her own handout material—the audience reading when it should be listening.

An easel is almost essential in displaying pictures, posters, cartoons, charts, maps, scale drawings, and other flat materials. Various lettering sets, tracing and template outfits, and graphic supplies can be purchased in hobby or art supply stores and facilitate a neat visual.

For a more detailed discussion of flat materials, see pages 542–554 in Chapter 11, Visuals.

Exhibits. Visual materials such as demonstrations, displays, dramatizations, models, mock-ups, dioramas, laboratory equipment, and real objects comprise exhibits. These are usually shown on a table or stand.

Undoubtedly the demonstration is one of the best aids in an oral presentation. In fact, at times the entire presentation can be in the form of a demonstration. When performing a demonstration, be sure that all equipment is flawlessly operable and that everyone in the audience can see; if practical, allow the audience to participate actively. (The oral presentation on how a mushroom grows, pages 45–46, uses demonstration and real objects.)

For a more detailed discussion of exhibits, see pages 554–556 in Chapter 11, Visuals.

Projected Materials. Projected materials are those shown on a screen by use of a projector: pictures, slides, films, filmstrips, and transparencies. (A transparency is used in the oral description of the ColorPro Plotter, pages 75–77.) When using projected materials, a long pointer is essential, and an assistant often is needed to operate the machine.

For a more detailed discussion of projected materials, see pages 556–557 in Chapter 11, Visuals.

The most effective use of visual materials occurs when the most appropriate kind of visual is selected and when the visual is prepared and shown well.

Preparation of Visuals. Once you have chosen specific kinds of visuals from the general types of flat materials, exhibits, and projected materials, careful attention should be given to their presentation. The following should assist you.

1. *Determine the purpose of the visual.* Select visuals that will help the audience understand the subject. Adapt them to your overall objective and to your audience.
2. *Organize the visual.* Information and its arrangement should be geared to quick visual comprehension.
3. *Consider the visibility of the aid: its size, colors used, and typography.* The size of the visual aid is determined largely by the size of the presentation room and the size of the audience. Visuals should be large enough to be seen by the entire audience.
4. *Keep the visual simple.* Do not include too much information.
5. *In general, portray only one concept or idea in each visual.*
6. *Make the visual neat and pleasing to the eye.* Clean, bold lines and an uncrowded appearance contribute to the visual's attractiveness.
7. *Select and test needed equipment.* If you need equipment to show your visuals—an overhead projector, a filmstrip projector, a movie projector—select the equipment and test it to be sure it is operable. Check the room for locations and types of electrical outlets; these may affect the placement of the visual equipment. Perhaps a long extension cord will be needed. Determining needs and setting up equipment ahead of time allow you to make your presentation in a calm, controlled manner.

Showing of Visuals. Visual materials should be shown with natural ease, avoiding awkwardness. This is the basic principle in showing visuals in any kind of oral communication. The following suggestions are simply aspects of that basic principle.

1. Place the visual so that everyone in the audience can see it.
2. Present the visual at precisely the correct time. If an assistant is needed, rehearse with the assistant. The showing of a visual near the beginning of a presentation often helps the speaker to relax and to establish contact with the audience.
3. Face the audience, not the visual, when talking. In using a chalkboard, for instance, be sure to talk to the audience, not the chalkboard.
4. Keep the visual covered or out of sight until needed. After use, cover or remove the visual, if possible. Exposed drawings, charts, and the like are distracting to the audience.
5. Correlate the visual with the verbal explanation. Make the relationship of visual and spoken words explicit.
6. When pointing, use the arm and hand next to the visual, rather than reaching across the body. Point with the index finger, with the other fingers loosely curled under the thumb; keep the palm of the hand toward the audience.
7. Use a pointer as needed, but don't make it a plaything.

Visuals should not be a substitute for the speaker, or a prop, or a camouflage for the speaker's inadequacies. Further, the use of visuals should not constitute a show, obviating the talk.

Appropriately used, visuals can decidedly enhance an oral presentation.

Evaluating an Oral Presentation

On the following page is a class evaluation form for oral presentations. The vertical spaces across the top are for students' names. The evaluative criteria are listed under two headings, "Delivery" and "Content & Organization"; to these is added "Overall Effectiveness." The members of the class are to evaluate one another on each criterion, using this scale:

4 = Outstanding

3 = Good

2 = Fair

1 = Needs improvement

Then the total number of points for each speaker is tabulated. Total scores can range from a high of 64 to a low of 16.

The evaluation procedure can be simplified, if desired, by using only the Overall Effectiveness criterion. The highest number of points for a speaker would then be 4; and the lowest, 1.

Job Interviews

Whether or not an applicant gets the job is usually a direct result of the interview. Certainly, the information in the application letter and resumé are important, but the *person* behind that information is the real focus (see pages 288–306 in Chapter 7, Memorandums and Letters, for a discussion of the application letter and resumé). The personal circumstance of the job interview allows the applicant to be more than written data. The impression that the applicant makes is often the deciding factor in whether he/she gets the job.

Following the suggestions below will help ensure your making the right impression. The suggestions are grouped according to the three steps in the job interview process: preparing for an interview, holding an interview, and following up an interview.

PREPARING FOR AN INTERVIEW

Careful attention should be given to preparing for an interview. This involves acquaintance with the job field, company analysis, job analysis, interviewer analysis, personal analysis, and preparation of a resumé. A procedure for rehearsing an interview is suggested at the end of this section.

EVALUATION OF ORAL PRESENTATIONS
(See page 516 for directions.)

Course and Section _____ Date _____ Evaluated by (Name) _____

Students' Names

DELIVERY

Forceful introduction

Poise

Eye contact

Sticking to mode of delivery

Zest (enthusiasm)

Voice control

Acceptable pronunciation and grammar

Avoidance of mannerisms

Ease in showing visuals

Clear-cut closing

Sticking to specified length

CONTENT & ORGANIZATION

Stating of main points at outset

Development of main points

Needed repetition and transitions

Effective kinds and sizes of visuals

OVERALL EFFECTIVENESS

TOTAL POINTS

4=Outstanding   2=Fair

3=Good   1=Needs improvement

COMMENTS:

Acquaintance with the Job Field. Part of preparing for a job interview involves learning as much as you can about the field in which you seek a job. Become familiar with the possible career choices, job opportunities, advancement possibilities, salaries, and the like. An excellent way of learning about your field is talking with persons who are currently employed in that field, especially in the kind of job that interests you.

Another excellent way of becoming acquainted with your job field is by reading in occupational guides, such as *The Occupational Outlook Handbook*, or by consulting brochures published by various professional societies. (All these reading materials usually are available in a college or public library or in a counselor's office.)

Company Analysis. Find out as much as you can about the company by which you are to be interviewed. Your investigation may lead you to conclude that you really don't want to work for that particular company. More likely, however, your investigation will provide you with information useful for your letter of application, for the interview, and for later if you get the job.

Your analysis of the company should take into account such items as the following:

Background. How old is the company? Who established it? What are the main factors or steps in its development?

Organization and management. Who are the chief executives? What are the main divisions?

Product (or service). What product does it manufacture or handle? What are its manufacturing processes? What raw materials are used and where do they come from? Who uses the product? Is there keen competition? How does the quality of the product compare with that of other companies?

Personnel. How many persons does the company employ? What is the rate of turnover? What is the range of skills required for the total work force? What are company policies concerning hiring, sick leave, vacation, overtime, retirement? What kinds of in-service training are provided? What is the salary range? Are there opportunities for advancement?

Obtaining answers to these and other pertinent questions may require consulting various sources. Remember that the enterprising applicant who really wants the job will overcome whatever difficulties or expend whatever energy is required to find needed information.

This information can be obtained from the Chamber of Commerce, trade and industrial organizations, local newspapers, interviews with company personnel, inspection tours through the company, company publications, and correspondence with the company.

Job Analysis. Just as you analyze the various aspects of the company with which you will interview, you also should analyze the particular job for which you will interview. That is, you should be as knowledgeable as possible about such job factors as the following:

Educational requirements
Necessary skills

Significance of experience

On-the-job responsibilities

Desirable personal qualities

Promotion possibilities

Salary range

Special requirements

This analysis can help you to see yourself and your qualifications more objectively and thereby contribute to self-confidence.

Sources of information concerning a particular job are likely to be the same as those for the company analysis.

Interviewer Analysis. Learn something about the person who will interview you; it usually is well worth the effort. Of course, the communication that gained you an interview provided some information about the interviewer; but more information would relieve undue anxiety and help to smooth the way and establish rapport.

Just as the salesperson analyzes a prospective customer—his or her wants, needs, and interests—before making a big sales effort, so you should analyze the person who will interview you.

Personal Analysis. In preparing for the job interview, analyze yourself. Think through your attitudes, your qualifications, and your career goals. Be prepared to answer such questions as the following:

Why did I apply for this job?

Do I really want this job?

Have I applied for jobs with other companies?

What do I consider my primary qualifications for this job?

Why should I be hired over someone else with similar qualifications?

What do I consider my greatest accomplishment?

Can I take criticism?

(If you have held other positions) Why did I leave other jobs?

Do I prefer to work with people or with objects?

How do I spend my leisure time?

What are my ambitions in life?

What are my salary needs?

Preparation of a Resumé. If your letter of application did not include a resumé, or data sheet, prepare one to take with you to the interview. This orderly listing of information about yourself will help you organize your qualifications. Further, having the resumé with you at the interview will help you to present all significant information and will provide the interviewer, at a glance, with an outline of pertinent information about your ability.

See pages 288–290 in Chapter 7, Memorandums and Letters, for specific directions in preparing a resumé.

Rehearsing an Interview. With the help of friends, practice an interview. Set up an office situation with desk and chairs. Ask a friend to assume the role of the interviewer and to ask you questions about yourself (see the questions listed in the section on Personal Analysis) and questions about the information on your resumé. Ask your friends to comment honestly on the rehearsal. Then swap roles; you be the interviewer and a friend, the applicant. If possible, tape the interview and play it back for critical analysis.

Another way to rehearse is by yourself, in front of a full-length mirror. Dress in the attire you will wear for the interview. Study yourself impartially; go over aloud the points you plan to discuss in the interview. Keep your gestures and facial expressions appropriate.

Rehearsals can help you gain self-confidence and organize your thoughts.

HOLDING AN INTERVIEW

Ordinarily, you cannot know exactly how an interview will be conducted. Thus it is impossible to be prepared for every situation that can arise. It is possible, however, to become acquainted with the usual procedure so that you can more easily adapt to the particular situation.

The usual procedure for an interview includes observance of business etiquette, establishing the purpose of the interview, questions and answers, and closing the interview.

Observance of Business Etiquette. Dressing appropriately is extremely important. Good business manners, if not common sense, demand that you arrive on time, be pleasant and friendly but businesslike, avoid annoying actions (such as chewing gum or tapping your feet), let the interviewer take the lead, and listen attentively.

Establishing the Purpose of the Interview. At the outset of the interview, establish why you are there. If you seek a specific position, say so. If you seek a job within a general area of a company, let that be known. Be flexible, but give the interviewer a clear idea of your job preferences.

Questions and Answers. During most of the interview, the interviewer will ask questions to which you respond. Your responses should be frank, brief, and to the point—yet complete.

Be honest in discussing your qualifications, neither exaggerating nor minimizing them. The interviewer's questions and comments can help you determine the type of employee sought; then you can emphasize your suitable qualifications. For instance, if you apply for a business position and the interviewer mentions that the job requires some customer contact, present your qualifications that show you have dealt with many people. That is, emphasize any work, experience, or courses that pertain to direct contact work.

In the course of the interview, you will likely be asked if you have any questions. Don't be afraid to ask what you need to know concerning the company or the job and its duties (such as employee insurance programs, vacation policy, overtime work, travel).

Closing the Interview. Watch the interviewer for clues that it is time to end the interview. Express appreciation for the time and courtesies given you, say goodbye, and leave. Lingering or prolonging the interview usually is an annoyance to the interviewer.

At the close of the interview, you may be told whether or not you have the job; or the interviewer may tell you that a decision will be made within a few days. If the interviewer does not definitely offer you a job or indicate when you will be informed about the job, ask when you may telephone to learn the decision.

Job Interview Tips*

Preparation:
- ☐ Learn something about the company
- ☐ Have specific job or jobs in mind
- ☐ Review in your mind your qualifications for the job
- ☐ Be prepared to answer broad questions about yourself
- ☐ Review your resumé
- ☐ Be there a few minutes before the scheduled time of your interview

Personal Appearance:
- ☐ Well groomed
- ☐ Suitable dress
- ☐ No chewing gum
- ☐ Only smoke when invited

The Interview:
- ☐ Answer each question as well as you can
- ☐ Be prompt in giving responses
- ☐ Be well mannered
- ☐ Use good English and avoid the use of slang
- ☐ Be cooperative and enthusiastic
- ☐ Don't be afraid to ask questions

Test (if employer gives one):
- ☐ Listen carefully to instructions
- ☐ Read each question carefully
- ☐ Write legibly and clearly
- ☐ Budget your time wisely and don't stay on one question too long

Information to Take With You:
- ☐ Social Security number
- ☐ Driver's license number
- ☐ Resumé. Although not all positions require job applicants to bring a resumé, you should be able to furnish the interviewer with information about your education and previous employment
- ☐ Usually an employer requires three references. Get permission from people before using their names. If you can avoid it, do not use the names of relatives. For each reference, give the following information: Name, address, telephone number, and occupation

*From *Occupational Outlook Handbook* 1986–1987 ed. U.S. Department of Labor.

Following up an interview reflects good manners and good business. Whether the follow-up is in the form of a telephone call, a letter, or another interview usually depends on whether you got the job, you did not get the job, or no decision has been made.

If you got the job, a telephone call or a letter can serve as confirmation of the job, its responsibilities, and the time for reporting to work. Too, the communication can express appreciation for the opportunity to become connected with the company.

If you did not get the job, thank the interviewer by telephone or by letter for his or her time and courtesy. This kind of goodwill is essential to business success.

Perhaps a final decision has not yet been reached concerning your employment. A favorable decision may well hinge on a wisely executed follow-up. If the interview ended with "Keep in touch with us" or "Check with us in a few days," you have the go-ahead. Have another interview. A telephone call or a letter would be less effective, for they would not be as forceful in maintaining the good impression that has been made. If the interview ended with "We'll keep your application on file" or "We may need a person with your qualifications a little later," follow up the lead. After a few days, write a letter. Mention the interview, express appreciation for it, include any additional credentials or emphasize credentials that since the interview seem to be particularly significant, and state your continued interest in the position.

Informational Interviews

A student writing a report about careers in sociology visits a sociologist, a social worker, and an occupational counselor. An employee who has been promoted to a higher position talks with associates concerning new job responsibilities. A member of a service organization polls the membership for suggested projects to undertake. The prospective buyer of a used car questions the owner about its condition. All these situations call for informational interviews—that is, conversing with another person to gain needed information. In addition, the preparer of an observation report (see pages 387–399 in Chapter 8, Reports) and of a research report (see Chapter 9, The Research Report) frequently derives information from knowledgeable people. In seeking such information, especially from those not obliged to give it, you must prepare carefully for the interview.

PREPARING FOR AN INTERVIEW

Knowledgeable people usually are willing to share their knowledge, provided the time required to do so seems worthwhile to them. Follow these steps to ensure that time is well spent for both interviewee and interviewer:

1. Ask for an interview. A frank, informal request usually is sufficient. Identify yourself, and explain briefly why the interview is important. State the kind of information being sought.
2. Set a convenient time and place for the interview. Accommodation of the person who must grant the interview is especially important.
3. Carefully plan the questions to be asked. Think through the reason for the interview and the kind of information desired.

4. The parties to the interview should be aware of each other's knowledge and resources so that a valuable exchange of information results.

THE INTERVIEW

After making the arrangements for the interview and planning specifically your contribution, you are prepared for the interview itself.

Follow this guide for a more satisfactory interview:

1. Be persuasive in explaining the significance of the interview and the value of the information sought. Make the other person feel that the time and knowledge shared can be of value to both parties.
2. Explain how you will use the information. Assure the interviewee that the information will be treated honestly and that confidential information will remain confidential.
3. Ask well-planned questions. Refer to your notes if necessary.
4. Take notes on what is said. One technique is to bring to the interview written questions. Space can be left to write in the interviewee's responses. Instead, if the interviewee does not object, you may want to tape the interview.
5. Review the information, clarifying where necessary. Especially if you intend to quote, read the quotation back to the interviewee as a check for accuracy.
6. Close the interview by sincerely thanking the person giving the information.

COURTESY FOLLOW-UP

After the interview, a brief letter of appreciation should be written to the person who granted the interview. This courteous gesture is both good manners and good business.

General Principles in Oral Communication

1. *An oral presentation differs from a written presentation.* Differences occur in level of diction, amount of repetition, kind of transitions, and kind and size of visuals.
2. *Modes of delivery include extemporaneous, memorized, and read from a manuscript.* The speaker selects a mode appropriate for a given situation.
3. *The general purpose of a presentation may be to entertain, to persuade, or to inform.*
4. *An effective presentation requires preparation.* The steps in preparation for a presentation include determining the specific purpose; analyzing the type of audience; gathering the material; organizing the material; determining the mode of delivery; outlining the speech, writing it out if necessary; preparing visual materials; and rehearsing.
5. *The delivery of the presentation is very important.* Effective delivery includes walking to the podium with poise and self-confidence; capturing the audience's attention and interest; looking at the audience; sticking to an appropriate mode of delivery; putting some zest in expression; getting words out clearly and distinctly; adjusting the volume and pitch of voice; varying the

rate of speaking to enhance meaning; standing naturally; avoiding mannerisms; showing visuals with natural ease; and closing appropriately.
6. *Visuals for oral presentations include flat materials, exhibits, and projected
   materials.*
7. *Evaluating a presentation requires considering the purpose and audience,*
   the delivery, and the content and organization of the presentation.

Application 1   Oral Communication

Select five advertisements and bring them to class. To what needs, wants, and
desires do the ads appeal? Decide for each one whether the major appeal is emotional
or rational.

Application 2   Oral Communication

Assume that you are in *one* of the situations below. What would you say?

a. Gather your supporting material.
b. Organize the material.
c. Outline the presentation.
d. Prepare appropriate visuals.
e. Rehearse.
f. Give the presentation.

Assume:

1. You have been employed in a firm for six months. You are asking your
   employer for a raise.
2. You have a plan for facilitating the flow of traffic for campus events. Present
   your plan to the local police chief (or other appropriate official).
3. You have an automobile (or some other piece of personal property) for sale.
   A prospective buyer is coming to talk with you.
4. You are a football coach getting ready to talk to your team at halftime. Your
   team is losing 17–0.
5. You want your parents to buy an automobile (or some other expensive item)
   for you.
6. Select any other situation in which you present an idea, plan, or product.

Ask your classmates to evaluate your speech by filling in an Evaluation of Oral
Presentations from pages 527–535.

Application 3   Oral Communication

As directed by your instructor, prepare one of the topics below for an oral
presentation.

a. Analyze the type of audience.
b. Gather the material.
c. Organize the material. Fill in an appropriate Plan Sheet.
d. Determine the mode of delivery.
e. Outline the presentation.
f. Prepare appropriate visuals.
g. Rehearse.
h. Give the presentation.

Prepare:

1. A set of instructions or description of a process. (Review Chapter 1 in Part I.)
2. A description of a mechanism. (Review Chapter 2 in Part I.)
3. An extended definition. (Review Chapter 3 in Part I.)
4. An analysis through classification or partition. (Review Chapter 4 in Part I.)
5. An analysis through effect and cause or comparison and contrast. (Review Chapter 5 in Part I.)
6. A report. (Review Chapter 8 in Part I.)

Ask your classmates to evaluate your speech by filling in an Evaluation of Oral Presentations from pages 527–535.

Application 4 Oral Communication

Team up with a classmate for practice job interviews. Alternate roles of interviewer and interviewee. Be constructively critical of each other. If you have access to a tape recorder, record your practice sessions and play them back for study.

Application 5 Oral Communication

Assume that your instructor is the personnel manager of a company you would like to work for. (*a*) Prepare for the interview by making a written job field analysis, company analysis, job analysis, interviewer analysis, personal analysis, and resumé. (*b*) Write a letter of application, including a resumé, to the personnel manager. See pages 288–306. If the letter is successful, you will be granted an interview. (*c*) Attend the interview at the scheduled time in the office of the personnel manager. (*d*) Use an appropriate follow-up.

Application 6 Oral Communication

Arrange for an interview (conference) with an instructor in whose class you would like to be making better grades. Then write a one-page report on the interview.

Application 7   Oral Communication

Arrange for an interview with your program adviser, a school counselor, an employment counselor, or some other person knowledgeable in your major field of study. Try to determine the availability of jobs, pay scale, job requirements, promotion opportunities, and other such pertinent information for the type of work you are preparing for. Present your findings jn an oral report.

    a. Gather the material.
    b. Organize the material.
    c. Outline the presentation.
    d. Prepare appropriate visuals.
    e. Rehearse.
    f. Give the presentation.

Ask your classmates to evaluate your speech by filling in an Evaluation of Oral Presentations from pages 527–535.

USING PART II: SELECTED READINGS

Application 8   Oral Communication

Read the speech "To Serve the Nation: Life Is More Than a Career," by Jeffrey R. Holland, pages 632–638. As directed by your instructor, answer the questions following the selection.

Application 9   Oral Communication

Read "Are You Alive?" by Stuart Chase, pages 611–615. Under "Suggestions for Response and Reaction," present your response to Number 1c as a speech. Ask your classmates to evaluate your speech by filling in an Evaluation of Oral Presentations from pages 527–535.

EVALUATION OF ORAL PRESENTATIONS

(See page 516 for directions.)

Course and Section _____ Date _____ Evaluated by (Name) _____

Students' Names

DELIVERY

Forceful introduction
Poise
Eye contact
Sticking to mode of delivery
Zest (enthusiasm)
Voice control
Acceptable pronunciation and grammar
Avoidance of mannerisms
Ease in showing visuals
Clear-cut closing
Sticking to specified length

CONTENT & ORGANIZATION

Stating of main points at outset
Development of main points
Needed repetition and transitions
Effective kinds and sizes of visuals

OVERALL EFFECTIVENESS

TOTAL POINTS

4=Outstanding   2=Fair

3=Good   1=Needs improvement

COMMENTS:

EVALUATION OF ORAL PRESENTATIONS
(See page 516 for directions.)

Course and Section _____   Date _____   Evaluated by (Name) _____

Students' Names

DELIVERY

Forceful introduction

Poise

Eye contact

Sticking to mode of delivery

Zest (enthusiasm)

Voice control

Acceptable pronunciation and grammar

Avoidance of mannerisms

Ease in showing visuals

Clear-cut closing

Sticking to specified length

CONTENT & ORGANIZATION

Stating of main points at outset

Development of main points

Needed repetition and transitions

Effective kinds and sizes of visuals

OVERALL EFFECTIVENESS

TOTAL POINTS

4=Outstanding   2=Fair

3=Good   1=Needs improvement

COMMENTS:

EVALUATION OF ORAL PRESENTATIONS
(See page 516 for directions.)

Course and Section _____ Date _____ Evaluated by (Name) _____

Students' Names

DELIVERY

Forceful introduction

Poise

Eye contact

Sticking to mode of delivery

Zest (enthusiasm)

Voice control

Acceptable pronunciation and grammar

Avoidance of mannerisms

Ease in showing visuals

Clear-cut closing

Sticking to specified length

CONTENT & ORGANIZATION

Stating of main points at outset

Development of main points

Needed repetition and transitions

Effective kinds and sizes of visuals

OVERALL EFFECTIVENESS

TOTAL POINTS

4=Outstanding   2=Fair

3=Good   1=Needs improvement

COMMENTS:

EVALUATION OF ORAL PRESENTATIONS
(See page 516 for directions.)

Course and Section _____ Date _____ Evaluated by (Name) _____

Students' Names

DELIVERY

Forceful introduction

Poise

Eye contact

Sticking to mode of delivery

Zest (enthusiasm)

Voice control

Acceptable pronunciation and grammar

Avoidance of mannerisms

Ease in showing visuals

Clear-cut closing

Sticking to specified length

CONTENT & ORGANIZATION

Stating of main points at outset

Development of main points

Needed repetition and transitions

Effective kinds and sizes of visuals

OVERALL EFFECTIVENESS

TOTAL POINTS

4=Outstanding   2=Fair

3=Good   1=Needs improvement

COMMENTS:

EVALUATION OF ORAL PRESENTATIONS
(See page 516 for directions.)

Course and Section _____   Date _____   Evaluated by (Name) _____

Students' Names

DELIVERY

Forceful introduction

Poise

Eye contact

Sticking to mode of delivery

Zest (enthusiasm)

Voice control

Acceptable pronunciation and grammar

Avoidance of mannerisms

Ease in showing visuals

Clear-cut closing

Sticking to specified length

CONTENT & ORGANIZATION

Stating of main points at outset

Development of main points

Needed repetition and transitions

Effective kinds and sizes of visuals

OVERALL EFFECTIVENESS

TOTAL POINTS

4 = Outstanding   2 = Fair

3 = Good   1 = Needs improvement

COMMENTS:

Chapter 11
Visuals: Seeing Is Convincing

537

Objectives

Upon completing this chapter, the student should be able to:

- List ways in which visuals can be helpful
- Explain how to use visuals effectively
- Explain how computers are influencing the use and the production of visuals
- Explain the special characteristics of tables, charts, graphs, photographs, drawings, diagrams, exhibits, and projected materials
- Use visuals in written and in oral presentations

Introduction

Visuals can help immeasurably in your communication needs. Both in written and in oral presentations, visuals can clarify the information and impress it upon the minds of your audience.

The various kinds of visuals have unique qualities that make each type the "right" one to use in certain circumstances. All kinds require careful planning and preparation. Some are usable only in written or only in oral presentations, although most types are adaptable to either; some require a great deal of technical knowledge to prepare—others can be rather easily constructed by an amateur. In order to select and show the visuals that best serve your purpose, you should consider both the reasons for using visuals and the best way to use them effectively. Further, you should become familiar with the most frequently used types (tables, charts, graphs, photographs, drawings, diagrams, exhibits, and projected materials), their specific contributions, and their special characteristics. (See also pages 513–516 in Chapter 10, Oral Communication.)

Advantages of Visuals

Visuals can be helpful in several ways:

1. Visuals can capitalize on *seeing*. For most people, the sense of sight—more so than hearing, smell, touch, or taste—is the most highly developed of the senses.
2. Visuals can convey some kinds of messages better than words can. Ideas or information difficult or impossible to express in words may be communicated more easily through visuals.
3. Visuals can simplify or considerably reduce textual explanation. Accompanying visuals often clarify words.
4. Visuals can add interest and focus attention.

Of course, visuals can work against you as well as for you. So guard against an overreliance on visuals, against poorly planned visuals, and, most of all, against snafus in timing or presentation.

Remember, however, that appropriately used, visuals can richly enhance your presentation.

Using Visuals Effectively

Visuals can be a simple, effective way of presenting information that will make a lasting, positive impression on your audience. Following are suggestions that will help ensure your using visuals to the best advantage.

1. *Study the use of visuals by others.* Analyze their use in books and periodicals and by speakers and lecturers, especially in your field of study. Note such things as intended audience, the kinds of information presented or supplemented, the kind of visual selected for a particular purpose, the design and layout of the visual, the amount of accompanying textual explanation, and the overall effectiveness of the visual.
2. *Select the kinds of visuals that are most suitable.* Consider the purpose of your presentation, the needs of your audience, and the specific information or idea to be presented.
3. *Prepare the visual carefully.* Organize information logically, accurately, completely, and consistently. Include all needed labels, symbols, titles, and headings.
   a. *Do not include too much in a visual.* Plan one overall focus. Information should be easy to grasp visually and intellectually.
   b. *Make the visual pleasing to the eye.* It should be neat, uncrowded, attractive, and should have sufficient margins on all sides.
   c. *Use lettering to good advantage.* Avoid carelessly mixing styles of lettering or typefaces and mixing uppercase (capital) letters with lowercase ones. Space consistently between letters and between words. Suggestion: Using a pencil, lightly rule guidelines and block out the words. Then do the actual lettering. Erase the pencil lines with art gum after the lettering has dried. Examples of lettering:

POOR	GOOD
EX pensez for the Y EAR	Expenses for the year
	Expenses for the Year
	EXPENSES FOR THE YEAR

   d. *Give each visual a caption.* State clearly and concisely what the viewer is looking at. For tables, place the caption above the visual; in all other instances, place the caption below the visual.
4. *Decide whether to make the visual "run on" in the text or to separate it from the text.* The run-on visual is a part of the natural sequence of information within a paragraph. It is not set apart with a title, a number, or lines (rules). Usually the run-on visual is short and the information contained is uncomplicated. The separate visual is "dressed up" with a number (unless the communication contains only one visual), a caption, and rule lines. Such a visual usually is more complicated and requires more space than the run-on visual. In addition, the separate visual is movable; that is, it can be located other than at the place where it is mentioned in the text (see number 5 below).
5. *Determine where to place the visual.* Ideally a visual is placed within the text at the point where it is discussed. The visual may be more practically placed,

however, other than at the point of reference (such as on a following page, on a separate page, or at the end of the communication) if (1) a visual too large for the remaining space on a page unavoidably causes a noticeable blank space; (2) a visual that merely supplements verbal explanation interferes with reader comprehension; or (3) a number of visuals are used and seem to break the content flow of a presentation. (When visuals form a significant part of a report, they are *listed,* together with page numbers, under a heading such as "List of Illustrations." This list appears on a separate page immediately following the table of contents.)

6. *Refer to the visual in the textual explanation.* The audience should never be left to wonder. "Why is this visual included?" It is essential that you establish a proper relationship between the visual and the text. The extent of textual explanation is determined largely by the complexity of the subject matter, the purpose of the visual, and the completeness of labels on the visual. In referring to the visual, use such pointers as "see Figure 1," "as illustrated in the following diagram," or "Table 3 indicates the pertinent factors."

7. *Use correct terminology in referring to visuals.* Tables are referred to as tables; all other visuals are usually referred to as figures. Examples: "Study the amounts of salary increases shown in Table 2." "Note the position of the automobile in Figure 6." "As the graph in Figure 4 indicates. . . ."

8. *Give credit for borrowed material.* The credit line, usually in parentheses, typically is placed immediately following the title of the visual or just below the visual. For bibliographical forms other than those illustrated below, see pages 473–474 in Chapter 9, The Research Report.

   a. *If you have borrowed or copied an entire visual, give the source.* Although there is no one standard format for giving the source, the following examples use acceptable formats.

EXAMPLE 1

```
                        TABLE 4
              CAUSES OF INDUSTRIAL ACCIDENTS

                 [The table goes here.]

    Source: John R. Barnes, Industry on Trial (New York: Harper & Row,
    1988), p. 221.
```

EXAMPLE 2

```
    Figure 6. Automobile troubleshooting using the charging system
    (Source: Ford Division—Ford Motor Co.)
```

EXAMPLE 3

```
    Table 2. Projected College Enrollments for 1990 (Education Almanac
    [New York: Educational Associates, Inc., 1988], p. 76.)
```

(Note the brackets where parentheses ordinarily are used; another pair of parentheses would be confusing.) For other examples, see page 76 and, in this chapter, Table 1 (page 543), figures 3 and 4 (page 546), Figure 6 (page 548), Figure 7 (page 549), Figure 10 (page 551), Figure 13 (page 553), and the table on page 561.

b. *If you have devised the visual yourself but gotten the information from another source, give the source.* The following examples use acceptable formats.

EXAMPLE 1

> Figure 5. Per Capita Income in Selected States. (Source of Information: U.S. Dept. of Urban Affairs)

EXAMPLE 2

> Figure 1. Average size of American families. (Data from U.S. Bureau of the Census)

For another example, see Figure 7 in this chapter (page 549).

9. *If necessary, mount the visual.* Photographs, maps, and other visuals smaller than the regular page should be mounted. Attach the visual with dry mounting tissue, spray adhesive, or rubber cement (glue tends to wrinkle the paper).

Computer Graphics

The computer is revolutionizing both the role and the production of visuals in communication. This computer revolution can be largely attributed to three conditions: (1) the increasing availability of terminals for mainframe systems, (2) the widespread availability and downward pricing of microcomputers, and (3) the continuous additions to a growing array of graphic software, that is, commercially produced programs for easily and swiftly transforming data into various kinds of graphics. For the novice or intermediate level computer user, the software aspect is of most interest.

Graphics software provides programs for creating basic kinds of visuals and programs for creating more complex visuals.

BASIC VISUALS—CHARTING

Persons in business and industry—regardless of the size of the company—are using computer graphics to enlarge upon the time-proved positive difference that visuals make. The general term for producing basic visuals on the computer is *charting*.

The computer user can create line graphs, pie graphs (pie charts), bar graphs (bar charts), consumer marketing maps, forecasting charts, and a host of other basic visuals. These can be produced in black-and-white hard copies, that is, on a sheet of paper; or, with a color plotter (a device that acts as a mechanical arm and that "draws" with a pen), the visuals can be produced in multiple colors and on clear polyester film for overhead projection.

What this means in terms of preparing a business report is that the data in the computer can be programmed to produce such visuals as a pie chart of percentages of sales for particular products, a graph displaying a sales matrix, a line graph depicting total sales and fixed expenses, or bar charts reflecting sales, gross profits, and overhead expenses. The visuals are generated as an integral aspect of the report.

Given in Figure 1 is an example of a computer-generated graph.

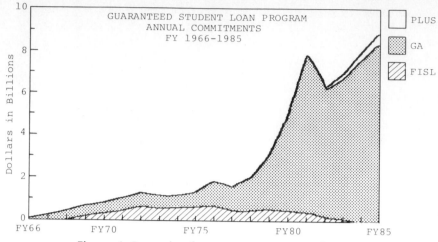

Figure 1. Example of a computer-generated graph.

See also the description of the ColorPro plotter in the sample presentation, pages 75–77.

MORE COMPLEX VISUALS

In addition to charting, other software programs provide more complex, more sophisticated kinds of visuals.

One kind of such graphics is the slide-show program. This program permits the user to select computerized charts for demonstration; the user sets up the charts in sequence, times the needed intervals, and programs the charts for projection on a screen. Some slide-show programs offer graphic editing and such design features as borders and multiple typefaces.

Other slide-show programs include those that splice such media as videotape, film, off-air broadcasts, and interactive computer programs and those that produce three-dimensional computerized graphics.

EFFECT ON MANUALLY PRODUCED VISUALS

Computer graphics are rapidly replacing manually produced graphs and charts. Manually produced visuals—whether by hand, with a rapidograph or other such aids, or through a company's graphic arts department—are increasingly being replaced by electronically produced visuals.

The computer-generated visuals can be produced quickly, accurately, and economically. Furthermore, they can be easily updated.

Tables

Tables are an excellent form for presenting large amounts of data concisely. (See the tables on pages 107, 143, 201, 385, 403, and 627–628.) Although tables lack the eye appeal and interest-arousing dramatization of such visuals as charts or graphs, they are unexcelled as a method of organizing and depicting statistical information compiled through research. In fact, information in most other visuals showing numerical amounts and figures derives from data originally calculated in tables.

Study Table 1, noting the quantity of information given and its arrangement.

TABLE 1 **Single (Never-Married) Persons 18 Years Old and Over as Percent of Total Population, by Age and Sex: 1960 to 1984**
[**1960**, as of **April;** thereafter based on Current Population Survey as of **March.**]

Age	Male					Female				
	1960	1970	1975	1980	1984	1960	1970	1975	1980	1984
Total	**17.3**	**18.9**	**20.8**	**23.8**	**25.5**	**11.9**	**13.7**	**14.6**	**17.1**	**18.4**
18 years	94.6	95.1	96.8	97.4	98.2	75.6	82.0	83.7	88.0	91.4
19 years	87.1	89.9	89.3	90.9	95.3	59.7	68.8	71.4	77.6	82.9
20–24 years	53.1	54.7	59.9	68.8	74.8	28.4	35.8	40.3	50.2	56.9
25–29 years	20.8	19.1	22.3	33.1	37.8	10.5	10.5	13.8	20.9	25.9
30–34 years	11.9	9.4	11.1	15.9	20.9	6.9	6.2	7.5	9.5	13.3
35–39 years	8.8	7.2	8.6	7.8	11.6	6.1	5.4	5.0	6.2	7.5
40–44 years	7.3	6.3	7.2	7.1	6.9	6.1	4.9	4.8	4.8	5.4
45–54 years	7.4	7.5	6.3	6.1	6.2	7.0	4.9	4.6	4.7	4.6
55–64 years	8.0	7.8	6.5	5.3	5.4	8.0	6.8	5.1	4.5	4.2
65 years old and over	7.7	7.5	4.7	4.9	5.0	8.5	7.7	5.8	5.9	5.6

Source: U.S. Bureau of the Census, *Current Population Reports,* series P-20, No. 399, and earlier reports.

Tables may be classified as informal (such as those on pages 199 and 375–376) or as formal (such as Table 1 above). Informal tables are incorporated as an integral part of a paragraph and thus are not given an identifying number or title. Formal tables are set up as a separate entity (with identifying number and title) but are referred to and explained as needed in the pertinent paragraphs.

GENERAL DIRECTIONS FOR CONSTRUCTING TABLES

As you prepare tables, observe the following basic practices:

1. Number each table (the number may be omitted if only one table is included) and give it a descriptive title (number and title are omitted in informal tables). Center the number and the title *above* the table. Often the word *table* is written in all capitals, the number written in arabic numerals, and this label centered above the title (caption) written in all capitals. Example:

TABLE 2

LEADING CAUSES OF DEATH AMONG AMERICANS

Just as acceptable:

Table 2. Leading Causes of Death among Americans

Table 2. Leading causes of death among Americans

Other acceptable practices include giving the table number in roman numerals or as a decimal sequence.

2. Label each column accurately and concisely. If a column shows amounts, indicate the unit in which the amounts are expressed; example: Wheat (in metric tons).

3. To save space, use standard symbols and abbreviations. If items need clarification, use footnotes, placed immediately below the table. (Table footnotes are separate from ordinary footnotes placed at the bottom of the page.)

4. Generally use decimals instead of fractions, unless it is customary to use fractions (as in the size of drill bits or hats).

5. Include all factors or information that affect the data. For instance, omission of wheat production in a table "Production of Chief United States Crops" would make the table misleading.

6. Use ample spacing and rule lines (straight lines) to enhance the clarity and readability of the table. Caution: Generally, use as few rule lines as necessary.

7. If a table is divided for continuation, repeat the column headings.

8. If a table is more than a page long and must be continued to another page, use the word *continued* at the bottom of each page to be continued and at the top of each continuation page. If column totals are given at the end of the table, give subtotals at the end of each page; and at the top of each continuation page, repeat the subtotals from the preceding page.

Charts

Although the term *chart* is often used as a synonym for *graph*, a chart is distinguished by the various shapes it can take, by its use of pictures and diagrams, and by its capacity to show nonstatistical as well as statistical relationships. More importantly, a chart can show relationships better than other types of visuals can. Frequently used types of charts are the pie chart, the bar chart, the organization chart, and the flowchart.

GENERAL DIRECTIONS FOR CONSTRUCTING CHARTS

Manually constructing charts, as well as other visuals, requires careful attention to details, a bit of arithmetic, and a few basic materials: ruler, pen or pencil, and paper.

In the construction process:

1. Number each chart as Figure 1 or Fig. 1, Figure 2, and so on (this may be omitted if only one visual is included), and give the chart a descriptive title. Center the number and title (both on the same line) *below* the chart.

2. Label each segment concisely and clearly.

3. Use lines or arrows if necessary to link labels to segments.

4. Place all labels and other information horizontally for ease in reading.

PIE CHART

The pie chart, also called circle chart or circle graph, is a circle representing 100 percent. It is divided into segments, or slices, that represent amounts or proportions. The pie chart is especially popular for showing monetary proportions and is often used to show proportions of expenditures, income, or taxes. Although not the most accurate form for presenting information, it has strong pictorial impact. More than any other kind of visual, the pie chart permits simultaneous comparison of the parts to one another and comparison of one part to the whole.

Constructing a pie chart is relatively simple if you will follow the general directions given above and these additional suggestions:

1. Begin the largest segment in the "twelve o'clock" position; then, going clockwise, give the next largest segment, and so on.
2. Lump together, if practical, items that individually would occupy very small segments. Label the segment "Other," "Miscellaneous," or a similar title, and place this segment last.
3. Put the label and the percentage or amount on or near each segment.

Pie Chart

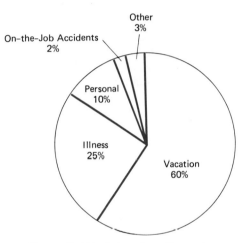

Figure 2. Work loss for all employees.

BAR CHART

The bar chart, also called column chart or bar graph, is one of the simplest and most useful of the visual aids, for it allows the immediate comparison of amounts. (Bar charts are shown on pages 76 and 135.)

The bar chart consists of one or more vertical or horizontal bars of equal width, scaled in length to represent amounts. (When the bar is vertical, the visual is called a column chart; when the bar is horizontal, the visual is called a bar chart.) The bars are often separated to improve appearance and readability.

To give multiple data, a bar may be subdivided, or multiple bars may be used, with crosshatching, colors, or shading to indicate different divisions.

Note the difference in the two accompanying examples of bar charts, Figure 3 and Figure 4.

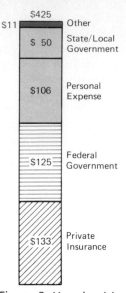

Figure 3. How health care was paid for in 1985. (*Source:* Department of Health and Human Services)

Horizontal Bar Chart with Multiple Bars

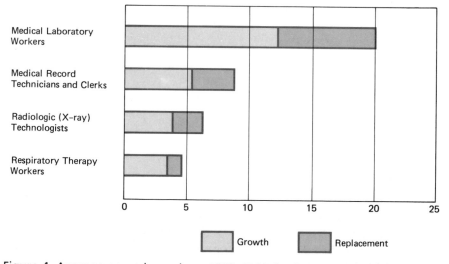

Figure 4. Average annual openings, 1980–1989 (in thousands), selected medical technologist, technician, and assistant occupations. (*Source:* Bureau of Labor Statistics)

ORGANIZATION CHART

The organization chart is effective in showing the structure of such organizations as businesses, institutions, and governmental agencies. (An organization chart appears in Figure 5 and in a sample report in Chapter 4, page 156.) Unlike most other charts, the organization chart does not present statistical information. Rather, it reflects lines of authority, levels of responsibility, and kinds of working relationships.

Organization Chart

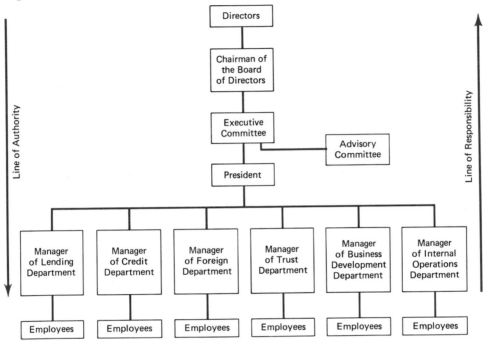

Figure 5. A line organization plan of a medium-sized bank. In this plan, the department head must perform highly specialized functions and at the same time direct or supervise subordinates.

Organization charts depict the interrelationships of (1) staff, that is, the personnel; (2) administrative units, such as offices or departments; or (3) functions, such as sales, production, and purchasing.

A staff organization chart shows the position of each individual in the organization, to whom each is responsible, over whom each has control, and the relationship to others in the same or different divisions of the organization. The administrative unit organization chart shows the various divisions and subdivisions. The administrative units of a large supermarket, for instance, include the produce department, the meat department, the grocery department, and the interrelationships of different activities, operations, and responsibilities. A college organization chart, for instance, might show its structure by functions: teaching, community service, research, and the like.

An organization chart must be internally consistent; that is, it should not jump randomly from, say, depicting personnel to depicting functions.

Blocks or circles containing labels are connected by lines to indicate the organizational arrangement. Heavier lines are often used to show chain of authority, while broken lines may show coordination, liaison, or consultation. Blocks on the same level generally suggest the same level of authority.

FLOWCHART

The flowchart, or flow sheet, shows the flow or sequence of related actions. The flowchart pictorially presents events, procedures, activities, or factors and shows how they are related. (Sample presentations, pages 42, 45, and 575, contain flowcharts.)

The flowchart is an effective visual for showing the flow of a product from its beginning as raw material to its completed form, or the movement of persons in a process, or the steps in the execution of a computer operation. Labeled blocks, triangles, circles, and the like (or simply labels) represent the steps, although sometimes simplified drawings that suggest the actual appearance of machines and equipment indicate the various steps. Usually, arrowhead lines show the direction in which the activity or product moves.

The flowchart in Figure 6 uses labeled blocks and a circle, arrowhead lines, and screened and unscreened lettering to explain the work process in a laundry and dry cleaning plant. The flowchart in Figure 7 depicts a natural process: nerve supply from the brain to the teeth.

Flowchart

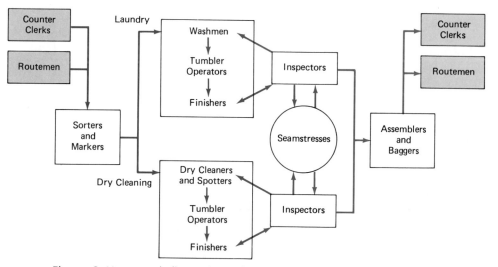

Figure 6. How work flows through a laundry and dry cleaning plant.
(*Source:* Bureau of Labor Statistics)

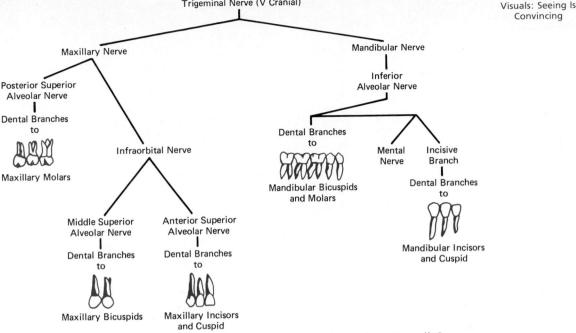

Figure 7. Nerve supply to the teeth. (Redrawn by permission from Russell C. Wheeler, *Dental Anatomy and Physiology* [Philadelphia: W. B. Saunders, 1985], p. 94)

Graphs

Graphs present numerical data in easy-to-read form. They are often essential in communicating statistical information, as a glance through a business periodical, a report, or an industrial publication will attest.

Graphs are especially helpful in identifying trends, movements, relationships, and cycles. Production or sales graphs, temperature and rainfall curves, and fever charts are common examples. Graphs simplify data and make their interpretation easier. But whatever purpose a specific graph may serve, all graphs emphasize *change* rather than actual amounts.

Consider, for instance, the graphs on pages 579 and 630 or the graph on page 550, showing the changes over a ten-year period in the cost of stock in a company.

At a glance you can see the change in cost over the years. Presented verbally or in another visual form, such as in a table, the information would be less dramatic and would require more time for study and analysis. But presented in a graph, the information is immediately impressed upon the eyes and the mind.

GENERAL DIRECTIONS FOR CONSTRUCTING GRAPHS

Consider the following as you prepare graphs:

1. A graph is labeled as Figure 1 or Fig. 1, Figure 2, etc. (this may be omitted

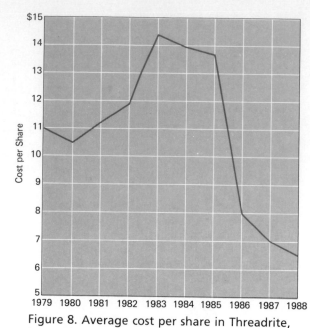

Figure 8. Average cost per share in Threadrite,
Inc., 1979–1988.

if only one visual is included), and is given a descriptive caption, or title. The label and caption (same line) are placed below the graph.

2. A graph has a horizontal and a vertical scale. The vertical scale usually appears on the left side (the same scale may also appear on the right side if the graph is large), and the horizontal scale appears underneath the graph.

3. Generally, the independent variable (time, distance, etc.) is shown on the horizontal scale; the dependent variable (money, temperature, number of people, etc.) is shown on the vertical scale.

4. The horizontal scale increases from left to right. If this scale indicates a value other than time, labeling is necessary. The vertical scale increases from bottom to top; it should always be labeled. Often this scale starts at zero, but it may start at any amount appropriate to the data being presented.

5. The scales on a graph should be planned so that the line or curve creates an accurate impression, an impression justified by facts. To the viewer, sharp rises and falls in the line mean significant changes. Yet the angle at which the line goes up or down is controlled by the scales. If a change is important, the line indicating that change should climb or drop sharply; if a change is unimportant, the line should climb or drop less sharply. See the graph in Figure 9.

6. A graph may have more than one line. The lines should be amply separated to be easily recognized and distinguished, yet close enough for clear comparison. Each line should be clearly identified either above or to the side of the line or in a legend. Often, variations of the solid unbroken line are used, such as a dotted line or a dot/dash line. Note the kinds of lines and their identification in Figure 9 and Figure 10.

Graph

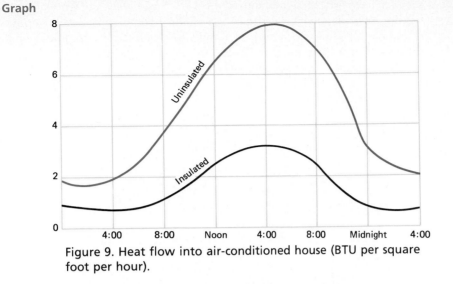

Figure 9. Heat flow into air-conditioned house (BTU per square foot per hour).

Multiple-Line Graph

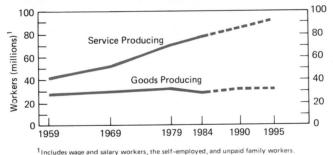

[1] Includes wage and salary workers, the self-employed, and unpaid family workers.

Figure 10. Industries providing services will continue to employ many more people than those providing goods. (*Source:* Bureau of Labor Statistics)

7. The line connecting two plotted points may be either straight, as in Figure 8, or smoothed out (faired, curved), as in Figure 9. A straight line is usually desired if the graph shows changes that occur at stated intervals; a faired line, if the graph shows changes that occur continuously.

Photographs

Photographs provide a more exact impression of actual appearance than other visual aids. Photographs on pages 105 and 399 are used to validate reports. They supply far more concreteness and realism than drawings. Though helpful in supplementing verbal description and for giving information, photographs are of the greatest value as evidence in proving or showing what something is.

Photographs have certain limitations, however. Since they present only appearance (except, of course, for such specialties as X-ray photography or holography), internal or below-the-surface exposure is impossible, and drawings or diagrams might be necessary. Further, unless retouched (or cropped) or taken with proper layout for a given purpose, photographs may unavoidably present both significant and insignificant elements of appearance with equal emphasis; or they may even miss or misrepresent important details.

The photograph in Figure 11 was used in filing an insurance claim.

Photograph

Figure 11. Damage to a car.

Drawings and Diagrams

DRAWINGS

Drawings are especially helpful in all kinds of technical communication. (Among the many drawings in the text are those on pages 20, 21, 22, 32, 33, 42, 67, 69, 70, 72, 79, 149, and 150.)

A drawing, though sometimes suggestive or interpretive, ordinarily portrays the actual appearance of an object. Like a photograph, it can picture what something looks like; but unlike a photograph—and herein lies one of its chief values—it can picture the interior as well as the exterior. A drawing makes it possible to place the emphasis where needed and thus omit the insignificant. Furthermore, a drawing can show details and relationships that might be obscured in a photograph. And a drawing can be tailored to fit the need of the user: it can show, for example, a cutaway view (see pages 42, 69, 79, 97, 177, and 178) or an enlarged view of a particular part (see Figure 12).

Making a simple drawing is relatively uncomplicated and usually is much easier and less expensive than preparing a photograph. If the object being drawn has more than one part, the parts should be proportionate in size (unless enlargement is indicated). The name of each part that is significant in the drawing should be clearly

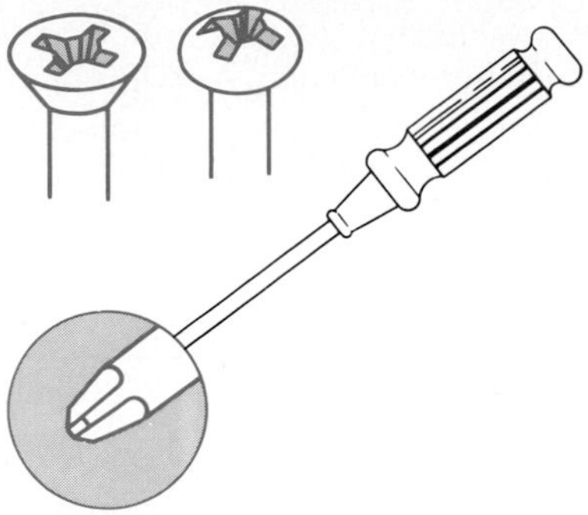

Figure 12. Phillips screwdriver.

given, either on the part or near it, connected to it by a line or arrow. If the drawing is complex and shows a number of parts, symbols (either letters or numbers) may be used with an accompanying key.

DIAGRAMS

A diagram is a plan, sketch, or outline, consisting primarily of lines and symbols. A diagram is designed to demonstrate or explain a process, object, or area, or to clarify the relationship existing among the parts of a whole. Diagrams are especially valuable for showing the shape and relative location of items and the manner in which equipment functions or operates (see the diagrams on page 82). Too, diagrams are helpful in showing the principles involved in an operation or concept (as in defining horsepower, page 110).

Diagrams are indispensable in modern construction, engineering, and manufacturing. A typical example is the design for a fireplace in Figure 13.

Diagram

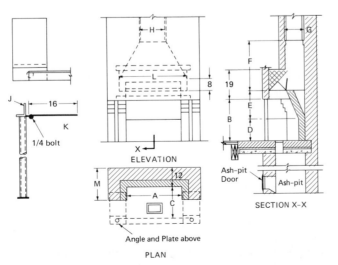

Figure 13. Proven design for a three-way, conventionally built fireplace. (Courtesy of Donley Brothers Co.)

SCHEMATIC DIAGRAM

The schematic diagram, a specialized diagram, is an invaluable aid in various mechanical fields, particularly in electronics. As with all visuals, the schematic diagram, like that on page 383 and that below of an electronic device, has standard symbols, terminology, and procedures that should be followed in its preparation.

Schematic Diagram

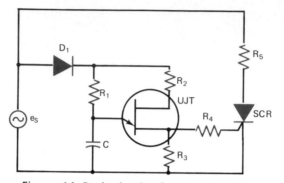

Figure 14. Basic circuit of an SCR controller.

Exhibits

Exhibits, particularly valuable in oral presentations, are designed as learning experiences that involve people, objects, or representations presented in orderly sequence. Among the most frequently used kinds of exhibits are demonstrations, displays, real objects, models, dioramas, posters, and the chalkboard. (See Chapter 10, Oral Communication, especially page 514.)

DEMONSTRATION

A demonstration describes, explains, or illustrates a procedure or idea. A demonstration provides realism, for it is an enactment of actual steps or aspects using real objects. For instance, an explanation of how to apply a tourniquet, a description of what happens when acid is applied to copper, or a discussion of Newton's law of gravity is more realistic to an audience if it is demonstrated. In the oral description of a color plotter on pages 75–77, the speaker uses demonstration to good advantage.

DISPLAY

A display is an arrangement of materials, such as photographs, news clippings, mobiles, or three-dimensional objects, designed to dramatize significant information or ideas. Often displays are presented on a bulletin board, flannel board, table, or similar area.

For an effective showing of displays, observe the following suggestions:

1. Plan the display around a specific theme.

2. Arrange the material so that it tells a story.
3. Use neat, clear lettering.
4. Show the material in an attractive setting with pleasing backgrounds and accessories.

REAL OBJECT

Real objects are especially effective in a presentation. More so than pictures, models, or other representations, real objects give immediacy. Both animate and inanimate, real objects provide an opportunity to show actual size, weight, sound, movement, and texture.

MODEL

A model is a three-dimensional representation of a real object. A model can be used when the real object is not available, is too large or too small, or is otherwise unsuitable for use with the presentation. Although models are of various types, generally they permit easy handling and convenient observation; they can provide interior views of objects; they can be stripped of some details so that other details can be easily observed; and they can be disassembled and put together to show the interrelationship of parts.

DIORAMA

A diorama is a three-dimensional scene of proportionately scaled objects and figures in a natural setting. A diorama is framed by a box, pieces of cardboard fastened together at right angles, or other such delineating devices. The diorama gives an in-depth, realistic view.

Dioramas are commonly used in museums, advertising displays, and instructional materials.

POSTER

A poster is a very versatile visual aid, for it can contain a wide range of information and include a number of other visuals such as charts, photographs, pictures, drawings, and diagrams. In addition, anyone can design a good poster. (A poster is used in the oral presentation describing the ColorPro plotter, pages 75–77.)

In preparing a poster, review Using Visuals Effectively, pages 539–541, especially number 3, on preparing a visual. Remember that everything on a poster must be large enough for *all* viewers to see.

CHALKBOARD

The chalkboard, though not really an exhibit in itself, is one of the most convenient means for visually communicating information. The chalkboard can be erased quickly, and new material can be added as the learning sequence progresses. (The chalkboard is used in the outlining of steps in the oral presentation of how a mushroom grows, pages 45–46.)

When using a chalkboard, be sure to write large enough for all material to be visible from the rear of the room; develop one point at a time; remove or cover distracting material; and stand to one side when pointing out material.

Projected Materials

Projected materials are uniquely effective in oral presentations. (See Chapter 10, Oral Communication, especially page 514.) Among the most common projected materials are films, filmstrips, slides, and overhead-projected transparencies. These materials require a certain room illumination, a screen, and specialized projectors.

FILM

Films, or more accurately motion pictures, are excellent, of course, for portraying the action and movement inherent in a subject. They present a sense of continuity and logical progression. And sound can be added easily. However, since the preparation of films requires specialized knowledge and is expensive in terms of time, equipment, and materials, commercially prepared 8-millimeter or 16-millimeter films are often borrowed, rented, or purchased.

FILMSTRIP

A filmstrip is a series of still pictures photographed in sequence on 35-millimeter film. They may be supplemented by captions on the frames (pictures), recorded narration, or script reading. Filmstrips are compact, easily handled, and, unlike slides, always in proper sequence. Although rather difficult to prepare locally, filmstrips from commercial producers are inexpensive and cover a wide array of topics.

SLIDE

Slides, usually 2 inches by 2 inches, are taken with a simple 35-millimeter camera; they provide colorful, realistic reproductions of original subjects. Exposed film is sent to a processing laboratory, which then returns the slides mounted and ready for projection. Also, sets of slides on particular topics can be purchased; these often are accompanied by taped narrations.

Slides are quite flexible: They are easily rearranged, revised, handled, and stored; and automatic and remotely controlled projectors are available for greater efficiency and effectiveness. If handled individually, however, slides can get out of order, be misplaced, or even be projected incorrectly.

TRANSPARENCY

Transparencies have become quite popular in conveying information to groups of people. They are easy and inexpensive to prepare and to use, and the projector is simple to operate. (A transparency is used in the oral presentation describing the ColorPro plotter, pages 75–77.)

The overhead projector permits a speaker to stand facing an audience and project transparencies (sheets of acetate usually 8½ inches by 10 inches) onto a screen behind the speaker. The projection may be enlarged to fit the screen, so the speaker may more easily point to features or mark on the projected image. Room light is at a moderate level.

In preparing a transparency, you may write on the film with a grease pencil, India ink, or special acetate ink; or you may use a special copying machine.

General Principles in Using Visuals

1. *Visuals can be extremely helpful.* They can capitalize on seeing, convey information, reduce textual explanation, and add interest.
2. *Effective use of visuals requires careful planning.* Appropriate selection, placement, and reference to visuals are crucial to their effective use. The visual should not include too much, should be pleasing to the eye, and should use lettering to good advantage.
3. *Credit should be given for borrowed material.* Whether an entire visual or simply the information is borrowed, credit should be given for the source.
4. *Layout of text on the page can visually aid the reader.* Such layout devices as headings, paragraphing, indentation, list formats, double columns, and ample white space can enhance a report.
5. *Computers have both simplified and expanded the production of graphic illustrations.*
6. *Tables present large amounts of data concisely.* Although tables lack eye appeal and interest-arousing dramatization, they are unexcelled as a method of organizing and depicting research data.
7. *Charts show relationships.* Common types of charts are the pie chart, the bar chart, the organization chart, and the flowchart.
8. *Graphs show change.* Graphs portray numerical data helpful in identifying trends, movements, and cycles.
9. *Photographs show actual appearance.* Photographs are of particular value as evidence in proving or showing what something is.
10. *Drawings and diagrams show isolated or interior views, exact details, and relationships.* Drawings and diagrams can be easily tailored to fit the needs of the user.
11. *Exhibits dramatize a concept or objects before an audience.* Frequently used kinds of exhibits are demonstrations, displays, real objects, models, dioramas, posters, and the chalkboard.
12. *Projected materials show pictures and other forms of information on a screen before an audience.* Common projected materials include films, filmstrips, slides, and overhead-projected transparencies.

Application 1   Using Visuals

From periodicals, reports, brochures, government publications, etc., make a collection of ten visuals pertaining to your major field of study. If necessary, copy the visual from the source, and write a two- to three-sentence comment concerning each visual.

Application 2 Using Visuals

Make a survey of 20 persons near your age concerning their preferences in automobiles: make, color, accessories. Present your findings in a table.

Application 3 Using Visuals

From a bank or home finance agency, obtain the following information concerning a $50,000 house loan, a $60,000 house loan, and a $70,000 house loan:

Interest rate

Total monthly payment for a 30-year loan

Total interest for 30 years

Total cost of the house after 30 years

Cost of insurance for 30 years

Estimated taxes for 30 years

Present this information in a table.

Application 4 Using Visuals

Present the following information in the form of a pie chart.

From Steer to Steak

Choice steer on hoof	1000 lbs
Dresses out 61.5%	615 lbs
Less fat, bone, and loss	183 lbs
Salable beef	432 lbs

Application 5 Using Visuals

Prepare a pie chart depicting your expenses at college for a term. Indicate the total amount in dollars. Show the categories of expenses in percentages.

Application 6 Using Visuals

Roy McGowan is transportation maintenance supervisor for Denis County Public Schools. His salary is $30,000 per year. McGowan participates in a retirement plan into which annually he pays 5 percent of his salary, Denis County pays 4 percent of his salary, and the state pays 2 percent of his salary.

Prepare a bar chart showing the total amount in dollars paid annually in McGowan's retirement plan and the individual amounts paid by McGowan, by Denis County, and by the state.

Application 7   Using Visuals

Make a bar chart showing the sources of income and areas of expenditure for a household, your college or some other institution, or a firm.

Application 8   Using Visuals

Make an organization chart of the administrative personnel of your college or of some other institution or of the personnel in a firm.

Application 9   Using Visuals

Make an organization chart of a club or organization with which you are familiar.

Application 10   Using Visuals

Make a flowchart of the registration procedures or some other procedure in your college or of a procedure in your place of employment.

Application 11   Using Visuals

Make a flowchart depicting from beginning to completion the flow of a process or product in your field of study.

Application 12   Using Visuals

Construct a line graph depicting your growth in height *or* weight, or someone else's growth, for a ten-year period. (Make estimates if actual amounts are unknown.)

Application 13   Using Visuals

Prepare a multiple-line graph showing the enrollment in your college and in another college in your area for the past ten years. If you are unable to obtain the exact enrollment figures, use your own estimates.

Application 14   Using Visuals

Find two photographs (from your own collection; in newspapers, periodicals, etc.) that present evidence as only a photograph can. Write a brief comment about each photograph.

Application 15   Using Visuals

Visualize a situation pertinent to your major field of study in which a photograph would be essential. Then find or make a suitable photograph. Finally, write a paragraph that describes the situation and include the photograph.

Application 16   Using Visuals

Make a drawing of a piece of equipment used in your major field of study and indicate the major parts. Then write a paragraph identifying the piece of equipment and include the drawing.

Application 17   Using Visuals

The ancient Greek philosopher Aristotle thought that the heavier something is, the faster it falls to the ground. The Renaissance physicist Galileo discovered that this is not true. Rather, everything falls to the ground at the same rate; for example, a baseball and a lead cannon ball dropped from a tower would both hit the ground at the same time.

Make a drawing or diagram that illustrates this concept.

Application 18   Using Visuals

Make a floor plan (diagram) of your living quarters, including the location of all pieces of furniture.

Application 19   Using Visuals

Prepare an oral presentation in which you use exhibits and/or projected materials.

Application 20   Using Visuals

Present the following information in an appropriate visual or visuals: Government service, one of the nation's largest fields of employment, provided jobs for 15 million civilian workers in 1986, about one out of six persons employed in the United States. Nearly four-fifths of these workers were employed by state or local governments, and more than one-fifth worked for the federal government.

Application 21   Using Visuals

Select information from Table 1, page 543. Present the selected information in a nontable visual.

Improper Driving as Factor in Accidents, 1982

Kind of improper driving	Fatal accidents			Injury accidents			All accidents[a]		
	Total	Urban	Rural	Total	Urban	Rural	Total	Urban	Rural
Improper driving	61.2	62.3	60.6	67.8	70.5	63.6	66.6	67.7	64.5
Speed too fast[b]	30.5	31.5	30.2	24.4	20.7	30.2	21.0	17.6	28.1
Right of way	12.1	17.8	9.6	24.3	30.7	14.3	23.7	27.6	15.7
Drove left of center	11.8	6.0	14.3	4.2	2.4	7.1	3.5	2.1	6.3
Improper overtaking	1.5	1.0	1.6	1.4	1.1	1.9	2.1	1.7	2.7
Made improper turn	0.7	0.8	0.6	2.2	2.5	1.8	3.7	4.2	2.7
Followed too closely	0.5	0.7	0.4	6.2	7.8	3.6	6.5	7.6	4.4
Other improper driving	4.1	4.5	3.9	5.1	5.3	4.7	6.1	6.9	4.6
Total	**100.0%**	**100.0%**	**100.0%**	**100.0%**	**100.0%**	**100.0%**	**100.0%**	**100.0%**	**100.0%**

[a]Principally property damage accidents, but also includes fatal and injury accidents.
[b]Includes "speed too fast for conditions."
Source: Urban and rural reports from ten state traffic authorities to National Safety Council.
NOTE: Figures are latest available.

Application 22 Using Visuals

Select information from the above table "Improper Driving as Factor in Accidents, 1982." Present the selected information in a nontable visual.

Application 23 Using Visuals

Select any 15 consecutive pages in this book. Analyze the visuals (including layout and design).

a. List the selected page numbers.
b. List the various techniques used in layout and design.
c. List the kinds of visuals included.
d. Write a paragraph evaluating the effectiveness of the layout and design and the visuals.

Application 24 Using Visuals

Look up your major field of study in the latest edition of *Occupational Outlook Handbook* (published biennially by the U.S. Department of Labor's Bureau of Labor Statistics). From the information given, prepare a report, using visuals where appropriate. Create your own visuals; do not merely copy the ones in the *Handbook*.

USING PART II: SELECTED READINGS

Application 25 Using Visuals

Read "American Labor at the Crossroads," by Steven M. Bloom and David E. Bloom, pages 583–588. Select any one paragraph of statistical information. Present the information in one or more visuals.

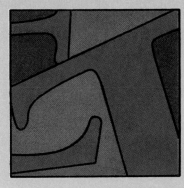

Part II
Selected Readings

PART I SUGGESTS WAYS TO IMPROVE writing and speaking, two levels of language usage. Part II suggests ways to improve reading, a third level of language usage.

Part II of *Technical English* is a collection of readings included for several reasons. First, the readings are intended to complement the material in Part I by illustrating, in part or as a whole, the kinds of writing discussed in Part I. They are intended to stimulate thinking and writing with at least one assignment for each reading correlated to a specific chapter in Part I.

The collection is also intended to stimulate reading. Hopefully, the selections will be of interest to you, the student and the worker, and reading an article or articles will encourage you to read additional materials.

Without question, reading skill is essential to the technician. In a recent yearbook of the Association for Supervision and Curriculum Development, the following comments appeared:

Reading is . . . the process of interrelating many varied experiences, drawing meanings with symbols that are almost infinitely varied in their combinations and permutations. It is, therefore, not a simple process that is mastered once and for all. As students move into the organized bodies of knowledge with their own technical terminologies and special vocabularies, in short their languages, they must to a degree learn to read again. Each special field has its own language and one who would succeed in the field must learn its language.

The printed word is a method of communication used daily by business, industry, and service institutions. Many large companies with multiple locations have a division that does nothing but prepare written communications for distribution to company personnel throughout the United States and around the world. These communications vary; they may be statements of company policy on such subjects as pay, working hours, holidays, and benefits; reports on research into new methods, techniques, processes, and uses of material; or periodic bulletins about company personnel who, for example, have made significant suggestions on how to improve or increase production. Of course, intercompany communications are also numerous and varied, the number and the variety increasing proportionally with the size of the company. To function as an employee, you will no doubt have to read and understand these communications.

Reading skill is essential to the technician who keeps up with what is happening in this rapidly changing technological age. Institutions and organizations publish numerous periodicals and professional journals for almost all possible areas of interest. Some of these are available to the general public; others are available only through membership in professional organizations. Reading such periodicals and journals keeps the skilled worker aware of current developments.

You will perhaps find that you cannot possibly read all the available materials related to your professional interest; so you must be selective. The following study-reading method offers one way to help you decide whether an article is worthy of careful reading and study.

This method will help you read and study assigned material more efficiently and carefully. By following each step as you read the selections in Part II, you will be better prepared to carry out the assignments following each selection.

The procedure includes scanning, reading, reviewing, responding and reacting—activities that will help you become familiar with the content of a selection and thus provide you with a basis for selective or assigned reading.

Scanning

Scanning (prereading, previewing, surveying) a selection should reveal pertinent information, such as the subject or topic, tone or mood, and depth or detail of content. You may discover such information by noting specific parts of a selection, such as its title and subtitle, headings, annotation, visuals, introduction and conclusion, and questions or problems.

TITLE AND SUBTITLE

Note any title or subtitle. It may indicate the subject or the topic. For example, the title of the selection "To Serve the Nation: Life Is More Than a Career" very clearly identifies the subject, whereas titles such as "Quality" and "Are You Alive?" require the reader to find more information before learning exactly what the selections are about.

The title or subtitle may also indicate the tone or the mood of the selection—humorous, sarcastic, persuasive, and so forth.

HEADINGS

Note headings that may be used throughout a selection. These will identify the subjects of the sections within a selection. "Videotex: Ushering in the Electronic Household," "American Labor at the Crossroads," "Holography: Changing the Way We See the World," "Clear Writing Means Clear Thinking Means . . . ," and "Starch-Based Blown Films" use headings. They may help you see the overall outline of a selection more quickly.

ANNOTATION

Sometimes an annotation or a headnote suggests the subject, content, or main point of a selection. Read any such annotation for the information on content it may give. Or it may give information about the author: background, education, experience.

Each of the selections in Part II has an annotation.

VISUALS

Look at any visuals included in a selection—pictures, sketches, graphs, diagrams, and the like. (Review Chapter 11 in Part I.) Visuals may give more specific clues about content. Note the use of visuals in the article "Videotex: Ushering in the Electronic Household" and in the report "Starch-Based Blown Films."

INTRODUCTION AND CONCLUSION

Read the first paragraph (or first few paragraphs); it usually introduces the topic of the selection. Of the 13 selections in Part II, the topic is identified in the first paragraph of five selections and in the first few paragraphs in five.

Read the concluding paragraph (or paragraphs); it may contain various kinds of information. The author may restate the central idea (see pages 684–685) or relist the main points used in developing or explaining the central idea. The author may summarize the content or reach a conclusion(s) supported or justified by the content.

Sometimes sections will be labeled "Summary" or "Conclusion." A reading of these sections will identify details the author considers important; look for these details as you read the entire selection.

Although no labels are used, of the selections in Part II, nine are summarized or a conclusion is drawn in either the last paragraph or in the last two or more paragraphs ("The Science of Deduction" does not include a summary).

Other selections in Part II, including a short story, a poem, and a section from a novel, because of their form, neither introduce topics in opening paragraphs nor summarize themes in final paragraphs.

QUESTIONS OR PROBLEMS

Sometimes questions or problems will follow a selection. Read these carefully. They will give you some idea about the content of the selection. Look for the answers as you read.

Reading

On the basis of information gained through scanning a selection, you may decide to read it, if, of course, the decision to read or not to read is yours to make. As a student or a worker you may be assigned a selection; scanning the selection, then, provides a background for your reading.

INITIAL READING

Read a selection through quickly for an overall view.

FURTHER READING

Read a selection a second (and if necessary, a third and a fourth) time, reading carefully and thoroughly for details.

VOCABULARY

Sometimes in reading a selection you find a word that is unfamiliar. You may be able to determine the meaning of such a word from other words in the sentence and in surrounding sentences. Or you may have to look up the word in a dictionary. There you will probably find several meanings; of these, you will have to decide which meaning is more logical as it is used in the selection.

In the following selections, vocabulary study words are printed in **boldface** type.

Reviewing

After reading a selection, think about what you have read. List the main ideas presented and summarize each briefly. You may want to list and summarize these ideas as an oral review of your reading, or you may want to write the review. Possibly you will do both an oral and a written review, especially if a selection contains information of immediate or future importance.

Think about the relationship among the parts of the article. Try to understand the relationship between the parts to the article as a whole (see pages 211–213, 217–222, 355–357).

Review words that are new to you. Define the words according to their use in the selection; then try using the words in your own sentences.

Responding and Reacting

Rarely does anyone read anything without responding or reacting in some way. The reaction may be simply a fleeting thought such as "I don't believe this," "I don't like this," or "Pretty good article." Or it may be more involved, depending on the reason you read the material or on your interest in or need for the information the material contains.

Procedure for Comprehending

1. Scan a selection. Note specifically:
   a. Title and subtitle(s)
   b. Headings
   c. Annotation
   d. Visuals
   e. Introduction and conclusion
   f. Questions or problems
2. Read a selection at least twice, once quickly for an overall view and again carefully and thoroughly for details. Learn the meanings of unfamiliar words in the selection.
3. Review what you have read. If necessary, write down main ideas and summarize each briefly; then review these orally. Think about the relationship among the parts of the article.
4. Respond and react, either as you choose or as you are directed.

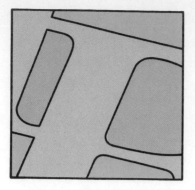

Clear Only If Known

EDGAR DALE

Edgar Dale of Ohio State University ponders the question: Why do people give directions poorly and sometimes follow excellent directions inadequately?

For years I have puzzled over the **inept** communication of simple directions, especially those given me when touring. I ask such seemingly easy questions as "Where do I turn off Route 40 for the by-pass around St. Louis? How do I get to the planetarium? Is this the way to the Federal Security Building?" The individual whom I hail for directions either replies, "I'm a stranger here myself" or gives you in kindly fashion the directions you request. He finishes by saying pleasantly, "You can't miss it."

But about half the time you do miss it. You turn at High Street instead of Ohio Street. It was six blocks to the turn, not seven. Many persons who give directions tell you to turn right when they mean left. You carefully count the indicated five stoplights before the turn and discover that your guide meant that blinkers should be counted as stoplights. Some of the directions exactly followed turn out to be inaccurate. Your guide himself didn't know how to get there.

Now education is the problem of getting our bearings, of developing orientation, of discovering in what direction to go and how to get there. An inquiry into the problem of giving and receiving directions may help us discover something important about the educational process itself. Why do people give directions poorly and sometimes follow excellent directions inadequately?

First of all, people who give directions do not always understand the complexity of what they are communicating. They think it a simple matter to get to the Hayden Planetarium because it is simple for them. When someone says, "You can't miss it," he really means, "I can't miss it." He is suffering from what has been called the COIK fallacy—Clear Only If Known. It's easy to get to the place you are inquiring about if you already know how to get there.

We all suffer from the COIK fallacy. For example, during a World Series game a recording was made of a conversation between a rabid Brooklyn baseball fan and an Englishman seeing a baseball game for the first time.

The Englishman asked, "What is a pitcher?"

"He's the man down there pitching the ball to the catcher."

"But," said the Englishman, "all of the players pitch the ball and all of them catch the ball. There aren't just two persons who pitch and catch."

Later the Englishman asked, "How many strikes do you get before you are out?"

The Brooklyn fan said, "Three."

"But," replied the Englishman, "that man struck at the ball five times before he was out."

These directions about baseball, when given to the uninitiated, are clear only if known.

Try the experiment sometime of handing a person a coat and ask him to explain how to put it on. He must assume that you have lived in the tropics, have never seen a coat worn or put on, and that he is to tell you *verbally* how to do it. For example, he may say, "Pick it up by the collar." This you cannot do, since you do not know what a collar is. He may tell you to put your arm in the sleeve or to button up the coat. But you can't follow these directions because you have no previous experience with either a sleeve or a button.

The communication of teachers to pupils suffers from the COIK fallacy. An uninitiated person may think that the decimal system is easy to understand. It is— if you already know it. Some idea of the complexity of the decimal system can be gained by teachers who are asked by an instructor to understand his explanation of the duo-decimal system—a system which some mathematicians will say is even simpler than the decimal system. It is not easy to understand with just one verbal explanation, I assure you.

A teacher of my acquaintance once presented a group of parents of first-grade children with the shorthand equivalents of the first-grade reader and asked them to read this material. It was a frustrating experience. But these parents no longer thought it was such a simple matter to learn how to read in the first grade. Reading, of course, is easy if you already know how to do it.

Sometimes our directions are over-complex and introduce unnecessary elements. They do not follow the law of **parsimony.** Any unnecessary element mentioned when giving directions may prove to be a distraction. Think of the directions given for solving problems in arithmetic or for making a piece of furniture or for operating a camera. Have all unrelated and unnecessary items been eliminated? Every unnecessary step or statement is likely to increase the difficulty of reading and understanding the directions. There is no need to overelaborate or **labor** the obvious.

In giving directions it is also easy to overestimate the experience of our questioner. It is hard indeed for a Philadelphian to understand that anyone doesn't know where the City Hall is. Certainly if you go down Broad Street, you can't miss it. We know where it is: why doesn't our questioner? . . .

Another frequent reason for failure in the communication of directions is that explanations are more technical than necessary. Thus a plumber once wrote to a research bureau pointing out that he had used hydrochloric acid to clean out sewer pipes and inquired, "Was there any possible harm?" The first reply was as follows: "The **efficacy** of hydrochloric acid is indisputable, but the corrosive residue is incompatible with metallic permanence." The plumber then thanked them for the information approving his procedure. The dismayed research bureau tried again, saying, "We cannot assume responsibility for the production of toxic and noxious residue with hydrochloric acid and suggest you use an alternative procedure." Once more the plumber thanked them for their approval. Finally, the bureau, worried about the New York sewers, called in a third scientist who wrote: "Don't use hydrochloric acid. It eats hell out of the pipes."

Some words are not understood and others are misunderstood. For example, a woman confided to a friend that the doctor told her she had "very close veins." A patient was puzzled as to how she could take two pills three times a day. A little girl told her mother that the superintendent of the Sunday school said he would drop them into the furnace if they missed three Sundays in succession. He had said that he would drop them from the register.

We know the vast difference between knowing how to do something and being able to communicate that knowledge to others, or being able to verbalize it. We know how to tie a bow knot but have trouble telling others how to do it.

Another difficulty in communicating directions lies in the unwillingness of a person to say that he doesn't know. Someone drives up and asks you where Oxford Road is. You realize that Oxford Road is somewhere in the vicinity and feel a sense of guilt about not even knowing the streets in your own town. So you tend to give poor directions instead of admitting that you don't know.

Sometimes we use the wrong medium for communicating our directions. We make them entirely verbal, and the person is thus required to keep them in mind until he has followed out each of the parts of the directions. Think, for example, how hard it is to remember Hanford 6-7249 merely long enough to dial it after looking it up.

A crudely drawn map, of course, would serve the purpose. Some indication of distance would also help, although many people seem unable to give adequate estimates of distances in terms of miles. A chart or a graph can often give us an idea in a glance that is communicated verbally only with great difficulty.

But we must not put too much of the blame for inadequate directions on those who give them. Sometimes the persons who ask for help are also at fault. Communication, we must remember, is a two-way process.

Sometimes an individual doesn't understand directions but thinks he does. Only when he has lost his way does he realize that he wasn't careful enough to make sure that he really did understand. How often we let a speaker or instructor get by with such mouth-filling expressions as "emotional security," "audiovisual materials," "self-realization," without asking the questions which might clear them up for us. Even apparently simple terms like "needs" or "interests" have hidden many confusions. Our desire not to appear dumb, to be presumed "in the know," prevents us from really understanding what has been said.

We are often in too much of a hurry when we ask for directions. Like many tourists we want to get to our destination quickly so that we can hurry back home. We don't bother to savor the trip or the scenery. So we impatiently rush off before our informant has really had time to catch his breath and make sure that we understand.

Similarly, we hurry through school subjects, getting a bird's-eye view of everything and a closeup of nothing. We aim to cover the ground when we should be uncovering it, looking for what is underneath the surface.

It is not easy to give directions for finding one's way around in a world whose values and directions are changing. Ancient landmarks have disappeared. What appears to be a lighthouse on the horizon turns out to be a mirage. But those who do have genuine expertness, those who possess tested, authoritative data, have an obligation to be clear in their explanations. Whether the issue is that of atomic energy, UNESCO, the UN, or conservation of human and natural resources, clarity in the presentation of ideas is a necessity.

We must neither overestimate nor underestimate the knowledge of the inquiring traveler. We must avoid the COIK fallacy, realize that many of our communications are clear only if already known.

Questions on Content

1. What is the COIK fallacy?
2. List five of the author's reasons why people give directions poorly.
3. The author puts some of the blame for lack of communication involving directions on the person who requested help. What two reasons did he describe?

Vocabulary

Define the following words as they are used in the article (the words are printed in **boldface** type):

1. inept
2. parsimony
3. labor
4. efficacy

Suggestions for Response and Reaction

1. Relate an experience when you failed to give or receive instructions or directions clearly.
2. Following the explanation for giving instructions (see pages 18–19), give directions that would explain to a campus visitor how to walk from the visitors' parking lot to your technical writing classroom.
3. Write a descriptive (see pages 214–217) and an informative (see pages 217–222) summary of this article.
4. Write instructions (see pages 19–33) for putting on a coat, tying shoelaces, or some other "simple" procedure.

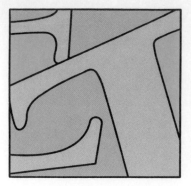

The Unknown Citizen

W. H. AUDEN

Even though Auden apparently describes a
desirable, well-received citizen, he gives a clue to
his reason for such a description in the title
"The Unknown Citizen."

TO JS/07/M/378
THIS MARBLE MONUMENT IS ERECTED BY THE STATE

He was found by the Bureau of Statistics to be
One against whom there was no official complaint,
And all the reports on his conduct agree
That, in the modern sense of an old-fashioned word, he was a saint,
For in everything he did he served the Greater Community.
Except for the War till the day he retired
He worked in a factory and never got fired,
But satisfied his employers, Fudge Motors Inc.
Yet he wasn't a **scab** or odd in his views,
For his Union reports that he paid his dues,
(Our report on his Union shows it was sound)
And our Social Psychology workers found
That he was popular with his mates and liked a drink.
The Press are convinced that he bought a paper every day
And that his reactions to advertisements were normal in every way.
Policies taken out in his name prove that he was fully insured,
And his Health card shows he was once in hospital but left it cured.
Both Producers Research and High-Grade Living declare
He was fully sensible to the advantages of the Installment Plan
And had everything necessary to the Modern Man,
A gramophone, a radio, a car and a frigidaire.
Our researchers into Public Opinion are content
That he held the proper opinions for the time of year;
When there was peace, he was for peace; when there was war, he went.
He was married and added five children to the population,
Which our **Eugenist** says was the right number for a parent of his generation,
And our teachers report that he never interfered with their education.
Was he free? Was he happy? The question is **absurd:**
Had anything been wrong, we should certainly have heard.

Questions on Content

1. Identify the "he" in the poem. The "we" in the last line.
2. List the activities that made the unknown citizen ordinary or acceptable, supposedly.

Vocabulary

Define the following words as they are used in the poem (the words are printed in **boldface** type):

1. scab   2. Eugenist   3. absurd

Suggestions for Response and Reaction

1. What seems to be Auden's attitude toward the society he describes?
2. What is the significance of the word *unknown* in the title? Of "TO JS/07/M/ 378" and "THIS MARBLE MONUMENT IS ERECTED BY THE STATE"?
3. What attitudes common among twentieth-century people does Auden apparently disapprove of?
4. If you were to design an abstract picture of the unknown citizen, what color would you use primarily? Why?
5. Answer the question in the next-to-last line: "Was he free?" Explain your answer using cause/effect analysis (see pages 170–184).
6. What does "free" mean to you? Write an extended definition of *free*, following the suggestions on pages 96–99, 101–112.

Videotex: Ushering in the Electronic Household

JOHN TYDEMAN

Home information services will transform the house of the future—and society. People will be able to learn, work, and shop at home. But the changes will raise important issues that society must face.

The microelectronic revolution is everywhere. Electronic substitutes appear daily for products and services. They range from digital watches and programmable microwave ovens to automated teller machines. Electronic games are billed as America's fastest growing sport. Car manufacturers talk confidently of fully automated robot-operated assembly lines in the near future. Businesses are reorganized around minicomputers, communicating word processors linked to local and national data networks, and electronic mail systems.

Electronic technology has the potential to transform our society. The business world is already beginning to experience this change. At home, we are fast approaching the threshold of the electronic household.

A technology that could help us cross this threshold and contribute to major changes in our lifestyle—from the design of our dwellings to the organization of activities performed within our daily lives—is the class of home information services known as teletext and videotex.

The terms teletext and videotex refer to systems that primarily disseminate verbal and pictorial information by wholly electronic means, for visual display or printing under the control of the user.

Teletext is a one-way service in which the information is broadcast as part of a television signal. The teletext material is mixed into the TV signal in such a way that a user with a keypad can select the information to be displayed on the screen (see Figure 1).

Videotex refers to systems with two-way information flows. The transmission lines between the user and the computer can be the public telephone network, a cable TV system with two-way capabilities, or hybrids, such as a one-way cable into the home with a normal telephone link out. Users are thus able to send as well as receive signals (see Figure 2).

Videotex is not of itself a new technology, but is a product of the merging of communication and computing. It is made up of a new combination of **hardware,** such as computers, data bases, communication networks, television sets, and terminal devices, and **software,** including frame design, graphics and text, gateways, retrieval mechanisms, and billing. If it becomes widespread, videotex will significantly change household use and acquisition of various information services and products. For the sake of simplicity, in this article we will use the term videotex as a generic term to include one-way as well as two-way services.

Reprinted, with permission, from *The Futurist*, February 1982, published by the World Future Society, 4916 St. Elmo Avenue, Bethesda, Maryland 20814.

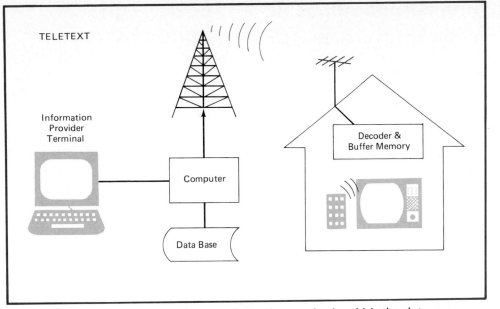

Figure 1. Teletext is a one-way home information service in which the data are broadcast as part of a television signal. A home user can select the information desired, which is displayed on the TV screen.

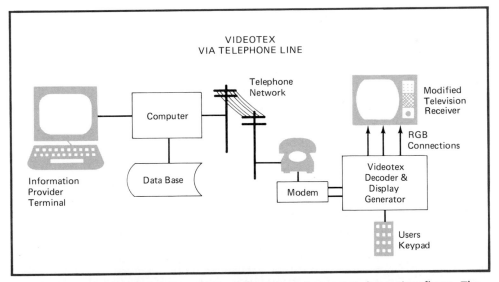

Figure 2. Videotex home information systems have two-way information flows. The user can receive information and respond through the telephone network, a cable TV system, or a combination of the two.

Home Applications

Videotex systems are still at the level of technical and market trials in the United States. Because the basic technologies are still evolving, the potential applications for videotex systems are still taking shape, but some of the major services now being offered in test trials and commercial services in the U.S. and other countries give some idea of what to expect.

Most current videotex trials offer some news, weather, and sports information. Newspapers and wire services have become both information providers and videotex service providers. In some instances, only selected features or stories are offered. In other cases, the entire editorial content of the paper is available for electronic retrieval.

Videotex systems can be used for advertising and shopping. Current videotex tests include display advertising (e.g., food market specials of the week), classifieds (e.g., help wanted, houses for sale), and promotional ads (e.g., "This weather information brought to you by. . .").

Adams-Russell Company and J. Walter Thompson are offering cable viewers in Peabody, Massachusetts, a "cable shop." Viewers can select a three-to-seven-minute video segment on a particular product or service, but cannot purchase the items. In the "Viewtron" test in Coral Gables, Florida, the at-home user not only has the chance to see the supermarket's specials of the week but also to order them.

Several organizations currently provide some kind of on-line electronic purchasing. Times-Mirror Cable, in conjunction with Comp-U-Card, is conducting an at-home buying service in Los Angeles in which the information is transmitted by cable, and purchasing is accomplished by a phone call on an "800" number.

Videotex is being explored for the services it might add to video programs. During election coverage, for example, teletext pages could be continuously updated for various contests, even while the viewer watches other programs or other election coverage. For people watching a sporting event, scores from other games could be available on demand. For hearing-impaired or multilingual audiences, videotex can provide subtitles for television programs.

Though often thought of as an inexpensive means of delivering "pages" of information to the home, both teletext and videotex have entertainment and game potential as well. In the early trials, entertainment, such as cartoons, quizzes, and horoscopes, has become quite popular. A cable service might send new game software to the home computer user, allowing him or her to interact with it or with others who might also wish to play.

Specialized libraries and information banks already provide electronic retrieval services. OCLC, a nationwide on-line library cataloging service, had a test in Columbus, Ohio, in which participants had access to the card catalog of the Columbus library system; they could request a desired book and receive it in the mail.

Videotex users can form closed user groups (CUG) that share information among them. CUGs provide a forum for clubs and associations to communicate and offer the opportunity for private, or at least restricted, access to information.

Electronic directories could be provided by any institution or corporation with frequently changing names and listings. Of course, both the white and yellow pages of the telephone directory would be prime candidates for an electronic delivery system. A major drawback of the yellow pages for advertisers has been its year-long time frame between updating. An electronic yellow page (EYP) service could be

Application Areas	Personal storage files	Electronic messaging	Home computer support	Banking transactions	Teleshopping	Games/entertainment	Electronic directories/yellow pages	Library information	Financial information	Education	Program captioning	Advertising	Electronic newspaper	News, weather, sports
KSL-TV (Salt Lake City)						■						■		■
CBS-Extravision										■		■		■
NCI/Closed Captioning											■			■
WETA-TV (Washington, D.C.)									■					■
Time Inc.*										■				■
WFLD-TV (Chicago)						■				■				■
WTBS (Atlanta)														■
Datacast*								■						■
WGN (Chicago)														■
WHA* (Madison)											■			■
KPIX-Westinghouse												■		■
KNBC* (Los Angeles)														■
Landmark														■

Application Areas	Personal storage files	Electronic messaging	Home computer support	Banking transactions	Teleshopping	Games/entertainment	Electronic directories/yellow pages	Library information	Financial information	Education	Program captioning	Advertising	Electronic newspaper	News, weather, sports
QUBE/Warner-Amex (Columbus, OH)						■				■				■
Indax/CoxCable (San Diego)				■	■	■				■				■
Belo/Dallas Morning News/Sammons Cable				■		■				■				■
Times Mirror Cable (Orange County, CA)*				■	■	■		■		■				■
Viewtron/Knight-Ridder/AT & T (Coral Gables, FL)		■		■	■	■		■	■	■		■		■
EIS/AT & T (Albany, NY)	■	■		■	■	■	■		■					■
OCLC/Channel 2000 (Columbus, OH)	■					■	■	■	■					■
The Source	■	■	■			■		■	■					■
MicroNet/Compu-Serv	■	■	■			■		■	■					■
Green Thumb (Kentucky)			■						■	■				■
Comp-U-Star/Comp-U-Card					■									■
Dow Jones									■					■
Associated Press														■
United American Bank				■	■									■
First Interstate Bank (Los Angeles)	■	■		■	■									■
Citibank (New York)				■										■
Chemical Bank (New York)			■	■		■								■
Continental Telephone*														■

Teletext and videotex systems are undergoing trials in many areas of the United States, for a variety of potential applications. The outcomes of these trials will help determine how the technologies will eventually be used.

updated almost instantaneously. In addition to its updating capability, an EYP service could also give the advertiser a direct link to the consumer. In other words, browsing in the EYP could lead to purchasing or ordering via the system.

Some services, such as The Source and CompuServe, offer mail and message sending capabilities at a relatively low cost. In others, with hand-held keypads, preformatted messages can be sent to other people who access the system. Alternatively, users might be able to access a community bulletin board, much as personal computer users do now for receiving information from other users. The bulletin board has been a feature of the test in Coral Gables.

The Viewtron trial includes educational features such as health tips. The U.S. Department of Agriculture-sponsored Green Thumb test in Kentucky provided information on weather, agricultural commodities markets, and related subjects to farmers in two rural counties. In some tests, the educational material has been linked with entertainment features, much as Sesame Street pioneered the development of entertaining educational television programs for preschool children.

In 1981, Dow Jones announced a financial information service oriented toward the home computer user. In this data base, the user can access current news from the pages of the *Wall Street Journal*, *Barron's*, and the Dow Jones News Source. In addition, current quotes in stocks, bonds, options, mutual funds, and U.S. Treasury issues are available (historical quotes are also available for stocks).

Automatic teller machines (ATMs) and the computerization of bank records have brought the possibility of home banking closer to reality. Videotex banking trials are now underway in many states on both cable and telephone networks. Services include information on account balances, transfers among accounts, and payments of certain bills.

The Household Market

Of these potential users, a market survey by the Institute for the Future found that in the case of teletext (one-way) the most popular use would probably involve information retrieval. Other popular applications could include computing services that also provide games and entertainment.

The Many Uses of Videotex

The potential uses of videotex span many aspects of our daily activities. They fit into five classes of likely applications:

Information retrieval: The most fundamental videotex service, information retrieval involves linking users with one or more data bases from which they can select material of interest. The number of specific applications within this category is very large. They range from general interest electronic newspapers to specialized data bases and directories.

Transactions: This category includes such interactive services as making reservations, paying bills, transferring funds, and teleshopping. Because transactions require the user to interact with an external computer, the service depends on having a two-way capability.

Messaging: Here a videotex computer acts as a switchboard to store and forward messages from one user to another. A messaging service can provide either one-to-one communication (electronic mail) or one-to-many communication (community bulletin board and computer conferencing).

Computing: At the most basic level, a videotex system can permit a person to answer yes or no or to choose a response to multiple-choice questions by using a simple numerical keyboard like the one on today's push-button telephones. Such a keyboard could be used to play games, do financial analyses, or take a multiple-choice test on any subject. At a more advanced level, videotex systems could be used to transmit computer programs (software) directly from a large central computer to the memory unit of a small personal computer.

Once transmitted, the large computer's program would be available for use on the small computer at any time.

Telemonitoring: This is a different kind of service from the others, but one that will be offered as part of the videotex package. A continuous link between a host computer and a user's terminal can result in two kinds of automated service. The first is for security—remote sensing for the detection of fire or intruders. The second kind of service would automatically control home devices, e.g., an energy management system might optimize efficiency by switching appliances off and on at appropriate times.

Two-way videotex systems are a different story. Not only is the technology for providing these services undergoing significant changes, but precisely which applications will provide the driving force for the spread of the technology is uncertain. In the short term, it is likely that information retrieval, computerbased gaming, and transactions (including home-based shopping) will be the driving applications. In the longer term, transactions and teleshopping will take on an even greater degree of prominence, and services such as electronic messaging and telemonitoring will become significant. This will open the market for home-based education and training.

Basic teletext or one-way broadcast services may have the potential to saturate the household market by the end of the century. The cost for the simple decoder is dropping significantly, and the text and graphics information services may be accepted as being basic TV services of the future.

Two-way videotex services, including systems offering some or all of the above generic applications, are expected to be in between 2 percent and 5 percent of households by 1985 and between 30 percent and 40 percent of households by the year 2000. Given that there are likely to be in excess of 100 million households in the United States by that time, this represents a sizable market (see Figure 3).

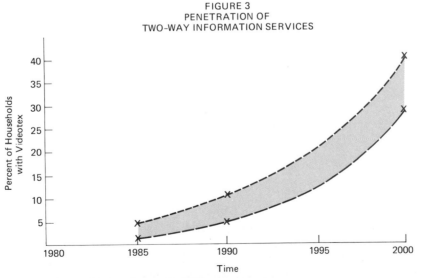

FIGURE 3
PENETRATION OF
TWO-WAY INFORMATION SERVICES

*Range of Forecasts Prepared by IFTF

Figure 3. Videotex is a technology all set to take off. For the next 20 years it will show phenomenal growth as it appears in more and more households.

Of course, if videotex does take off, it will not be because of a passive desire by the consumer to "go electronic." The pressure will come from manufacturers, bankers, wholesalers, or retailers who realize that the electronic option offers them a more economical way to provide their services.

Finding electronic alternatives cheaper and more convenient, businesses will begin to charge consumers the full cost of services such as paper-based bill paying.

Where To Next?

Today's videotex systems offer a glimpse into the electronic household of the future. They are relatively primitive in design and have a long way to go before being as acceptable to the consumer as the postal service, the newspaper, or the shopping mall. In fact, the evolutionary potential of these early systems is so great that videotex as we know it today may disappear. It will evolve considerably with printers added to systems, voice-activated systems, touch-sensitive screens, and animation and full-motion video services. Some home-based electronic interface to a wide variety of information services, however, will certainly emerge.

This interface will bring with it a number of major **societal** changes.

☐ The dwelling unit will also become a place of employment. This not only affects the type of structure, in terms of architecture and layout, but also the geographical location.

☐ The electronic household will facilitate a new home-based cottage industry in electronic products.

☐ Home-based shopping, along with computer-aided production, will allow consumers to control the manufacturing process. Consumers will be able to order exactly what they want, for "production on demand."

☐ The family will determine the electronic schooling (or, more correctly, the education) required for children and for the retraining of adults. There will be a shift away from the traditional school and work socialization process to ones in which peer groups and alliances are electronically determined.

☐ Stereotypical male-female roles will be replaced by roles determined on the basis of contribution to the household.

☐ New skills and career paths associated with the management of information will emerge. These will range from information brokers who provide the "best" deal on a used car to gatekeepers who monitor politicians and corporate activities and selectively release this information to interested parties.

☐ There will be an increase in opportunities to participate in educational, social, and political arenas. The interactive capacity of the new media allows greater involvement.

These are but a few of the social changes that will result from widespread use of this technology.

However videotex and teletext are used—or even if some successor technology linking the home to outside resources appears—some key impacts and trade-offs will face society.

To the question of whether widespread implementation of videotex systems will be "good" or "bad," the answer is, of course, "Yes."

. . .

Videotex: Ready or Not?

Radio took 20 years before it became a mass-communication product. During the years in the wilderness, it was used as a point-to-point or ship-to-shore transmission medium. It was not until Westinghouse understood that the radio offered *broadcasting* as its major product that programming emerged, an industry structure unfolded, and advertising support drove the technology toward household saturation.

On the other hand, television (a radio with a significant add-on—motion pictures) took off immediately. It had been ready to be released in 1939 by RCA at the World's Fair, but its debut was postponed due to the advent of World War II. When it arrived, it did so with a pent-up demand, post-war affluence, and a single national standard.

Further, the radio **infrastructure**—organization, licensing, allocation of frequency, advertising—was in place and there was little learning to do by the relevant actors in the television industry to have a commercial product.

Consider also cable TV, which was regulated out of the market until as late as 1977. Its penetration since that time has been nothing less than spectacular—currently running around 1% of households per month. The telephone, on the other hand, took 35 years to achieve 50% penetration.

Videotex resembles radio more closely than TV in some ways. There is no infrastructure in place. One only has to look at the new alliances that are coming together—telephone companies and broadcasters, banks and cable companies, newspapers and cable operators—to realize that the roles of information provider, videotex system provider, and to a lesser extent communications provider are not clear cut or well defined. The videotex product (or products) is not well defined. Is it information retrieval, is it transactions, is it messaging, or is it all of these things?

There is no single videotex standard, and there is no clearly identifiable pent-up demand by consumers. Further, there are a lot of electronic products competing for the householder's next dollar—pay and subscription TV, video and audio cassettes, videodiscs, personal computers, electronic games, and stand-alone security systems.

It would seem that there is still a lot of "shaking out" before videotex, or a videotexlike system, finally arrives. The only certainty is that the move toward the electronic household is irrevocable.

Questions on Content

1. Define the terms *teletext* and *videotex*.
2. Give four examples of home applications for videotex systems.
3. Where will videotex get its push for growth?
4. As videotex systems expand and develop, they will bring about major changes in society. List seven of these social changes.

Vocabulary

Define the following words as they are used in the article (the words are printed in **boldface** type):

1. electronic technology
2. hardware
3. software

4. societal
5. infrastructure

Suggestions for Response and Reaction

1. Using information from this article, write an extended definition (see pages 96–99, 101–112) of the term *videotex*.
2. Following the guidelines for setting up a classification system (see pages 122–139), explain the classification of the uses of videotex.
3. Choose two of the societal changes described on page 580. Contrast (see pages 193–202) the way of life described with the same aspects in today's society.

American Labor at the Crossroads

STEVEN M. BLOOM AND DAVID E. BLOOM

Organized labor in America is in a transition period. Following decades of popular support, the labor system is now facing many changes in the workplace.

Fifty years ago the American labor movement was plagued by riots, murders, endless court battles, and widespread and bloody strikes. Since then, it has enjoyed decades of popular support and has been at the cutting edge of institutional innovation. Today, however, **organized labor** is suffering a dramatic and accelerating reversal of its fortunes.

Union membership as a percent of total employment has fallen from 33 percent in 1960 (18 million members) to 20 percent in 1984 (21 million members). Union membership in manufacturing and mining—historic union strongholds—has fallen both in absolute numbers and as a fraction of employment in those industries, from more than one in two workers in 1960 to fewer than one in three workers in 1985. With these declines in membership, the political strength of organized labor has been shattered.

Public faith in trade union leaders has fallen over time. Only 55 percent of respondents in a 1981 nationwide Gallup Poll (and 73 percent of respondents belonging to union households) approved of labor unions, the lowest the approval rating has been since Gallup first asked the question in 1936. Only 26 percent of respondents (39 percent of those in union households) had "quite a lot" of confidence in organized labor as an institution, less than that for big business and government, and about the same as that for used-car salesmen.

Beyond the weakening of organized labor, the American **workplace** is changing. While unionized companies were once the principal site of innovative labor practices, now changes are occurring in the workplace regardless of the union status of firms. These include employee stock ownership plans, early retirement programs, profit-sharing, quality of work life programs, job security agreements, and **cafeteria benefit plans.**

The New American Worker

Behind the eroding power of the labor unions and the continuing evolution of the American workplace are the changing tastes and preferences of American workers, the growing technological sophistication of the workplace, and the increasing competition American industries are facing from foreign producers and from industrial deregulation.

American workers are different than they used to be, and the ways in which they have changed spell continued trouble for the unions. Young, educated and female workers have historically been less likely to join a union than other workers; yet these have been the most rapidly growing segments of the labor force for the past twenty years.

In 1960, one in three workers was a woman. By 1984, 44 percent of the labor force was female. At the same time, the share of the labor force aged 16 to 34 jumped from 37 percent in 1960 to fully 50 percent in 1984. One in four workers today is a college graduate, up from one in ten in 1959. The rise in the number of people who live alone and in the share of married couples with two earners also has changed the tastes and preferences of workers, not only making the labor force different than it once was, but making workers increasingly different from one another.

Technologically, the workplace has changed in ways that were not foreseen even ten years ago. The "second industrial revolution" includes microprocessors and microcomputers (about 2 million desktop computers are in use in the United States), automated production processes (roughly 5,000 industrial robots were in operation in the United States in the first half of 1984, mainly in the automobile, metalworking, and electronics industries—up from just 1,300 in 1979), lasers and satellite communications equipment, and data-handling and information systems.

The demands of the workplace have also changed as the economy moves away from continuous process shift work and toward service-sector employment. Between 1955 and 1984, employment in private sector goods-producing industries dropped from 41 percent of total nonagricultural employment to 26 percent. At the same time, employment in private sector service-producing industries grew from 46 percent to 57 percent of employment. Job schedules in the service sector are not as rigidly controlled as those in goods-producing industries, allowing individual employees greater flexibility to devise their own work schedules.

Increasing international competition has led to the permanent contraction of several key U.S. industries in which unions are heavily represented. Foreign steelmakers' share of the U.S. steel market has grown to 26 percent. Steel imports grew by nearly 30 percent while U.S. steel production declined by over 20 percent between 1970 and 1985. During the same period membership in the United Steelworkers and the United Auto Workers, two of the nation's largest international unions, declined 24 percent and 33 percent respectively.

Unionized industries have also faced stiffer competition from industrial deregulation. In the trucking and railroad industries in 1980, for example, deregulation dismantled the route assignment and freight rate controls of the Interstate Commerce Commission. The deregulated markets have attracted smaller, nonunion companies which are competing with larger, unionized operations.

The weakening power of unions is evident on a number of fronts. Concession bargaining during the past recession placed the union movement on the defensive. Fifty-six percent of the manufacturing workers covered by major private-sector labor agreements that were settled in 1983 received either no wage increase or a wage decrease in the first year of their contracts. The comparable figure in 1980 was 1 percent.

Union financial positions are eroding. After adjusting for inflation, consolidated union net assets per member declined by 8 percent during the 1970s to below 1962 levels. In 1969, consolidated union assets would have ranked organized labor 18th among the *Fortune* 500 industrial corporations. By 1982 its ranking would have fallen to 37th.

U.S. strike activity has reached a post-World War-II low. Fewer strikes involving 1,000 or more workers occurred between 1975 and 1984 than in any other ten-year period since 1945. There were under 300 strikes each year, with only 81 in 1983.

Seeing their influence waning, unions are regrouping. Union membership in the U.S. has become more concentrated in big unions, and the pace of union merger activity has quickened. Thirty-eight union mergers took place from 1975 to 1984, more than in any other ten-year period since the first recorded merger in 1859 between the machinists and the blacksmiths.

In an effort to stem further erosion, unions are going after public-sector workers. The share of employees who are union members in the public sector is now about twice that in the private sector. Nevertheless, the proportion of union workers in the public sector peaked in the late 1970s and has been steadily declining since then. On the other hand, the number of public-sector strikes has been increasing for more than two decades.

Piecemeal Innovation

The improvement of wages and working conditions for nonunion workers reflects, to some extent, the desire of employers to remain free of union constraints in an increasingly competitive economy. One way for companies to avoid unionization is to beef up personnel policies and practices.

While enforcement of the National Labor Relations Act has become bogged down in conflicting and changing interpretations of various legal aspects of **collective bargaining,** such as what constitutes an appropriate bargaining unit, piecemeal public policies and programs have been put into place.

In 1984 President Reagan signed into law a set of sweeping new tax incentives promoting employee ownership of business. There are now approximately 7,500 employee stock ownership plans in the U.S., covering roughly 6 million employee-shareholders.

The traditional employment-at-will doctrine, holding that employment involves a voluntary agreement between two parties, either of whom is free to terminate it at will, has been somewhat modified in a recent series of court rulings. Courts in at least 13 states have taken the position that there are circumstances under which an employee can regard a company manual, handbook, or employment interview as an implicit contract to which an employer is legally bound, even in the absence of a written labor agreement.

Between 1979 and 1984, an estimated 5.6 million workers were displaced as a result of plant shutdowns or relocations. Successive versions of a National Employment Priorities Bill, which would require companies to provide advance notice, severance pay, and retraining/relocation for displaced workers, have been introduced in Congress since 1974. By 1984, plant closing legislation existed in Maine, Michigan, and Wisconsin, and bills had been introduced in at least 38 other states.

The growing number of two-earner families and the increasingly diverse benefit preferences of the labor force have resulted in the development of cafeteria benefit plans under which workers can choose among various job benefits. Virtually nonexistent before 1970, cafeteria benefit plans are now being offered by over 200 major companies, including TRW, American Can, Johnson Controls, Pepsico, Minnesota Power and Light, and the State of Alaska.

The growth of two-earner families has also led to more flexible work scheduling and child-care benefits as employers and the government seek ways to aid working parents. An estimated 15 percent of all full-time nonfarm wage and salary workers were on flexible work schedules in 1985 (7.6 million employees), up from less than 5 percent in 1974. In 1976 the federal tax code was revised to permit working parents with a dependent child to take a tax credit for work-related child-care expenses.

Increasing competition is forcing management to involve workers in decision making through employee suggestion plans. In 1983, for example, employee suggestions saved 700 U.S. companies an estimated $800 million. Workers are also bearing more of the risk of doing business. An estimated 31 percent of all corporations with 100 or more employees had some form of profit-sharing plan in 1983.

Participatory management and quality of work life programs are being instituted by a growing number of companies. An estimated one-sixth of all U.S. corporations and one-third of all corporations with 500 or more employees had such programs in 1984.

Employee board representation has spread from the Chrysler Corporation in 1980 to Eastern Airlines, Pan Am, Western Airlines, Weirton Steel, and Hyatt Clark Industries, among others. This trend has been confined principally to financially troubled companies in exchange for wage concessions, and the representation is usually accompanied by some degree of employee ownership.

Two-tier contracts are another innovative response to increased competition. Under two-tier contracts, the wages and/or benefits of new workers are lower than those of **incumbent** jobholders. In some cases the new workers can eventually earn what the older employees earn, but in other cases equal pay for equal work is ruled out for the life of the agreement. Two-tier compensation was mentioned in 8 percent of contract settlements negotiated in the first half of 1984.

Employers and employees are also coming up with a variety of innovative job security arrangements. In its landmark agreement with the United Auto Workers in 1984, General Motors Corporation established a $1 billion job security fund to compensate workers who are laid off because of technological change, transfer of work to an outside supplier, or changes in the work process.

Work sharing—another way to ensure job security—was recently adopted by Motorola, Inc. and by steelworkers at Jones and Laughlin. Worksharing is most popular in Arizona, California, Maryland, Oregon, and in Canada, all of which have laws allowing workers who participate in such programs to receive unemployment benefits for the time they do not work. Florida, Illinois, and Washington adopted similar laws in 1984.

A Needed Transition

The changing labor market is pressing the American labor system to the limits of its ability to function efficiently. The original intent of the National Labor Relations Act—to offer workers a free choice about belonging to unions for purposes of **collective bargaining**—is being increasingly circumvented by administrative procedures and court rulings.

The void created by the decline of a national labor policy is being filled by piecemeal public policies, and by new labor practices that are being privately adopted by employers and employees. But the fact that longstanding labor institutions are yielding to innovative arrangements in certain companies and for certain workers does not mean that even more change on a larger scale is unnecessary.

By the year 2000, increasingly expensive and technologically-advanced equipment will find its way to the shop floor. There is a growing belief that complex equipment is used most productively when the individuals operating it are encouraged to exercise judgment and act independently. This flexibility is incompatible with the philosophy of scientific management and the accompanying body of rigid work rules that continue to dominate the American labor system.

The increasing age of the labor force over the next two decades is also likely to have a significant impact on the American labor system. Older workers are less **mobile** than younger ones, and it is less cost-effective to retrain older workers than to train their younger counterparts. As it ages, the work force may become less able to adapt to new economic conditions and new labor institutions. This suggests that many of today's opportunities for companies, workers, and the government to modify the American labor system may disappear in the future.

The American labor system is changing, but more like the tortoise than the hare. Public and private neglect are beginning to take their toll. It is time for policymakers to take stock of the system because the opportunities to make all of us better off are too good to miss.

Questions on Content

1. List at least three ways in which the American workplace is changing.
2. Why aren't more workers joining labor unions?
3. Cite ways in which unions are trying to prevent membership erosion.

Vocabulary

Define the following words as they are used in the article (the words are printed in **boldface** type):

1. organized labor
2. cafeteria benefit plans
3. workplace
4. collective bargaining
5. incumbent
6. mobile

Suggestions for Response and Reaction

1. Give the pros and cons from the employer's standpoint of employees belonging to a labor union.
2. Give the pros and cons from an employee's standpoint of employees belonging to a labor union.

3. Write a paper comparing and contrasting (see pages 193–202) the position of the employee and of the employer concerning organized labor.

4. Take a segment of statistical information and present it as a visual, such as a chart or graph (see pages 544–549 and 549–551).

5. Write a report (see pages 348–349) in which you compare (see pages 193–202) the role of unions in the United States today with the role of unions in the 1950s and 1960s. Use visuals wherever they are appropriate.

Holography: Changing the Way We See the World

JONATHAN BACK AND SUSAN S. LANG

Artists, scientists, entrepreneurs, and homemakers are all finding new uses for holograms.

"The greatest advance in imaging since the eye," as Isaac Asimov once described holography, is changing the way we see—and think about—the world. At once an artist's medium, a surgeon's guide, and a child's plaything, holography promises a multitude of applications in science, industry, communications, and art.

Holograms (from the Greek, "whole message") are three-dimensional images created by exposing a piece of film or glass with **laser light** reflected by an object.

Holography As Art

Spanish surrealist painter Salvador Dali predicted more than 12 years ago: "The possibility of a new renaissance in art has been realized with the use of holography. The doors have been opened for me into a new house of creation."

Holography may be viewed as a synthesis of all the arts. Like sculpture, holography employs and generates three-dimensional information. Like architecture, it creates and manipulates space. Like photography, its medium is light. Like painting, its colors are controlled by the artist, not by nature. Like other graphics, holograms can be one-of-a-kind or printed by the millions. Like dance or cinematography, holography can present every kind of motion.

At a more pragmatic level, holography currently serves the arts through archival recording. Holograms of ancient relics can be examined anywhere in the world while the real object remains stored in a safe place. Russian art treasures already have been holographed and sent to other nations in a traveling exhibit.

While one hologram of a masterpiece might be used to direct a laser in a restoration of the original, another perfect copy can be displayed on the gallery wall and a third can be sent overseas on an authentication mission to weed out a forgery. Holographers in Besançon, France, have made a life-sized image of the Louvre's "Venus de Milo." Mass produced, such holograms would allow you to turn your living room into a gallery.

Holography at Home

From the architect's projection of a future house to the stress-testing of its beams, the home of the future will utilize many holographic innovations. Because holography affords microscopic detail, home security will be foolproof and home

Reprinted, with permission, from *The Futurist*, December 1985, published by the World Future Society, 4916 St. Elmo Avenue, Bethesda, Maryland 20814.

maintenance will be trouble-free. Enter the robot that can see with holographic eyes.

Equipped with holographic vision and template-matching spatial memory, the household robot of the future won't miss a speck of dust. Companion robots with holographic **artificial intelligence** could reason by association, and even their knowledge need not be program-based. Sonar holography will allow the robot to navigate even after you move the furniture.

These robots will not only be good for housework, but they will also make great chess opponents. And the mix of optical-array processors, holographic optical element eyes with microscopic mapping, and an uncanny humanlike artificial intelligence will create a competent combatant who functions at the speed of light.

Other games: ATARI has already advertised a holographic video game called Cosmos, while Holosystems, Inc., is developing the first holographic jigsaw puzzle. This is especially challenging since in a hologram the whole image is in each piece.

Motorists will see holographic images of their instruments projected directly in front of them, just as pilots can now. Similar technology will position game opponents in your visual field for any sport. The holographic image can be teleported on earth thanks to a communications system using **optical fibers** like the ones recently developed at AT&T/Bell Labs. Since light waves aren't limited to our atmosphere, transmission may allow us to play with friends on other planets.

Holograms will be used in "designed experiences" . . . Holographics with interactive computer programming will allow you to "plot" against Napoleon, who spontaneously responds to your every strategy.

For anything from *t'ai chi* to tennis, from portrait painting to engine tune-ups, the step-by-step perfect way to learn would be from a hologram. Holographic-projection films of masters in each craft would play upon your fledgling efforts and be filmed. Examining these films would pinpoint exactly where you reached too far or bent too slowly. The more expensive version would continuously check your progress, signaling weak points like a biofeedback machine.

For the energy-conscious, holographic optical elements already manufactured in the United States can take sunlight and shape it around a room, concentrate its transformation into electricity, or direct its heat to save fossil fuels.

Holography in Commerce

A city planner sells a proposed building by showing how its image is compatible with the landscape. Game packages with holograms on the cover compete against flat and inert graphics. Books and record albums have hologram covers. Hologram jewelry, keychains, belt buckles, and paperweights are given as promotional specialties.

These innovations have set the stage for rampant emulation, along with some surprises. Department store catalogs, now on laser disk, might be more effective if they used holograms to showcase products; hologram posters on buses or subway cars would certainly be eye-catching. Billboards, too, could become giant composite holograms, controlled by satellite to change their message with differing traffic flow and audience profile.

"Wherever you see photographs today, you're going to see holograms . . . three-dimensional, and in color," says Ron Erickson, educational consultant for the Museum of Holography.

Communicating with Holography

591

Jonathan Back
and Susan S. Lang
Holography:
Changing the Way
We See the World

Fiber optics not only allow holograms to be made around corners or under water, but they also can transmit numerous signals because of the infinite variability of light wavelengths. Holograms filter and control wavelengths better than glass lenses since they can be lighter, cheaper, and tougher than glass. And they can be reproduced; there is no limit to how many holograms you can make from a lens master, while the grinding of glass is a costly one-of-a-kind affair.

Optical fibers also may open the way to 3-D television, a dream with at least 30 years of history. But transmission bandwidth restrictions are a stumbling block, since a hologram consists of an incredible amount of information. Fred Unterseher, former education director for the Museum of Holography, points out that it would take two hours to send a single TV frame—that's one-thirtieth of a second in duration! "Some world-class inventions are called for if the reconstruction of holographic images . . . is to be grappled with," writes Robert De Marrais in *Holosphere*, a newsletter published by the Museum of Holography.

There's still the hope that computers may unlock the secret to true 3-D television, according to H. John Caulfield, senior scientist at Aerodyne Research. Computers can analyze and digitize a visual field, using sonar, perhaps, to plot distances, and make the calculations needed to make the holograms. Many kinds of hologram writers already exist with computers directing lasers.

Holographic cinema? Considering that no two people see the same image, how would you synchronize the sound? Considering that a hologram contains information about an uncountably large number of depths of field, what screen could replay them and be highly reflective while also being highly transparent for the pass-through of other fields? Despite these problems, the Soviets and the Japanese have reported good results with projected holograms, and small showings of holographic films have taken place in the Soviet Union and in France.

3-D Security

Look again at your new credit cards. MasterCard and Visa now have a holographic 3-D rainbow image of a globe and dove that are devilishly hard to forge. Holographic embossing is beginning to put a dent in the multimillion-dollar counterfeiting business. The U.S. Mint is considering holographic money to cut losses to forgers; before long, there will be holograms on stock certificates, passports, licenses, and birth certificates.

One company in Switzerland has developed a telephone credit card on which the holographic element not only defeats forgers but also encodes your credit status. Other European firms have employed holographic security passes because they cannot be duplicated. The security pass of the future may be a holographic memory of your own eye, which has features that are not duplicated by any other eye.

Mechanical locks could one day be operated by holographic keys since holographic images constitute definitive identification. Such a lock could never be picked or duplicated without authorization, making the principle useful for restricting illegal access into critical computer programs. Finally, classified documents could be holographically encoded with trillions of possible solutions.

Holography in Engineering

In engineering, advances in imaging made possible by holography may make your next car the sleekest, swiftest, most efficient one you've ever driven.

A car's aerodynamic properties could be improved thanks to holograms of fluid dynamics. Critical stress points will be strengthened thanks to holographic **interferometry,** or **nondestructive testing.** Holograms of combustion allow General Motors engineers to develop more efficient and pollution-free engines. The car of the future also may be more attractive thanks to holographic hard-copy of a computer-aided design.

Since 1966, when Lockheed first tested the bonding of wings, nondestructive testing made holography invaluable to engineers. Trans World Airlines estimates that millions of dollars have been saved through holographic tests of airplane tires, and the Central Electricity Generating Board of the United Kingdom considers holograms the best test device for detecting flaws in gas-cooled nuclear reactors.

The Geologist's Tool

With lasers already used to measure continental drift (plate tectonics), it would be natural to make holographic models of the plates and their motion. It also would be reasonable to use holographic interferometry to detect minute surface deformations. Rugged coastlines and ocean floors already are mapped by radar and sonar holography, although this application is not yet widespread.

If a geological sample cannot be removed for study, a hologram can be used to study all the optical properties of the original. Many copies of rare samples can be made from a master hologram, leaving fragile or unique specimens in safety. Looking for minerals by airplane or satellite is easy now because of holographic remote sensing.

Samples that *can* be removed make holography even more useful, since stress testing already uses holographic interferometry. Holographic comparisons in laser spectroscopy would greatly enhance soil analysis, since holographic optical elements are already used to find trace elements in stellar spectroscopy. And oil companies can use superimposed holograms to avoid drilling dry wells.

Holography and Health

Hospitals won't have to store tons of plaster now that holograms of orthodontic molds can be stored flat. Nondestructive testing methods have been used successfully to design prosthetic implants that resist stress-breaks better, yet can articulate the same as bone. Scans made with tomography, ultrasound, or nuclear magnetics can be combined into a 3-D image with motion parallax, so a surgeon can see the whole problem at a glance. These are just the start of a holographic revolution, in and out of the hospital.

The analysis of body fluids will be faster, more thorough, and need less blood, thanks to the unique "light signature" of a molecule in suspension. A holographic record of a small amount of blood, for example, could reveal its insulin count, then go on to a battery of other tests since no chemical change has taken place.

The body's structural integrity could be evaluated by a direct comparison with a base-line or prototype hologram; minute changes would be obvious, so subtle diagnoses would become commonplace.

Lasers are already used to do microsurgery, repair torn retinas, and remove birthmarks. These procedures could be standardized and performed with holograms guiding the light path. The ultimate extension of this notion is laser surgery: a complete face change—bloodless, painless, and totally automated. Pick the hologram face you'd like, and lasers follow its instructions with sub-cell precision.

The man who discovered holography, Dennis Gabor, was trying to increase the usefulness of electron microscopes. With the perfection of an X-ray laser, a greater boon than even that visionary sought might be realized.

If X rays could be made to "lase" or emit laser light, holograms might catch a DNA molecule or even study a virus in action. Doctors could see human cells in three dimensions. Anything that X rays could penetrate would be exposed for minutest inspection, which would be an invaluable tool for gene-splicing. Recent studies show that X rays can be made to lase as a by-product of controlled nuclear fusion.

Holography and the Mind

In 1966, neuroscientist Karl Pribram first proposed a connection between the physics of holography and his studies in human memory. Years of experiments on salamanders and primates showed that memory, while apparently concentrated locally, could not be destroyed locally. Pribram suggested that memory may be stored holographically, in wave/phase relationships duplicated and distributed across the brain.

Later experiments further the analogy, showing that brain waves operate in a frequency domain like light waves and may be mathematically transforming, decoding, and reconstructing primary reality the same way a hologram does.

In the early 1970s, these startling notions gained powerful support from the seminal thinking of physicist David Bohm, whose work arose just in time to answer Pribram's query, "What is the hologram the hologram of?" Bohm's reply: "According to the view we're proposing, the world itself is constructed or is structured on the same general principle as the hologram."

This dialogue of geniuses from different fields spurred lively debates, and, according to Marilyn Ferguson, editor of *Brain/Mind Bulletin*, a new paradigm was born. She describes the holographic supertheory: "Our brains mathematically construct 'hard' reality by interpreting frequencies from a dimension transcending time and space. The brain is a hologram, interpreting a holographic universe."

Where will these developments lead? Perhaps toward a dramatic shift in the way humans think about themselves and the world.

Questions on Content

1. Define the term *holography*.
2. List nine areas in which holography has significant implications.
3. From each of the nine areas, give at least one specific example of how holography is changing the way we see the world.

593

Jonathan Back
and Susan S. Lang
Holography:
Changing the Way
We See the World

Vocabulary

Define the following words as they are used in the article (the words are printed in **boldface** type):

1. laser light
2. artificial intelligence
3. optical fibers
4. interferometry
5. nondestructive testing

Suggestions for Response and Reaction

1. Write a general description (see pages 62–70, 78–85) of the hologram as a mechanism.
2. Using information from this article, write an extended definition (see pages 96–99, 101–112) of the term *holography*.
3. Following the guidelines for setting up a classification system (see pages 123–139), explain the classification of the uses of holography.
4. Research any one of the nine areas of holography development. Write a research paper on your findings (see pages 454–473).

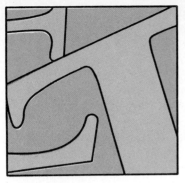

Quality

JOHN GALSWORTHY

The short story "Quality" portrays the unrelenting struggle of the craftsman who takes pride in the quality of his work but who is slowly starved to death by mass production.

Descriptive Summary

I knew him from the days of my extreme youth, because he made my father's boots; inhabiting with his elder brother two little shops let into one, in a small by-street—now no more, but then most fashionably placed in the West End.

That tenement had a certain quiet distinction; there was no sign upon its face that he made for any of the Royal Family—merely his own German name of Gessler Brothers; and in the window a few pairs of boots. I remember that it always troubled me to account for those unvarying boots in the window, for he made only what was ordered, reaching nothing down, and it seemed so inconceivable that what he made could ever have failed to fit. Had he bought them to put there? That, too, seemed inconceivable. He would never have tolerated in his house leather on which he had not worked himself. Besides, they were too beautiful—the pair of pumps, so inexpressibly slim, the patent leathers with cloth tops, making water come into one's mouth, the tall brown riding-boots with marvellous sooty glow, as if, though new, they had been worn a hundred years. Those pairs could only have been made by one who saw before him the Soul of Boot—so truly were they **prototypes** incarnating the very spirit of all footgear. These thoughts, of course, came to me later, though even when I was promoted to him, at the age of perhaps fourteen, some inkling haunted me of the dignity of himself and brother. For to make boots—such boots as he made—seemed to me then, and still seems to me, mysterious and wonderful.

I remember well my shy remark, one day, while stretching out to him my youthful foot:

"Isn't it awfully hard to do, Mr. Gessler?"

And his answer, given with a sudden smile from out of the **sardonic** redness of his beard: "Id is an Ardt!"

Himself, he was a little as if made from leather, with his yellow crinkly face, and crinkly reddish hair and beard, and neat folds slanting down his cheeks to the corners of his mouth, and his guttural and one-toned voice; for leather is a sardonic substance, and stiff and slow of purpose. And that was the character of his face, save that his eyes, which were grey-blue, had in them the simple gravity of one secretly possessed by the Ideal. His elder brother was so very like him—though watery, paler in every way, with a great industry—that sometimes in early days I was not quite sure of him until the interview was over. Then I knew it was he, if the words, "I will ask my brudder," had not been spoken; and that, if they had, it was his elder brother.

When one grew old and wild and ran up bills, one somehow never ran them up with Gessler Brothers. It would not have seemed becoming to go in there and stretch out one's foot to that blue iron-spectacled glance, owing him for more than— say—two pairs, just the comfortable reassurance that one was still his client.

For it was not possible to go to him very often—his boots lasted terribly, having something beyond the temporary—some, as it were, essence of boot stitched into them.

One went in, not as into most shops, in the mood of: "Please serve me, and let me go!" but restfully, as one enters a church; and, sitting on the single wooden chair, waited—for there was never anybody there. Soon, over the top edge of that sort of well—rather dark, and smelling soothingly of leather—which formed the shop, there would be seen his face, or that of his elder brother, peering down. A guttural sound, and tip-tap of bast slippers beating the narrow wooden stairs, and he would stand before one without coat, a little bent, in leather apron, with sleeves turned back, blinking—as if awakened from some dream of boots, or like an owl surprised in daylight and annoyed at this interruption.

And I would say: "How do you do, Mr. Gessler? Could you make me a pair of Russia leather boots?"

Without a word he would leave me, retiring **whence** he came, or into the other portion of the shop, and I would continue to rest in the wooden chair, inhaling the incense of his trade. Soon he would come back, holding in his thin, veined hand a piece of gold-brown leather. With eyes fixed on it, he would remark: "What a beaudiful biece!" When I, too, had admired it, he would speak again. "When do you wand dem?" And I would answer: "Oh! As soon as you conveniently can." And he would say: "To-morrow fordnighd?" Or if he were his elder brother: "I will ask my brudder!"

Then I would murmur: "Thank you! Good-morning, Mr. Gessler." "Goot-morning!" he would reply, still looking at the leather in his hand. And as I moved to the door, I would hear the tip-tap of his base slippers restoring him, up the stairs, to his dream of boots. But if it were some kind of footgear that he had not yet made me, then indeed he would observe ceremony—**divesting** me of my boot and holding it long in his hand, looking at it with eyes at once critical and loving, as if recalling the glow with which he had created it, and rebuking the way in which one had disorganized this masterpiece. Then, placing my foot on a piece of paper, he would two or three times tickle the outer edges with a pencil and pass his nervous fingers over my toes, feeling himself into the heart of my requirements.

I cannot forget that day on which I had occasion to say to him: "Mr. Gessler, that last pair of town walking-boots creaked, you know."

He looked at me for a time without replying, as if expecting me to withdraw or qualify the statement, then said:

"Id shouldn'd 'ave greaked."

"It did, I'm afraid."

"You godden wed before dey found demselves?"

"I don't think so."

At that he lowered his eyes, as if hunting for memory of those boots, and I felt sorry I had mentioned this grave thing.

"Zend dem back!" he said; "I will look at dem."

A feeling of compassion for my creaking boots surged up in me, so well could I imagine the sorrowful long curiosity of regard which he would bend on them.

"Zome boods," he said slowly, "are bad from birdt. If I can do noding wid dem, I dake dem off your bill."

Once (only once) I went absent-mindedly into his shop in a pair of boots bought in an emergency at some large firm's. He took my order without showing me any leather, and I could feel his eyes penetrating the inferior **integument** of my foot. At last he said:

"Dose are nod my boods."

The tone was not one of anger, nor of sorrow; not even of contempt, but there was in it something quiet that froze the blood. He put his hand down and pressed a finger on the place where the left boot, endeavoring to be fashionable, was not quite comfortable.

"Id 'urds you dere," he said. "Dose big virms 'ave no self-respect. Drash!" And then, as if something had given way within him, he spoke long and bitterly. It was the only time I ever heard him discuss the conditions and hardships of his trade.

"Dey get id all," he said, "dey get id by adverdisement, nod by work. Dey dake it away from us, who lofe our boods. Id gomes to this—bresently I haf no work. Every year id gets less—you will see." And looking at his lined face I saw things I had never noticed before, bitter things and bitter struggle—and what a lot of grey hairs there seemed suddenly in his red beard!

As best I could, I explained the circumstances of the purchase of those ill-omened boots. But his face and voice made so deep an impression that during the next few minutes I ordered many pairs. **Nemesis** fell! They lasted more terribly than ever. And I was not able conscientiously to go to him for nearly two years.

When at last I went I was surprised to find that outside one of the two little windows of his shop another name was painted, also that of a boot-maker—making, of course, for the Royal Family. The old familiar boots, no longer in dignified isolation, were huddled in the single window. Inside, the now contracted well of the one little shop was more scented than ever. And it was longer than usual, too, before a face peered down, and the tip-tap of the bast slippers began. At last he stood before me, and gazing through those rusty iron spectacles, said:

"Mr—, isn'd it?!"

"Ah, Mr. Gessler," I stammered, "but your boots are really *too* good, you know! See, these are quite decent still!" And I stretched out to him my foot. He looked at it.

"Yes," he said, "beople do nod wand good boods, id seems."

To get away from his reproachful eyes and voice I hastily remarked: "What have you done to your shop?"

He answered quietly: "Id was too exbensif. Do you wand some boods?"

I ordered three pairs, though I had only wanted two, and quickly left. I had, I do not know quite what feeling of being part, in his mind, of a conspiracy against him; or not perhaps so much against him as against his idea of boot. One does not, I suppose, care to feel like that; for it was again many months before my next visit to his shop, paid, I remember, with the feeling: "Oh! well, I can't leave the old boy—so here goes! Perhaps it'll be his elder brother!"

For his elder brother, I knew, had not character enough to reproach me, even dumbly.

And, to my relief, in the shop there did appear to be his elder brother, handling a piece of leather.

"Well, Mr. Gessler," I said, "how are you?"

He came close, and peered at me.

"I am breddy well," he said slowly, "but my elder brudder is dead."

And I saw that it was indeed himself—but how aged and wan! And never before had I heard him mention his brother. Much shocked, I murmured: "Oh! I am sorry!"

"Yes," he answered. "He was a good man, he made a good bood; but he is dead." And he touched the top of his head, where the hair had suddenly gone as thin as it had been on that of his poor brother, to indicate, I suppose, the cause of death. "He could nod ged over losing de oder shop. Do you wand any boods?" And he held up the leather in his hand: "Id's a beaudiful biece."

I ordered several pairs. It was very long before they came—but they were better than ever. One simply could not wear them out. And soon after that I went abroad.

It was over a year before I was again in London. And the first shop I went to was my old friend's. I had left a man of sixty; I came back to one of seventy-five, pinched and worn and **tremulous**, who genuinely, this time, did not at first know me.

"Oh! Mr. Gessler," I said, sick at heart, "how splendid your boots are! See, I've been wearing this pair nearly all the time I've been abroad; and they're not half worn out, are they?"

He looked at my boots—a pair of Russian leather, and his face seemed to regain steadiness. Putting his hand on my instep, he said:

"Do dey vid you here? I 'ad drouble wid dat bair, I remember."

I assured him that they had fitted beautifully.

"Do you wand any boods?" he said. "I can made dem quickly; id is a slack dime."

I answered: "Please, please! I want boots all round—every kind!"

"I will make a vresh model. Your food must be bigger." And with utter slowness, he traced round my foot, and felt my toes, only once looking up to say: "Did I dell you my brudder was dead?"

To watch him was painful, so feeble had he grown; I was glad to get away.

I had given those boots up, when one evening they came. Opening the parcel, I set the four pairs out in a row. Then one by one I tried them on. There was no doubt about it. In shape and fit, in finish and quality of leather, they were the best he had ever made me. And in the mouth of one of the town walking-boots I found his bill. The amount was the same as usual, but it gave me quite a shock. He had never before sent it in till quarter day. I flew downstairs and wrote a cheque, and posted it at once with my own hand.

A week later, passing the little street, I thought I would go in and tell him how splendidly the new boots fitted. But when I came to where his shop had been, his name was gone. Still there, in the window, were the slim pumps, the patent leathers with cloth tops, the sooty riding boots.

I went in, very much disturbed. In the two little shops—again made into one—was a young man with an English face.

"Mr. Gessler in?" I said.

He gave me a strange, **ingratiating** look.

"No, sir," he said, "no. But we can attend to anything with pleasure. We've taken the shop over. You've seen our name, no doubt, next door. We make for some very good people."

"Yes, yes," I said; "but Mr. Gessler?"

"Oh!" he answered; "dead."

"Dead!! But I only received these boots from him last Wednesday week."

"Ah!" he said; "a shockin' go. Poor old man starved 'imself."

"Good God!"

"Slow starvation, the doctor called it! You see he went to work in such a way! Would keep the shop on; wouldn't have a soul touch his boots except himself. When he got an order, it took him such a time. People won't wait. He lost everybody. And there he'd sit, goin' on and on—I will say that for him—not a man in London made a better boot. But look at the competition! He never advertised! Would 'ave the best leather, too, and do it all 'imself. Well, there it is. What could you expect with his ideas?"

"But starvation—!"

"That may be a bit flowery, as the sayin' is—but I know myself he was sittin' over his boots day and night, to the very last. You see I used to watch him. Never gave 'imself time to eat; never had a penny in the house. All went in rent and leather. How he lived so long I don't know. He regular let his fire go out. He was a character. But he made good boots."

"Yes," I said. "He made good boots."

Questions on Content

1. How does Galsworthy describe the central character?
2. Explain the procedure for ordering a pair of boots from Mr. Gessler.
3. What was Gessler's attitude toward boots? Point out lines in the story to support your answer.

Vocabulary

Define the following words as they are used in the article (the words are printed in **boldface** type):

1. prototypes
2. sardonic
3. whence
4. divesting
5. integument
6. nemesis
7. tremulous
8. ingratiating

Suggestions for Response and Reaction

1. What *really* caused Mr. Gessler's death?
2. Evaluate the effectiveness of the story's ending.
3. **a.** For the skilled worker or technician, explain one advantage in working under the old system of individual craftsmanship and one advantage in working under the system of mass production.
   b. What is a disadvantage of working in each system?
4. As a skilled worker in a technical field, defend or attack the assertion that mass production has eliminated personal pride in craftsmanship.
5. Following the principles for cause/effect explanation (see pages 170–184), show how the general public has benefited and suffered from the shift from one production system to the other.
6. Write a documented paper (see pages 454–473) on some aspect of mass production.

Excerpt from *Zen and the Art of Motorcycle Maintenance*

ROBERT M. PIRSIG

The narrator, in explaining the views of his friends John and Sylvia and their dislike for maintaining their own motorcycles, gets at the larger issues of one's relationship to technology.

Disharmony I suppose is common enough in any marriage, but in John and Sylvia's case it seems more tragic. To me, anyway.

It's not a personality clash between them; it's something else, for which neither is to blame, but for which neither has any solution, and for which I'm not sure I have any solution either, just ideas.

The ideas began with what seemed to be a minor difference of opinion between John and me on a matter of small importance: how much one should maintain one's own motorcycle. It seems natural and normal to me to make use of the small tool kits and instruction booklets supplied with each machine, and keep it tuned and adjusted myself. John **demurs.** He prefers to let a competent mechanic take care of these things so that they are done right. Neither viewpoint is unusual, and this minor difference would never have become magnified if we didn't spend so much time riding together and sitting in country roadhouses drinking beer and talking about whatever comes to mind. What comes to mind, usually, is whatever we've been thinking about in the half hour or forty-five minutes since we last talked to each other. When it's roads or weather or people or old memories or what's in the newspapers, the conversation just naturally builds pleasantly. But whenever the performance of the machine has been on my mind and gets into the conversation, the building stops. The conversation no longer moves forward. There is a silence and a break in the continuity. It is as though two old friends, a Catholic and Protestant, were sitting drinking beer, enjoying life, and the subject of birth control somehow came up. Big freeze-out.

And, of course, when you discover something like that it's like discovering a tooth with a missing filling. You can never leave it alone. You have to probe it, work around it, push on it, think about it, not because it's enjoyable but because it's on your mind and it won't get off your mind. And the more I probe and push on this subject of cycle maintenance the more irritated he gets, and of course that makes me want to probe and push all the more. Not deliberately to irritate him but because the irritation seems **symptomatic** of something deeper, something under the surface that isn't immediately apparent.

When you're talking birth control, what blocks it and freezes it out is that it's not a matter of more or fewer babies being argued. That's just on the surface. What's underneath is a conflict of faith, of faith in empirical social planning versus faith in the authority of God as revealed by the teachings of the Catholic Church. You can prove the practicality of planned parenthood till you get tired of listening to yourself and it's going to go nowhere because your **antagonist** isn't buying the assumption that anything socially practical is good per se. Goodness for him has other sources which he values as much as or more than social practicality.

So it is with John. I could preach the practical value and worth of motorcycle maintenance till I'm hoarse and it would make not a dent in him. After two sentences on the subject his eyes go completely glassy and he changes the conversation or just looks away. He doesn't want to hear about it.

Sylvia is completely with him on this one. In fact she is even more emphatic. "It's just a whole other thing," she says, when in a thoughtful mood. "Like garbage," she says, when not. They want *not* to understand it. Not to *hear* about it. And the more I try to fathom what makes me enjoy mechanical work and them hate it so, the more elusive it becomes. The ultimate cause of this originally minor difference of opinion appears to run way, way deep.

Inability on their part is ruled out immediately. They are both plenty bright enough. Either one of them could learn to tune a motorcycle in an hour and a half if they put their minds and energy to it, and the saving in money and worry and delay would repay them over and over again for their effort. And they *know* that. Or maybe they don't. I don't know. I never confront them with the question. It's better to just get along.

But I remember once, outside a bar in Savage, Minnesota, on a really scorching day when I just about let loose. We'd been in the bar for about an hour and we came out and the machines were so hot you could hardly get on them. I'm started and ready to go and there's John pumping away on the kick starter. I smell gas like we're next to a refinery and tell him so, thinking this is enough to let him know his engine's flooded.

"Yeah, I smell it too," he says and keeps on pumping. And he pumps and pumps and jumps and pumps and *I* don't know what more to say. Finally, he's really winded and sweat's running down all over his face and he can't pump anymore, and so I suggest taking out the plugs to dry them off and air out the cylinders while we go back for another beer.

Oh my God no! He doesn't want to get into all that stuff.

"All what stuff?"

"Oh, getting out the tools and all that stuff. There's no reason why it shouldn't start. It's a brand-new machine and I'm following the instructions perfectly. See, it's right on full choke like they say."

"Full *choke!*"

"That's what the instructions say."

"That's for when it's *cold!*"

"Well, we've been in there for a half an hour at least," he says.

It kind of shakes me up. "This is a hot day, John," I say. "And they take longer than that to cool off even on a freezing day."

He scratches his head. "Well, why don't they tell you that in the instructions?" He opens the choke and on the second kick it starts. "I guess that was it," he says cheerfully.

And the very next day we were out near the same area and it happened again. This time I was determined not to say a word, and when my wife urged me to go over and help him I shook my head. I told her that until he had a real felt need he was just going to resent help, so we went over and sat in the shade and waited.

I noticed he was being superpolite to Sylvia while he pumped away, meaning he was furious, and she was looking over with a kind of "Ye gods!" look. If he had asked any single question I would have been over in a second to diagnose it, but he wouldn't. It must have been fifteen minutes before he got it started.

Later we were drinking beer again over at Lake Minnetonka and everybody was talking around the table, but he was silent and I could see he was really tied up in knots inside. After all that time. Probably to get them untied he finally said, "You know . . . when it doesn't start like that it just . . . really turns me into a *monster* inside. I just get **paranoic** about it." This seemed to loosen him up, and he added, "They just had this *one* motorcycle, see? This *lemon*. And they didn't know what to do with it, whether to send it back to the factory or sell it for scrap or what . . . and then at the last moment they saw *me* coming. With eighteen hundred bucks in my pocket. And they knew their problems were over."

In a kind of singsong voice I repeated the plea for tuning and he tried hard to listen. He really tries hard sometimes. But then the block came again and he was off to the bar for another round for all of us and the subject was closed.

He is not stubborn, not narrow-minded, not lazy, not stupid. There was just no easy explanation. So it was left up in the air, a kind of mystery, that one gives up on because there is no sense in just going round and round and round looking for an answer that's not there.

It occurred to me that maybe I was the odd one on the subject, but that was disposed of too. Most touring cyclists know how to keep their machines tuned. Car owners usually won't touch the engine, but every town of any size at all has a garage with expensive lifts, special tools and diagnostic equipment that the average owner can't afford. And a car engine is more complex and inaccessible than a cycle engine so there's more sense to this. But for John's cycle, a BMW R60, I'll bet there's not a mechanic between here and Salt Lake City. If his points or plugs burn out, he's done for. I *know* he doesn't have a set of spare points with him. He doesn't know what points are. If it quits on him in western South Dakota or Montana I don't know what he's going to do. Sell it to the Indians maybe. Right now I know what he's doing. He's carefully avoiding giving any thought whatsoever to the subject. The BMW is famous for not giving mechanical problems on the road and that's what he's counting on.

I might have thought this was just a peculiar attitude of theirs about motorcycles but discovered later that it extended to other things. . . . Waiting for them to get going one morning in their kitchen I noticed the sink faucet was dripping and remembered that it was dripping the last time I was there before and that in fact it had been dripping as long as I could remember. I commented on it and John said he had tried to fix it with a new faucet washer but it hadn't worked. That was all he said. The **presumption** left was that that was the end of the matter. If you try to fix a faucet and your fixing doesn't work then it's just your lot to live with a dripping faucet.

This made me wonder to myself if it got on their nerves, this drip-drip-drip, week in, week out, year in, year out, but I could not notice any irritation or concern about it on their part, and so concluded they just aren't bothered by things like dripping faucets. Some people aren't.

What it was that changed this conclusion, I don't remember . . . some intuition, some insight one day, perhaps it was a subtle change in Sylvia's mood whenever

Robert M. Pirsig
Excerpt from *Zen
and the Art of
Motorcycle
Maintenance*

the dripping was particularly loud and she was trying to talk. She has a very soft voice. And one day when she was trying to talk above the dripping and the kids came in and interrupted her, she lost her temper at them. It seemed that her anger at the kids would not have been nearly as great if the faucet hadn't also been dripping when she was trying to talk. It was the combined dripping and loud kids that blew her up. What struck me hard then was that she was *not* blaming the faucet, and that she was *deliberately* not blaming the faucet. She wasn't ignoring that faucet at all! She was *suppressing* anger at that faucet and that goddamned dripping faucet was just about *killing* her! But she could not admit the importance of this for some reason.

Why suppress anger at a dripping faucet? I wondered.

Then that patched in with the motorcycle maintenance and one of those light bulbs went on over my head and I thought Ahhhhhhhh!

It's not the motorcycle maintenance, not the faucet. It's all of technology they can't take. And then all sorts of things started tumbling into place and I knew that was it. Sylvia's irritation at a friend who thought computer programming was "creative." All their drawings and paintings and photographs without a technological thing in them. Of course she's not going to get mad at that faucet, I thought. You always suppress momentary anger at something you deeply and permanently hate. Of course John signs off every time the subject of cycle repair comes up, even when it is obvious he is suffering for it. That's technology. And sure, of course, obviously. It's so simple when you see it. To get away from technology out into the country in the fresh air and sunshine is why they are on the motorcycle in the first place. For me to bring it back to them just at the point and place where they think they have finally escaped it just frosts both of them, tremendously. That's why the conversation always breaks and freezes when the subject comes up.

Others things fit in too. They talk once in a while in as few pained words as possible about "it" or "it all" as in the sentence, "There is just no escape from it." And if I asked, "From what?" the answer might be "The whole thing," or "The whole organized bit," or even "The system." Sylvia once said defensively, "Well, *you* know how to *cope* with it," which puffed me up so much at the time I was embarrassed to ask what "it" was and so remained somewhat puzzled. I thought it was something more mysterious than technology. But now I see that the "it" was mainly, if not entirely, technology. But, that doesn't sound right either. The "it" is a kind of force that gives rise to technology, something undefined, but inhuman, mechanical, lifeless, a blind monster, a death force. Something hideous they are running from but know they can never escape. I'm putting it way too heavily here but in a less emphatic and less defined way this is what it is. Somewhere there are people who understand it and run it but those are technologists, and they speak an inhuman language when describing what they do. It's all parts and relationships of unheard-of things that never make any sense no matter how often you hear about them. And their things, their monster keeps eating up land and polluting their air and lakes, and there is no way to strike back at it, and hardly any way to escape it.

That attitude is not hard to come to. You go through a heavy industrial area of a large city and there it all is, the technology. In front of it are high barbed wire fences, locked gates, signs saying NO TRESPASSING, and beyond, through sooty air, you see ugly strange shapes of metal and brick whose purpose is unknown, and whose masters you will never see. What it's for you don't know, and why it's there,

there's no one to tell, and so all you can feel is alienated, estranged, as though you didn't belong there. Who owns and understands this doesn't want you around. All this technology has somehow made you a stranger in your own land. Its very shape and appearance and mysteriousness say, "Get out." You know there's an explanation for all this somewhere and what it's doing undoubtedly serves mankind in some indirect way, but that isn't what you see. What you see is the NO TRESPASSING, KEEP OUT signs and not anything serving people but little people, like ants, serving these strange, incomprehensible shapes. And you think, even if I were a part of this, even if I were not a stranger, I would be just another ant serving the shapes. So the final feeling is hostile, and I think that's ultimately what's involved with this otherwise unexplainable attitude of John and Sylvia. Anything to do with valves and shafts and wrenches is a part of *that* dehumanized world, and they would rather not think about it. They don't want to get into it.

If this is so, they are not alone. There is no question that they have been following their natural feelings in this and not trying to imitate anyone. But many others are also following their natural feelings and not trying to imitate anyone and the natural feelings of very many people are similar on this matter; so that when you look at them collectively, as journalists do, you get the illusion of a mass movement, an **antitechnological** mass movement, an entire political antitechnological left emerging, looming up from apparently nowhere, saying, "Stop the technology. Have it somewhere else. Don't have it here." It is still restrained by a thin web of logic that points out that without the factories there are no jobs or standard of living. But there are human forces stronger than logic. There always have been, and if they become strong enough in their hatred of technology that web can break.

Clichés and stereotypes such as "beatnik" or "hippie" have been invented for the antitechnologists, the antisystem people, and will continue to be. But one does not convert individuals into mass people with the simple coining of a mass term. John and Sylvia are not mass people and neither are most of the others going their way. It is against being a mass person that they seem to be revolting. And they feel that technology has got a lot to do with the forces that are trying to turn them into mass people and they don't like it. So far it's still mostly a passive resistance, flight into the rural areas when they are possible and things like that, but it doesn't always have to be this passive.

I disagree with them about cycle maintenance, but not because I am out of sympathy with their feelings about technology. I just think that their flight from and hatred of technology is self-defeating. The Buddha, the Godhead, resides quite as comfortably in the circuits of a digital computer or the gears of a cycle transmission as he does at the top of a mountain or in the petals of a flower. To think otherwise is to demean the Buddha—which is to demean oneself.

Questions on Content

1. What is the narrator's attitude toward a person's being able to maintain his or her own motorcycle?
2. What is John and Sylvia's attitude toward maintaining their own motorcycles?
3. Why do some people become antitechnological?

Vocabulary

Robert M. Pirsig
Excerpt from *Zen
and the Art of
Motorcycle
Maintenance*

Define the following words as they are used in the article (the words are printed in **boldface** type):

1. demurs
2. symptomatic
3. antagonist
4. paranoic
5. presumption
6. antitechnological

Suggestions for Response and Reaction

1. Are you more like the narrator or like John and Sylvia; that is, if you had a motorcycle, would you prefer maintaining it yourself or taking it to a cycle shop? Or if you had a leaky faucet, would you repair it or put up with the drip? (Other questions: Do you change the oil in your car or take the car to a service station? If mowing the lawn and the mower breaks down, do you attempt to repair the mower yourself or do you take it to a shop for repairs?)

2. Write an extended definition (see pages 96–99, 101–112) of *technology*.

3. Do you think that computer programming can be "creative"?

4. What is the relationship between the title *Zen and the Art of Motorcycle Maintenance* and the reference to Buddha in the last paragraph?

5. Do you agree or disagree with this statement from the last paragraph: "The Buddha, the Godhead, resides quite as comfortably in the circuits of a digital computer or the gears of a cycle transmission as he does at the top of a mountain or in the petals of a flower"?

6. Reread the last paragraph of the selection. Think about it for a while. Then describe an enjoyable activity in which you can "lose" yourself to the point that you have a new insight into life. (Examples: Pride in a newly washed and waxed automobile, satisfaction in a more efficient rearrangement of furniture, the contentment of knowing that you gave a patient or client your complete attention.) In your description, explain the situation, the activity, the outcome, your overall assessment, and the impact on your thinking and feelings.

The Science of Deduction

SIR ARTHUR CONAN DOYLE

Sherlock Holmes, the world's only unofficial consulting detective, illustrates his powers of observation, deduction, and knowledge.

"I abhor the dull routine of existence. I crave for mental **exaltation.** That is why I have chosen my own particular profession—or rather created it, for I am the only one in the world."

"The only unofficial detective?" I said, raising my eyebrows.

"The only unofficial consulting detective," [Sherlock Holmes] answered. "I am the last and highest court of appeal in detection. When Gregson or Lestrade or Athelney Jones are out of their depths—which, by the way, is their normal state—the matter is laid before me. I examine the data, as an expert, and pronounce a specialist's opinion. I claim no credit in such cases. My name figures in no newspaper. The work itself, the pleasure of finding a field for my peculiar powers, is my highest reward. But you have yourself had some experience of my methods of work in the Jefferson Hope case."

"Yes, indeed," said I, cordially. "I was never so struck by anything in my life. I even embodied it in a small brochure with the somewhat fantastic title of 'A Study in Scarlet.'"

He shook his head sadly. "I glanced over it," said he. "Honestly, [Dr. Watson,] I cannot congratulate you upon it. Detection is, or ought to be, an exact science, and should be treated in the same cold and unemotional manner. You have attempted to tinge it with romanticism, which produces much the same effect as if you worked a love-story or an elopement into the fifth proposition of Euclid."

"But the romance was there," I remonstrated. "I could not tamper with the facts."

"Some facts should be suppressed, or at least a just sense of proportion should be observed in treating them. The only point in the case which deserved mention was the curious analytical reasoning from effects to causes by which I succeeded in unravelling it."

I was annoyed at this criticism of a work which had been specially designed to please him. I confess, too, that I was irritated by the egotism which seemed to demand that every line of my pamphlet should be devoted to his own special doings. More than once during the years that I had lived with him in Baker Street I had observed that a small vanity underlay my companion's quiet and **didactic** manner. I made no remark, however, but sat nursing my wounded leg. I had had a Jezail bullet through it some time before, and, though it did not prevent me from walking, it ached wearily at every change of the weather.

"My practice has extended recently to the Continent," said Holmes, after a while, filling up his old brier-root pipe. "I was consulted last week by François Le Villard, who, as you probably know, has come rather to the front lately in the French detective service. He has all the Celtic power of quick intuition, but he is deficient in the wide range of exact knowledge which is essential to the higher developments of his art. The case was concerned with a will, and possessed some features of interest. I was able to refer him to two parallel cases, the one at Riga in 1857, and the other at St. Louis in 1871, which have suggested to him the true solution. Here is the letter which I had this morning acknowledging my assistance." He tossed over, as he spoke, a crumpled sheet of foreign note-paper. I glanced my eyes down it, catching a profusion of notes of admiration, with stray "magnifiques," "coup-de-maîtres," and "tours-de-force," all testifying to the ardent admiration of the Frenchman.

"He speaks as a pupil to his master," said I.

"Oh, he rates my assistance too highly," said Sherlock Holmes, lightly. "He has considerable gifts himself. He possesses two out of the three qualities necessary for the ideal detective. He has the power of observation and that of deduction. He is only wanting in knowledge; and that may come in time. He is now translating my small works into French."

"Your works?"

"Oh; didn't you know?" he cried, laughing. "Yes, I have been guilty of several monographs. They are all upon technical subjects. Here, for example, is one 'Upon the Distinction between the Ashes of the Various Tobaccoes.' In it I enumerate a hundred and forty forms of cigar-, cigarette-, and pipe-tobacco, with colored plates illustrating the difference in the ash. It is a point which is continually turning up in criminal trials, and which is sometimes of supreme importance as a clue. If you can say definitely, for example, that some murder has been done by a man who was smoking an Indian lunkah, it obviously narrows your field of search. To the trained eye there is as much difference between the black ash of a Trichinopoly and the white fluff of bird's-eye as there is between a cabbage and a potato."

"You have an extraordinary genius for **minutiae,**" I remarked.

"I appreciate their importance. Here is my monograph upon the tracing of footsteps, with some remarks upon the uses of plaster of Paris as a preserver of impresses. Here, too, is a curious little work upon the influence of a trade upon the form of the hand, with lithotypes of the hands of slaters, sailors, cork-cutters, compositors, weavers, and diamond-polishers. That is a matter of great practical interest to the scientific detective,—especially in cases of unclaimed bodies, or in discovering the antecedents of criminals. But I weary you with my hobby."

"Not at all," I answered, earnestly. "It is of the greatest interest to me, especially since I have had the opportunity of observing your practical application of it. But you spoke just now of observation and deduction. Surely the one to some extent implies the other."

"Why, hardly," he answered, leaning back luxuriously in his arm-chair, and sending up thick blue wreaths from his pipe. "For example, observation shows me that you have been to the Wigmore Street Post-Office this morning, but deduction lets me know that when there you despatched a telegram."

"Right!" said I. "Right on both points! But I confess that I don't see how you arrived at it. It was a sudden impulse upon my part, and I have mentioned it to no one."

"It is simplicity itself," he remarked, chuckling at my surprise,—"so absurdly simple that an explanation is **superfluous**; and yet it may serve to define the limits of observation and of deduction. Observation tells me that you have a little reddish mould adhering to your instep. Just opposite the Seymour Street Office they have taken up the pavement and thrown up some earth which lies in such a way that it is difficult to avoid treading in it in entering. The earth is of this peculiar reddish tint which is found, as far as I know, nowhere else in the neighborhood. So much is observation. The rest is deduction."

"How, then, did you deduce the telegram?"

"Why, of course I knew that you had not written a letter, since I sat opposite to you all morning. I see also in your open desk there that you have a sheet of stamps and a thick bundle of post-cards. What could you go into the post-office for, then, but to send a wire? Eliminate all other factors, and the one which remains must be the truth."

"In this case it certainly is so," I replied, after a little thought. "The thing, however, is, as you say, of the simplest. Would you think me **impertinent** if I were to put your theories to a more severe test?"

"On the contrary," he answered. . . . "I should be delighted to look into any problem which you might submit to me."

"I have heard you say that it is difficult for a man to have any object in daily use without leaving the impress of his individuality upon it in such a way that a trained observer might read it. Now, I have here a watch which has recently come into my possession. Would you have the kindness to let me have an opinion upon the character or habits of the late owner?"

I handed him over the watch with some slight feeling of amusement in my heart, for the test was, as I thought, an impossible one, and I intended it as a lesson against the somewhat dogmatic tone which he occasionally assumed. He balanced the watch in his hand, gazed hard at the dial, opened the back, and examined the works, first with his naked eyes and then with a powerful convex lens. I could hardly keep from smiling at his crestfallen face when he finally snapped the case to and handed it back.

"There are hardly any data," he remarked. "The watch has been recently cleaned, which robs me of my most suggestive facts."

"You are right," I answered. "It was cleaned before being sent to me." In my heart I accused my companion of putting forward a most lame and impotent excuse to cover his failure. What data could he expect from an uncleaned watch?

"Though unsatisfactory, my research has not been entirely barren," he observed, staring up at the ceiling with dreamy, lack-lustre eyes. "Subject to your correction, I should judge that the watch belonged to your elder brother, who inherited it from your father."

"That you gather, no doubt, from the H. W. upon the back?"

"Quite so. The W. suggests your own name. The date of the watch is nearly fifty years back, and the initials are as old as the watch: so it was made for the last generation. Jewelry usually descends to the eldest son, and he is most likely to have the same name as the father. Your father has, if I remember right, been dead many years. It has, therefore, been in the hands of your eldest brother."

"Right, so far," said I. "Anything else?"

"He was a man of untidy habits,—very untidy and careless. He was left with good prospects, but he threw away his chances, lived for some time in poverty with occasional short intervals of prosperity, and finally, taking to drink, he died. That is all I can gather."

I sprang from my chair and limped impatiently about the room with considerable bitterness in my heart.

"This is unworthy of you, Holmes," I said. "I could not have believed that you would have descended to this. You have made inquiries into the history of my unhappy brother, and you now pretend to deduce this knowledge in some fanciful way. You cannot expect me to believe that you have read all this from his old watch! It is unkind, and, to speak plainly, has a touch of **charlatanism** in it."

"My dear doctor," said he, kindly, "pray accept my apologies. Viewing the matter as an abstract problem, I had forgotten how personal and painful a thing it might be to you. I assure you, however, that I never even knew that you had a brother until you handed me the watch."

"Then how in the name of all that is wonderful did you get these facts? They are absolutely correct in every particular."

"Ah, that is good luck. I could only say what was the balance of probability. I did not at all expect to be so accurate."

"But it was not mere guess-work?"

"No, no: I never guess. It is a shocking habit,—destructive to the logical faculty. What seems strange to you is only so because you do not follow my train of thought or observe the small facts upon which large inference may depend. For example, I began by stating that your brother was careless. When you observe the lower part of that watch-case you notice that it is not only dented in two places, but it is cut and marked all over from the habit of keeping other hard objects, such as coins or keys, in the same pocket. Surely it is no great feat to assume that a man who treats a fifty-guinea watch so cavalierly must be a careless man. Neither is it a very far-fetched inference that a man who inherits one article of such value is pretty well provided for in other respects."

I nodded, to show that I followed his reasoning.

"It is very customary for pawnbrokers in England, when they take a watch, to scratch the number of the ticket with a pin-point upon the inside of the case. It is more handy than a label, as there is no risk of the number being lost or transposed. There are no less than four such numbers visible to my lens on the inside of this case. Inference,—that your brother was often at low water. Secondary inference,—that he had occasional bursts of prosperity, or he could not have redeemed the pledge. Finally, I ask you to look at the inner plate, which contains the key-hole. Look at the thousands of scratches all round the hole—marks where the key has slipped. What sober man's key could have scored those grooves? But you will never see a drunkard's watch without them. He winds it at night, and he leaves these traces of his unsteady hand. Where is the mystery in all this?"

"It is as clear as daylight," I answered. "I regret the injustice which I did you. I should have had more faith in your marvellous faculty."

Questions on Content

1. According to Sherlock Holmes, what are the three qualities necessary for the ideal detective?
2. What does Sherlock Holmes observe and deduce concerning Dr. Watson's watch?

Vocabulary

Define the following words as they are used in the article (the words are printed in **boldface** type):

1. exaltation
2. didactic
3. minutiae

4. superfluous
5. impertinent
6. charlatanism

Suggestions for Response and Reaction

1. Compare and contrast (see pages 193–202) *observation* and *deduction*.
2. Do you think it is really possible for a person to develop his or her powers of observation, deduction, and knowledge to the extent of those of Sherlock Holmes?
3. What is the relationship between analysis through effect-cause (see pages 170–184) and detective work?
4. In what occupations, professions, or types of work (other than that of being a detective) are the powers of observation, deduction, and knowledge essential? Helpful?
5. Select an object about which you do not have prior knowledge and try out your powers of observation and deduction. Suggested objects: a pair of shoes, a billfold, a suitcase, a notebook, an automobile, a belt. To record your findings and inferences, divide a sheet of paper into two columns, one headed *Observations* and the other, *Deductions*.

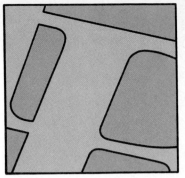

Are You Alive?

STUART CHASE

Chase has written interestingly and convincingly on a fundamental question—the enjoyment of life. By a simple process of analysis and classification, he has divided life into two categories, living and existing. He has then proceeded to show specifically the differences between the two conditions by labeling accordingly phases of his own life.

I have often been perplexed by people who talk about "life." They tend to strike an attitude, draw a deep breath, and wave their hands—triumphantly but somehow vaguely. Americans they tell me do not know how to live, but the French—ah, the French!—or the Hungarians, or the Poles, or the Patagonians. When I ask them what they mean by life they look at me with pitying silence for the most part, so that their long thin eyes haunt me. Sometimes they say: "Life is the next emotion. . . . Life is—well, you know what Shaw says . . . Life is a continual becoming." Unfortunately this is all Babu to me. These people do not advance me an inch in my quest of the definition of life.

Edgar Lee Masters is more helpful. In *Spoon River* he talks of a woman who dies at the age of ninety-six. She speaks of her hard yet happy existence on the farm, in the woods by the river gathering herbs, among the neighbors. "It takes life to love life," she says in a ringing climax. This is **cryptic,** but I get a flood of meaning from it.

What does it mean to be alive, to live intensely? What do social prophets mean when they promise a new order of life? Obviously they cannot mean a new quality of life never before enjoyed by anyone, but rather an extension of vitality for the masses of mankind in those qualities of "life" which have hitherto been enjoyed only by a few individuals normally, or by large numbers of individuals rarely.

What is it which is enjoyed, and how is it to be shared more extensively? Behind the phrase and the gesture, can we catch the gleam of **verihood** and hold life on a point for a moment while we examine it?

Initially, we must differentiate between "life" in the sense of not being physically dead, and "life" in the sense of awareness and vitality. Obviously everybody who is not dead is alive—"lives" in a sense. And all experience of undead people is a part of life in its broadest interpretation. But there seems to be an ascending scale of values in life, and somewhere in this scale there is a line—probably a blurred one—below which one more or less "exists," and above which one more or less "lives."

Secondly, we must examine the distinction that is drawn between physical and mental living. While this distinction is of no consequence so far as the joy of living is concerned, it is of profound consequence so far as the techniques of progress are concerned. If all well-being proceeds from a state of mind with little reference to physical environment, it is evident that all people have to do is to create well-being mentally—to think "life." If well-being on the other hand proceeds primarily from

Reprinted by permission of Stuart Chase.

611

physical causes, it is necessary to clean up slums, produce more usable goods and distribute them more widely, broaden the arts, look into **eugenics,** establish comprehensive systems of sanitation, release the creative impulse, and what not. Here we collide with a problem which is destined to agitate the race for centuries to come. Probably both approaches have their place, but I for one would stress the engineer against the **metaphysician.**

What, concretely, is this "awareness," this "well-being"? These words are shorthand words, meaningless in themselves unless we know the longhand facts for which they are a symbol. I want in a rather personal way to tell you the facts as I have found them. I want to tell you when I think I live in contradistinction to when I think I "exist." I want to make life very definite in terms of my own experience, for in matters of this nature about the only source of data one has is oneself. I do not know what life means to other people—I can only guess—but I do know what it means to me, and I have worked out a method of measuring it.

I get out of bed in the morning, bathe, dress, fuss over a necktie, hurry down to breakfast, gulp coffee and headlines, demand to know where my raincoat is, start for the office—and so forth. These are the crude data. Take the days as they come, put a plus beside the living hours and a minus beside the dead ones; find out just what makes the live ones live and the dead ones die. Can we catch the verihood of life in such an analysis? The poet will say no, but I am an accountant and only write poetry out of hours.

My notes show a classification of eleven states of being in which I feel I am alive, and five states in which I feel I only exist. These are major states, needless to say. In addition I find scores of sub-states which come from Heaven knows where and are too obscure for me to analyze. The eleven "plus" reactions are these:

I seem to live when I am creating something—writing this article for instance; making a sketch, working on an economic theory, building a bookshelf, making a speech.

Art certainly vitalizes me. A good novel—"The Growth of the Soil," for instance—some poems, some pictures, operas (not concerts), many beautiful buildings and particularly bridges affect me as though I took the artist's blood into my own veins. But I do not only have to be exposed to get this reaction. The operation is more subtle, for there are times when a curtain falls over my perceptions which no artist can penetrate.

In spite of those absurd Germans who used to perspire over Alpine passes shouting "Co-lo-sall!," I admit that mountains and the sea and stars and things—all the old subjects of a thousand poets—renew life in me. As in the case of art, the process is not automatic—I hate the sea sometimes—but by and large the chemistry works, and I feel the line of existence below me when I see these things.

Love, underneath its middle-class manners, and its frequent hypocrisies, is life, vital and intense. Very real to me also is the love one bears one's friends.

I feel very much alive in the presence of a genuine sorrow.

I live when I am stimulated by good conversation, good argument. There is a sort of vitality in just dealing in ideas—that to me at least is very real.

I live when I am in the presence of danger—rock-climbing, or being shadowed by an agent of the Department of Justice.

I live when I play—preferably out-of-doors. Such things as diving, swimming, skating, skiing, dancing, sometimes driving a motor, sometimes walking. . . .

One lives when one takes food after genuine hunger, or when burying one's lips in a cool mountain spring after a long climb.

One lives when one sleeps. A sound, healthy sleep after a day spent out-of-doors gives one the feeling of a silent, whirring dynamo. In vivid dreams I am convinced one lives.

I live when I laugh—spontaneously and heartily.

These are the eleven states of well-being which I have checked off from the hours of my daily life. Observe they only represent causes for states of being which I accept as "life." The definition of what state of being is in itself—the physical and mental chemistry of it, the feeling of expansion, of power, of happiness, or whatever it may be that it is—cannot be attempted at this point. The Indian says: "Those who tell do not know. Those who know do not tell." I deal here only in definite states of being which I recognize by some obscure but infallible sign as "life." *Why* I know it, or what it is, intrinsically, cannot be told.

In contradistinction to "living," I find five main states of "existence," as follows:

I exist when I am doing drudgery of any kind—adding up figures, washing dishes, writing formal reports, answering most letters, attending to money matters, reading newspapers, shaving, dressing, riding on street cars, or up and down in elevators, buying things.

I exist when attending the average social function. The whole scheme of middle-class manners bores me—a tea, a dinner, a call on one's relatives, listening to dull people talk, being polite, discussing the weather.

Eating, drinking, or sleeping when one is already replete, when one's senses are dulled, are states of existence, not life. For the most part, I exist when I am ill, but occasionally pain gives me a **lucidity** of thought which is near to life.

Old scenes, old monotonous things—city walls, too familiar streets, houses, rooms, furniture, clothes—drive one to the existence level. Even a scene that is beautiful to fresh eyes may grow intolerably dull. Sheer ugliness, such as one sees in the stockyards or in a city slum, depresses me intensely.

I retreat from life when I become angry. I feel all my handholds slipping. It is as though I were a deep-sea diver in a bell of compressed air. I exist through rows and misunderstandings and in the blind alleys of "getting even."

So in a general way I locate my line and set life off from existence. It must be admitted of course that "living" is often a mental state quite independent of physical environment or occupation. One may feel—in springtime for instance—suddenly alive in old, monotonous surroundings. Then even dressing and dishwashing become eventful and one sings as one shaves. But these outbursts are on the whole abnormal. By and large there seems to be a definite cause for living and a definite cause for existing. So it is with me at any rate. I am not at the mercy of a blind fate in this respect. I believe that I could deliberately "live" twice as much—in hours—as I do now, if only I could come out from under the chains of an artificial necessity—largely economic—which bind me.

I have indeed made some estimates of the actual time I have spent above and below the "existence" line. For instance, let me analyze a week I spent in the city during the past fall—an ordinary busy week of one who works at a desk. Of the 168 hours contained therein my notes show that I only "lived" about forty of them, or 25 per cent of the total time. This allowed for some creative work, a Sunday's hike, some genuine hunger, some healthy sleep, a little stimulating reading, two acts of a play, part of a moving picture, and eight hours of interesting discussion with various friends, including one informal talk on the technique of industrial administration which stimulated me profoundly.

Twenty-five per cent is not a very high life ratio, although if I may be permitted to guess, I guess it is considerably over the average of that of my fellow Americans. It is extremely doubtful if my yearly average is any higher than that which obtained in this representative week. Some weeks—in vacation time, or when the proportion of creative work is high—may be individually better, but the never-ceasing grind of income-getting, and the never-ceasing obscenity of city dwelling, undermine too ruthlessly the hope of a higher average ratio.

I can conceive of a better ratio in another social order. In such a society I do not see why one should not have the opportunity to do creative work at least three hours a day—allowing two hours for the inevitable drudgery—nor why I should not be out-of-doors playing for two or three hours a day; nor why I should not have a great deal more opportunity than I do now to hear good music, and go to good plays, and see good pictures; while reading, discussion, being with one's friends, sound sleep, and the salt of danger should all have an increased share of my time. I do not see why I should not laugh a lot more in that society. Adding these possibilities up—I will not bother you with the mathematical details—the total shows some one hundred hours, or a life ratio of 60 per cent as against the 25 per cent which is now my portion.

If this is true of me, it may be true of you, of many others perhaps. It may be that the states of being which release life in me release it in most human beings. But this I know and to this I have made up my mind: my salvation is too closely bound up with that of all mankind to hope for any great personal advance. In the last analysis—despite much beating against the bars—my ratio of living can only grow with that of the mass of my fellow-men.

Questions on Content

1. Chase divides life into two categories. Name them.
2. In examining the distinction between physical and mental living, which approach does the author favor? Write the sentence in the essay that tells you so.
3. List five of the eleven states of being in which the author feels he is alive. List five states of existence for him.

Vocabulary

Define the following words as they are used in the article (the words are printed in **boldface** type):

1. cryptic
2. verihood
3. eugenics
4. metaphysician
5. lucidity

Suggestions for Response and Reaction

1. **a.** List and explain at least three states of being that occur during your average work (school) day in which you are truly alive in the author's sense of the term.
    b. List at least three states of existing. Be very specific in describing times or activities.
    c. Present the material in *a* and *b* in an organized paper or speech, using the Chase essay as a model.
2. Excluding hours asleep, for an average work (school) day, what percentage of your time is spent in living and what in existing? Following the principles for cause/effect explanations (see pages 170–184), explain why you do or do not believe you could increase the percentage of time living. Construct a pie chart (see pages 544–545) to accompany your explanation.
3. For an ideal holiday or weekend day (just one day), list and explain five states of living.
4. Write an informative summary (see pages 217–222) of this selection.

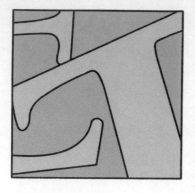

Clear Writing Means Clear Thinking Means . . .

MARVIN H. SWIFT

Associate professor of communication at the General Motors Institute, Marvin H. Swift analyzes the way in which a manager reworks and rethinks a memo of minor importance, to point up a constant management challenge of major importance—the clear and accurate expression of a well-focused message.

If you are a manager, you constantly face the problem of putting words on paper. If you are like most managers, this is not the sort of problem you enjoy. It is hard to do, and time consuming; and the task is doubly difficult when, as is usually the case, your words must be designed to change the behavior of others in the organization.

But the chore is there and must be done. How? Let's take a specific case.

Let's suppose that everyone at *X* Corporation, from the janitor on up to the chairman of the board, is using the office copiers for personal matters; income tax forms, church programs, children's term papers, and God knows what else are being duplicated by the gross. This minor piracy costs the company a pretty penny, both directly and in employee time, and the general manager—let's call him Sam Edwards—decides the time has come to lower the boom.

Sam lets fly by dictating the following memo to his secretary:

To: All Employees
From: Samuel Edwards, General Manager
Subject: Abuse of Copiers

It has recently been brought to my attention that many of the people who are employed by this company have taken advantage of their positions by availing themselves of the copiers. More specifically, these machines are being used for other than company business.

Obviously, such practice is contrary to company policy and must cease and desist immediately. I wish therefore to inform all concerned—those who have abused policy or will be abusing it—that their behavior cannot and will not be tolerated. Accordingly, anyone in the future who is unable to control himself will have his employment terminated.

If there are any questions about company policy, please feel free to contact this office.

Now the memo is on his desk for his signature. He looks it over; and the more he looks, the worse it reads. In fact, it's lousy. So he revises it three times, until it finally is in the form that follows:

To: All Employees
From: Samuel Edwards, General Manager
Subject: Use of Copiers

We are revamping our policy on the use of copiers for personal matters. In the past we have not encouraged personnel to use them for such purposes because of the costs involved. But we also recognize, perhaps belatedly, that we can solve the problem if each of us pays for what he takes.

We are therefore putting these copiers on a pay-as-you-go basis. The details are simple enough. . . .

Samuel Edwards

This time Sam thinks the memo looks good, and it *is* good. Not only is the writing much improved but the problem should now be solved. He therefore signs the memo, turns it over to his secretary for distribution, and goes back to other things.

From Verbiage to Intent

I can only speculate on what occurs in a writer's mind as he moves from a poor draft to a good revision, but it is clear that Sam went through several specific steps, mentally as well as physically, before he had created his end product:

☐ He eliminated wordiness.
☐ He modulated the tone of the memo.
☐ He revised the policy it stated.

Let's retrace his thinking through each of these processes.

ELIMINATING WORDINESS

Sam's basic message is that employees are not to use the copiers for their own affairs at company expense. As he looks over his first draft, however, it seems so long that this simple message has become diffused. With the idea of trimming the memo down, he takes another look at his first paragraph:

It has recently been brought to my attention that many of the people who are employed by this company have taken advantage of their positions by availing themselves of the copiers. More specifically, these machines are being used for other than company business.

He edits it like this:

Item: "recently"
Comment to himself: Of course; else why write about the problem? So delete the word.

Item: "It has been brought to my attention"
Comment: Naturally. Delete it.

Item: "the people who are employed by this company"
Comment: Assumed. Why not just "employees"?

Item: "by availing themselves" and "for other than company business"
Comment: Since the second sentence repeats the first, why not **coalesce?**

And he comes up with this:

> Employees have been using the copiers for personal matters.

He proceeds to the second paragraph. More confident of himself, he moves in broader swoops, so that the deletion process looks like this:

> Obviously, such practice is contrary to company policy ~~and must cease and desist immediately, I wish therefore to inform all concerned—those who have abused policy or will be abusing it—that their behavior cannot and will not be tolerated. Accordingly, anyone in the future who is unable to control himself will have his employment terminated.~~

The final paragraph, apart from "company policy" and "feel free," looks all right, so the total memo now reads as follows:

> To: All Employees
> From: Samuel Edwards, General Manager
> Subject: Abuse of Copiers
>
> Employees have been using the copiers for personal matters. Obviously, such practice is contrary to company policy and will result in dismissal.
>
> If there are any questions, please contact this office.

Sam now examines his efforts by putting these questions to himself:

> *Question:* Is the memo free of deadwood?
> *Answer:* Very much so. In fact, it's good, tight prose.
>
> *Question:* Is the policy stated?
> *Answer:* Yes—sharp and clear.
>
> *Question:* Will the memo achieve its intended purpose?
> *Answer:* Yes. But it sounds foolish.
>
> *Question:* Why?
> *Answer:* The wording is too harsh; I'm not going to fire anybody over this.
>
> *Question:* How should I tone the thing down?

To answer this question, Sam takes another look at the memo.

CORRECTING THE TONE

What strikes his eye as he looks it over? Perhaps these three words:

□ Abuse . . .
□ Obviously . . .
□ . . . dismissal . . .

The first one is easy enough to correct: he substitutes "use" for "abuse." But "obviously" poses a problem and calls for reflection. If the policy is obvious, why are the copiers being used? Is it that people are outrightly dishonest? Probably not. But that implies the policy isn't obvious; and whose fault is this? Who neglected to clarify policy? And why "dismissal" for something never publicized?

These questions impel him to revise the memo once again:

To: All Employees
From: Samuel Edwards, General Manager
Subject: Use of Copiers

Copiers are not to be used for personal matters. If there are any questions, please contact this office.

REVISING THE POLICY ITSELF

The memo now seems courteous enough—at least it is not discourteous—but it is just a blank, perhaps overly simple, statement of policy. Has he really thought through the policy itself?

Reflecting on this, Sam realizes that some people will continue to use the copiers for personal business anyhow. If he seriously intends to enforce the basic policy (first sentence), he will have to police the equipment, and that raises the question of costs all over again.

Also, the memo states that he will maintain an open-door policy (second sentence)—and surely there will be some, probably a good many, who will stroll in and offer to pay for what they use. His secretary has enough to do without keeping track of affairs of that kind.

Finally, the first and second sentences are at odds with each other. The first says that personal copying is out, and the second implies that it can be arranged.

The facts of organizational life thus force Sam to clarify in his own mind exactly what his position on the use of copiers is going to be. As he sees the problem now, what he really wants to do is put the copiers on a pay-as-you-go basis. After making that decision, he begins anew:

To: All Employees
From: Samuel Edwards, General Manager
Subject: Use of copiers

We are revamping our policy on the use of copiers . . .

This is the draft that goes into distribution and now allows him to turn his attention to other problems.

The Chicken or the Egg?

What are we to make of all this? It seems a rather lengthy and **tedious** report of what, after all, is a routine writing task created by a problem of minor importance. In making this kind of analysis, have I simply labored the obvious?

To answer this question, let's drop back to the original draft. If you read it over, you will see that Sam began with this kind of thinking:

□ "The employees are taking advantage of the company."
□ "I'm a nice guy, but now I'm going to play Dutch uncle."
   . . ."I'll write them a memo that tells them to shape up or ship out."

In his final version, however, his thinking is quite different:

□ "Actually, the employees are pretty mature, responsible people. They're capable of understanding a problem."
□ "Company policy itself has never been crystallized. In fact, this is the first memo on the subject."
□ "I don't want to overdo this thing—any employee can make an error in judgment."
   . . . "I'll set a reasonable policy and write a memo that explains how it ought to operate."

Sam obviously gained a lot of ground between the first draft and the final version, and this implies two things. First, if a manager is to write effectively, he needs to isolate and define, as fully as possible, all the critical variables in the writing process and **scrutinize** what he writes for its clarity, simplicity, tone, and the rest. Second, after he has clarified his thoughts on paper, he may find that what he has written is not what has to be said. In this sense, writing is feedback and a way for the manager to discover himself. What are his real attitudes toward that **amorphous,** undifferentiated gray mass of employees "out there"? Writing is a way of finding out. By objectifying his thoughts in the medium of language, he gets a chance to see what is going on in his mind.

In other words, *if the manager writes well, he will think well*. Equally, the more clearly he has thought out his message before he starts to dictate, the more likely he is to get it right on paper the first time round. In other words, *if he thinks well, he will write well*.

Hence we have a chicken-and-the-egg situation: writing and thinking go hand in hand; and when one is good, the other is likely to be good.

REVISION SHARPENS THINKING

More particularly, rewriting is the key to improved thinking. It demands a real openmindedness and objectivity. It demands a willingness to cull **verbiage** so that ideas stand out clearly. And it demands a willingness to meet logical contradictions head on and trace them to the premises that have created them. In short, it forces a writer to get up his courage and expose his thinking process to his own intelligence.

Obviously, revising is hard work. It demands that you put yourself through the wringer, intellectually and emotionally, to squeeze out the best you can offer. Is it worth the effort? Yes, it is—if you believe you have a responsibility to think and communicate effectively.

Questions on Content

1. List the three steps manager Sam went through in revising the memo.

2. Explain the chicken-and-the-egg analogy used in the article as it relates to writing and thinking.
3. What conclusion does Swift reach about the worth of revision?

Vocabulary

Define the following words as they are used in the article (the words are printed in **boldface** type):

1. coalesce
2. tedious
3. scrutinize

4. amorphous
5. verbiage

Suggestions for Response and Reaction

1. From you own experience, describe a situation in which a piece of correspondence written or received by you caused confusion, embarrassment, or anger. This could have taken place in a school, job, or social situation.
2. Explain why the tone of a business letter or a memo is significant.
3. Write an interoffice memorandum (see pages 247–253) to call attention to and suggest an alternate plan to correct an undesirable situation.
4. Define (see pages 96–108) *revision*. Also, consult handbooks and other composition texts. Then make your own list of suggestions to follow in revising your writing.

The World of Crumbling Plastics

STEPHEN BUDIANSKY

Cornstarch and sunlight may help clean up landfills, says a United States Department of Agriculture chemist.

Ordinary cornstarch may be the solution to a major environmental headache: How to dispose of plastic. Fifty billion pounds of plastics are used in the U.S. each year. Burning is out because of air pollution, and landfills are filling up. Carelessly discarded plastic is a danger to pets and wild animals, which get **entangled** in six-pack rings or choke on other plastic junk.

The problem would be considerably eased if plastic could be made to **decompose** as wood and paper do. A discarded plastic soda bottle can last for decades, for example. Chemists at the U.S. Department of Agriculture's Northern Regional Research Center in Peoria, Ill., are mixing new plastics with up to half starch, which is greedily devoured by soil **microorganisms.**

The major target is a substitute for the black plastic sheeting used as a mulch to control weeds, warm the soil and retain moisture. Farmers spend $100 to $200 an acre at the end of each growing season to dispose of the 125 million pounds of plastic **mulch** that they put down on their fields each year. If it weren't removed, it would become entangled in tilling equipment.

Felix Otey, a chemist who heads the USDA research effort, has spent 30 years studying starch. He got the idea of a starch-based plastic when he noticed that starch **molecules** could readily be tacked onto the chemical structure of a newly introduced form of **polyethylene.**

Disposable Laundry Bags

By varying the amount of starch, the researchers can control the length of time the plastic takes to decay after it's placed on the ground. The more starch, the faster the soil microbes chew their way through. Ideally, the process would take three or four months, just long enough for the plastic mulch to last through the growing season. Otey says there is evidence that once the microbes have devoured the starch, the tiny leftover pieces of **synthetic** chemicals also are eaten fairly quickly.

Otey's group is only turning out the mulch in experimental batches of a half pound or so. But another starch-based plastic he developed already is being marketed: Water-soluble hospital laundry bags. The filled bags are sealed and put directly into washing machines, where they dissolve—cutting staff and patient exposure to germs.

Plastics that fall apart when exposed to sunlight may offer another solution to the mulch problem. EcoPlastics, Ltd., in Ontario is selling 150,000 pounds a year of a plastic-film mulch containing a chemical that reacts with ultraviolet light by breaking the plastic's molecular bonds. The self-destruct timer can be set more predictably than with starch-based plastics. The soil microorganisms that eat starch are affected by variables such as soil moisture and temperature, which can vary from year to year. But total sunlight hitting the ground during a growing season is quite constant.

These plastics are also being enlisted to fight roadside garbage. "One characteristic of litter is that it's exposed to sunlight," says Anthony Redpath, president of EcoPlastics. The company has sold a million pounds of self-destructing shopping bags to Italy, where public concern about plastic waste runs high.

The USDA hopes that starch-based plastics might eventually dent the American farmers' huge corn surplus. It wouldn't be much of a dent, however. Even if the 50 billion pounds of plastic made every year were half starch—an unlikely possibility—that would only eat up a quarter of the surplus, currently about 100 billion pounds a year.

Question on Content

1. Why is a decomposing plastic needed?
2. How did Felix Otey get the idea for a starch-based plastic?
3. What three methods for disposing of plastic are mentioned in the article?
4. a. Which starch-based plastic is in the experimental stage only?
   b. Which starch-based plastic is already on the market?

Vocabulary

Define the following words as they are used in the article (the words are printed in **boldface** type):

1. entangled
2. decompose
3. microorganisms
4. mulch
5. molecules
6. polyethylene
7. synthetic

Suggestions for Response and Reaction

1. Explain the relationship between the title and the article.
2. Write an extended definition (pages 96–99, 101–112) of cornstarch.
3. Write a paragraph comparing (see pages 672–673) decomposing plastic and nondecomposing plastic.

Starch-Based Blown Films

FELIX H. OTEY, RICHARD P. WESTHOFF, AND WILLIAM M. DOANE

Northern Regional Research Center, Agricultural Research, Science and Education Administration, U.S. Department of Agriculture,[1] Peoria, Illinois 61604

Chemists have developed a cornstarch film that can help alleviate the problem of nonbiodegradable plastics.

Formulations containing high levels of gelatinized starch and various levels of poly(ethylene-co-acrylic acid) and, optionally, polyethylene can be readily blown into biodegradable films having the feel and general appearance of conventional plastic films. Ammonia and 2–10% moisture were essential ingredients for obtaining uniform films. Inclusion of polyethylene in the film formulation improved the economics and increased UV stability and rate of biodegradability of the blown films. The films have potential application as agricultural mulch and packaging, especially where biodegradation is important.

Introduction

The U.S. produces more than 30 billion pounds of plastics each year. Virtually all of this plastic is made from petroleum-based raw materials. Rapidly increasing prices, dwindling supplies of petroleum, and the need for **biodegradable** plastics have intensified interest in natural products as alternate sources of raw materials for making plastics. Starch, especially from corn, is probably the most abundant and lowest cost natural **polymer** available and its use in plastics production would greatly reduce the demand for petrochemicals and the negative impact on the environment now caused by discarding plastics that do not biodegrade.

Farmers especially need degradable plastic items, including mulch, coatings for the controlled release of herbicides and insecticides, seedling containers, films for protecting roots during transplanting, and containers for toxic agricultural chemicals. Approximately 140 million lb of polyethylene (PE) film is used annually in the U.S. as mulch to improve crop yields 50 to 350% by controlling weeds, retaining soil moisture, and reducing nutrient leaching (Carnell, 1978).

Since PE mulch cannot be reused and does not degrade between growing seasons, it must be removed from the field and disposed of at an estimated cost of $100 per acre. Hence, the development of a biodegradable film for mulch application would significantly benefit agricultural technology.

[1]The mention of firm names or trade products does not imply that they are endorsed or recommended by the U.S. Department of Agriculture over other firms or similar products not mentioned.

Published in *I&EC Product Research & Development*, Dec. 1980. pp. 592–595, by the American Chemical Society. Purchased by the U.S. Department of Agriculture for official use.

Since starch alone forms a brittle film that is sensitive to water, it is generally understood that starch must be combined with other materials in order to produce a satisfactory film. PE is the most widely used material for producing films that have desirable physical properties for packaging and mulch applications, and it is available at a relatively low cost. It is, therefore, a particularly desirable material to combine with starch to achieve the desired flexibility, strength, and water resistance. However, our previous attempts to produce blown films from compositions containing high levels of starch combined with PE were unsuccessful. Griffin developed techniques for producing blown films from compositions containing 8% predried starch, 80% PE, 1.6% ethyl oleate, and less than 1% oleic acid. However, the films became paperlike as the starch level was increased to about 15% (Griffin, 1977).

In earlier studies we developed technology for casting quality films from **aqueous dispersions** of starch and poly(ethylene-*co*-acrylic acid) (EAA) (Otey et al., 1977). These films are especially promising for applications where up to 90% starch is needed for rapid biodegradation but where strength, flexibility, and water resistance are not critical. When lower levels of starch were used, good physical properties and water resistances were achieved; however, the cost could prohibit certain large-scale applications because casting is a relatively slow process and aqueous dispersions of EAA are considerably more expensive than pelletized EAA.

In the present study, we prepared and evaluated films from various combinations of starch and pelletized EAA, using the less expensive **extrusion** blowing technique. Extrusion blowing is a common, economical method for producing film in which a tubular extrudate is expanded and shaped by internal air pressure to form a bubble several times the size of the die opening. The process also allowed the incorporation of low density PE (LDPE) as a partial replacement of the EAA, which further reduced the film cost and in some instances improved properties.

Experimental Section

FILM PREPARATION

Slight variations in methods and equipment were used to produce films reported in this paper; however, the following procedure is representative.

A mixture of air-dried corn starch (11% moisture) and enough water to equal the total solids in the final composition were blended for 2–5 min at 95 to 100 °C in a steam-heated Readco mixer (type: 1-qt lab made by Read Standard Div., Capitol Products Corp.) to partially gelatinize the starch. EAA pellets (type: 2375.33 manufactured by Dow Chemical Co.) were added while heating and mixing were continued. When LDPE was used, it was added about 15 min after addition of the EAA; but since it did not melt at this temperature, it can be added any time prior to the first extrusion. After the mixture was stirred and heated for a total of about 45 min, aqueous ammonia was added and heating and stirring continued for at least another 5 min. Immediately upon addition of the ammonia, the **viscosity** of the **matrix** increased rapidly. During the mixing, enough water evaporated around the loosely fitted cover so that the final product usually contained about 35% moisture. To further improve blending and reduce moisture content, the matrix was extrusion processed with an extrusion head attached to a Brabender Plasti-Corder (type: PL-V300 manufactured by C. W. Brabender Instruments, Inc.). The screw of the

Felix H. Otey, Richard P. Westhoff, and William M. Doane
Starch-Based Blown Films

extruder was $\frac{3}{4}$ in. in diameter, had a length to diameter ratio of 12, and a compression ratio of 2:1. If LDPE was present, the mix was first extruded 2–3 times through a $\frac{1}{4}$-in. die at a barrel temperature of 130° C to flux the LDPE pellets. All samples were then extruded through a die having 24 holes of $\frac{1}{32}$-in. diameter until the moisture content of the extrudate was about 5 to 10%. Usually 2 to 3 passes were sufficient to achieve the desired moisture level. The extrudate was transparent, flexible, and strong plastic strands. In some instances there appeared to be an improvement in blowing characteristics when these strands were further blended 2–3 min on cold rolls of a differential rubber mill. The extrudate was blown into a film by passing it through the same extruder, except that the die was replaced with a heated $\frac{1}{2}$-in. blown film die. The screw rpm was about 70–80, torque reading was 400–500 mg, barrel temperature was 120–130° C, and the die temperature was set in the range of 125 to 145° C. Higher levels of LDPE required the upper end of the temperature range.

In most instances, a clear film was obtained during the first pass through the blown film die. However, if excess moisture was present, one or more passes were required to obtain a uniform film. If the extrudate was too dry or contained insufficient ammonia, it could be exposed to ammonia for a few minutes and re-blown to yield a clear film.

BIODEGRADABLE TESTS

Although some preliminary degradability tests were conducted by exposing film samples to outdoor and indoor soil contact, more controlled data were obtained by placing the specimens on solid agar growth medium and inoculating them with test microorganism commonly found in the soil according to ASTM D1924-70. Each sample was sprayed with a mixed fungus spore suspension containing *Aspergillus niger*, *Penicillium funiculosum*, *Trichoderma* sp., and *Pullularia pullulans*. The amount of mold coverage on each specimen after 1 to 4 weeks was reported as: 1, less than 10% coverage; 2, 10 to 30%; 3, 30 to 60%; and 4, 60 to 100%.

PHYSICAL TEST METHODS

Films were evaluated for the following physical properties: tensile strength and percent **elongation** using the Scott tester; folding endurance, ASTM D2176-69 using the MIT tester with 0.5 kg of tension; and burst factor, TAPPI Standard T220 os-71. All tests were conducted at 50% relative humidity and 25° C.

Results and Discussion

FILM PREPARATION

Addition of ammonia was especially important to achieve a clear uniform film. Attempts to blow film without ammonia usually yielded films with streaks, indicating incompatibility between the starch and EAA resin. However, when such films were exposed to NH_3 for a few minutes, they could again be passed through the blown film die to achieve uniform films. Whether aqueous NH_3 was added to the formulation of NH_3 to the extruded product, the films showed an infrared absorption peak at 1550 cm^{-1} that is in the range for a carbonyl of a carboxylate group. The exact amount of ammonia used did not appear critical since any excess was driven off during the processing steps. The ammonia probably improved the dispersibility of EAA in water and reduced the rate of starch retrogration, both of which would improve compatibility between starch and EAA molecules.

627

Felix H. Otey,
Richard P.
Westhoff, and
William M. Doane
Starch-Based
Blown Films

Moisture content also appeared important to achieving quality films. We arbitrarily chose to incorporate about 50% water initially into the formulations to assure a high degree of starch gelatinization during the early mixing. The moisture content was reduced to about 35% during the first mixing by using a loosely fitted cover on the mixing head. To achieve adequate mixing while removing additional moisture, the samples were passed through the extruder several times. **Formulations** were easiest to extrusion blow, and the resulting films had the best appearance when the moisture content was 5–8% as determined by weight loss when subjecting film samples to a reduced pressure of 1 mmHg at 100° C for 2 h.

In commercial equipment, probably many of the processing steps could be simplified by using a vented extruder. Regardless of the equipment used, the composition should be kept above the gelatinization temperature of starch long enough to significantly gelatinize the starch, and at some time the temperature must be above the melting temperature of EAA and LDPE.

BLOWN FILM PROPERTIES

Formulating, blowing, and testing small batches of films inevitably involves certain uncontrolled variables and causes inconsistencies in data from one preparation to another. However, data reported in Table I reflect general trends in film properties as the levels of starch, EAA, and LDPE were changed.

Properties of film samples from runs 1–5 reveal effects of increasing the starch level from 10% to 50% without any PE in the formulation. All of the films containing up to 50% starch were transparent, flexible, self-supporting, and generally were

TABLE I. Effect on Film Properties of Starch and LDPE Levels

Run No.	Formulation, %[a,b] Starch	EAA	PE	Tensile Strength, psi	Elongation %	NH$_3$[c]	MIT Fold, No. Folds	Burst Factor	Weather-Ometer, h	Fungi Susceptibility, Weeks[d] 1	2	3	4
1	10	90	0	3470	260	4.9	—	—	402	0	0	0	0
2	20	80	0	4140	120	4.3	—	—	212	0	0	0	0
3	30	70	0	3225	150	3.8	—	—	168	0	0	0	0
4	40	60	0	3870	92	3.3	—	—	90	1	1	1	1
5	50	50	0	3940	61	2.7	—	—	90	1	2	3	3
6	40	50	10	3570	80	3.6	3800	24	111	2	3	4	4
7	40	40	20	3477	66	2.2	7000	24	134	1	2	3	4
8	40	30	30	3150	85	1.7	2700	21	151	1	2	3	4
9	40	20	40	2920	34	2.8	4800	19	199	—	—	—	4
10	40	10	50	1840	10	2.8	470	9	559	—	—	—	4
11[e]	30	32.5	32.5	2000	62	1.6	—	—	710	—	—	—	—

[a]Based on combined dry weight of starch, EA, and PE, exclusive of water and NH$_3$.
[b]Formulations of examples 4, 6, 9, and 11 additionally contained about 1% antioxidant (Irganox 1035, Ciba Geigy Corp.).
[c]Parts of Ammonia per 100 parts of formulation dry weight.
[d]ASTM D 1924-70, fungus growth; 0 = none, 1 = 10%; 2 = 10–30%; 3 = 30–60%; 4 = 60–100%.
[e]Formulation contained 5% carbon black (Industrial Reference Black, No. 3).

considered to have good physical properties. However, the degree of transparency and flexibility decreased slightly as the level of starch was increased. All of the samples were uniform and indicated that good compatibility existed between the starch and EAA. It was apparent from the general appearance of these films, their blowing rate, and flow characteristics in the extruder that the maximum level of starch that could be incorporated to achieve blown films was about 60%, with the preferred level at about 40%. As the starch level increased, there was a significant decrease in film resistance to artificial weathering in a Weather-Ometer. Deterioration was attributed to UV instability that caused small cracks or tears in the film. More significant was the lack of fungal attack under controlled conditions with up to 30% starch and a very slow attack with 40% starch present. Although it is expected that all the films are biodegradable, the ASTM method used for measuring fungal attack did not extend beyond 4 weeks.

The relatively high level needed and price of EAA could prevent the application of starch-EAA film in certain large-scale areas. Although starch and LDPE alone did not produce quality films, it was possible to incorporate relatively high levels of LDPE into the starch-EAA compositions.

Films made with 40% starch and up to about 40% LDPE (runs 6–9, Table I) were clear, flexible, and uniform, indicating good compatibility. Above about 40% levels of LDPE, the films were less transparent and in some instances translucent, and were observed to have less tear resistance. In contrast to the film samples without LDPE, these films reflected a substantial increase in resistance to Weather-Ometer exposure as increasing amounts of LDPE were incorporated into formulations. Furthermore, the addition of LDPE greatly increased the fungal attack on the samples, showing that the film would biodegrade more readily when exposed to outdoor soil contact.

Films corresponding to those prepared in runs 4 and 8 were subjected to a 35-day outdoor exposure test. The film with 40% starch and 60% EAA developed cracks within 11–13 days, whereas that containing 40% starch, 30% EAA, and 30% LDPE did not develop any cracks.

Compositions containing more than the preferred level of either starch or LDPE for producing blown films could be extruded into sheets and tubing that could have application where rigid biodegradable plastics are needed.

Other materials, either polymeric or **monomeric,** may be added to the composition to **impart** specific properties to the film. Films (Table II) with good physical properties and improved rates of fungi attack were prepared by incorporating 5 to 25% polyvinyl alcohol into the formulation.

TABLE II. Effect on Film Properties of Polyvinyl Alcohol (PVA) Level. [Film Formulation (Dry Basis): 3.5% NH$_3$; 0.5% Irganox 1035; EAA and Starch Varied]

Run No.	PVA, %	Starch, %	EAA, %	LDPE, %	Tensile Strength, psi	Elongation,	Weather-Ometer, h	Fungi[a] Susceptibility
1	5	40	55	0	4650	97	75	2
2	10	40	50	0	3320	40	140	3
3	20	40	40	0	5765	59	95	4
4	10	40	25	25	1970	4	77	3
5	10	30	30	30	2100	26	140	4
6[b]	25	25	30	—	3500	300	—	4

[a]Mold growth after 4 weeks; see Table I, footnote d.
[b]18% Sorbitol and 2% glycerol included in formulation.

629

Felix H. Otey,
Richard P.
Westhoff, and
William M. Doane
Starch-Based
Blown Films

Plasticizers can be added to significantly increase percentage elongation and rate of fungi attack as illustrated in run 6, Table II. However, the **optimum** level and types of plasticizers were not determined in this study.

Also, ultraviolet (UV) stabilizers would be essential for most applications where the film would be exposed to extended periods of sunlight. No effort was made to establish the preferred type and level of UV stabilizer. However, a few blown films were prepared and evaluated that contained various levels of carbon black (Industrial Reference Black, No. 3). One such example (run 11), is given in Table I to illustrate the significant improvement in UV stability that can be gained from adding carbon black. Our preliminary experience suggested that to obtain a uniform film the carbon black must be blended into the EAA on a rubber mill prior to blending the other components.

COST CALCULATIONS

The plastic film reported here would have potential application for the most part where LDPE is now used. Their principal advantages over LDPE film would depend upon their faster rate of degradation. LDPE pellets now sell for about 47¢/lb. Calculated material costs for various starch-based films along with their approximate specific gravity are listed in Table III. No attempt was made to estimate the additional costs required to formulate and pelletize the starch products suitable for blowing films.

Although the replacement of a synthetic resin with low-cost starch substantially reduces the cost of a film formulation, part of this cost advantage is offset by the increased specific gravity imparted by starch. Starch has a specific gravity of 1.5 compared to about 0.92 for LDPE and 0.954 for EAA. The starch-based films have specific gravity ranging from about 1 to 1.2.

The effect of film specific gravity on the film cost for mulching 80% of a 1-acre field with 1.5-mil film is plotted in Figure 1. The graph is a plot of the equation

$$\text{cost/acre, \$} = 272 \times \text{film specific gravity} \times \text{film cost per lb, \$}$$

where the 272 lb is the weight of 80% of one acre of film (specific gravity, 1) 1.5 mil thick. This figure illustrates that the resin cost of mulching one acre with 47¢/lb LDPE (0.92 specific gravity) would be $118, whereas a starch-based film at the same

TABLE III. Calculated Raw Material Costs for Various Starch-EAA-LDPE Formulations

Starch %	EAA, %	LDPE, %	Density g/cm^{3a}	Cost, ¢/lb^b
20	80	0	1.0632	82.0
40	60	0	1.1724	64.0
50	50	0	1.2270	55.0
40	40	20	1.1656	53.4
40	30	30	1.1622	48.1
40	20	40	1.1588	42.8
50	25	25	1.2185	41.8

[a]Calculated based on following specific gravities: starch, 1.5; EAA, 0.954; and LDPE 0.92.
[b]Calculated chemical cost based on following costs per pound: starch at 10¢; EAA at 100¢; and LDPE at 47¢.

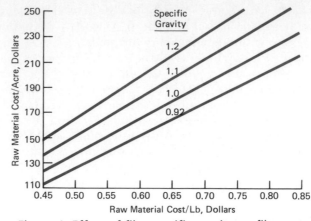

Figure 1. Effect of film specific gravity on film cost for mulching 80% of a one-acre field with 1.5-mil film.

resin cost would be about $153 assuming a specific gravity of 1.2. However, a nondegradable film cannot be reused and must be removed from the field and buried or burned. Based on an estimated removal and disposal cost of $100 per acre, a degradable film that need not be removed from the field could theoretically have a resin cost, after processing into pellets, of about 66.8¢/lb at 1.2 specific gravity or 72.9¢/lb at 1.1 specific gravity ($218 per acre) and yet be competitive with LDPE at 47¢/lb. The approximate raw material cost for the experimental films that have acceptable properties range from about 42¢ to 53¢/lb.

As EAA becomes more widely used and priced more competitively with conventional resins, these raw material costs could go much lower.

LITERATURE CITED

Carnell, D., "Photodegradable Plastic Mulch in Agriculture," presented at the National Agricultural Plastics Association, 14th Agricultural Plastics Congress, Miami Beach, Fla., Nov 10–13, 1978.

Griffin, G. J. L. (to Coloroll Limited), U. S. Patent 4016 117 (Apr 5, 1977).

Otey, F. H., Westhoff, R. P., Russell, C. R., *Ind. Eng. Chem. Prod. Res. Dev.*, 16, 305–308 (1977).

Questions on Content

1. **a.** What are the three first-level (major) headings in the report?
   b. What do these headings tell you about the content of the report?
2. What is the most abundant and lowest cost natural polymer available?
3. Why is it necessary to combine starch with other material to produce a satisfactory film?
4. List the physical properties used to evaluate the film.
5. How many different combinations of starch, EAA, and PE were tried for effect on film properties?
6. What element was added to achieve a clear, uniform film?

Vocabulary

Felix H. Otey,
Richard P.
Westhoff, and
William M. Doane
Starch-Based
Blown Films

Define the following words as they are used in the article (the words are printed in **boldface** type):

1. biodegradable
2. polymer
3. aqueous
4. dispersions
5. extrusion
6. viscosity

7. matrix
8. elongation
9. formulations
10. monomeric
11. impart
12. optimum

Suggestions for Response and Reaction

1. Read "The World of Crumbling Plastics," pages 622–623, and read "Starch-Based Blown Films," pages 624–630. What observations can you make about audience, purpose, and content?
2. Find a paragraph that clearly illustrates process explanation (see pages 41–51) using passive voice (see pages 41, 698).
3. Write an extended definition (pages 96–99, 101–112) of *polymer*.
4. Write a letter (see pages 262–267) to Felix Otey, agricultural researcher, requesting additional information on the starch-based blown films.

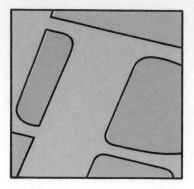

To Serve the Nation: Life Is More Than a Career

JEFFREY R. HOLLAND

A former vocational student, now a successful educator, looks at life from an educational, vocational, and leisure-time point of view.

I am pleased to stand before you as a vocational student of some years past—though a very poor one I am chagrined to admit. My high school industrial arts teacher would be so proud today because he, of all my high school teachers, quite rightfully thought I would not amount to much of anything—at least not in vocational education. He had been my brother's teacher and my brother was a genius in industrial arts. I remember envying all of his school projects. He made a magnificent gun cabinet. He made a hope chest for my mother. He made shutters for our windows and iron railings for our front steps. He made wall shelves and work benches and drawing tables. He made book ends and tool cabinets and tools to go in them. There was nothing he could not produce. He was a shop teacher's dream.

Then I came along. I think in two years of industrial arts the only things I concluded successfully were a tin quart cup (which I accidentally split down the side and which never held a drop of anything liquid), a funnel which had a spout on it so broad that all materials rushed through with the abandonment of a spring avalanche, and a frightfully shaped breadboard on which no respectable loaf of bread would have been caught dead—or sliced. It is hard to produce an unrecognizable breadboard but I did. The fact that, in spite of such a meager record, I am now invited to address your conference heartens me and would hearten that teacher. Thank you for a second chance to earn your respect after an obviously unsuccessful first attempt 20 years ago.

By the way, my mother never found a way to use any of these implements I made but she received them with the love that mothers are famous for. She saw what that shop teacher saw—that my brother and I were very different but that we both could, if we gave it our best effort, find some area in which we would not only contribute to our own well-being and development but also make some larger contribution to society. I think that is true of every mother's child and we commend you for offering many honest, honorable ways in which hundreds of young men and women in this state can, in ways of their own choosing, accept the challenge of your conference and indeed "serve the nation" with skills they have acquired under your direction.

A story is told of Ralph Waldo Emerson, one of the great men in modern history and perhaps the single most influential literary and cultural mind of mid-nineteenth century America. It seems he was out on his grounds coaxing, bullying, imploring, and shoving a balky calf toward the barn door but with no discernible success. For all of his New England wisdom and Yankee ingenuity, Emerson could not budge

Reprinted by permission from *Vital Speeches of the Day*, XLIV (June 15, 1978), 533–536. Headings have been added.

the animal, shove and tug and pull as he would. Nearby a bright young Irish milk-maid was watching "the Sage of Concord" and fought valiantly to hold back her audible laughter. Just when Mr. Emerson's frustration was at the boiling point, the young girl walked over to the calf, moistened her forefinger, stuck it in the calf's mouth, and led it gently into the barn.

Emerson watched in astonishment. He stood and gazed and said nothing. Then he marched into his home, washed his hands, sat down at his study, and picked up his pen. In his daily journal he wrote boldly, "I like people who can do things!"

The wonderful thing about education is that it prepares people to do things—and it teaches a respect among the family of men for the variety of talents and skills which lead us to accomplishment. Emerson admired the milkmaid for the practical, even ingenious skills which made her effective in her work. She, too, surely must have admired him for the poetry he wrote and the lectures he gave and the books he produced. In a world of education—education in its broadest and best meaning—there is room for that kind of mutual respect, all in the name of genuine ability, all in admiration for people who can "do things."

One of the beauties of the American tradition lies in its pluralism—in its willingness to entertain, even to encourage, diversity and breadth and multiplicity. And surely part of that American tradition is the freedom of choice we have within such attractive options. We enjoy that breadth and freedom in the educational programs of this nation. I represent a private, church-sponsored educational system, but there are other private institutions as well as a superb and influential public system. Most of you represent certain kinds of vocational and technical teaching, but there are others who teach the arts and the professions. Surely there is not only room enough for all but a need for all, to make our choices broad and own growth certain. No one school or one teacher or one classroom experience can cater to the complex diversified educational activity that we encourage as part of our national strength.

I was pleased to note this response from Chairman Donald B. Holbrook of the Utah State Board of Regents in a recent address at the University of Utah. He said,

> The Utah System of Higher Education is based on the **premise** that institutions can greatly differ in their roles and methods and still find harmony and cooperation which will complement the whole, while strengthening the need for difference . . . we have pledged our efforts to **nurture** opportunities in the technical vocational field [while preserving] differences in the institutions of higher learning . . . the spirit of individualism between institutions of higher learning has provided opportunity for uniqueness, not [necessarily] sameness or standardization.
>
> I want universities of excellence. I want the very best for technical colleges. I want each of our four- and two-year colleges to find total fulfillment of its role as defined by our Utah system of cooperative endeavor.

I am sure that same spirit and sentiment prevails with the secondary school administration of this state. It suggests an enlightened and encouraging time for vocational educators, not at the expense of other programs but alongside and in harmony with them. I join the thousands of parents in this state who are grateful for the many agencies, schools, and programs which teach our sons and daughters "to do things" whether that be write a poem, repair an automobile, paint a picture, sew a dress, program a computer, milk a cow, balance an account ledger, turn a lathe, write a book, or perform open-heart surgery. All of our lives are enhanced and our nation is best served when each individual is led to the broadest possible opportunity for meaningful service.

Against this backdrop of service and fulfillment, let me suggest just three principles I would hope every educator would teach. These are, I think, three which you are particularly prepared and able to teach in vocational and technical programs. I speak to you now as colleagues—brothers and sisters if you will—in the work of education in this nation. And perhaps I speak now not only as a fellow professional but also as a parent. My sons and daughter will undoubtedly be in some of your classes soon and they will sit at your feet to learn. While you develop their skills and provide technical, vocational training for these children—mine and yours—I invite you to teach them at least three values that will be important to them and this nation whatever their personal career choices may be.

VALUE OF WORK

Please teach them the need for and the great value in hard, productive, faithfully pursued work. No mortal force has been more powerful in our personal and collective progress than clean, clear-cut, honest effort. Work. Industry. Perspiration. Some of our first technical learning may have come by accident but not much and not often. Thomas Edison once said that nothing he ever did in his life that was worthwhile came by accident. It is ultimately long hours of effort that pay off in the steady advancement of civilization. Past and present generations in this nation have, by continuous and purposeful labor, converted the resources of this country into a quantity and variety of goods and services never before known in the history of mankind. We must teach the rising generation to pay that same price in order to reap the same rewards. I include in these rewards the blessings of heart and spirit as well as head and hand.

President Spencer W. Kimball said recently at the dedication of the beautiful new Utah Technical College campus in Orem,

> I for one worry lest we value leisure too highly in our society along with material things. Leisure and materialism can drain our society of strength and purpose, but work and morality will permit us to progress and grow as a people and to keep our nation strong.

It interests me that President Kimball links "work" with "morality," for I believe there is an important and explicit relationship between the two. Our integrity should demand of us an honest day's effort for an honest day's pay. Thoroughness and reliability and the will to stand by our product or service are qualities for which the best men and women have always been known. These are "moral" issues. When the switch is thrown, the lights should go on. When the entries are made, the books should balance. When the medication is applied, the disease should waste away. When the plowing is finished, the furrows should be straight. Since time began, the quality and the quantity of his work has spoken volumes about a man.

I am reminded of the young boy who watched and waited and waited and watched while his grandfather spent long, painstaking hours on a boat he was building. Great care and concern was given to every cut of the wood, to every angle and curve and connection. Great care was given to the quality of the mast and to the sealing and stretching and reinforcements. In weariness and impatience the child pled with his grandfather, "Why do you take so long? I want to go out to sea." His grandfather replied with the wisdom of age. "When you finally sail your ship, it has to float, son. You can't 'explain' anything to the ocean."

Please teach my children the value of hard work and the positive moral meaning of work when it is performed to the best and most faithful of their abilities.

PRIDE AND SELF-ESTEEM IN CAREER CHOICE

The second principle which I would hope you could teach my children—and it is closely related to the first—is a sense of appropriate pride and self-esteem in the work they do and the career choices they make. I think it is fair to say that technical, vocational skills have not always been held in the kind of esteem that honest labor well performed deserves. As a nation we may have consciously or unconsciously held back our commendation for the work done inside a blue collar or a pair of rubber boots or a domestic apron. I have loved for a long time the telling comment by John W. Gardner, former Secretary of Health, Education, and Welfare, on some of those misspent attitudes. He said,

> An excellent plumber is infinitely more admirable than an incompetent philosopher. The society which scorns excellence in plumbing because plumbing is a humble activity and tolerates shoddiness in philosophy because it is an exalted activity will have neither good plumbing nor good philosophy. Neither its pipes nor its theories will hold water.

Consider what that may mean for us today—this week, this month. Economics, the availability and cost of goods and services, dominate much of our lives. No one in the room, I presume, is a coal miner. But coal mines can capture the nation's attention. No one in the room, I presume, is a farmer. But farms can capture the nation's attention. The list of battle fronts goes on, inevitably touching every other front until all in this room *are* affected. Ultimately the success of this nation's economy and technical accomplishment depends in no small part on the peace and proficiency of the nation's workers. I use that designation "worker" to include everyone from the president of the United States down to the minimum wage earner—philosophers, plumbers, laborers, lawyers. And once again that is where we look to you who teach. Whatever else two or three thousand years of western civilization has taught us, it has taught us that every citizen should be as well equipped as possible to contribute effectively for the welfare of that individual, his family, and the community as a whole. The logic of that suggests that the highest possible welfare is achieved only when each individual produces to the limit of his or her capacity. Adequate training and assignments for those varied talents and skills have been recognized by even the most primitive societies. I think finally we are now recognizing them in America.

I, for one, will take good plumbing over poor philosophy anytime. But I also prefer good philosophy to poor plumbing. Ultimately good philosophy and good plumbing are necessary and we have a right to expect both. What we really ask for here is the **dissipation** of old prejudices, which all of us have, and a turn to appreciation for excellence, whatever the task. When we do that, then any **stigma** inappropriately attached to certain walks of life and career choices in the past disappear. What we will have, I believe, is the best of the pluralism, the best of the freedoms and choices we referred to earlier. Then even the distinctions Mr. Gardner was forced to use fifteen years ago—distinctions such as "exalted activity" and "humble activity"—will carry less social force. We will praise or criticize the quality and not simply the image of the endeavor. Then both our pipes *and* our theories will hold water.

Give my child and yours appropriate pride and self-esteem in the work he or she does, when it is excellently done. This nation needs the self-esteem again. To have it will aid a kind of moral unity we have managed to lose in the last twenty-five or thirty years. Strikes, and wars, and droughts, and inflation constantly remind us that we are on this planet together and we must make it work—together. As Thomas Jefferson did two centuries ago, I hope we can set aside gender, ancestral lines, and inheritance, dispel false prejudices and arrogance for a return to mutual respect and common admiration for a job well done, whatever the background of the labor or the laborer. Help my children believe in that kind of professional world.

LIFE IS MORE THAN A CAREER

The third and concluding principle I hope you will teach my children, and it stems from the first two, is that finally life is more than a career, more than a vocation, whatever that choice may be and however fulfilling it should become. Whether secretary or scientist, mechanic or musician, farmer or physiologist—those employments are not all there is to this mortal experience. We owe it to our students to teach not only the breadth *within* careers but the immensely full and rewarding life outside of careers. One hundred years ago Henry Ward Beecher said,

> Education is the knowledge of how to use the whole of oneself. Many men use but one or two faculties out of the score with which they are endowed. A man is educated who knows how to make a tool of every faculty, how to open it and keep it sharp and how to apply it to all practical purposes.

Every person needs training. I define training as that which teaches us to perform a task, to make a living. Secretaries and scientists, mechanics and musicians, farmers and physiologists need training. But every person needs education, too. I define education as that which not only helps us make a "living" but helps us make a "life." I believe in the eternal possibilities of the human soul and I am committed to a philosophy of training *and* education which pushes and stretches and expands and broadens life forever. We ought to do as Beecher said—make a tool of *every* faculty and use the *whole* of ourselves. I think secretaries ought to understand something of Newtonian physics. On the other hand I think physicists ought to know how to type and find something they have filed. I think mechanics ought to recognize the difference between Bach's music and Beethoven's. And I certainly think a musician ought to be able to connect a battery cable on his own car. I think a farmer ought to know a little of the skeletal structure of man. I think a physiologist ought to know what to spray on his fruit trees. Such learning and experience and enjoyment outside our work-a-day world has both practical *and* spiritual meaning in terms of what I believe mortal men and women can someday become.

Along with these personal achievements there are, of course, larger and more public issues which should stretch us. Recently President David Gardner of the University of Utah said,

We must equip our students to cope with the complexities of the global world, whether their life's work finds them here in our valley or in Tehran or Taiwan. We must develop in our students a capacity for clear thinking, for living with uncertainty and **ambiguity,** for understanding [themselves] and others, for communicating with effect, for tolerance and respect for those whose views and opinions may be at odds with our own and for an appreciation of the physical, natural, social, political, and economic world in which we live.

I agree and I assume you do also.

I was pleased to note that former Secretary of Labor, Willard W. Wirtz, had gone on record regarding the need to fulfill lives not only within the world of work but also in those dimensions that are outside it. He said,

> When I realize that work occupies only one-third as much of the waking hours of the human being today as it did in 1900 . . . I wonder what kind of mistake is about to be made when we put the measure of the value of education [only] in terms of the earnings, the wages, [the job] of a particular [person] . . . it seems to me that the evidence is mounting that here is danger of our taking a shortcut across the quicksand.

You can help our children—mine and yours—avoid that quicksand. Even as you give them both the skills and self-esteem of their career choices, so can you give them a vision of skill and self-esteem that goes out and up and beyond anything they have ever done before. You can ignite the flame within these students' souls— and watch it burn far beyond the lathe or the laboratory or your classroom. You can watch it leap into a consuming passion for the best and brightest that life has to offer. And chances are, if we do it right, you will meet that student years hence and the flames of fulfillment will still be dancing. If there is one joy and reward for teaching that ranks above all the others, it is that one—the stunning, awesome realization, sometimes many years after the fact, that you have elevated forever, not only the vocation but also the vision, of a child of God. May you be encouraged and faithful in that worthiest of all the nation's professional services—teaching. Thank you.

Questions on Content

1. What use does the speaker make of comparison and contrast? What point does each comparison and contrast make?
2. Explain the speaker's view of the educational process.
3. According to Holland, every educator should teach three principles. List the three principles.
4. Give Holland's definition of (a) training and (b) education.

Vocabulary

Define the following words as they are used in the article (the words are printed in **boldface** type):

1. premise
2. nurture
3. dissipation

4. stigma
5. ambiguity

Suggestions for Response and Reaction

1. What can you assume about the content of the speech from reading only the title and the subtitle?
2. Do you agree or disagree that educators should teach the three principles identified by Holland? Explain your answer.
3. Do you agree or disagree with Holland's view of the educational process? Explain your answer.
4. Review Holland's definition of "training" and "education." Then, following the principles for writing a sentence definition (see pages 99–101), write your own definition of "training" and "education."
5. Holland says, "Part of the American tradition is freedom of choice." Following the principles for effect-cause explanations (see pages 170–184), develop this statement into a full explanation.
6. Following the principles for summarizing (see page 225), write an evaluative summary (see pages 222–224) of this speech.

Part III
Handbook

ART III OF *TECHNICAL ENGLISH* CONSISTS of a handbook to be used as a guide to accepted practices in writing and in matters of usage. These practices, developed over a period of time, are conventions through which communication of meaning is made easier.

Chapter 1
The Sentence

Objectives

Upon completing this chapter, the student should be able to:

- Write sentences using sentence patterns common to the English language
- Write sentences with expanded meaning by adding modifiers or by compounding
- Write sentences to emphasize certain material within the sentence
- Combine the ideas of several sentences into one or more sentences
- Write sentences free of the common problems in sentence construction

Introduction

As human beings, we communicate with one another through patterns of speech that we call sentences, although we do not always speak or write what is traditionally called the complete sentence. For example, if you and your friend Joe were discussing going to the drag races, your conversation might be:

Joe, wanna go to the drag strip tonight?
Sure. What time?
About eight.
OK.
See you then.

There would be no misunderstanding between you and Joe even though both of you spoke mostly in fragments and each supplied mentally and unconsciously the missing parts of the sentences. Joe was ready to go at 8:00 P.M. Such omission is more common in conversation than in writing, except perhaps when conversation is reproduced in writing.

Naturally most of us speak much more easily than we write, mainly because we have had a great deal more experience with speech than with the written word. Almost anyone, however, can improve writing skills by studying generally accepted principles of writing and by practicing writing according to these principles, just as almost anyone can develop skill at welding through understanding the principles of welding and seriously practicing welding.

This chapter will discuss the sentence as the basic element of the English language and will show the basic sentence patterns used in the English language.

Key to the English Sentence: Word Order

The key to the English sentence is word order. If you read or heard, "He the machine operated," you would think that you had misread or heard incorrectly. The expected order for this group of words is "He operated the machine."

Through study and analysis of many, many sentences, linguists have recognized that the words forming English sentences generally are arranged in one of several orders. Linguists, then, attempting to provide a way to talk about sentences, have classified these several orders as patterns. In making these classifications, linguists did not come to an agreement on any set number of possible sentence patterns, nor did they identify a single sentence pattern in the same way. Some linguists list as few as three sentence patterns while others list as many as nine or more. This chapter shows five basic sentence patterns used by almost all speakers of English, although very few people are aware that they are using any particular sentence pattern.

Pattern 1: Subject + Verb

Pattern 2: Subject + Verb + Direct Object

Pattern 3: Subject + Verb + Direct Object + Objective Complement

Pattern 4: Subject + Verb + Indirect Object + Direct Object

Pattern 5: Subject + Verb (linking) + Subjective Complement

Each of these five basic patterns can be expanded by the addition of modifiers in the form of words, phrases, or clauses and through the process of compounding. (For further discussion of expansion, see pages 648–650.)

Basic Sentence Patterns

Pattern 1: Subject + Verb

Pattern 1 requires at least two units: a unit to function as a subject (identifies who or what is acting in some way) and a unit to function as a verb (identifies the action done by the who or what of the subject).

EXAMPLES

      S   V

Airplanes fly.

         S   V

Some ores melt.

      S   V

Students learn.

Pattern 1 may be expanded by adding modifiers (adjective and adverb: word, phrase, or clause—for detailed explanation, see pages 648–650). The basic pattern, however, remains unchanged; it must contain a subject unit and a verb unit.

EXAMPLES

MODIFIER   S   V MODIFIER

    Jet planes fly swiftly.

               (Prepositional phrase)

MODIFIER   S   MODIFIER   V   MODIFIER

    Some students in math courses learn easily.

MODIFIER S   V   MODIFIER (Clause)

    Some ores melt when heated to a specified temperature.

Pattern 1 may also be expanded by compounding either or both of the essential units.

EXAMPLES

(Subject compounded)

S   S   V

Airplanes and spacecraft fly.

(Verb compounded)

S   V   V

Some ores melt and resolidify.

Pattern 2: Subject + Verb + Direct Object

Pattern 2 requires at least three units: a unit to function as a subject (identifies who or what is acting in some way), a unit to function as a verb (identifies the action done by the who or what of the subject), and a third unit following the verb to function as a direct object (word, phrase, or clause that identifies to whom or to what the action is done).

EXAMPLES

S   V   DO (Word)

Experiments can illustrate theories.

S   V   DO (Clause)

We know that the Bessemer process is named for Sir Henry Bessemer.

S   V   DO (Infinitive phrase)

The frayed wires began to burn.

To determine if the word, phrase, or clause following the verb is a direct object, read the subject and the verb of the sentence and add "What?" or "Whom?" to form a question. If the unit following the verb answers the question but does not rename or describe the subject, it is the direct object.

EXAMPLE

S   V   DO

Insulators prevent a ready current flow.

Question: Insulators prevent what?
Answer: "a flow" (the direct object), or "a ready current flow" (the direct object and all modifiers)

EXAMPLE

S   V   DO

You ask the supervisor.

Question: You ask whom?
Answer: "the supervisor" (Direct object)

Pattern 2 may also be expanded by adding modifiers (adjective or adverb: word, phrase, or clause), but it still must have the basic units of subject + verb + direct object.

EXAMPLE

645

The Sentence

MODIFIER   S   MODIFIER (Participial phrase)
The frayed wire coming from the motor of the air conditioning unit

MODIFIER   V   DO (Infinitive phrase)
suddenly began to burn.

Pattern 2 may be expanded by compounding.

EXAMPLE

S   V   V   DO   DO
You may call or write the personnel manager or the president.

Pattern 3: Subject + Verb + Direct Object + Objective Complement

Pattern 3 requires four units: a unit to function as subject (identifies who or what is acting in some way), a unit to function as a verb (identifies the action done by the who or what of the subject), a unit following the verb to function as the direct object (identifies to whom or to what the action is done), and a unit (word or phrase) following the direct object to function as an objective complement (tells more information abut the direct object by renaming or describing).

EXAMPLES

S   V   DO OC (Objective complement renames direct object)
The employees chose him supervisor.

S   V   DO   OC (Objective complement describes direct object)
Shaping made the point sharp.

Modifiers (adjective and adverb: word, phrase, and clause) may be added to the basic units of Pattern 3, or the basic units may be compounded to expand meaning.

EXAMPLE USING MODIFIERS

S   MODIFIER (clause)   MODIFER   V   DO
The persons who represent the company's board of directors unanimously chose her
OC
district supervisor in charge of sales.

EXAMPLE USING COMPOUNDING

S   S   V   DO   DO
The president and the board of directors chose Janet Herron and William Strickland,
OC   OC
vice-president and district director, respectively.

Pattern 4: Subject + Verb + Indirect Object + Direct Object

Pattern 4 requires four units: a unit to function as a subject (identifies who or what is acting in some way), a unit to function as a verb (identifies the action done by the who or what of the subject), a unit to function as an indirect object (specifies what or who receives the direct object), and a fourth unit to function as a direct object (identifies to what or to whom the action expressed in the verb is done).

      S   V   IO   DO
The shop instructor gave the students their assignment.

      S   V   IO   DO
The drafter has taken the engineer the drawings.

Notice that a sentence pattern which includes an indirect object *must* include a direct object. A test of this pattern is to read the subject and the verb followed by "Whom?" and "What?" For example, the drafter has taken whom what? The answer to "Whom?" is "engineer," the indirect object; the answer to "What?" is "drawings," the direct object.

If these sentences were rewritten as follows, they would be changed to Pattern 2: Subject + Verb + Direct Object. Notice that the indirect object becomes a prepositional phrase.

      S   V   DO   (Prepositional phrase)
The shop instructor gave the assignments to the students.

      S   V   DO   (Prepositional phrase)
The drafter has taken the drawings to the engineer.

Pattern 5: Subject + Verb (Linking) + Subjective Complement

Pattern 5 requires three units. The first unit functions as the subject, which names who or what is being described or identified more specifically (renames). The second unit functions as a verb, which in this pattern is a linking verb, a verb that "links" or "joins" the subject to the subjective complement. The third unit functions as a subjective complement, a unit that tells something about the subject by describing the subject unit (perhaps you identify this unit as a predicate adjective) or renaming it (sometimes identified as a predicate nominative, a predicate noun, or a predicate pronoun).

EXAMPLES

      S   V   SC (Renames subject)
Sir Henry Bessemer was an English engineer.

      S   V   SC (Renames subject)
Metals are good conductors.

               S   V SC (Describes subject)
Micrometer depth gauges are obtainable.

Notice this linking verb has a special "joining" characteristic. In English, the more common verbs that have this characteristic include: forms of *be (am, is, are, was, were, be, being, been), seem, appear, become,* verbs of senses *(smell, taste, look, feel),* and *grow.*

Pattern 5 may be expanded by adding modifiers (adjective and adverb: word phrase, or clause) and by compounding.

MODIFIER (Participial phrase)   S   V   SC   MODIFIER (Clause)

Born in 1813, Sir Henry Bessemer was an English engineer who discovered a process

for manufacturing steel.

         S   MODIFIER (Participial phrase)   V   SC

Micrometer depth gauges having a range up to 9 inches are obtainable.

EXAMPLES USING COMPOUNDING

            S   V   SC   SC   SC   SC

Frequently used hand tools are the hammer, chisel, file, and wrench.

    S   S   S V   SC

The steel rule, the narrow rule, and the tape rule are semiprecision measuring tools.

Building Sentences

In speaking, few of us have to stop and plan how we will organize a particular sentence. The sentence just comes out because, somewhere within, the sentence is inexplicably put together and generated through speech. When we write, on the other hand, the process is a bit more difficult. More often than not we spend some time in planning how to write certain sentences, and for many people this process is difficult. It might be helpful if we realize how sentences are constructed.

The process is very similar to that used by a boy playing with Tinker Toys. He knows what he wants to build, so he adds or removes sticks and wheels until he gets just the shape he has in his mind. The same procedure occurs in constructing sentences. We know what we want to say; we have all the material there. The problem is how to say it, how to get it out. We need to start with our main idea and add to or take away until we get the sentence just the way we want it.

BASIC SENTENCE UNITS

From the preceding section on the basic patterns that sentences form, it is evident that every sentence has two basic units: a unit to function as the subject of the sentence (identifies who or what is acting in some way) and a unit to function as the verb (identifies the action done by the who or what of the subject). Every English sentence, from the simplest to the most complicated, may be divided into two units: the subject and the verb. The natural arrangement of these two units is subject followed by verb.

The sentence may be one word, "Wait." Its two parts are:

   S   V

(You) wait. (The subject "you" is understood.)

Or it may be as complicated as:

              S   V

The new machinist-trainee in the shop began operating the drill press before he put

on goggles, thus creating a situation in which possible eye injury could occur.

Sometimes the sentence may need to be rearranged into its natural order (S + V) before this division can be made.

> Has the machinist been instructed to start the chuck on the lathe spindle by hand, not power? (Inverted order)
>
> The machinist has been instructed to start the chuck on the lathe spindle by hand, not power. (Natural order)

To the subject and verb, generally required units of the sentence, you can add direct objects (Pattern 2), direct objects and objective complements (Pattern 3), indirect objects and direct objects (Pattern 4), or subjective complements (Pattern 5); each of these then becomes a unit of a basic sentence pattern.

EXPANDING THE SENTENCE

The basic technique for expansion of the English sentence patterns is the addition of modifiers: words, phrases, or clauses that function as adjectives or adverbs. The sentence keeps its basic units (S + V, S + V + DO, etc.) and adds meaning through the adjective and/or adverb units. Another technique for expansion is compounding.

Adjective Modifiers. The adjective modifier may be a word, phrase, or clause; it will always modify words that we will classify as nouns or pronouns, for want of better terms.

EXAMPLE

The machine broke. (Basic sentence)

SENTENCE EXPANDED BY ADJECTIVE MODIFIERS

> (Word)
> The *old* milling machine broke.

> (Prepositional phrase)
> The milling machine *in the machine shop* broke.

> (Participial phrase)
> The milling machine *used in Shop 141* broke.

> (Infinitive phrase)
> The milling machine *to be used in the shop demonstration* broke.

> (Clause)
> The milling machine *that was most recently assembled* broke.

Notice that every adjective modifier in the preceding examples tells which machine. The adjective modifier, whether word, phrase, or clause, answers one of three questions: "Which?" "What kind of?" "How many?" Study the sentences that follow.

EXAMPLES

> (Word)
> The *broken* equipment was repaired.
> (*Which* equipment? The *broken* equipment)

(Word)   (Prepositional phrase)
Heavy blocks *of copper* are used to make ammeter shunts.
   (*What kind of* blocks? *Heavy* blocks *of copper*)

(Word)   (Clause)
Metal pieces *that are triangularly shaped* should be used.
   (*What kind of* pieces? *Metal* pieces *that are triangularly shaped*)

                                      (Word)
Trainees in aircraft mechanics have overhauled *six* engines.
   (*How many* engines? *Six*)

                   (Word)   (Word)
Job Placement receives *numerous* calls about *employment* opportunities.
   (*How many* calls? *Numerous* calls. *What kind of* opportunities? *Employment* opportunities)

Adjective modifiers most often come just before or just after the word modified.

Adverb Modifiers. The adverb modifier may also be a word, a phrase, or a clause, but it is more movable than the adjective modifier. Although the adverb modifier may be moved to almost any point within its sentence, the adjective modifier can be placed only before or after (usually before) the word it relates to.

EXAMPLES

Helium is a *light, inert,* and *colorless* gas. (Adjective modifiers before "gas")
Helium is a gas, *light, inert,* and *colorless.* (Adjective modifiers after "gas")

Now look at examples showing the movability of the adverb modifier.

Initially, the project seemed worthless.
The project, *initially,* seemed worthless.
The project seemed, *initially,* worthless.
The project seemed worthless, *initially*.

Notice that this one-word adverb, "initially," may be placed in common adverb positions: at the beginning of the sentence, before the verb, after the verb, or at the end of the sentence.

Moving the adverb shifts the emphasis of the sentence and sometimes changes the meaning, as in the following:

Only the supervisors received the $100 per month raise.
The supervisors *only* received the $100 per month raise.
The supervisors received *only* the $100 per month raise.
The supervisors received the *only* $100 per month raise.

See also Squinting Modifiers, pages 704–705.

Adverb modifiers (words, phrases, or clauses) answer: "When?" (time), "Where?" (place), "Why?" (reason), "How?" (manner), "How much?" (degree).

EXAMPLES

In machine shop work (prepositional phrase: answers *When?* or *Where?*), workers read decimals *in thousandths of an inch* (prepositional phrase: answers *How?*).

When they are not in use (clause: answers *When?*), calipers should be hung *in the tool closet* (prepositional phrase: answers *Where?*).

To ensure safety (infinitive phrase: answers *Why?*), arc welders need to wear safety glasses *under their hoods* (prepositional phrase: answers *Where?*).

Compounding. Compounding is a technique for expanding sentences in which the writer doubles a sentence unit (word, phrase, or clause), a complete sentence pattern, or both. Required in this process of compounding are words called coordinating conjunctions. In the English language these conjunctions are *and, so, for, but, or, nor,* and *yet.* Also, *both . . . and, not only . . . but also, either . . . or,* and *neither . . . nor* are coordinating conjunctions used in pairs and called correlatives. The conjunction joins, links, or connects units of the same value, such as two or more subjects, two or more adjective clauses, two or more prepositional phrases, two or more sentences, and so forth.

EXAMPLES

Professors (and) *business executives* have recognized the need for a thorough study of the management process. (Compound subject)

The *roughness* (or) *smoothness* of a material causes it to *absorb* (or) *reflect light.* (Compound subject) (Compound infinitive phrase)

The horizontal line naturally suggests *repose, steadiness,* (and) *duration.* (Compound direct object)

The analysis of a product deals with the question *of size, of capacity,* (and) *of product output.* (Compound prepositional phrase)

Because the corners of the nut may be rounded or damaged (and) *because the wrench may slip off the nut and cause an accident,* it is important that a wrench be of a size and type suited to the nut. (Compound adverb clause)

The lathe *that was most recently assembled* (and) *that was used in the class demonstration yesterday* has broken down. (Compound adjective clause)

Apprehending a suspect, taking him into custody, (and) *questioning him,* the police officer was careful to inform the suspect of his rights. (Compound participial phrase)

(Neither) *the steel rule, the rule depth gauge,* (nor) *the slide caliper rule* is classified as a precision measuring instrument. (Compound subject)

Dumping garbage out to sea is an economical means of disposal, (but) *other factors outweigh the advantages in many instances.* (Compound sentence)

The most commonly used coordinating conjunction is *and.*

Emphasis in Sentences

In sentence construction, important parts of the sentence should stand out and less important parts should be subordinate. In the English sentence, word order controls emphasis.

1. Put material to be emphasized either at the beginning or at the end of the sentence.

EXAMPLES

The *tape rule* is a common measuring device in the machine shop.

A common measuring device in the machine shop is the *tape rule.*

In the machine shop a common measuring device is the *tape rule.*

2. Emphasize important details by putting them in independent clauses and subordinate the less important details in dependent clauses and phrases.

EXAMPLES

The dimensions on a blueprint, which are given in terms of a fraction, *are called scale dimensions*. (Dependent clause)

Note: A dependent clause or phrase that gives additional information not essential to the basic sentence meaning is set off by commas. (See Nonrestrictive Modifiers, page 727.)

The generator acts like an electron pump, causing electrons to flow through the circuit provided by the wire in the direction indicated by the arrow. (Participial phrase)
One method for treating posts, called double diffusion, *is fairly inexpensive and can be done on the farm*. (Participial phrase)
The roping type of saddletree, as the name implies, *is used by professional ropers*. (Dependent clause)
The vernier caliper, one of the machinist's most versatile precision instruments, *is used to take either inside or outside measurements accurate to 0.001 inch*. (Phrase used as appositive)

3. Invert the normal order of the sentence to gain emphasis.

EXAMPLES

Unexpected was the reaction to the new plant rules.
Fragile and brittle is the tap tool.

4. Add emphasis by placing the adjective after the noun modified instead of before it.

EXAMPLES

Helium is a gas, *light, inert,* and *colorless.*
The drill press, *dangerous when not properly operated,* is commonly employed for drilling holes.
The personnel manager, *tired and hungry,* interviewed the final applicant.
The tap tool, *fragile and brittle,* is easily broken.

5. Add emphasis by using verbs in the active voice.

EXAMPLES

The Greeks first *noticed* that lodestones had a peculiar and invisible quality they named magnetism.
The machinist *values* highly the rule depth gauge.
The column *supports* the head and the table of the drill press.
Use of an iron centerpiece or core greatly *increases* the magnetizing effect of a wire loop.
The resistance of most conductors *increases* if the temperature of the conductor increases.
To convert from Fahrenheit to Celsius, *subtract* 32 from the Fahrenheit value and *multiply* the remainder by 5/9.

Using active voice verbs creates an effectiveness in writing because the reader knows immediately the subject of the discussion; the writer mentions who or what is being talked about first. Passive voice verbs, however, sometimes may be used more emphatically. (The passive verb includes a form of *be* plus the past participle of the main verb.)

a. When the who or what is not as significant as the action or the result, use passive voice.

EXAMPLE

The lathe *has been broken* again. (More emphatic than "Someone has broken the lathe again.")

b. When the who or what is unknown, preferably unnamed, or relatively insignificant, use passive voice.

EXAMPLES

The blood sugar test *was made* yesterday.
The transplant operation *was performed* by an outstanding heart surgeon.
Light *is provided* for technical drawing classrooms by windows in the north wall.

For further discussion of active and passive voice, see pages 12–13, 698–699.

Combining Ideas

The preceding pages have illustrated how adjective modifiers, adverb modifiers, and compounding allow the writer to expand sentence meaning. These techniques also offer the writer other advantages; they allow the writer to economize on words and, at the same time, make clearer the relationship of words. For example, the two sentences

1. Mr. Jones built a patio.
2. It is made of brick.

could be written as a single sentence

1. Mr. Jones built a brick patio.

By combining the two sentences, the writer eliminates "It is made of," and by placing the adjective modifier *brick* before the noun *patio*, the relationship of the adjective *(brick)* and the word modified *(patio)* is clearer.

The three sentences

1. Edmund Fonche developed the first oxyacetylene torch.
2. It was developed in 1900.
3. He was a Frenchman.

could be written as a single sentence:

1. Edmund Fonche, a Frenchman, developed the first oxyacetylene torch in 1900.

The words "It was developed" and "He was" are eliminated; the appositive, "a Frenchman," is placed after "Edmund Fonche," and the adverb phrase, "in 1900," is placed at the end of the sentence.

The ideas in sentences may be combined to make meaning clearer by adding a word, usually a subordinate conjunction or a coordinate conjunction, to show the relationship between the ideas.

From the two sentences

1. The company did not hire him.
2. He was not qualified.

a single sentence

1. The company did not hire him *because* he was not qualified.

makes the relationship of ideas clearer with the addition of the subordinate conjunction "because." The second sentence is changed to an adverb clause, "because he was not qualified," telling why the company did not hire him.

The two sentences

1. John applied for the job.
2. He did not get the job.

might be stated more clearly in

1. John applied for the job, but he did not get it.

The addition of the coordinate conjunction "but" indicates that the idea following is in contrast to the idea preceding it.

Study the following examples.

1. Magnetic lines of force can pass through any material.
2. They pass more readily through magnetic materials.
3. Some magnetic materials are iron, cobalt, and nickel.

The ideas in these three sentences might be combined into a single sentence.

1. Magnetic lines of force can pass through any material, but they pass more readily through magnetic materials such as iron, cobalt, and nickel.

The addition of the coordinate conjunction "but" indicates the contrasting relationship between the two main ideas; sentence 3 has been reduced to "such as iron, cobalt, and nickel." The three sentences might also be combined as follows:

1. Although magnetic lines of force can pass through any material, they pass more readily through magnetic materials: iron, cobalt, and nickel.

Almost any group of ideas can be combined in several ways. The writer chooses the arrangement of ideas that best "fits in with" preceding and following sentences.

The following ideas are combined into a single sentence through the use of all three techniques: adjective modifiers, adverb modifiers, and compounding.

1. A manager cannot eliminate grapevines, even if she tries.
2. She should make good use of the inevitable.
3. She can learn what the grapevine is saying.
4. She can scotch false rumors.
5. She can spread information informally.

Ideas combined:

1. Since a manager cannot eliminate grapevines, even if she tries, she should make good use of the inevitable by learning what the grapevine is saying, by scotching false rumors, and by spreading information informally.

Sentence 1 becomes an adverb clause with the addition of the subordinate conjunction "since"; sentence 2 remains an independent clause. Sentence 3 becomes a series of adverb modifiers in the form of prepositional phrases—"by learning what the grapevine is saying," "by scotching false rumors," and "by spreading information informally"—joined by the coordinate conjunction, "and."

Ideas may also be combined effectively into two or more sentences. The ideas expressed in the following five sentences

1. The mission's most important decision came.
2. It was early on December 24.
3. Apollo was approaching the moon.
4. Should the spacecraft simply circle the moon and head back toward earth?
5. Should it fire the Service Propulsion System engine and place the craft in orbit?

might be combined:

1. As Apollo was approaching the moon early on December 24, the mission's most important decision came.
2. Should the spacecraft simply circle the moon and head back toward earth or should it fire the Service Propulsion System engine and place the craft in orbit?

In the following example, ideas in the five sentences

1. More than 90 percent of the electrical energy generated today is generated as alternating current rather than direct current.
2. There are several reasons for generating energy as alternating current.
3. Alternating current may be generated in large quantities more cheaply than direct current.
4. Alternating current may be generated and transmitted at high voltages.
5. These high voltages can be reduced efficiently at the receiving end of the transmission line to voltages suitable for lights or motors with small loss.

might be combined into three sentences:

1. More than 90 percent of the electrical energy generated today is generated as alternating current rather than direct current.
2. One reason is that alternating current can be generated in large quantities more cheaply.
3. It may be generated and transmitted at high voltages that can be reduced efficiently at the receiving end of the transmission line, with small loss, to voltages suitable for lights or motors.

See also pages 11–12 and 349 for discussion of conciseness.

Common Problems in Writing Sentences

Common problems in writing sentences are sentence fragments, parallelism in sentences, run-on or fused sentences, comma splices, and shift in focus. Following is a discussion of these problems. See also Problems of Usage, pages 700–705.

SENTENCE FRAGMENTS

A group of words containing a subject and a verb and standing alone as an independent group of words is a sentence. If a group of words lacks a subject or a verb or cannot stand alone as an independent group of words, it is called a sentence fragment. Sentence fragments generally occur because:

■ A noun (subject) followed by a dependent clause or phrase is written as a sentence. The omitted unit is the verb.

FRAGMENT   The engineer's scale, which is graduated in the decimal system.
FRAGMENT   Colors that are opposite.

To make the sentence fragment into an acceptable sentence, add the verb and any modifiers needed to complete the meaning.

SENTENCE   The engineer's scale, which is graduated in the decimal system, is often called the decimal scale.
SENTENCE   Colors that are opposite on the color circle are used in a complementary color scheme.

■ Dependent clauses or phrases are written as complete sentences. These are introductory clauses and phrases that require an independent clause.

FRAGMENT   While some pencil tracings are made from a drawing placed underneath the tracing paper.
FRAGMENT   Another development that has promoted the recognition of management.

To make the fragments into sentences, add the independent clause.

SENTENCE   While some pencil tracings are made from a drawing placed underneath the tracing paper, most drawings today are made directly on pencil tracing paper, cloth, vellum, or bond paper.
SENTENCE   Another development that has promoted the recognition of management is the separation of ownership and management.

Acceptable Fragments. There are occasions when types of fragments are acceptable.

Emphasis

Open the window. Right now.

Transition between ideas in a paragraph or composition

Now to the next point to be discussed.

Dialogue

"How many different meals have been planned for next month?" "About ten."

PARALLELISM IN SENTENCES

Parallel structure involves getting like ideas into like constructions. A coordinate conjunction, for example, joins ideas that must be stated in the same grammatical form. Other examples: an adjective should be parallel with an adjective, a verb with a verb, an adverb clause with an adverb clause, and an infinitive phrase with an infinitive phrase. Parallel structure in grammar helps to make parallel meaning clear.

A *typewriter,* a *table,* and a *filing cabinet* were delivered today. (Nouns in parallel structure)

Whether you accept the outcome of the experiment or *whether I accept it* depends on our individual interpretation of the facts. (Dependent clauses in parallel structure)

Failure to express each of the ideas in the same grammatical form results in faulty parallelism.

FAULTY   This study should help the new spouse *learn* skills and *to be knowledgeable* about sewing and cooking. (Verb and infinitive phrase)

REVISED   This study should help the new spouse *learn* skills and *become* knowledgeable about sewing and cooking. (Verbs in parallel structure)

FAULTY   Three qualities of tungsten steel alloys are *strength, ductility,* and *they have to be tough.* (Noun, noun, and independent clause)

REVISED   Three qualities of tungsten steel alloys are *strength, ductility,* and *toughness.* (Nouns in parallel structure)

FAULTY   The best lighting in a study room can be obtained *if windows are placed in the north side, if tables are placed so that the light comes over the student's left shoulder,* and *by painting the ceiling a very light color.* (Dependent clause, dependent clause, phrase)

REVISED   The best lighting in a study room can be obtained *if windows are placed in the north side, if tables are placed so that the light comes over the student's left shoulder,* and *if the ceiling is painted a very light color.* (Dependent clauses in parallel structure)

See also, in Part I, Chapter 4, Analysis Through Classification and Partition, especially Coordination, page 126.

RUN-ON OR FUSED SENTENCE

The run-on or fused sentence occurs when two sentences are written with no punctuation to separate them. (See also Comma Splice below.)

The evaluation of the Automated Drafting System is encouraging we expect to operate the automated system in an efficient and profitable manner.

Obviously the sentence above needs some punctuation to make it understandable. Several methods could be used to make it into an acceptable form:

a. Period between "encouraging" and "we"

The evaluation of the Automated Drafting System is encouraging. We expect to operate the automated system in an efficient and profitable manner.

b. Semicolon between "encouraging" and "we"

The evaluation of the Automated Drafting System is encouraging; we expect to operate the automated system in an efficient and profitable manner.

c. Comma plus "and" between "encouraging" and "we"

The evaluation of the Automated Drafting System is encouraging, and we expect to operate the automated system in an efficient and profitable manner.

d. Recasting the sentence

> Since the evaluation of the Automated Drafting System is encouraging, we expect to operate the automated system in an efficient and profitable manner.

See also Punctuation, especially usage of the period, the comma, and the semi-colon, pages 725, 726–727, 730–731.

COMMA SPLICE

The comma splice occurs when a comma is used to connect, or splice together, two sentences. The comma splice, a common error in sentence punctuation, is also called the comma fault. (See also Run-on or Fused Sentence above.)

> Mahogany is a tropical American timber tree, its wood turns reddish brown at maturity.

The comma splice in the above sentence may be corrected by several methods:

a. Replacing the comma with a period

> Mahogany is a tropical American timber tree. Its wood turns reddish brown at maturity.

b. Replacing the comma with a semicolon

> Mahogany is a tropical American timber tree; its wood turns reddish brown at maturity.

c. Adding a coordinate conjunction, such as "and"

> Mahogany is a tropical American timber tree, and its wood turns reddish brown at maturity.

d. Recasting the sentence

> Mahogany is a tropical American timber tree whose wood turns reddish brown at maturity.

SHIFT IN FOCUS

The writer may begin a sentence that expresses one thought but, somehow, by the end of the sentence has unintentionally shifted the focus. Consider this sentence:

> There are several preparatory steps in spray painting a house, which is easy to mess up if you are not careful.

The focus in the first half of the sentence is on preparatory steps in spray painting a house; the focus in the second half is on how easy it is to mess up when spray painting a house. Somehow, in the middle of the sentence, the writer shifted the focus of the sentence. The result is a sentence that is confusing to the reader.

To correct such a sentence, the writer should think through the intended purpose and emphasis of the sentence and should review preceding and subsequent sentences. The poorly written sentence above might be divided into two sentences:

> There are several preparatory steps in spray painting a house. Failure to follow these steps may result in a messed-up paint job.

or might be recast as one sentence:

> Following several preparatory steps in spray painting a house will help to keep you from messing up.

or

> To help assure satisfactory results when spray painting a house, follow these preparatory steps.

Application 1 Writing Sentences

After studying the preceding pages on basic sentence patterns and thinking about your own sentence-pattern habits, write at least five original sentences to illustrate each of the five basic patterns. Try to use material from your major field as subject matter for the sentences.

Pattern 1: S + V

1.

2.

3.

4.

5.

Pattern 2: S + V + DO

1.

2.

3.

4.

5.

Pattern 3: S + V + DO + OC

1.

2.

3.

4.

5.

Pattern 4: S + V + IO + DO

1.

2.

3.

4.

5.

Pattern 5: S + V (Linking) + SC

1.

2.

3.

4.

5.

Application 2 Writing Sentences

1. Using material from your major field, write five sentences, each containing at least one adjective modifier. Try to include a sentence containing a one-word adjective modifier, a sentence containing an adjective phrase, and a sentence containing an adjective clause.
2. Write five sentences containing adverb modifiers, attempting to write these sentences just as you were instructed to write the sentences containing adjective modifiers.
3. Write five sentences containing compound units. Try to write each sentence so that a different unit is compounded.

Application 3 Writing Sentences

Each of the following groups of sentences contains related information stated in simple sentences. Take this information and combine it into one sentence or, if you think necessary, two or more sentences. Be careful to punctuate each sentence in an acceptable way.

GROUP A SENTENCES

1. **a.** All drawings are projections.
   b. Two main types of projections are perspective and parallel projection.
2. **a.** Some lathes are built with two lead screws.
   b. One is used for thread cutting.
   c. The other is used for general lathe-turning operations.
3. **a.** Btu is the abbreviation for British thermal unit.
   b. It is a unit of heat equal to 252 calories.
   c. It is the quantity of heat required to raise the temperature of one pound of water from 62°F to 63°F.
4. **a.** Machining is shaping by the removal of excess metal.
   b. It includes many operations.
   c. Some of these operations are grinding, sawing, drilling, and milling.
5. **a.** Fusion welding involves three processes.
   b. One of these processes is casting.
   c. Another process is heat treating.
   d. A third process is dilution.
6. **a.** Management is a social process.
   b. It is a process because it comprises a series of actions that leads to the accomplishment of objectives.
   c. It is a social process because these actions are principally concerned with relations between people.
7. **a.** Drafting students need special equipment.
   b. They need a drawing board.
   c. They need a T square.
   d. They need a set of drawing tools.
8. **a.** The operating tasks each company must perform are affected by many factors.
   b. One factor is changes in market.
   c. Another factor is new technology.
   d. A third factor is shifts in competition.
   e. A final factor is government regulations.
9. **a.** One type of mass production is manufacturing parts in large quantities by specially designed machine tools.
   b. Machine tools have helped to change our economy from principally agricultural to industrial.
10. **a.** The vernier height gauge is a measuring tool.
    b. It is used for measuring heights.
    c. It is used for laying out height dimensions for work resting on a surface plate.
    d. It is used for locating centers for holes that are to be drilled or bored.

GROUP B SENTENCES

1. **a.** In interior decoration straight lines are considered intellectual, not emotional.
   b. They are considered classic, not romantic.
   c. They are sometimes considered severe and masculine.
2. **a.** The three basic circuits are series, parallel, and series-parallel.

      b. They are distinguished by the way the electrical equipment is connected.

      c. Their names describe their connection.

3. a. Water from newly constructed wells will normally have a very high bacterial count.

      b. This count will not "settle down" for a matter of weeks or even months unless the supply is treated.

      c. This treatment reduces the bacterial population that found its way into the water during construction operations.

4. a. The Edison storage battery is the only nickel-iron-alkaline type on the American market.

      b. It was developed by Thomas A. Edison.

      c. It is built in many sizes and for many uses.

      d. It was put on the market in 1908.

5. a. Tool designers are specialists.

      b. One duty of the tool designer is to design drill jigs, reaming and tapping jigs, etc.

      c. Another duty is to choose correct cutting tool material.

      d. Sometimes she must establish cutting speeds, feeds, and depths of cut.

6. a. The engineer's scale is often referred to by other terms.

      b. It is called the decimal scale because the scale is graduated in the decimal system.

      c. It is called the civil engineer's scale because it was originally used in civil engineering.

      d. It is called the chain scale because it was derived from the surveyor's chain of 100 links.

7. a. Micrometer calipers are measuring instruments.

      b. They are designed to use the decimal divisions of the inch.

      c. They are used to take very precise measurements.

      d. They are made in different styles and sizes.

8. a. One essential part of an air damper box is a vane.

      b. The vane is made of a thin sheet of lightweight alloy.

      c. It is stiffened by ribs stamped into it.

      d. It is stiffened by the bending of its edges.

      e. Its edges conform to the surfaces of the damping chamber.

9. a. Soil may contain unusually high numbers of *Clostridium tetani*.

      b. This is the etiological agent in lockjaw.

      c. Soil that has been fertilized with manure, particularly horse manure, has high numbers of this agent.

10. a. Every time a manager delegates work to a subordinate, three actions are either expressed or implied.

      b. She assigns duties.

      c. She grants authority.

      d. She creates an obligation.

GROUP C SENTENCES

1. a. Landscape architects employ texture as a valuable tool.

      b. Repetition of dominant plant texture unifies a plan.

      c. Contrast of texture at corners and focal points gives emphasis.

2. a. There are three basic factors in an electrical circuit.

    b. These are pressure, current, and resistance.

    c. To be usable, these factors must be defined in terms of units that can be handled in measurements.

3. **a.** Any material employed to grow bacteria is called a culture medium.

    b. This medium, to be satisfactory, must have the proper moisture content.

    c. It must contain readily available food materials.

    d. It must have the correct acid–base balance.

4. **a.** Metallurgy is the science of metals.

    b. Metallurgy has two divisions: extractive and physical.

    c. Extractive metallurgy is the study of extracting metals from their ores.

    d. Physical metallurgy is the study of the properties of metals.

5. **a.** Of all the microbiologists none was so accurate as Leeuwenhoek.

    b. None was so completely honest.

    c. None had such common sense.

    d. He died in 1723.

    e. He was 91.

    f. Microbiology went into a dormant stage for almost 150 years.

6. **a.** In ammeter shunts, the shunt is usually made of a manganin strip.

    b. This strip is brazed in fairly heavy copper blocks.

    c. These heavy copper blocks serve two purposes.

    d. They carry heat away from the manganin strip.

    e. They keep all parts of each copper block at nearly the same potential.

7. **a.** Gaspard Mongé advanced the theory of projection by introducing two planes of projection at right angles.

    b. He was French.

    c. He was a mathematician.

    d. This development provides the basis of descriptive geometry.

8. **a.** Flat surfaces are measured with common tools.

    b. The steel rule and the depth gauge are common tools.

    c. Round work is measured by feel with nonprecision tools.

    d. The spring caliper and the firm-joint caliper are nonprecision tools.

    e. They have contact points or surfaces.

9. **a.** Ohm's law states that the current in an electrical circuit varies directly as the voltage and inversely as the resistance.

    b. This law was developed by George Simon Ohm.

    c. He was born in Germany.

    d. He was born in 1787.

    e. He was a physicist.

10. **a.** Vitruvius wrote a treatise on architecture.

    b. In the treatise he referred to projection drawings for structures.

    c. He wrote the treatise in 30 B.C.

    d. The theory of projections was well developed by Italian architects.

    e. The Italian architects were Brunelleschi, Alberti, and others.

    f. They developed the theory in the early part of the fifteenth century.

Application 4 Writing Sentences

Examine the pieces of writing you have done in all of your courses this term.

1. Make a list of the sentences indicated by the instructor (or now discovered by yourself) that require(d) revision.
2. For each sentence, explain why you originally wrote the sentence as you did.
3. Explain what changes are needed to make the sentences clear.
4. Rewrite the sentences.

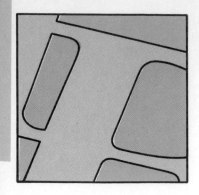

Chapter 2
The Paragraph

Objectives

Upon completing this chapter, the student should be able to:

- Select a subject and limit it so that it can be developed in a paragraph
- Write a topic statement
- Explain and illustrate the various methods of development and various orders in paragraph development
- Explain the function of and identify transitional words and phrases in a paragraph
- Write and revise a paragraph of development
- Write a paragraph of introduction, of transition, and of closing

Introduction

Page after page of continuous print or handwriting, each line beginning at the same left margin, creates problems for most readers and writers. A reader may look at such a page and think, "I don't want to read *that*." "How will I know where divisions of content occur?" "What parts of the overall content are to be emphasized?" "Where can I pause to be sure I understand what I've read before reading further?"

Writers who write line after line with no indentations may lose sight of the central idea and ramble away from their point. They cannot quickly review the major points already developed by looking at paragraph beginnings.

Paragraphing is the method used by writers to divide material into units (paragraphs) to make writing more clearly understood and more readable and to make the writing process easier.

A paragraph may be a single sentence, but more often it is a group of major and minor support sentences combined to develop some central idea.

Planning the Paragraph

Before putting pen or pencil to paper, any writer should do some thinking and planning. Just as a contractor needs a blueprint before beginning construction or a coach and a team need a plan before beginning a game, a writer, whether experienced or inexperienced, needs a plan before beginning to write. This plan includes selecting a subject, limiting the subject, making a topic statement, adding details, and deciding on transition, order, and method of development.

SELECTING A SUBJECT

The first step in planning a paragraph is selecting a subject. If an instructor assigns a subject, you may bypass this step; if not, you must make this decision. Try to choose a subject that interests you, that you know something about or would like to find out about through investigation and research, and that you can tell about in some detail. Logically, a student planning a future in engineering is interested in engineering and related fields, has some knowledge about the field, and could tell about areas in the field in some detail. This student would find research on engineering profitable. On the other hand, such a student might have little interest, if any, in a subject such as hotel-motel restaurant management.

Most of the subjects that you might think of immediately would probably be too broad to write about specifically. Therefore you may want to limit or restrict the subject to a manageable topic.

LIMITING THE SUBJECT

Limiting or restricting a subject to a manageable topic allows you to write about the topic in specific, concrete details. For example, you could not adequately cover the subject "books" in one paragraph. The subject is too broad. You could limit "books" to a subdivision such as "books in libraries," but this is still too broad. Better, you could limit the subject to the topic "systems of classifying books." An investigation of material to be covered shows that there are only two systems for classifying books. Thus, this topic can be adequately treated in a paragraph.

The subject "books," of course, could have been limited in any number of ways and in any number of steps:

Books > books in libraries > library systems of classifying books

Books > publication of books > copyright laws > how to copyright a book

Books > books by American writers > books by contemporary American writers > books by Ralph Ellison > Ellison's *Invisible Man* > implications of the name "Trueblood" in Ellison's *Invisible Man*

For almost any subject there are many, many possible topics for writing. Limiting a paragraph by subject, as described above, is usually only the first step. The limiting process may be continued through limitation by attitude, time, space, or a combination of these.

(Subject)   (Attitude)   (Time)
A *wound* should be cleansed *gently* but *thoroughly before stitches are taken*.

Almost any single idea sentence is limited by subject and attitude. The subject identifies what will be written about and the attitude pinpoints the writer's point of view, or feelings, toward the subject.

The important thing to remember in narrowing a subject for a paragraph is that the topic finally decided upon must be so limited that it can be satisfactorily developed within *one* paragraph. The term *satisfactorily developed* is relative, of course, depending on the purpose of the paragraph, the intended audience, and the writer's knowledge of the topic.

FORMING A TOPIC STATEMENT

In the preceding step, the subject "books" was limited to library systems of classifying books; the limited idea can now be stated in a sentence:

Libraries classify books according to one of two systems: the Library of Congress Classification or the Dewey Decimal Classification.

This sentence is the *topic statement,* the central idea to be developed in the paragraph. The topic statement is sometimes expressed in a phrase within a longer sentence; more frequently it appears in a main clause, with additional information. But most often—and usually most effectively—the topic statement is a short, simple sentence.

Two Parts to the Topic Statement. Every topic statement has these two parts:

- ☐ Introduction of the subject of the paragraph
- ☐ Specific indication of the kind of information that will follow

Examine, for example, the topic statement *Libraries classify books according to one of two systems: the Library of Congress Classification or the Dewey Decimal Classification.* The subject of the paragraph is *Libraries classify books.* That is the overriding concern of every sentence in the paragraph.

But the subject of the paragraph, *Libraries classify books,* is lacking—unless the writer gives specific indication of the kind of information that will follow. It is a statement, yes; but by itself, it is not a usable topic statement. There is no clue, no directional signal, no indication of what information will be given about libraries classifying books. Will the information that follows be concerned with the random shelving of books in some homes, with the problems in classifying nonprint materials, with difficulties that libraries have in storing periodicals? The subject *Libraries classify books* does not indicate the kind of information that will follow.

Now add the remainder of the sentence. The statement then is *Libraries classify books according to one of two systems: the Library of Congress Classification or the Dewey Decimal Classification.* There is no question now of what kind of specific information will follow in the paragraph. The information will concern the two classification systems: the Library of Congress and the Dewey Decimal.

Everything in the paragraph must discuss these two systems by which libraries classify books. The writer could not discuss how to check out a book or the arrangement of books in a personal library.

Following are other sample topic statements. The specific indication of the kind of information that would follow in the paragraph is italicized.

Maintaining a healthy physical condition offers *personal advantages.*
A successful marriage results from a *proper mix of several qualities.*
Three major mistakes have hampered rehabilitation of disabled veterans.
To change the air filter on a home central air conditioning unit, *follow these four basic steps.*
The plots of the various TV soap operas have many *similarities.*
The automobile needs *better safety features.*

Notice that each topic statement requires details to support or develop each statement. These added details fill out the statement and create a paragraph.

If a proposed topic statement requires no details to make it clear and understood, it is not a usable topic statement. For example, the statement "I am an animal technician" is not a usable topic statement because it requires no details; it is a statement of fact, complete in itself. It introduces the subject of the paragraph, being an animal technician, but it does not indicate the kind of information to be given in the paragraph. On the other hand, the statement "I like being an animal technician" could logically be developed: It introduces the subject of the paragraph and it indicates that the kind of information to follow will be the reasons for the preference.

Topic Statement as the Thesis or Purpose Statement for Entire Paper.

The topic statement announces the central idea to be developed in the paragraph. It lets the reader know what to expect. The topic statement is similar to the thesis statement or purpose statement for an entire paper. The thesis statement or purpose statement lets the reader know what to expect in the paper.

EXAMPLES OF THESIS STATEMENTS (also called key statement, controlling idea, central idea):

Second-generation personal computers have more capability, are easier to use, and are less expensive than first-generation personal computers.

Replacing a component in a PC board requires four steps.

Soaring costs for hospital care are tied to high-technology advances in medicine.

EXAMPLES OF PURPOSE STATEMENTS:

The purpose of this paper is to explore teenage emotional and physical conditions that may lead to anorexia nervosa.

This report discusses the feasibility of three possible locations for the proposed branch bank.

In this report I will evaluate current research concerning no-till planting of soybeans.

Placement of the Topic Statement.

Although the topic statement may be placed at several different positions in a paragraph, most often it appears at the beginning. As the paragraph develops, frequent rereading of the topic statement will help the writer stick to the limited topic and avoid wandering into a discussion of related ideas that have no place within the paragraph being developed. The paragraph must include only material related to the central idea given in the topic statement. In the following paragraph, the central idea is in the opening sentence:

Certain areas of the world are particularly exposed to minor or major disasters for geologic and geographic reasons. The potential earthquake disaster zones are Central and Southeast Asia, including Japan, the entire West Coast of the North American Continent, and the Mediterranean region, as well as Asia Minor. These areas also include more than 500 volcanoes, which are still active today and whose eruptions in some instances have led to serious disasters. Tropical storms, too, cause serious disasters in specific areas: Typhoons are predominant in the Philippines, Formosa, Japan, and the coastal regions of China and Indochina; cyclones occur primarily in the Bay of Bengal; and hurricanes strike mainly in the West Indies and along the coastal areas of North and Central America. Flood disasters occur in almost every country of the world. Generally, however, they are a serious threat to human life only in densely populated plains of large rivers.

The skillful writer often moves the topic statement from the beginning to other positions within the paragraph—for example, to the end:

The bones of dogs have been found in numerous campsites of the New Stone Age, dating back more than 10,000 years. The domestication of dogs, then, probably began many thousands of years earlier than that. From these campsite remains, at least five different kinds of dogs comparable to modern dogs have been identified. *Thus, one can assume that the dog, undoubtedly used in hunting, was one of the first animals domesticated by humans.*

Sometimes the skillful writer might omit the topic statement and imply rather than state the main idea. However, having read a paragraph with an implied topic statement, the reader should be able to state it without doubt. The following paragraph has no stated topic sentence, yet a reader knows immediately that the idea developed is the procedure for using bondo:

> In using bondo to smooth a dent in a car body, the body and fender repairman first washes the car carefully, making sure the dented area is clean. He then straightens the dented area to the desired position and, using a disk sander, removes all the old paint so the bondo can be applied against the bare metal. The repairman next applies the bondo in upward strokes with a plastic spreader, taking care to spread it as evenly as possible. After allowing time for the bondo to dry until it can be worked properly, he smooths the rough places even with the car body, using a rough bondo file, and sands the area to a smooth finish, using a 400 sander. Where there was once a dented area in the car body, there is now a smooth, shiny finish.

So, while the beginning and the end of a paragraph are common positions for the topic statement, it is not unusual to find a topic statement implied. The beginning writer, however, might find it helpful to place the topic statement at the beginning of the paragraph.

ADDING DETAILS

Once a subject has been limited and the central idea stated in a topic statement, the next step is to think of possible details that might develop the topic statement. Details *support* general statements and help make them more specific.

It is quite helpful to jot down on scratch paper possible supporting details and then select from these the ones to include in the paragraph.

In adding details, consider what the topic statement asks for; especially look at the key word(s) or phrase in the statement. The key words in *"Libraries classify books* according to *one of two systems:* the Library of Congress Classification or the Dewey Decimal Classification" give clues to possible details for supporting the idea.

Here is a topic statement with a list of possible supporting details.

TOPIC STATEMENT   Libraries classify books according to one of two systems: the Library of Congress Classification or the Dewey Decimal Classification.

POSSIBLE DETAILS
1. Library of Congress (LC) system has 20 basic groups.
2. List of basic groups of LC.
3. Dewey Decimal Classification (DD) has ten basic groups.
4. List of basic DD groups.
5. Reasons a library chooses one system over another.
6. LC uses letters and numbers.
7. LC letters and numbers used to form many subdivisions.
8. Origin of the LC system.

9. Origin of the DD system: Melvil Dewey thought of the idea while sitting in church.
10. DD uses numbers.
11. The DD numbers are divided to form many subdivisions.
12. Most libraries use the DD system.
13. Smaller libraries use the DD.
14. In LC, six unused letters—for future classifications.
15. Book *Lasers and Their Applications* in LC: TK7872.L3S7.
16. Same book in DD: 621.329.
17. Subdivisions in DD designated by numeral-decimal combination.

Once all details that come to mind are jotted down, analyze the topic statement carefully to determine the kind of detail that can best be used to develop the main idea. Then carefully consider your list of details, modifying and combining where necessary, and eliminate those that do not seem valuable.

A check of the preceding list, for example, might eliminate those details crossed out below.

TOPIC STATEMENT   Libraries classify books according to one of two systems: the Library of Congress Classification or the Dewey Decimal Classification.

POSSIBLE DETAILS
1. Library of Congress (LC) system has 20 basic groups.
2. ~~Listing of basic groups of LC.~~
3. Dewey Decimal Classification (DD) has ten basic groups.
4. ~~List of basic DD groups.~~
5. ~~Reasons a library chooses one system over another.~~
6. LC uses letters and numbers.
7. LC letters and numbers used to form many subdivisions.
8. ~~Origin of the LC system.~~
9. ~~Origin of the DD system. Melvil Dewey thought of the idea while sitting in church.~~
10. DD uses numbers.
11. The DD numbers are divided to form many subdivisions.
12. Most libraries use the DD system.
13. Smaller libraries use the DD.
14. In LC, six unused letters—for future classifications.
15. Book *Lasers and Their Applications* in LC: TK7872.L3S7.
16. Same book in DD: 621.329.
17. Subdivisions in DD designated by numeral-decimal combination.

These remaining details can now be combined to develop a satisfactory paragraph. You will notice the crossed-out details are about the two systems libraries use in classifying books, but they do not help to explain the two systems, as the topic statement requires.

Developing the Paragraph

Now you are ready to plan how to put topic statement and details together. At this stage in planning, consider at least three points: method (pattern) of development, order, and transition.

To be considered before writing a paragraph is the method or pattern of development. Some common methods are fact, fact, fact; examples or illustrations; comparison (similarity) and contrast (differences); cause and effect (things happen and the results of their happening); definition (defining a term, a process, etc.); and partition (division into parts). More than likely, any one paragraph will contain a combination of these methods of development.

Fact, Fact, Fact. Probably the most common method of paragraph development is simply the compilation of facts. These facts support, substantiate, or expand the topic statement.

In the following sample paragraph, the topic statement is the first sentence. The details in the remainder of the paragraph give factual information.

> The job outlook in the medical/health area is very promising. Not just physicians and nurses are needed; technicians, aides, and specialists are much in demand. Although many of the jobs are in hospitals, about half of the jobs are in such institutions as nursing homes, doctors' offices, public health agencies, psychiatric clinics, research institutes, and neighborhood clinics. The salaries for jobs in the medical/health field vary widely. Affecting salaries are such factors as job location, type of employment, required skill and training, and the employee's overall experience.

Examples or Illustrations. A single, lengthy illustration, or several examples or illustrations, can effectively develop a topic statement into a paragraph. The following sample paragraph includes several brief examples.

> The decade between 1920 and 1930 is often referred to as sports' golden age, primarily because of the caliber of performance of many noted competitors. In baseball, for instance, there were Ty Cobb and George Herman (Babe) Ruth, who set a major league home run record of 60 in 1927. Jack Dempsey, the heavyweight boxing champion, became well known for his quick victories. Lawn tennis supplied William T. (Big Bill) Tilden, considered by many the finest all-round player in the game's history. Golf's Robert T. (Bobby) Jones, Jr., football's "Red" Grange, swimming's Johnny Weissmuller, figure skating's Sonja Henie—all are examples from the impressive list of competitors in sports' golden '20s.

This pattern might also be used in writing definitions (Chapter 3 in Part I) or in analyzing through classification or partition (Chapter 4 in Part I).

Comparison and Contrast. Comparison is a method of showing similarity between two or more people, ideas, or objects; and contrast is a method of showing the difference or differences between two or more people, ideas, or objects.

Comparison emphasizes the similarity between apparently unlike things, especially when one is unfamiliar; by comparing the unfamiliar with the familiar, the unfamiliar becomes clearer. Contrast, on the other hand, emphasizes the differences between apparently like things.

There are three usual ways to arrange material for comparison and contrast: point by point, subject by subject, and similarities/differences.

One way to arrange material for comparison and contrast is to consider points one by one, completing each point before going to the next. The following sample paragraph, which contrasts the alligator and the crocodile, uses this arrangement.

TOPIC STATEMENT	The alligator is a close relative of the crocodile. The alligator, how-
POINT 1	ever, has a broader head and blunter snout. Alligators are usually
POINT 2	found in fresh water; crocodiles prefer salt water. The alligator's
POINT 3	lower teeth, which fit inside the edge of the upper jaw, are not
	visible when the lipless mouth is closed. The crocodile's teeth
	are always visible.

The following paragraph, which discusses similarities in the care of the caliper and the micrometer, uses the subject-by-subject arrangement; the first part discusses the care of the caliper and the second part discusses the care of the micrometer. Notice the transition from the first subject to the second subject with the word *similarly*.

TOPIC STATEMENT	Although the caliper is a semiprecision measuring tool and the mi-
	crometer is a precision measuring tool, both must receive proper
SUBJECT 1	care to ensure accuracy of measurement when they are used. The
	caliper has slender legs that can be sprung easily. When not in
	use, it should be hung in the tool cabinet, never placed on a
	bench where heavy tools can be laid on it. When stored, the
	caliper should be protected from rusting by a light coat of oil.
SUBJECT 2	*Similarly*, the micrometer should be kept in a protected place
	away from emery dust or chips and away from heavier tools. It
	should be checked for accuracy often and necessary adjustments
	should be made. When not in use, the spindle and anvil should
	not be in contact. Further, the micrometer should not be com-
	pletely closed and allowed to remain closed because rust may
	form on the ends of the spindle and on the anvil. Finally, the
	micrometer should be oiled occasionally with a light-grade oil to
	prevent rust and corrosion.

A third way to arrange details for comparison-contrast is to give the similarities of the subjects first, followed by the differences in the subjects. The paragraph below illustrates this arrangement.

TOPIC STATEMENT	Football and soccer, although alike in some respects, are different
	in others. Both games are alike in that they require two teams
SIMILARITIES	of 11 players each, that an inflated ball is used, that the games
	are played on a rectangular field, and that scoring is achieved by
	advancing the ball through the opponent's territory into a goal
DIFFERENCES	area. The games differ, however, in the kind of clothing the
	players wear, the shape of the ball, the way in which the field is
	divided, and the manner in which the ball is advanced into the
	goal area.

See Chapter 5 in Part I for an extended discussion of analysis through comparison and contrast.

Cause and Effect. A third method or pattern of arranging ideas into paragraph form is cause and effect. Using this pattern, you may move in one of two directions: from cause to effect, or, reversing this movement, from a discussion of effects to the identification of the cause. In the following paragraph, movement is from cause (passage of the Stamp Act) to effect (action of the Stamp Act congress in the American colonies).

The passage of the Stamp Act in 1765 by the British Parliament aroused great excitement in the American colonies. Interpreting the act as an attempt at taxation without representation, the colonies assembled a Stamp Act congress. This congress formulated an address to the king, petitions to Parliament, and a declaration of the rights and grievances of the colonies. The congress also asserted that the colonies could be taxed only by their own representatives in the colonial assemblies, claimed the inherent right of trial by jury, and proclaimed the Stamp Act an expression of the subversion of the rights and liberties of the colonies.

To gain suspense or to make a point dramatically, you may choose to give the effects first and then identify the cause.

Then rose a frightful cry—the hoarse, hideous, indescribable cry of hopeless fear— the despairing animal cry man utters when suddenly brought face to face with Nothing- ness, without preparation, without consolation, without possibility of respite. . . . *Sauve qui peut!* Some wrenched down the doors; some clung to the heavy banquet tables, to the sofas, to the billiard tables: during one terrible instant, against fruitless heroisms, against futile generosities, raged all the frenzy of selfishness, all the brutalities of panic. And then, then came, thundering through the blackness, the giant swells, boom on boom! . . . One crash! The huge frame building rocks like a cradle, seesaws, crackles. What are human shrieks now? The tornado is shrieking! Another! Chandeliers splinter; lights are dashed out; a sweeping cataract hurls in: the immense hall rises, oscillates, twirls as upon a pivot, crepitates, crumbles into ruin. Crash again! The swirling wreck dissolves into the wallowing of another monster billow; and a hundred cottages overturn, spin in sudden eddies, quiver, disjoint, and melt into the seething. . . . So the hurricane passed.*

See Chapter 5 in Part I for an extended discussion of analysis through cause and effect.

Definition. A fourth method or pattern of paragraph development is defini- tion. This arrangement gives details to explain the nature and characteristics of the term being defined and often begins with a sentence definition. A sentence definition names the term, states the class or group to which it belongs, and gives its essential, distinguishing characteristics. This sentence definition is then expanded in the re- mainder of the paragraph through such details as a description of the item, an explanation of its function, and an analysis of its parts. The following paragraph defines "adolescence."

Adolescence is the period of physical and emotional transition when a person discards childish ways and prepares for the duties and responsibilities of adulthood. During this period the adolescent is neither child nor adult, although having the characteristics of both. "To grow to maturity," the meaning of the derivative Latin word *adolescere*, de- scribes the state of the adolescent. This period of transition is customarily divided into three phases: preadolescence, or puberty, the approximately two-year period of sexual maturation; early adolescence, extending from the time of sexual maturity to the age of 16½; and late adolescence, extending from the age of 16½ to 21.

See Chapter 3 in Part I for an extended discussion of definition.

* From *Chita*, by Lafcadio Hearn (New York: Harper & Brothers, 1889).

Partition. Partition divides a subject into parts, steps, or aspects. Each division is discussed, showing its individual function and its relationship to the whole.

> The United States government has three branches: the legislative, the executive, and the judicial. The legislative branch is responsible for enacting laws. Legislators, or congresspersons, from each state are elected by the popular vote of that state. Members of the House of Representatives have a two-year term; members of the Senate, a six-year term. The executive branch is responsible for carrying out (executing) the laws enacted by the Congress. The chief executives, the President and the Vice-President, are chosen by the Electoral College for a four-year term. The third branch of the government, the judicial, is entrusted with interpreting the laws. This branch is headed by the Supreme Court, composed of nine persons appointed for life by the President. Because of the different methods of selecting the chief members of each branch and their different lengths of term in office, each branch serves as a check and balance for the other branches.

See Chapter 4 in Part I for an extended discussion of partition.

ORDER

Method of development in a paragraph relates closely to the order, or arrangement, of details. Some common kinds of orders are time order, space order, general to particular, particular to general, and climax. You must decide which order best arranges your selected details.

Time Order. In a paragraph using time, or chronological, order, arrange the events according to *when* they happened. For example, you might use time order to report your progress in completing course requirements in a major field; you could also report the completed courses in this order. You would use time order to explain a particular operation or narrate an experience. Study the following paragraph; in it the writer gives directions for designing a furnace. The order is time.

> The first step in designing a furnace is to measure the fireplace to determine the length of pipe to buy. Start at the front of the fireplace and measure from there to the back wall. Then, measure from the floor of the fireplace to the top wall. Finally, measure from the back wall along the top part of the fireplace. Subtract 7 inches from the measurement made from the floor to the top wall so that the furnace can rest 2 inches off the floor and still have 5 inches of clearance between the furnace and the top of the fireplace. Add all three of these measurements together to find the length of each of the smaller pipes needed.

Time order would be used especially to develop instructions and process descriptions (Chapter 1 in Part I); it might also be used in certain letters (Chapter 7 in Part I) in which the time that events or steps occurred is of major importance.

Space Order. In a paragraph using spatial order, discuss the items according to *where* they are. For example, in describing the physical layout of a college, you might show certain buildings located to the south of a main building, others to the north, and still others directly to the east. Also, in describing the luxurious features of a house room by room, you might use space order. In the following paragraph, which explains a shop layout, the details are arranged in this order.

> Entering the front door of the machine shop at Westhaven Area Technical School, an industrialist would see a well-organized shop. To his left he would see a group of lathes, turret lathes and engine lathes. In the left back corner he would see a tool room from which students check out materials and tools. Just to the right of the tool room is a cutting saw. In the center of the shop he would see a line of grinding machines. Included are surface grinders, tool-cutter grinders, cylindrical grinders, and centerless grinders. Nearer the center back wall is the heat-treating section where the industrialist would see furnaces used in heat treatment. To his right he would see two other types of machines: the milling machines and the drill presses. Milling machines are of the vertical and the horizontal types; drill presses included are the multispindle and the radial drill.

Space order may also be used to describe a mechanism (Chapter 2 in Part I).

General to Particular (Specific) Order.

General to Particular (Specific) Order. In a paragraph using general to particular order, the topic statement, the most general statement within the paragraph, appears near the beginning. Following the topic statement is the particular, or the specific, material of the paragraph. This material makes up the body of the paragraph and includes details giving additional information, examples, comparisons, contrasts, reasons, and so forth. This is the order most often used by students and with good reason. Placing the topic statement near the beginning of the paragraph allows you to reread it often as you write your paragraph and helps you stick to the main idea, which is stated in it. The following paragraph illustrates general to particular order.

> Lures for bass fishing should be selected according to the time of day the fishing will take place. For early morning and late afternoon fishing, use top-water baits that make a bubbling or splashing sound. For midday fishing, select deep-running lures and plastic worms. And for night fishing try any kind of bait that makes a lot of noise and has a lot of action. The reason for selecting these different baits is that bass feed on top of shallow water in the early morning, late afternoon, and at night; the only way to get their attention is to use a noisy, top-water lure. During the middle of the day, however, bass swim down to cooler water; then a deep-running lure is best.

Particular (Specific) to General Order. A paragraph using particular to general order, of course, reverses the general to particular order. The most general statement, or the topic statement, is near the end of the paragraph, and the specific material—reasons, examples, comparisons, details, and the like—precedes and leads up to it, as in the following paragraph.

> The human body requires energy for growth and activity; this energy is provided through calories in food. The weight of an adult is determined in general by the balance of the intake of energy in food with the expenditure of energy in activity and growth. When intake and outgo of energy are equal, weight will stay the same. Similarly, weight is reduced when the body receives fewer calories from food than it uses, and weight is increased when the body receives more calories than it uses. Weight can be controlled therefore by regulating either the amount of food eaten, or the extent of physical activity, or both.

Climax. In a paragraph using the order of climax, arrange the ideas so that each succeeding idea is more important than the preceding. The most important idea comes as the last statement of the paragraph.

> People often drink water without knowing for sure that it is safe to drink. They drink water from driven wells into which potentially impure surface water has drained. They drink water from springs without checking around the outlet of the spring to be sure that no surface pollution from human or animal wastes has occurred. Without knowing the rate of flow, the degree of pollution, or the temperature, they drink running water, thinking it safe because of a popular notion that running water purifies itself. These are dangerous acts; polluted water can kill!

Other Orders. Among other orders of paragraph development are alphabetical and random listings, least important to most important, least pertinent to most pertinent, least likely to most likely, simple to complex, known to unknown.

TRANSITION

Transition aids in the smooth flow of thought from sentence to sentence and from paragraph to paragraph. It relates thoughts so there is a clear, coherent progression from one to the next. Transition is achieved through *transitional words or phrases*. As the name implies, these are words or phrases—even sentences and groups of sentences—that relate a preceding thought or topic to a succeeding thought or topic. Transition may also be achieved through the use of other techniques.

Transition Through Conjunctions. Conjunctions are familiar transitional words. To show that several items are of equal value, use the word *and*: item 1, item 2, item 3, *and* item 4. *And*, of course, is the conjunction that means "also," or "in addition," and it indicates a relationship of sameness or similarity between ideas.

To show difference or contrast between items, use *but*.

Other conjunctions and the relationships they show include: *for* (condition), *nor* (exclusion), *or* (choice), and *yet* (condition).

Transition Through Conventional Words or Phrases. In our language there is a large group of transitional words or phrases that show relationships between ideas and facts. These words are signs that the material following is related in a specific way to the preceding material. For example, if in reading you find the phrase *in contrast*, you know the material following will be opposite in meaning to the material preceding the phrase.

Listed below and on the next page are some of the more commonly used words and phrases and the relationship they signal.

Relationship	Transitional Words/Phrases
addition	also, and, first, moreover, too, next, furthermore, similarly
contrast	but, however, nevertheless, on the contrary, on the other hand
passage of time	meanwhile, soon, presently, immediately, afterward, at last
illustration	for example, to illustrate, for instance, in other words
comparison	likewise, similarly

Relationship	Transitional Words/Phrases
conclusion	therefore, consequently, accordingly, as a result, thus
summary	to sum up, in short, in brief, to summarize
concession	although, of course, at the same time, after all

Transition Through Other Techniques. Other techniques used to ensure the smooth flow of sentences and thoughts are pronoun reference, repetition of key terms, continuity of sentence subjects, parallel sentence patterns, or a combination of these. In the following paragraph from Abraham Lincoln's "Gettysburg Address" italics have been added to indicate these transition techniques.

CONTINUITY OF SUBJECT "WE"
REPETITION OF KEY TERMS:
  "WAR," "NATION," AND
  "BATTLEFIELD"
PARALLEL SENTENCE PATTERNS

PRONOUN REFERENCE

Now *we* are engaged in a great civil *war*, testing whether that *nation*, or any *nation* so conceived and so dedicated, can long endure. *We* are met on a great *battlefield* of that *war*. *We* have come to dedicate a portion of *that field* as a final resting-place for those who *here* gave their lives that that *nation* might live. It is altogether fitting and proper that *we* should do *this*.

Transition and Continuity. Read the following paragraph.

The diesel engine, an increasingly popular engine in automobiles, has its disadvantages. The higher compression that makes the diesel more efficient and clean burning necessitates the use of heavier engine components such as pistons, rods, and cylinder heads. Diesels are heavier than gasoline engines of the same displacement. Diesel engines cost more than gasoline engines. Diesel engines in Oldsmobiles cost $850, $750 over the cost of the 260-cubic-inch V8 gasoline engine; in Volkswagen's Rabbit, the diesel costs an extra $170. Diesel engines are difficult to start in cold weather. "Glow plugs" are available to preheat the combustion chambers and engine block heaters to provide easier starting. A diesel engine has relatively poor acceleration compared to the gasoline engine. It takes the diesel longer to rev up to its peak horsepower. Diesel engines have been noted for their noise, vibration, and smoke. These three characteristics have been reduced to a minimum in recent diesel-powered autos.

Although the preceding paragraph contains information, it presents some difficulty in reading and understanding because the sentences are choppy and there are no transitional words or phrases, or any other techniques, to relate one idea to another. Read the paragraph revised.

The diesel engine, an increasingly popular engine in automobiles, has its disadvantages. **For one,** the higher compression that makes the diesel more efficient and clean burning necessitates the use of heavier engine components such as pistons, rods, and cylinder heads; **thus,** diesel engines are heavier than gasoline engines. **Also** diesel engines cost more than gasoline engines. **For example,** diesel engines in Oldsmobiles cost $850, a $750 increase over the cost of the 260-cubic-inch V8 gasoline engine; in Volkswagen's Rabbit, the diesel costs an extra $170. **Another disadvantage** is that diesel engines are difficult to start in cold weather. **However,** "glow plugs" are available to preheat the combustion chambers and engine block heaters to provide easier starting. **Further,** a diesel engine has relatively poor acceleration compared to the gasoline engine; **thus,** the diesel takes longer to rev up to its peak horsepower. **Finally,** diesel engines have been noted for their noise, vibration, and smoke, **but** these characteristics have been reduced to a minimum in recent diesel-powered autos.

Notice the words in boldface print; these added words, transitional markers, indicate clearly the relationship of ideas. The paragraph now reads smoothly and is easily understood because there is continuity, a smooth flow of sentences moving from one to another. The "stop-go" effect characterizing the first writing of the paragraph has been eliminated.

Writing the Paragraph

WRITING THE FIRST DRAFT

Once the preceding steps have been completed—selecting a subject, limiting the subject, forming a topic statement, adding details, and deciding on method of development, order, and transition—you are ready to write a first draft of the paragraph. In this first writing, it may be helpful to identify the selected details as *major* supporting details or as *minor* supporting details.

Major and Minor Supporting Details. Major supporting details give direct support to the key idea given in the topic statement.

TOPIC STATEMENT	Libraries classify books according to one of two systems: the Library of Congress Classification or the Dewey Decimal Classification.
MAJOR SUPPORT 1	The Library of Congress Classification divides all books into 20 basic groups.
MAJOR SUPPORT 2	The Dewey Decimal Classification divides all books into ten basic groups.
MAJOR SUPPORT 3	The designations for the book *Lasers and Their Applications* illustrate the differences in the two systems.

For a more exact discussion, the major ideas supporting the topic statement should be developed further with minor supporting details.

TOPIC STATEMENT	Libraries classify books according to one of the two systems: the Library of Congress Classification or the Dewey Decimal Classification.
MAJOR SUPPORT 1	The Library of Congress Classification divides all books into 20 basic groups.
MINOR SUPPORT	Each group is indicated by a letter of the alphabet.
MINOR SUPPORT	The six letters not currently used are for future classification expansion.
MINOR SUPPORT	The basic groups have many divisions.
MINOR SUPPORT	Subdivisions are designated by letter-number combinations.
MAJOR SUPPORT 2	The Dewey Decimal Classification divides all books into ten basic groups.
MINOR SUPPORT	Each group is indicated by a number.
MINOR SUPPORT	The basic groups are divided and subdivided numerous times.
MINOR SUPPORT	Subdivisions are designated by numeral-decimal combinations.
MAJOR SUPPORT 3	The designations for the book *Lasers and Their Applications* illustrate the differences in the two systems.
MINOR SUPPORT	The Library of Congress designation is TK7872.L3S7.
MINOR SUPPORT	The Dewey Decimal designation is 621.329.

There is no required number of major or minor supporting details in any one paragraph. You might write a paragraph containing only major supports for the topic statement or a paragraph with one major support and several minor supports. You must decide how to develop your selected topic statement.

Further, there may be statements in the paragraph other than the topic statement that would not give major or minor supporting details. For example, a final sentence stating a conclusion such as "It is not feasible to begin plant expansion at this time" gives neither major nor minor support; it simply concludes the paragraph. Sometimes the topic statement may need one or more additional statements to explain the central idea; this would be an explanatory statement, not a major or a minor support. Or, a paragraph may contain transition sentences that connect major details.

TOPIC STATEMENT	Libraries classify books according to one of the two systems: the Library of Congress Classification or the Dewey Decimal Classification.
MAJOR SUPPORT	The Library of Congress Classification divides all books into 20 basic groups.
TRANSITION SENTENCE	The Dewey Decimal Classification, however, the older and more common system, is the one used by most libraries, especially smaller libraries.
MAJOR SUPPORT	The Dewey Decimal Classification divides all books into ten basic groups.

Putting Sentences Together. A first draft of a paragraph on the two library systems of classifying books might look like this.

TOPIC STATEMENT	Libraries classify books according to one of two systems: the Library of Congress Classification or the Dewey Decimal Classification.
MAJOR SUPPORT 1	The Library of Congress Classification divides all books into 20 basic groups;
MINOR SUPPORT	each group is indicated by a letter of the alphabet.
MINOR SUPPORT	(The six letters not currently used are for future classification expansion.)
MINOR SUPPORT	These basic groups have many divisions,
MINOR SUPPORT	designated by a letter-number combination.
TRANSITION SENTENCE	The Dewey Decimal Classification, however, the older and more common system, is the one used by most libraries, especially smaller libraries.
MAJOR SUPPORT 2	The Dewey Decimal Classification divides all books into ten basic groups,
MINOR SUPPORT	using a number system.
MINOR SUPPORT	These basic groups are divided and then subdivided numerous times, all on the decimal (by tens) system.
MAJOR SUPPORT 3	The book *Lasers and Their Applications* has the following designations, with each letter and each numeral having a particular significance:
MINOR SUPPORT	Library of Congress, TK7872.L3S7;
MINOR SUPPORT	Dewey Decimal, 621.329.

REVISING

A major part of writing is revising, checking to see if the paragraph needs any revisions or corrections. If it does, make these changes before copying the paragraph a final time.

CHECKLIST FOR REVISING

Topic Statement
1. Carefully stated so main idea is evident
2. If implied, main idea evident upon reading the paragraph

Words
1. Words chosen carefully to convey exact meaning
2. Words used economically; no wordiness
3. Each word spelled correctly
4. Correct usage of words that sound alike: *to, too, two; its, it's; there, their, they're;* etc.
5. No words carelessly omitted

Sentences
1. All ideas expressed in complete sentences
2. Sentence structure varied to avoid monotony
3. No run-on sentences

Development
1. Main idea developed by including specific details of support
2. All details clearly related to main idea

Transition
1. Transitional techniques used to keep sentences moving smoothly from one to the next
2. Transitional words and phrases used to show relationship between ideas

Punctuation
1. Correct comma usage
2. Correct end punctuation

Grammatical Usage
1. Subject-verb agreement
2. Correct verb form
3. Correct case of pronoun

Content
1. Accurate
2. Complete

General

1. Legible writing
2. Acceptable manuscript form

(See also the inside front cover and the inside back cover of this textbook.)

MAKING THE FINAL COPY

Now make a copy of the revised draft. Read the final copy carefully to catch any careless mistakes, such as mistakes in spelling or punctuation. If the paragraph has been typed, check to be sure there are no typographical errors. The final paragraph might be like the following paragraph:

> Libraries classify books according to one of two systems: the Library of Congress Classification or the Dewey Decimal Classification. The Library of Congress Classification divides all books into 20 basic groups, each group indicated by a letter of the alphabet (the six letters not currently used are for future classification expansion). These basic groups have many divisions, designated by a letter-number combination. The Dewey Decimal Classification, however, the older and more common system, is the one used by most libraries, especially smaller libraries. The Dewey Decimal Classification divides all books into ten basic groups, using a number system. These basic groups are divided and then subdivided numerous times, all on the decimal (by tens) system. As an example of the differences in the two classifying systems, the book *Lasers and Their Applications* has the following designations, with each letter and each numeral having a particular significance: Library of Congress, TK7872.L3S7; Dewey Decimal, 621.329.

Paragraphs of Introduction, Transition, and Closing

The preceding pages of this chapter have dealt with paragraphs that are developmental, that is, paragraphs that can stand alone as single units developing an idea, or that can, in a series, develop a stated topic. There are, however, at least three other types of paragraphs—introductory, transitional, and closing. Although these paragraphs may add details, they are used primarily to introduce a topic, to connect sections of a paper, or to end a paper.

PARAGRAPH OF INTRODUCTION

The introductory paragraph is a very important paragraph. This is what a reader sees first, and it could well determine whether the reader continues reading or puts the selection aside. The introductory paragraph, then, should get the reader's attention and introduce the subject being discussed.

Some introductory paragraphs provide a guide for the report to follow by *stating the major points the writer will develop* to support the topic.

> The volt is defined in terms of two other electrical qualities, the ampere and the ohm. The volt is that pressure which will force 1 ampere of current through a resistance of 1 ohm. To understand this definition fully, one must study in detail the meaning of *ampere* and *ohm.*

It is evident that the major points to be developed in following paragraphs are the detailed meanings of *ampere* and *ohm*.

Other introductory paragraphs introduce the central idea *without giving any specific statements about ideas* to be used in development.

> The relatively thin upper layer of the earth's crust is called soil. One dictionary definition of *soil* is "finely divided rock material mixed with decayed vegetable or animal matter, constituting that portion of the surface of the earth in which plants grow, or may grow." This upper layer of the earth's surface, varying in thickness from 6 to 8 inches in the case of humid soils to 10 or 20 feet in the case of arid soils, possesses characteristic properties that distinguish it from the underlying rocks and rock ingredients.

A report following this introductory paragraph would list and discuss the properties of the upper layer of the earth's surface. At this point, however, these properties have not been identified.

Various techniques can be used to begin a report, as illustrated by the following examples from various reports in the text.

Identification of Object and Reason for Sharpening

> A major instrument in making a mechanical drawing is the ruling pen, which is used to ink in straight lines and noncircular curves with a T square, a triangle, a curve, or a straightedge as a guide. The shape of the nibs (blades resembling tweezers) is the most important aspect of the pen. The nibs must be rounded (elliptical) to create an ink space between the nibs. To assure neat, clean lines, the ruling pen must be kept sharp and in good condition. It must be sharpened from time to time after extensive use because the nibs wear down. (Page 32)

Effect of Subject and Division into Parts

> A flashlight is a common household item used often as a light source which requires no external energy source. Four parts of the flashlight work together to produce light: the battery, the switch, the bulb, and the crown. (Page 177)

Definition of Subject and Major Stages of Process

> When eating a mushroom, have you ever wondered how it grows into the umbrella shape? Mushrooms are fungi that grow in decaying vegetable matter. The common table mushroom grows in much the same way as other mushrooms. Belonging to a group of mushrooms called the agarics, the common mushroom grows wild, but it is also cultivated. The five stages of mushroom growth include the spawn, the pinhead, the button, the cap and gills, and the mature mushroom shaped like an opened umbrella. (Page 45)

PARAGRAPH OF TRANSITION

The paragraph of transition connects two paragraphs or two groups of paragraphs. Usually brief (often only one or two sentences), the transition paragraph signals a marked shift in thought between two sections of a paper. The transition paragraph looks back to the preceding paragraph, summarizes the preceding material, or indicates the material to come.

Not all writing, of course, requires transition paragraphs. Frequently a transitional word, phrase, or sentence is sufficient to connect the sections. You must decide, after considering the purpose and method of development in your report, whether it is more effective to show transition between paragraphs by means of the opening and closing sentences of the adjacent paragraphs or by means of a transition paragraph.

For instance, in a report discussing the advantages and disadvantages of living at home while attending college, you might decide that you need a transition paragraph to connect the "advantages" section of the report with the "disadvantages" section.

> Thus, the student who lives at home while attending college has fewer living expenses, is free from dormitory rules, and can enjoy the conveniences of family living. On the other hand, the student must consider that he or she misses the experience of dormitory life, has less opportunity to meet other students, spends a lot of time and money commuting, and has less opportunity to become independent in making decisions.

The first sentence summarizes the first section of the report; the second sentence looks forward to the next section by suggesting the material that is to come.

The following paragraph separates a discussion of general anesthetics and local anesthetics.

> Not all surgery requires that the patient be unconscious. For many minor operations, only a restricted, or local, area of the body need be made insensible to pain; thus a local anesthetic is administered. The local anesthetic prevents sensations of pain and touch from traveling through the nerves in the drugged area.

In "Clear Only If Known" (see pages 568–570), Edgar Dale uses the following transition paragraph as he moves from a discussion of why people give poor directions to one of why people do not follow good directions.

> But we must not put too much of the blame for inadequate directions on those who give them. Sometimes the persons who ask for help are also at fault. Communication, we must remember, is a two-way process. (Page 570)

PARAGRAPH OF CLOSING

The closing paragraph, which leaves a final impression on the reader, should be compatible with the purpose and the emphasis of the paper.

The closing paragraph may signal the drawing to a close of a report in several ways. A report discussing the pros and cons of advertising might have the following closing paragraph.

> Many other good and bad points of advertising could be given. But these are sufficient to suggest the relation of advertising to people's lives, the vast power it controls, and the social responsibilities it carries.

This paragraph *draws a conclusion* about the material discussed in the report.

A report explaining the primary differences between plants and animals might end with the following *summary* paragraph.

The primary differences between plants and animals, then, are these: Animals can move about, animals sense their surroundings, animals live on ready-made foods, and animals do not contain cellulose.

This paragraph *reemphasizes the main points* discussed in the report.

Remember that many short reports, three to four pages, do not need closing paragraphs. Actually there is no particular rule to guide you in deciding if a closing paragraph is needed. Read the report, and if it seems to end abruptly, consider adding a closing paragraph—usually one containing either a conclusion or a summary.

Application 1   Writing Paragraphs

Read the following topic statements. Draw one line under the words that introduce the subject of the paragraph. Draw two lines under the words that indicate the kind of information that will follow in the paragraph.

EXAMPLES

Compacted soil is bad for plants.

More electrical accidents occur outside than inside the home.

1. Four types of solder are used to connect refrigeration lines.
2. The process of flaring copper tubing requires little skill.
3. One of the most satisfactory pumps for positive delivery of fuel is the vane pump.
4. The type of current used is a major factor in selecting transformers.
5. The technician or the novice mechanic should consider the size and shape of a room before installing a set of audio speakers.
6. Pistons vary in size, according to their intended use.
7. Refrigerant gases must have several specific qualities.
8. To get the longest use from a Crescent wrench, follow these simple rules.
9. Open-pit mining is a fast, easy way to obtain minerals.
10. A fairly high-temperature flame can be made to produce sound waves.
11. Experimenting with electricity can be dangerous.
12. Turning a patient in the bed requires care and skill.
13. In selecting plants for grounds beautification, one must consider year-round climate.
14. The colors used in decorating a room may affect the owner's moods.
15. To get maximum efficiency from drawing instruments, a drafter should take care in handling and in storing them.
16. Nondestructive testing is more reliable than destructive testing.
17. The type and the amount of fertilizer used determine crop yield.
18. Profitable cattle raising requires keeping records systematically.
19. Hydraulic oil must have the proper viscosity index.
20. The distinction between a health technologist and a health technician lies in the complexity of the job.

Application 2 Writing Paragraphs

Of the following sentences, indicate those that are usable as topic statements. For the sentences that are not usable as topic statements, rewrite them so that they are usable.

1. I am majoring in computer science.
2. The first consideration when either buying or building a workbench should be your personal needs.
3. If we were to build another kit boat, we would do a few things differently.
4. Arcade, in architecture, is a series of arches supported by piers or columns, either standing free or attached to a wall.
5. The changes brought by computer technology are tremendous.
6. Most perennials are descendants of wild flowers.
7. If iron is permitted to contact water and air, it forms iron oxide (rust).
8. Each part is inspected to make certain that its specifications are within the allowed tolerances.
9. Automobiles are generally equipped with two independent brake systems.
10. An important characteristic of fuels is their knocking properties.

Application 3 Writing Paragraphs

Using material from your technical field, write five topic statements, each of which could be developed into a paragraph approximately 100–150 words in length. Draw one line under the words that introduce the subject and two lines under the words that indicate the kind of information that will follow in the paragraph.

Application 4 Writing Paragraphs

Choose one of the topic statements from Application 2 or Application 3 above and develop it into a paragraph.

Application 5 Writing Paragraphs

Following the example on page 671, eliminate any sentences from the following possible list of details for developing the topic statement into a paragraph. Then write the paragraph, adding any transitional words or phrases necessary to make a coherent paragraph.

TOPIC STATEMENT	Regular telephone service charges are determined by three kinds of telephone calls: local service, long-distance service, and overseas service.
POSSIBLE DETAILS	1. Local service offers several arrangements based on cost.
	2. Local service offers an unlimited number of calls for a specified charge.

3. A local telephone call goes into a central office, traveling over wires or by radio.
4. For a set amount, a specified number of local calls can be made; additional calls are extra charge.
5. Extended area calling refers to long-distance calls in nearby areas for a specified rate.
6. Ninety-five percent of all calls in the United States are local calls.
7. Long-distance service is either station-to-station or person-to-person.
8. Station-to-station call charges begin when the telephone is answered.
9. Person-to-person calls begin when the specific person called answers the phone.
10. Long-distance calls cost according to other factors: time of day, the distance, minutes spent in conversation.
11. Long-distance calls today rarely involve an operator.
12. Long-distance dialing requires an area code plus a seven-digit telephone number.
13. Overseas calls can be made from the United States to approximately 200 other countries and territories.
14. Charges for overseas calls vary according to time of day, distance, and number of minutes talked.
15. Overseas calls travel by undersea cable or by radio or to ships at sea.
16. Mobile telephone service is a special telephone service.

Application 6 Writing Paragraphs

Choose one of the following topics or supply one from your major field. Using this as your subject, write a topic statement and develop it into a paragraph. As you plan the paragraph, identify major and minor supports; use the sample paragraphs on pages 666–671 and 679–682 as guides.

1. How a magnetic compass works
2. Different kinds of files
3. The differences between "addiction" and "obsession"
4. Why rubber stretches
5. Definition of a term common to your technology
6. Parts of an instrument, tool, or piece of equipment common to your major field

Application 7 Writing Paragraphs

Assume that you are to write a paper on job opportunities in your major field of study. Write the introductory paragraph for such a paper.

Application 8 Writing Paragraphs

For Application 7 above, assume that the paper has four major sections. Write a transitional paragraph connecting any two sections of the report.

Application 9 Writing Paragraphs

For Application 7 above, write a closing paragraph.

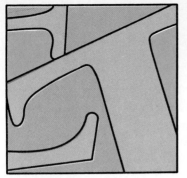

Chapter 3
Grammatical Usage

Objectives

Upon completing this chapter, the student should be able to:

- Recognize and revise nonstandard English for communication in business, industry, and government
- Solve typical usage problems: agreement of nouns; shifts in person, number, and gender; agreement of subject and verb; and use of modifiers
- Write sentences to emphasize certain material within the sentence
- Select an appropriate word for a specific communication need

Introduction

English language usage varies and constantly changes. Attempts made to classify language according to its usage in different dialects (the form of language spoken in different geographic areas), social statuses, educational levels or styles (formal and informal) are all incomplete and overlapping. Amid these differences in grammatical usage, however, there are generally accepted standard practices in business and industry.

Usage does make a difference. Select the usage most likely to get the desired results, recognizing that usages, like people, have different characteristics. For example, you say or write *I have known him a week* rather than *I have knowed him a week* because *I have known him a week* is more acceptable to most people. You base this decision, consciously or unconsciously, on the grammatical expressions of people you respect or of people with whom you want to "fit in," and on the reactions of people to these expressions.

Because judgments are based on knowledge, the following guides are designed to improve your judgment by increasing your knowledge of acceptable grammatical usages and word choice in standard English.

Parts of Speech: Definition and Usage

A customary way of examining the English language is to divide the vocabulary into eight major divisions called parts of speech: adjectives, adverbs, conjunctions, interjections, nouns, prepositions, pronouns, and verbs. Added to these divisions in the following discussion is another group, called verbals.

ADJECTIVES

Adjectives describe, limit, or qualify a noun or pronoun by telling "Which one?" "What kind?" or "How many?" An adjective may be a word, a phrase, or a clause.

DESCRIBING ADJECTIVE	the *gray* paint *with a yellow tint* (Tells "What kind?")
LIMITING ADJECTIVE	*three* pencils (Tells "How many?")
QUALIFYING ADJECTIVE	*younger* sister *who lives in Denver* (Tells "Which one?")

- Adjectives show degrees of comparison in quality, quantity, and manner by affixing *-er* and *-est* to the positive form or by adding *more* and *most* or *less* and *least* to the positive form.

POSITIVE	tall	beautiful	wise
COMPARATIVE	taller	more beautiful	less wise
SUPERLATIVE	tallest	most beautiful	least wise

The comparative degree is used when speaking of two things, and the superlative when speaking of three or more. One-syllable words generally add -*er* and -*est* while words of two or more syllables generally require *more* and *most*. To show degrees of inferiority, *less* and *least* are added.

■ A double comparison, such as *most beautifulest*, should not be used; either -*est* or *most* makes the superlative degree, not both.

■ Adjectives that indicate absolute qualities or conditions (such as *dead, round,* or *perfect*) cannot be compared. Thus: *dead, more nearly dead, most nearly dead* (not *deader, deadest*).

■ Adjective clauses are punctuated according to the purpose they serve. If they are essential to the meaning of the sentence, they are not set off by commas. If they simply give additional information, they are set off from the rest of the sentence. (See comma usage, item 6, page 727.) Adjective clauses beginning with *that* are essential and are not set off (with one exception: when *that* is used as a substitute for *which* to avoid repetition).

> Edward Jenner is the man *who experimented with smallpox vaccination*. (Essential to meaning)
> Bathrooms are often decorated in colors *that suggest water*. (Essential to meaning)
> My ideas about exploration of space, *which are different from most of my friends' ideas*, were formulated mainly through reading. (Gives additional information)

ADVERBS

Adverbs modify verbs, adjectives, or other adverbs by telling "When?" (time), "Where?" (place), "Why?" (reason), "How?" (manner), "How much?" (degree), or "How often?" (frequency). An adverb may be a word, a phrase, or a clause.

> The class began *after everyone arrived*. (Tells "When?")
> The recording session will take place *in Booth A*. (Tells "Where?")
> Medical technicians often go to developing nations *to help the people*. (Tells "Why?")
> Experienced salespeople can recognize a quality product *by inspection*. (Tells "How?")
> In our laboratories we have containers of *very* clear plastic. (Tells "How much?")
> The agricultural experts have a meeting *once a year* to exchange new knowledge. (Tells "How often?")

■ An adjective should not be used when an adverb is needed.

INCORRECT	The new employee did *good* on her first report. (Adverb needed)
CORRECT	The new employee did *well* on her first report.
INCORRECT	For a department to operate *smooth* and *efficient*, all employees must cooperate. (Adverbs needed)
CORRECT	For a department to operate *smoothly* and *efficiently*, all employees must cooperate.

■ Introductory adverb clauses are generally followed by a comma. Adverb clauses at the end of a sentence are not set off. (See comma usage, page 727.)

Although most people agree clothes do not make the person, they spend considerable time and money dressing themselves in the latest fashions.
When the investigation is completed, the company will make its decision. The company will make its decision *when the investigation is completed*.

CONJUNCTIONS

A conjunction connects words, phrases, or clauses. Conjunctions may be divided into two general classes: coordinating conjunctions and subordinating conjunctions.

1. COORDINATING CONJUNCTIONS—SUCH AS *AND*, *BUT*, *OR*—CONNECT WORDS, PHRASES, OR CLAUSES OF EQUAL RANK.

I bought nails, brads, staples, *and* screws. (Connects nouns in a series)
Nancy *or* Elaine will serve on the committee. (Connects nouns)

■ Conjunctive adverbs—such as *however, moreover, therefore, consequently, nevertheless*—are a kind of coordinating conjunction. They link the independent clause in which they occur to the preceding independent clause. The clause they introduce is grammatically independent, but it depends on the preceding clause for complete meaning. Note that a semicolon separates the independent clauses (see semicolon usage, pages 730–731) and that the conjunctive adverb is usually set off with commas (see comma usage, item 10, page 728).

The film was produced and directed by students; other students, *therefore*, should be interested in viewing the film.
Jane had a severe cold; *consequently*, she was unable to participate in the computer tournament.

■ Correlative conjunctions—*either . . . or, neither . . . nor, both . . . and, not only . . . but also*—are a kind of coordinating conjunction. Correlative conjunctions are used in pairs to connect words, phrases, and clauses of equal rank.

The movie was *not only* well produced *but also* beautifully filmed.

■ The units joined by coordinate conjunctions should be the same *grammatically*. (See also Coordination, page 126, and Parallelism in Sentences, pages 655–656.)

CONFUSING   Our store handles two kinds of drills—manual *and* electricity. (Adjective and noun)
REVISED   Our store handles two kinds of drills—manual *and* electric. (Adjective and adjective)
CONFUSING   I *either* will go today *or* tomorrow. (Verb and adverb)
REVISED   I will go *either* today *or* tomorrow. (Adverb and adverb)

■ The units joined by coordinate conjunctions should be the same *logically*. (See also Coordination, page 126, and Parallelism in Sentences, pages 655–656.)

CONFUSING   I can't decide whether I want to be a lab technician, a nurse, *or* respiratory therapy. (Two people and a field of study)
REVISED   I can't decide whether I want to be a lab technician, a nurse, *or* a respiratory therapist. (Three people)
REVISED   I can't decide whether I want to study lab technology, nursing, *or* respiratory therapy. (Three fields of study)

2. SUBORDINATE CONJUNCTIONS—SUCH AS *WHEN, SINCE, BECAUSE, ALTHOUGH, AS, AS IF*—INTRODUCE SUBORDINATE CLAUSES.

Because the voltmeter was broken, we could not test the circuit.
The instructor gave credit for class participation *since the primary purpose of the course was to stimulate thought*.

INTERJECTIONS

An interjection indicates sudden or strong feeling. It has no grammatical relationship to the sentence following, if there is one.

Ouch!
Darn!
Good grief, I dropped the thermometer.
Oh, you wouldn't believe what happened.

NOUNS

Nouns name persons *(Ms. Jones)*, places *(Portland, Oregon)*, things *(motorcycle)*, actions *(sterilization)*, and qualities *(mercy)*. A noun may be a word, a phrase, or a clause.

The *record* sold one million *copies*.
Making my first record was a satisfying *experience*.
The *group* believed *that the record would be a hit*.

Note: The second and third sentences above may be analyzed in two ways. Generally, the italicized sections may be identified simply as nouns. More specifically, they may also be analyzed on the basis of elements within them. For example, *Making my first record* includes *making*, a gerund used as subject; *my*, a pronoun functioning as an adjective; *first*, an adjective; *record*, a noun used as the object of *making*.

■ Nouns can also be characterized by their ability to have an affix added to show possession or plurality.

The *man's* overcoat protected him from the cold. *(man + 's forms a possessive)*
Experiences teach vivid *lessons*. *(experience and lesson add s to form plurals)*

■ Nouns have many uses in sentences:

SUBJECT   A noun identifying *who* or *what* the sentence is about.

Da Vinci invented the bicycle in 1493.

DIRECT OBJECT   A noun identifying *what* or *who* receives the action expressed by the verb.

Da Vinci's bicycle did not have *pedals*.

PREDICATE NOUN (also called PREDICATE NOMINATIVE and SUBJECTIVE COMPLEMENT)   A noun following a linking verb and referring to the subject of the sentence.

The inventor of the turbojet engine was *Sir Frank Whittle*.

INDIRECT OBJECT   The noun identifying *to whom (what, which)* or *for whom (what, which)* the action expressed by the verb is done. If the *to* or *for* appears in the sentence, the indirect object becomes a prepositional phrase.

INDIRECT OBJECT   The children gave *Spot* a bath.
                    He gave the *plan* a second chance.
                    They bought *him* a new leash.
PREPOSITIONAL PHRASE   They bought a new leash *for him*.

OBJECT OF PREPOSITION   The noun answering the question "What?" following a preposition.

The first step in the *procedure* is sanding the wood surface.

APPOSITIVE   The noun following another noun or pronoun and renaming that noun or pronoun.

My gift to John, my oldest *brother*, was a calculator.

OBJECTIVE COMPLEMENT   The noun (or adjective) following a direct object, referring to the direct object and completing its meaning.

The employees selected Sarah Smith *spokesperson*.

PREPOSITIONS

Traditionally a preposition (such as *in, by, to, with, at, into, over*) is a word that precedes a noun or a pronoun and shows its relationship to some other word in the sentence, generally a noun, adjective, or verb. A preposition is also a function word whose form does not change; the importance of the preposition lies in its grammatical function, not in its meaning. A prepositional phrase is composed of the preposition, its object, and any modifiers with them. It functions as an adjective, adverb, or noun.

The tools *on the shelves* need dusting. (Adjective)
Most commuting students live *at home*. (Adverb)
Under the mattress is our hiding place. (Noun)

■ A preposition may end a sentence when any other placement of the preposition would result in a clumsy, unnatural sentence.

UNNATURAL   Sex is a topic *about* which many people think.
NATURAL   Sex is a topic which many people think *about*.

PRONOUNS

A pronoun takes the place of or refers to a noun.

They worked overtime.
Who is *that?*
She gave the tube to *him*.
These belong to *someone*.

■ The personal pronouns and *who* have different forms, or cases, for different uses in the sentence. These pronoun forms divide into three groups—nominative case, objective case, and possessive case.

1. Nominative case forms

	Singular	*Plural*
FIRST PERSON	I	we
SECOND PERSON	you	you
THIRD PERSON	he, she, it	they
	who, whoever	who, whoever

Nominative case forms may be used in the following ways:

Subject of Dependent or Independent Clauses

We know about bacteria because of the microscope.
A student hands in an assignment when *it* is completed.
Ferdinand Cohn, *who* was a German botanist, worked out the first scheme for classifying bacteria as plants rather than as animals.

Subjective Complement (Predicate Pronoun) Following a Linking Verb

(Linking verbs include forms of "to be" and such verbs as *appear, seem, remain, become*.)

The delegates to the convention are *he* and *she*.
The most qualified applicant seemed to be *I*.

Appositive Following a Subject or a Subjective Complement

Two students, *Joan* and *I*, received awards for achievement.
The representatives to the DECA convention are club members *Dale* and *I*.

2. Objective case forms

	Singular	*Plural*
FIRST PERSON	me	us
SECOND PERSON	you	you
THIRD PERSON	her, him, it	them
	whom, whomever	whom, whomever

Objective case forms may be used in the following ways:

Direct Object of a Verb or of a Verbal (Infinitive, Participle, Gerund)

The flying chips hit *me* in the face. (Object of verb *hit*)
The artist carefully handled the clay, shaping *it* so that it did not crack. (Object of verbal *shaping*)

Indirect Object

The flight attendant told *them* the reason for the change in plans.

Object of Preposition

Persons like Marie and *him* will make excellent workers.
Give the plans only to Jack or *me*.

Subject of Infinitive

I believe *her* to be the best choice.
Let *me* help Tom.

*Appositive Following a Direct Object, Indirect Object, Object of a
Preposition, or Subject of an Infinitive*

We gave the article to the editors, James and *him*.

3. Possessive case forms

	Singular	Plural
FIRST PERSON	my, mine	our, ours
SECOND PERSON	your, yours	your, yours
THIRD PERSON	his, her, hers, its	their, theirs
	whose, whosever	whose, whosever

Possessive pronoun forms are used in the following way:

Show Ownership or Possession

Pasteur was also experimenting while Koch and *his* disciples were busy perfecting
  new techniques.
Whose guitar is this?

■ Compound personal pronouns are formed by combining personal pronoun forms
with -*self* or -*selves*: *myself, yourself, herself, himself, itself, ourselves, yourselves,*
and *themselves*.

■ *Myself* should not be used in the place of *I* or *me*.

INCORRECT	Mother or *myself* will be present.
CORRECT	Mother or *I* will be present.
INCORRECT	Reserve the conference room for the supervisor and *myself*.
CORRECT	Reserve the conference room for the supervisor and *me*.

■ *Themself, theirselves*, and *hisself* should never be used; they are not acceptable
forms.

CORRECT	The members of the team *themselves* voted not to go.
CORRECT	Henry cut *himself*.

VERBS

Verbs indicate state of being or express action.

That machine *is* a very expensive piece of equipment. (State of being)
The engineer *studied* the local water problem carefully. (Action)

■ Verb tense indicates the time of the state of being or action. There are six tenses.

1. Present tense verbs indicate action taking place now or continuing action.

There are three kinds of present tense: simple present, emphatic present, and progressive present.

> I *work* eight hours a day. (Simple present tense)
> I *do work* eight hours a day. (Emphatic present tense)
> I *am working* eight hours a day. (Progressive present tense)

2. Past tense verbs indicate action completed before the present time. There are three kinds of past tense: simple past, emphatic past, and progressive past.

> I *worked* eight hours a day. (Simple past tense)
> I *did work* eight hours a day. (Emphatic past tense)
> I *was working* eight hours a day. (Progressive past tense)

3. Future tense verbs indicate action that will occur sometime in the future, that is, sometime after the present. There are two kinds of future tense: simple future and progressive future.

> I *will work* eight hours a day. (Simple future tense)
> I *will be working* eight hours a day. (Progressive future tense)

4. Present perfect tense verbs indicate action completed prior to the present but connected in some way with the present or action begun in the past and continuing in the present. There are two kinds of present perfect tense: simple present perfect and progressive present perfect.

> I *have worked* eight hours a day. (Simple present perfect tense)
> I *have been working* eight hours a day. (Progressive present perfect tense)

5. Past perfect tense verbs indicate action completed prior to some stated past time. There are two kinds of past perfect tense: simple past perfect and progressive past perfect.

> I *had worked* eight hours a day. (Simple past perfect tense)
> I *had been working* eight hours a day. (Progressive past perfect tense)

6. Future perfect tense verbs indicate action to be completed prior to some stated future time. There are two kinds of future perfect tense: simple future perfect and progressive future perfect.

> I *will have worked* eight hours a day. (Simple future perfect tense)
> I *will have been working* eight hours a day. (Progressive future perfect tense)

■ The writer should not needlessly shift from one tense to another. Since verb tenses tell the reader when the action is happening, inconsistent verb tenses confuse the reader.

SHIFTED VERBS (CONFUSING)   As time *passed*, technology *becomes* more complex.
passed—past tense verb
becomes—present tense verb

Since the two indicated actions occur at the same time, the verbs should be in the same tense.

CONSISTENT VERBS   As time *passed*, technology *became* more complex.

CONSISTENT VERBS   As time *passes*, technology *becomes* more complex.

passed
became —past tense verbs

passes
becomes —present tense verbs

SHIFTED VERBS (CONFUSING)   While I *was installing* appliances last summer, I *had learned* that customers appreciate promptness.
was installing—past tense verb
had learned—past perfect tense verb

The verb phrase *was installing* indicates a past action in progress while the verb phrase *had learned* indicates an action completed prior to some stated past time. To indicate that both actions were in the past, the sentence might be written as follows:

CONSISTENT VERBS   While I *was installing* appliances last summer, I *learned* that customers appreciate promptness
was installing
learned —past tense verbs

■ Verbs have two voices, active and passive.

1. The active voice indicates action done by the subject. The active voice is forceful and emphatic.

Roentgen *won* the Nobel Prize for his discovery of X rays.
Meat slightly marbled with fat *tastes* better.
The scientist *thinks*.

2. The passive voice indicates action done to the subject. The recipient of the action receives more emphasis than the doer of the action. A passive voice verb is always at least two words, a form of the verb *to be* and the past participle (third principal part) of the main verb.

My husband's surgery *was performed* by Dr. Petro Vanetti.
Logarithm tables *are found* on page 210.

See also pages 12–13, 651–652.

■ The voice of the verb that permits the desired emphasis should be chosen.

ACTIVE VOICE   The ballistics experts *examined* the results of the tests. (Emphasis on *ballistics experts*)
PASSIVE VOICE   The results of the tests *were examined* by the ballistics experts. (Emphasis on *results of the tests*)
ACTIVE VOICE   Juan *gave* the report. (Emphasis on *Juan*)
PASSIVE VOICE   The report *was given* by Juan. (Emphasis on *report*)

■ The writer should avoid needlessly shifting from the active to the passive voice.

NEEDLESS SHIFT IN VOICE (CONFUSING)   Management and labor representatives *discussed* the pay raise, but no decision *was reached*.
discussed—active voice
was reached—passive voice

Management and labor representatives *discussed* the pay raise, but
they *reached* no decision.

$$\frac{discussed}{reached}$$—active voice

VERBALS

Verbals are formed from verbs and are used as modifiers or nouns.

The trees *planted on the slope* have prevented further erosion. (Modifier)
The persons who wanted *to scale the mountain* have done so. (noun)
Delegating authority is a simple and natural process. (noun)

Verbals are divided into three groups—participles, gerunds, and infinitives.

■ Participles are verb forms that function as adjectives, and they have three forms:
present participle (ends in *-ing*), past participle (ends in *-d, -ed, -n, -en,* or *-t*),
and present perfect participle (*having* plus the past participle).

Designing her own home, the architect used an ultramodern decor.
Pictures *produced by colored light* may replace painted pictures in future homes.
Having completed the experiment, I was convinced that silver was a better conductor
than copper.

Participles that are not necessary to identify the word modified are set off by
commas. (See comma usage, item 6, page 727.) Participles that identify the noun
modified are *not* set off by commas.

Mr. Owens, *presiding at the council meeting,* called for the committee's report.
A picture *improperly mounted and insecurely fastened* may come loose from the mounting and fall to the floor.

Participles introducing a sentence are set off by a comma. (See comma usage,
item 8, page 728.)

Representing the company, the salesmen can sign contracts.

■ Gerunds are verb forms that end in *-ing, -d, -ed, -n, -en,* or *-t* and that function
as nouns.

Measuring the pattern is necessary for proper fit.
The *unrecognized* are often the backbone of a company.

■ Infinitives are verb forms that function as nouns, adjectives, or adverbs, depending on their use in a sentence. There are two forms of the infinitive: present (*to*
plus first principal part of main verb) and present perfect (*to* plus *have* plus the
third principal part of the main verb).

My purpose is *to become involved in ecological discussion.* (Noun)
By the first of the year I hope *to have completed the tests.* (Noun)
This is the equipment *to be returned.* (Adjective)
To prevent injury to the worker or damage to materials, safe practice must be observed
at all times. (Adverb)

The *to* sign of an infinitive is sometimes omitted for the sake of conventional
sentence sense.

We saw him *perform*.
Let them *go* when they have finished.

Introductory infinitive phrases used as adjectives or adverbs are usually followed by a comma. (See comma usage, item 8, page 728.)

To secure the most satisfactory results, buy a good quality material.
To choose the best job, Lin first talked with an employment counselor.

Problems of Usage

Several common usage problems have been treated in Chapter 1 of the Handbook on pages 654–658: sentence fragments; parallelism in sentences; run-on or fused sentences; comma splice; and shift in focus. The following section is designed to help you solve some other typical usage problems: agreement of pronoun and antecedent; shifts in person, number, and gender; agreement of subject and verb; and placement of modifiers.

AGREEMENT OF PRONOUN AND ANTECEDENT

Pronouns take the place of or refer to nouns. They provide a good way to economize in writing. For example, writing a paper on Anthony van Leeuwenhoek's pioneer work in developing microscopes, you could use the pronoun forms *he, his,* and *him* to refer to Leeuwenhoek rather than repeat his name numerous times. Since pronouns take the place of or refer to nouns, there must be number (singular or plural) agreement between the pronoun and its antecedent (the word the pronoun stands for or refers to).

■ A singular antecedent requires a singular pronoun; a plural antecedent requires a plural pronoun.

NONAGREEMENT   Gritty ink *erasers* should be avoided because *it* invariably damages the working surface of the paper.

AGREEMENT   Gritty ink *erasers* should be avoided because *they* invariably damage the working surface of the paper.

■ Two or more subjects joined by *and* must be referred to by a plural pronoun.

The *business manager* and the *accountant* will plan *their* budget.

■ Two or more singular subjects joined by *or* or *nor* must be referred to by singular pronouns.

Ms. Pullen or *Ms. Merton* has left *her* purse.
Neither *Mr. Conway* nor *Mr. Freeman* will appear to give *his* report.

■ Nouns such as *part, rest,* and *remainder* may be singular or plural. The number is determined by a phrase following the noun.

The *remainder* of the *students* asked that *their* grades be mailed.
The *rest* of the *coffee* was stored in *its* own container.

■ The indefinite pronouns *each, every, everyone* and *everybody, nobody, either, neither, one, anyone* and *anybody,* and *someone* and *somebody* are singular and thus have singular pronouns referring to them.

> *Each* of the sorority members has indicated *she* wishes *her* own room.
> *Neither* of the machines has *its* motor repaired.

■ The indefinite pronouns *both, many, several,* and *few* are plural; these forms require plural pronouns.

> *Several* realized *their* failure was the result of not studying.
> *Both* felt that nothing could save *them.*

■ Still other indefinite pronouns—*most, some, all, none, any,* and *more*—may be either singular or plural. Usually a phrase following the indefinite pronoun will reveal whether the pronoun is singular or plural in meaning.

> *Most* of the *patients* were complimentary of *their* nurses.
> *Most* of the *money* was returned to *its* owner.

SHIFTS IN PERSON, NUMBER, AND GENDER

Avoid shifting from one person to another, from one number to another, or from one gender to another—such shifting creates confusing, awkward constructions.

SHIFT IN PERSON
: When *I* was just an apprentice welder, the boss expected *you* to know all the welding processes. (*I* is first person; *you* is second person)

REVISION
: When *I* was just an apprentice welder, the boss expected *me* to know all the welding processes.

SHIFT IN NUMBER
: *Each* person in the space center control room watched *his* dial anxiously. *They* thought the countdown to "zero" would never come. (*Each* and *his* are singular pronouns; *they* is plural)

REVISION
: *Each* person in the space center control room watched *his* dial anxiously and thought the countdown to "zero" would never come.

SHIFT IN GENDER
: The mewing of my cat told me *it* was hungry, so I gave *her* some food. (*It* is neuter; *her* is feminine)

REVISION
: The mewing of my cat told me *she* was hungry, so I gave *her* some food.

AGREEMENT OF SUBJECT AND VERB

A subject and verb agree in person and number. Person denotes person speaking (first person), person spoken to (second person), and person spoken of (third person). The person used determines the verb form that follows.

Present Tense

FIRST PERSON	I go	we go	I am	we are
SECOND PERSON	you go	you go	you are	you are
THIRD PERSON	he goes	they go	he is	they are
IMPERSONAL	one goes		one is	

- A subject and verb agree in number; that is, a plural subject requires a plural verb and a singular subject requires a singular verb.

  > The *symbol* for hydrogen *is* H. (Singular subject; singular verb)
  > Comic *strips are* often vignettes of real-life situations. (Plural subject; plural verb)

- A compound subject requires a plural verb.

  > *Employees* and *employers work* together to formulate policies.

- The pronoun *you* as subject, whether singular or plural in meaning, takes a plural verb.

  > *You were assigned* the duties of staff nurse on fourth floor.
  > Since *you are* new technicians, *you are* to be paid weekly.

- In a sentence containing both a positive and a negative subject, the verb agrees with the positive.

  > The *employees,* not the *manager, were asked* to give their opinions regarding working conditions. (Positive subject plural; verb plural)

- If two subjects are joined by *or* or *nor,* the verb agrees with the nearer subject.

  > *Graphs* or a *diagram aids* the interpretation of statistical reports. (Subject nearer verb singular; verb singular)
  > A *graph* or *diagrams aid* the interpretation of this statistical report. (Subject nearer verb plural; verb plural)

- A word that is plural in form but names a single object or idea requires a singular verb.

  > The *United States has changed* from an agricultural to a technical economy.
  > Twenty-five *dollars was offered* for any usable suggestion.
  > Six *inches is* the length of the narrow rule.

- The term *the number* generally takes a singular verb; *a number* takes a plural verb.

  > *The number* of movies attended each year by the average American *has decreased.*
  > A *number* of modern inventions *are* the product of the accumulation of vast storehouses of smaller, minor discoveries.

- When followed by an *of* phrase, *all, more, most, some,* and *part;* fractions; and percentages take a singular verb if the object of the *of* is singular and a plural verb if the object of the *of* is plural.

  > *Two-thirds* of his *discussion was* irrelevant. (Object of *of* phrase, *discussion,* is singular; verb is singular)
  > *Some* of the *problems* in a college environment *require* careful analysis by both faculty and students. (Object of *of* phrase, *problems,* is plural; verb is plural)

- Elements that come between the subject and the verb ordinarily do not affect subject-verb agreement.

  > *One* of the numbers *is* difficult to represent because of the great number of zeros necessary.
  > Such *factors* as temperature, available food, age of organism, or nature of the suspending medium *influence* the swimming speed of a given organism.

- Occasionally the subject may follow the verb, especially in sentences beginning with the expletives *there* or *here;* however, such word order does not change the subject-verb agreement. (The expletives *there* and *here*, in their usual meaning, are never subjects.)

> There *are* two common temperature *scales:* Fahrenheit and Celsius.
> There *is* a great *deal* of difference in the counseling techniques used by contemporary ministers.
> Here *are* several *types* of film.

- The introductory phrase *It is* or *It was* is always singular, regardless of what follows.

> *It is* these problems that overwhelm me.
> *It was* the officers who met.

- Relative pronouns *(who, whom, whose, which, that)* may require a singular or a plural verb, depending on the antecedent of the relative pronoun.

> She is one of those people *who keep* calm in an emergency. (Antecedent of *who* is *people*, a plural form; verb plural)
> The earliest lab *that is* offered is at 11:00. (Antecedent of *that* is *lab*, a singular form; verb singular)

MODIFIERS

Modifiers are words, phrases, or clauses, either adjective or adverb, that limit or restrict other words in the sentence. Careless construction and placement of these modifiers may cause problems such as dangling modifiers, dangling elliptical clauses, misplaced modifiers, and squinting modifiers.

Dangling Modifiers. Dangling modifiers or dangling phrases occur when the word the phrase should modify is hidden within the sentence or is missing.

WORD HIDDEN IN SENTENCE	Holding the bat tightly, the ball was hit by the boy. (Implies that the ball was holding the bat tightly)
REVISION	Holding the bat tightly, the boy hit the ball.
WORD MISSING	By placing a thermometer under the tongue for approximately three minutes, a fever can be detected. (Implies that a fever places a thermometer under the tongue)
REVISION	By placing a thermometer under the tongue for approximately three minutes, anyone can tell if a person has a fever.

- Dangling modifiers may be corrected by either rewriting the sentence so that the word modified by the phrase immediately follows the phrase (this word is usually the subject of the sentence) or rewriting the sentence by changing the phrase to a dependent clause.

DANGLING MODIFIER	*Driving down Main Street*, the city auditorium came into view.
SENTENCE REWRITTEN TO CLARIFY WORD MODIFIED	Driving down Main Street, I saw the city auditorium.
SENTENCE REWRITTEN WITH A DEPENDENT CLAUSE	As I was driving down Main Street, the city auditorium came into view.

Dangling Elliptical Clauses. In elliptical clauses some words are understood rather than stated. For example, the dependent clause in the following sentence is elliptical: *When measuring the temperature of a conductor, you must use the Celsius scale;* the subject and part of the verb have been omitted. The understood subject of an elliptical clause is the same as the subject of the sentence. This is true in the example: *When (you are) measuring the temperature of a conductor, you must use the Celsius scale.* If the understood subject of the elliptical clause is not the same as the subject of the sentence, the clause is a dangling clause.

DANGLING CLAUSE	*When using an electric saw*, safety glasses should be worn.
REVISION	*When using an electric saw*, wear safety glasses.

■ Dangling elliptical clauses may be corrected by either including within the clause the missing words or rewriting the main sentence so that the stated subject of the sentence and the understood subject of the clause will be the same.

DANGLING CLAUSE	*After changing the starter switch*, the car still would not start.
SENTENCE REVISED; MISSING WORDS INCLUDED	After *the mechanic* changed the starter switch, the car still would not start.
SENTENCE REVISED; UNDERSTOOD SUBJECT AND STATED SUBJECT THE SAME	*After changing the starter switch, the mechanic* still could not start the car.

Misplaced Modifiers. Place modifiers near the word or words modified. If the modifier is correctly placed, there should be no confusion. If the modifier is incorrectly placed, the intended meaning of the sentence may not be clear.

MISPLACED MODIFIER	The machinist placed the work to be machined in the drill press vise *called the workpiece*.
CORRECTLY PLACED MODIFIER	The machinist placed the work to be machined, *called the workpiece*, in the drill press vise.
MISPLACED MODIFIER	Where are the shirts for children *with snaps?*
CORRECTLY PLACED MODIFIER	Where are the children's shirts *with snaps?*

Squinting Modifiers. A modifier should clearly limit or restrict *one* sentence element. If a modifier is so placed within a sentence that it can be taken to limit or restrict either of two elements, the modifier is squinting; that is, the reader cannot tell which way the modifier is looking.

SQUINTING MODIFIER	When the student began summer work *for the first time* he was expected to follow orders promptly and exactly. (The modifying phrase could belong to the clause that precedes it or to the clause that follows it.)

Punctuation may solve the problem. The sentence might be written in either of the following ways, depending on the intended meaning.

REVISION	When the student began summer work *for the first time*, he was expected to follow orders promptly and exactly.
REVISION	When the student began summer work, *for the first time* he was expected to follow orders promptly and exactly.

■ Squinting modifiers often are corrected by shifting their position in the sentence. Sentence meaning determines placement of the modifier.

SQUINTING MODIFIER   The students were advised *when it was midmorning* the new class schedule would go into effect.

REVISION   *When it was midmorning,* the students were advised that the new class schedule would go into effect.

REVISION   The students were advised that the new class schedule would go into effect *when it was midmorning.*

■ Particularly difficult for some writers is the correct placement of *only, almost,* and *nearly.* These words generally should be placed immediately before the word they modify, since changing position of these words within a sentence changes the meaning of the sentence.

In the reorganization of the district the member of congress *nearly* lost a hundred voters. (Didn't lose any voters)

In the reorganization of the district the member of congress lost *nearly* a hundred voters. (Lost almost one hundred voters)

The customer *only* wanted to buy soldering wire. (Emphasizes *wanted*)

The customer wanted to buy *only* soldering wire. (Emphasizes *soldering wire*)

Words Often Confused and Misused

Here is a list of often confused and misused words, with suggestions for their proper use.

a, an   *A* is used before words beginning with a consonant sound; *an* is used before words with a vowel sound. (Remember: Consider sound, not spelling.)

EXAMPLES   This is *a* banana.
This is *an* orange.

accept, except   *Accept* means "to take an object or idea offered" or "to agree to something"; *except* means "to leave out" or "excluding."

EXAMPLES   Please *accept* this gift.
Everyone *except* Joe may leave.

access, excess   *Access* is a noun meaning "way of approach" or "admittance"; *excess* means "greater amount than required or expected."

EXAMPLES   The children were not allowed *access* to the laboratory.
Her income is in *excess* of $50,000.

ad   *Ad* is a shortcut to writing *advertisement.* In formal writing, however, write out the full word. In informal writing and speech, such abbreviated forms as *ad, auto, phone, photo,* and *TV* may be acceptable.

advice, advise   *Advice* is a noun meaning "opinion given," "suggestions"; *advise* is a verb meaning "to suggest," "to recommend."

EXAMPLES   We accepted the lawyer's *advice.*
The lawyer *advised* us to drop the charges.

affect, effect   *Affect* as a verb means "to influence" or "to pretend"; as a noun, it is a psychological term meaning "feeling" or "emotion." *Effect* as a verb means "to make something happen"; as a noun, it means "result" or "consequence."

EXAMPLES   The colors used in a home may *affect* the prospective buyer's decision to buy.
The new technique will *effect* change in the entire procedure.

aggravate, irritate   *Aggravate* means "to make worse or more severe." Avoid using *aggravate* to mean "to irritate" or "to vex," except perhaps in informal writing or speech.

ain't   A contraction for *am not, are not, has not, have not*. This form is still regarded as substandard; careful speakers and writers do not use it.

all ready, already   *All ready* means that everyone is prepared or that something is completely prepared; *already* means "completed" or "happened earlier."

EXAMPLES   We are *all ready* to go.
It is *already* dark.

all right, alright   Use *all right*. In time, *alright* may become accepted, but for now *all right* is generally preferred.

EXAMPLE   Your choice is *all right* with me.

all together, altogether   *All together* means "united"; *altogether* means "entirely."

EXAMPLES   We will meet *all together* at the clubhouse.
There is *altogether* too much noise in the hospital area.

almost, most   *Almost* is an adverb meaning "nearly"; *most* is an adverb meaning "the greater part of the whole."

EXAMPLES   The emergency shift *almost* froze.
The emergency shift worked *most* of the night.

a lot, alot, allot   *A lot* is written as two separate words and is a colloquial term meaning "a large amount"; *alot* is a common miswriting for *a lot*; *allot* is a verb meaning "to give a certain amount."

among, between   Use *among* when talking about more than two. Use *between* to express the relation between two things or the relation of a thing to many surrounding things.

amount, number   *Amount* refers to mass or quantity; *number* refers to items, objects, or ideas that can be counted individually.

EXAMPLES   The *amount* of money for clothing is limited.
A large *number* of people are enrolled in the class.

and/or   This pairing of coordinate conjunctions indicates appropriate alternatives. Avoid using *and/or* if it misleads or confuses the reader, or if it indicates imprecise thinking. In the sentence, "Jones requests vacation leave for Monday–Wednesday and/or Wednesday–Friday," the request is not clear because the reader does not know whether three days' leave or five days' leave is requested. Use *and* and *or* to mean exactly what you want to say.

EXAMPLES   She is a qualified lecturer *and* consultant.
For our vacation we will go to London *or* Madrid *or* both.

See also Virgule, page 733.

angel, angle   *Angel* means "a supernatural being"; *angle* means "corner" or "point of view." Be careful not to overuse *angle* meaning "point of view." Use *point of view, aspect,* and the like.

anywheres, somewheres, nowheres   Use *anywhere, somewhere,* or *nowhere.*

as if, like   *As if* is a subordinate conjunction; it should be followed by a subject-verb relationship to form a dependent clause. *Like* is a preposition; in formal writing *like* should be followed by a noun or a pronoun as its object.

EXAMPLES   He reacted to the suggestion *as if* he never heard of it.
Pines, *like* cedars, do not have leaves.

as regards, in regard to   Avoid these phrases. Use *about* or *concerning.*

average, mean, median   *Average* is the quotient obtained by dividing the sum of the quantities by the number of quantities. For example, for scores of 70, 75, 80, 82, 100, the *average* is 81⅖. *Mean* may be the simple average or it may be the value midway between the lowest and the highest quantity (in the scores above, 85). *Median* is the middle number (in the scores above, 80).

balance, remainder   *Balance* as used in banking, accounting, and weighing means "equality between the totals of two sides." *Remainder* means "what is left over."

EXAMPLES   The company's bank *balance* continues to grow.
Our shift will work overtime the *remainder* of the week.

being that, being as how   Avoid using either of these awkward phrases. Use *since* or *because.*

EXAMPLE   *Because* the bridge is closed, we will have to ride the ferry.

beside, besides   *Beside* means "alongside," "by the side of," or "not part of"; *besides* means "furthermore" or "in addition."

EXAMPLES   The tree stands *beside* the walk.
Besides the cost there is a handling charge.

bi-, semi-   *Bi-* is a prefix meaning "two," and *semi-* is a prefix meaning "half," or "occurring twice within a period of time."

EXAMPLES   Production quotas are reviewed *biweekly.* (Every two weeks)
The board of directors meets *semiannually.* (Every half year, or twice a year)

brake, break   *Brake* is a noun meaning "an instrument to stop something"; *break* is a verb meaning "to smash," "to cause to fall apart."

EXAMPLES   The mechanic relined the *brakes* in the truck.
If the vase is dropped, it will *break.*

can't hardly   This is a double negative; use *can hardly.*

capital, capitol *Capital* means "major city of a state or nation," "wealth," or, as an adjective, "chief" or "main." *Capitol* means "building that houses the legislature"; when written with a capital "C," it usually means the legislative building in Washington.

EXAMPLES Jefferson City is the *capital* of Missouri.
Our company has a large *capital* investment in preferred stocks.
The *capitol* is located on Third Avenue.

cite, sight, site *Cite* is a verb meaning "to refer to"; *sight* is a noun meaning "view" or "spectacle"; *site* is a noun meaning "location."

EXAMPLES *Cite* a reference in the text to support your theory.
Because of poor *sight*, he has to wear glasses.
This is the building *site* for our new home.

coarse, course *Coarse* is an adjective meaning "rough," "harsh," or "vulgar"; *course* is a noun meaning "a way," "a direction."

EXAMPLES The sandpaper is too *coarse* for this wood.
The creek followed a winding *course* to the river.

consensus *Consensus* means "a general agreement of opinion." Therefore, do not write *consensus of opinion*; it is repetitious.

contact *Contact* is overused as a verb, especially in business and industry. Consider using in its place such exact forms as *write to, telephone, talk with, inform, advise,* or *ask*.

continual, continuous *Continual* means "often repeated"; *continuous* means "uninterrupted" or "unbroken."

EXAMPLES The conference has had *continual* interruptions.
The rain fell in a *continuous* downpour for an hour.

could of The correct form is *could have*. This error occurs because of the sound heard in pronouncing such statements as: We *could have* (could've) completed the work on time.

council, counsel, consul *Council* is a noun meaning "a group of people appointed or elected to serve in an advisory or legislative capacity." *Counsel* as a noun means "advice" or "attorney"; as a verb, it means "to advise." *Consul* is a noun naming the official representing a country in a foreign nation.

EXAMPLES The club has four members on its *council*.
The *counsel* for the defense advised him to testify.
The *consul* from Switzerland was invited to our international tea.

criteria, criterion *Criteria* is the plural of *criterion*, meaning "a standard on which a judgment is made." Although *criteria* is the preferred plural, *criterions* is also acceptable.

EXAMPLES This is the *criterion* that the trainee did not understand.
On these *criteria*, the proposals will be evaluated.

device, devise *Device* is a noun meaning "a contrivance," "an appliance," "a scheme"; *devise* is a verb meaning "to invent."

EXAMPLES   This *device* will help prevent pollution of our waterways.
We need to *devise* a safer method for drilling offshore oil wells.

different from, different than   Although *different from* is perhaps more common, *different than* is also an acceptable form.

discreet, discrete   *Discreet* means "showing good judgment in conduct and especially in speech." *Discrete* means "consisting of distinct, separate, or unconnected elements."

EXAMPLES   The administrative assistant was *discreet* in her remarks.
Discrete electronic circuitry was the standard until the advent of integrated circuits.

dual, duel   *Dual* means "double." *Duel* as a noun means a fight or contest between two people; as a verb, it means "to fight."

EXAMPLES   The car has a *dual* exhaust.
He was shot in a *duel*.

due to   Some authorities object to *due to* in adverbial phrases. Acceptable substitutes are *owing to* and *because of*.

each and every   Use one or the other. *Each and every* is a wordy way to say *each* or *every*.

EXAMPLE   *Each* person should make a contribution.

except for the fact that   Avoid using this wordy and awkward phrase.

fact, the fact that   Use *that*.

field   Used too often to refer to an area of knowledge or a subject.

had ought, hadn't ought   Avoid using these phrases. Use *ought* or *should*.

EXAMPLE   He *should* not speak so loudly.

hisself   Use *himself*.

imply, infer   *Imply* means to "express indirectly." *Infer* means to "arrive at a conclusion by reasoning from evidence."

EXAMPLES   The supervisor *implied* that additional workers would be laid off.
I *inferred* from her comments that I would not be one of them.

in case, in case of, in case that   Avoid using this overworked phrase. Use *if*.

in many instances   Wordy. Use *frequently* or *often*.

in my estimation, in my opinion   Wordy. Use *I believe* or *I think*.

irregardless   Though you hear this double negative and see it in print, it should be avoided. Use *regardless*.

its, it's   *Its* is the possessive form; *it's* is the contraction for *it is*. The simplest way to avoid confusing these two forms is to think *it is* when writing *it's*.

EXAMPLES   The tennis team won *its* match.
It's time for lunch.

lay, lie   *Lay* means "to put down" or "place." Forms are *lay, laid,* and *laying*. It is a transitive verb; thus it denotes action going to an object or to the subject.

EXAMPLES   *Lay* the books on the table.
We *laid* the floor tile yesterday.

Lie means "to recline" or "rest." Forms are *lie, lay, lain,* and *lying*. It is an intransitive verb; thus it is never followed by a direct object.

EXAMPLES   The pearls *lay* in the velvet-lined case.
Lying in the velvet-lined case were the pearls.

lend, loan   *Lend* is a verb. *Loan* is used as a noun or a verb; however, many careful writers use it only as a noun.

loose, lose   *Loose* means "to release," "to set free," "unattached," or "not securely fastened"; *lose* means "to suffer a loss."

EXAMPLE   We turned the horses *loose* in the pasture, but we locked the gate so that we wouldn't *lose* them.

lots of, a lot of   In writing, use *many, much, a large amount*.

might of, ought to of, must of, would of   *Of* should be *have*. See *could of*.

off of   Omit *of*. Use *off*.

on account of   Use *because*.

one and the same   Wordy. Use *the same*.

outside of   Use *besides, except for,* or *other than*.

passed, past   *Passed* identifies an action and is used as a verb; *past* means "earlier" and is used as a modifier.

EXAMPLES   He *passed* us going 80 miles an hour.
In the *past*, bills were sent out each month.

personal, personnel   *Personal* is an adjective meaning "private," "pertaining to the person"; *personnel* is a noun meaning "body of persons employed."

EXAMPLES   Please do not open my *personal* mail.
He is in charge of hiring new *personnel*.

plain, plane   *Plain* is an adjective meaning "simple," "without decoration"; *plane* is a noun meaning "airplane," "tool," or "type of surface."

EXAMPLES   The *plain* decor of the room created a pleasing effect.
We worked all day checking the engine in the *plane*.

principal, principle   *Principal* means "highest," "main," or "head"; *principle* means "belief," "rule of conduct," or "fundamental truth."

EXAMPLES   The school has a new *principal*.
His refusal to take a bribe was a matter of *principle*.

proved, proven   Use either form.

put across   Use more exact terms, such as *demonstrate, explain, prove, establish*.

quiet, quite   *Quiet* is an adjective meaning "silence" or "free from noise"; *quite* is an adverb meaning "completely" or "wholly."

> EXAMPLES   Please be *quiet* in the library.
> It's been *quite* a while since I've seen him.

raise, rise   *Raise* means "to push up." Forms are *raise, raised, raising*. It is a transitive verb; thus it denotes action going to an object or to the subject.

> EXAMPLES   *Raise* the window.
> The technician *raised* the impedance of the circuit 300 ohms.

Rise means "to go up" or "ascend." Forms are *rise, rose, risen*. It is an intransitive verb and thus it is never followed by a direct object.

> EXAMPLES   Prices *rise* for several reasons.
> The sun *rose* at 6:09 this morning.

rarely ever, seldom ever   Avoid using these phrases. Use *rarely* or *seldom*.

read where   Use *read that*.

reason is because, reason why   Omit *because* and *why*.

respectfully, respectively   *Respectfully* means "in a respectful manner"; *respectively* means "in the specified order."

> EXAMPLES   I *respectfully* explained my objection.
> The capitals of Libya, Iceland, and Tasmania are Tripoli, Reykjavik, and Hobart, *respectively*.

sense, since   *Sense* means "ability to understand"; *since* is a preposition meaning "until now," an adverb meaning "from then until now," and a conjunction meaning "because."

> EXAMPLES   At least he has a *sense* of humor.
> I have been on duty *since* yesterday.

set, sit   *Set* means "to put down" or "place." Its basic form does not change: *set, set, setting*. It is a transitive verb; thus it denotes action going to an object or to the subject.

> EXAMPLES   Please *set* the test tubes on the table.
> The electrician *set* the breaker yesterday.

Sit means "to rest in an upright position." Forms are *sit, sat, sitting*. It is an intransitive verb and thus it is never followed by a direct object.

> EXAMPLES   Please *sit* here.
> The electrician *sat* down when the job was finished.

state   Use exact terms such as *say, remark, declare, observe. To state* means "to declare in a formal statement."

stationary, stationery   *Stationary* is an adjective meaning "fixed"; *stationery* is a noun meaning "paper used in letter writing."

> EXAMPLES   The workbench is *stationary*.
> The school's *stationery* is purchased through our firm.

their, there, they're   *Their* is a possessive pronoun; *there* is an adverb of place; *they're* is a contraction for *they are*.

> EXAMPLES   *Their* band is in the parade.
> *There* goes the parade.
> *They're* in the parade.

this here, that there   Avoid this phrasing. Use *this* or *that*.

> EXAMPLE   *This* machine is not working properly.

thusly   Use *thus*.

till, until   Either word may be used.

to, too, two   *To* is a preposition; *too* is an adverb telling "How much?"; *two* is a numeral.

try and   *Try to* is generally preferred.

type, type of   In writing, use *type of*.

used to could   Use *formerly was able* or *used to be able*.

where . . . at   Omit the *at*. Write "Where is the library?" (not "Where is the library at?")

who's, whose   *Who's* is the contraction for *who is; whose* is the possessive form of *who*.

> EXAMPLES   *Who's* on the telephone?
> *Whose* coat is this?

-wise   Currently used and overused as an informal suffix in such words as *timewise, safetywise, healthwise*. Better to avoid such usage.

would of   The correct form is *would have*. This error occurs because of the sound heard in pronouncing such a statement as this: He *would have* (would've) come if he had not been ill.

Application 1   Grammatical Usage

Think for a few minutes about your choice of career and the study necessary to prepare for that career. Then write five sentences about your thoughts. Analyze the sentences by answering the following questions.

1. Do your sentences contain any *adjectives, adverbs, conjunctions, interjections, nouns, prepositions, pronouns, verbs,* or *verbals?* Divide a page into nine columns; head column I "Adjectives," column II "Adverbs," and so on. Then, list the words from your sentences in the appropriate column.
2. Identify the subject(s) and verb(s) in each of the five sentences. Explain the agreement in number of each subject and verb.
3. For every pronoun listed in Question 1 above, identify an antecedent. Explain how each pronoun agrees with its antecedent.

Application 2 Grammatical Usage

Make a list of usage problems you have had in writing assignments. (Study this chapter and other available sources that explain how to solve your usage problems. Then, as you write, try to correct these problems.)

Application 3 Grammatical Usage

Rewrite the following sentences to correct errors in usage.

1. The team only managed to practice two days.
2. The glass face of the meter.
3. The workers were injured by the blast security officers were on the scene immediately to help.
4. The children's ward located in the hospital's east wing which houses patients from age 2 through age 10 was completed last year.
5. According to a report I read in last night's newspaper.
6. The project was quite complex, nevertheless I managed to complete it.
7. The instructors taught the new employees to serve customers efficiently, and while teaching them they seemed alert and knowledgeable.
8. If one understands a set of instructions, they will have little difficulty in following them.
9. We washed the windows to make the room clean and for more light.
10. Though tired, the suggestion that we go out to dinner was appealing to us.
11. Molly wanted to go to school and continuing with her work.
12. Because of the shortage of trained personnel.
13. While working for IBM, Walt received many honors one of them was being selected as the outstanding speaker at the annual convention.
14. After observing the assembly workers for a period of time, the supervisor made suggestions for improving production, then the suggestions were analyzed and implemented.
15. One of the workers have left their tools here.
16. The committee chairperson or myself will prepare the annual report.
17. My department manager insists on.
18. Tables or a graph aid the clear presentation of these statistics.
19. The number of calories are easily reduced.
20. The diesel engine is an increasingly popular engine in automobiles it does however have some disadvantages.

Application 4 Grammatical Usage

On pages 705–712 is a list of words often confused and misused. Study this list. Choose ten of the words that you confuse and/or misuse and write sentences using the words in an acceptable way.

Application 5 Grammatical Usage

Keep a list of words that you often confuse and/or misuse. Study definitions of these words; practice using the words in writing and in speaking.

Application 6 Grammatical Usage

Students should separate into two groups. From each group select a leader who will write on the board words from the list of words often confused and misused. The students in the first group will take turns using the words acceptably in sentences. If a student in that group cannot use a word acceptably, the opportunity passes to the other group. Count five points for each acceptable usage.

Application 7 Grammatical Usage

The following paragraph contains many misused and confused terms; for each unacceptable usage substitute an acceptable one.

Joe Schmoe was a good worker in regard to his actual performance on the job. But in many instances he complained about the access amount of responsibilities. He all ready had sought the council of alot of his fellow workers, trying to find an angel to use in devicing a duel plan to present to his superiors. He hisself wanted to try and establish a consensus of opinion, establish some principal, and then offer his advise to his superiors. Joe felt that altogether the workers between them could device a coarse of action that would be exceptable to management personal sense they had proved quiet agreeable to passed suggestions, irregardless of the fact that their had been too few personal added. The workers could of let there emotions guide them, but they didn't. Led by Joe, they sited plane facts, showing how in many instances workers were assigned to many responsibilities. Thusly they avoided a brake between management and workers. Its better generally to site facts!

Chapter 4
Mechanics

715

Objectives

Upon completing this chapter, the student should be able to:

- Use accepted forms of abbreviations, capitalization, numbers, plurals of nouns, punctuation, and spelling
- Explain the difficulty in using symbols
- Give examples to show how conventional use of mechanics (abbreviations, capitalization, etc.) helps make meaning clearer
- Identify the different punctuation marks and illustrate the various uses of each

Introduction

Threw the yrs. certin, conventions in the mechanics of written communication; have developed? These convention: Or generally accepted practicen;·eze the communication process for when. These conventional usages or followed! the (readers attention) can be rightly focused on the c. of the wtg.

Through the years certain conventions in the mechanics of written communication have developed. These conventions, or generally accepted practices, ease the communication process, for when these conventional usages are followed, the reader's attention can be rightly focused on the content of the writing.

Which is easier to follow—the first or the second paragraph? Undoubtedly the second, because it follows generally accepted practices in the mechanics of written English. The second paragraph permits you to concentrate on what is being said; further, it is thoughtful of you, the reader, in not requiring you to spend a great deal of time in simply figuring out the words and the sentence units before beginning to understand the subject material. The matter of mechanics, in short, is a matter of convention and of courtesy to the reader.

The following are accepted practices concerning abbreviations, capitalization, numbers, plurals of nouns, punctuation, spelling, and symbols. Applying these practices to your writing will help your reader understand what you are trying to communicate.

Abbreviations

Always consult a recent dictionary for forms you are not sure about. Some dictionaries list abbreviations together in a special section; other dictionaries list abbreviated forms as regular entries in the body of the dictionary.

Always acceptable abbreviations

1. Abbreviations generally indicate informality. Nevertheless, there are a few abbreviations always acceptable when used to specify a time or a person, such as a.m. (ante meridiem, before noon), BC (before Christ), AD (anno Domini, in the year of our Lord), Ms. or Ms (combined form of Miss and Mrs.), Mrs. (mistress), Mr. (mister), Dr. (doctor).

Dr. Ann Meyer and Mrs. James Brown will arrive at 7:30 p.m.

Titles following names

2. Certain titles following a person's name may be abbreviated, such as Jr. (Junior), Sr. (Senior), MD (Doctor of Medicine), SJ (Society of Jesus).

> Martin Luther King, Jr., was assassinated on April 4, 1968. (The commas before and after "Jr." are optional.)
> George B. Schimmet, MD, signed the report.

Titles preceding names

3. Most titles may be abbreviated when they precede a person's full name, but not when they precede only the last name.

> Lt. James W. Smith or Lieutenant Smith but *not* Lt. Smith
> The Rev. Arthur Bowman or the Reverend Bowman but *not* Rev. Bowman or the Rev. Bowman

Terms with numerals

4. Abbreviate certain terms only when they are used with a numeral.

> a.m., p.m., BC, AD, No. (number), $ (dollars)

(Careful writers place "BC" after the numeral and "AD" before the numeral: 325 BC, AD 597)

ACCEPTABLE   He arrives at 2:30 p.m.
             Julius Caesar was killed in 44 BC.
             The book costs $12.40.
UNACCEPTABLE He arrives this p.m.
             Julius Caesar was killed a few years B.C.
             The book costs several $.

Repeated term or title

5. To avoid repetition of a term or a title appearing many times in a piece of writing, abbreviate the term or shorten the title. Write out in full the term or title the first time it appears, followed by the abbreviated or shortened form in parentheses.

> Many high school students take the American College Test (ACT) in the eleventh grade; however, a number of students prefer to take the ACT in the twelfth grade.
> *A Portrait of the Artist as a Young Man (Portrait)* is an autobiographical novel by James Joyce. *Portrait* shows the struggle of a young man in answering the call of art.

Periods with abbreviations

6. Generally, place a period after each abbreviation. However, there are many exceptions:

☐ The abbreviations of organizations, of governmental divisions, and of educational degrees usually do not require periods (or spacing between the letters of the abbreviation).

> IBM (International Business Machines), DECA (Distributive Education Clubs of America), IRS (Internal Revenue Service), AAS (Associate of Applied Science) degree

☐ The U.S. Postal Service abbreviations for states are written without periods (and in all uppercase—capital—letters). For a complete list of these abbreviations, see pages 256–257 in Chapter 7, Memorandums and Letters.

☐ Roman numerals in sentences and contractions are written without periods.

> Henry VIII didn't let enemies stand in his way.

☐ Units of measure, with the exceptions of in. (inch) and at. wt. (atomic weight), are written without periods. See 9 below.

> He weighs 186 lb and stands 6 ft tall.

Plural terms

7. Abbreviations of plural terms are written in various ways. Add *s* to some abbreviations to indicate more than one; others do not require the *s*. See 9 below.

ABBREVIATIONS ADDING "s"	Figs. (or Figures) 1 and 2
	20 vols. (volumes)
ABBREVIATIONS WITHOUT "s"	pp. (pages)
	ff. (and following)

Lowercase letters

8. Generally use lowercase (noncapital) letters for abbreviations except for abbreviations of proper nouns.

> mpg (miles per gallon)   Btu (British thermal unit)
> c.d. (cash discount)   UN (United Nations)

Units of measure as symbols

9. Increasingly, the designations for units of measure are being regarded as symbols rather than as abbreviations. As symbols, the designations have only one form—regardless of whether the meaning is singular or plural—and are written without a period.

> 100 kph (kilometers per hour)   200 rpm (revolutions per minute)
> 50 m (meters)   1 T (tablespoon)

Capitalization

There are very few absolute rules concerning capitalization. Many reputable businesses and publishers, for instance, have their own established practices of capitalization. However, the following are basic conventions in capitalization and are followed by most writers.

Sentences

1. Capitalize the first word of a sentence or of a group of words understood as a sentence (except a short parenthetical sentence within another).

> After the party no one offered to help clean up. Not one person.

Quotations

2. Capitalize the first word of a direct quotation.

> Melissa replied, "Tomorrow I begin."

Proper nouns

3. Capitalize proper nouns:

People	☐ Names of people, and titles referring to specific persons

    Frank Lloyd Wright   Mr. Secretary
    Aunt Marian   the Governor

Places   ☐ Places (geographic locations, streets), but not directions

    Canada   Golden Gate Bridge
    Canal Street   the Smoky Mountains
    the South   the Red River

    Go three blocks south; then turn west.

Groups   ☐ Nationalities, organizations, institutions, and members of each

    Indian   Bear Creek High School
    British   League of Women Voters
    a Rotarian   International Imports, Inc.

Calendar divisions   ☐ Days of the week, months of the year, and special days, but not seasons of the year

    Monday   Halloween   New Year's Day
    January   spring   summer

Historic occurrences   ☐ Historic events, periods, and documents

    World War II   the Industrial Revolution
    the Magna Carta   Battle of San Juan Hill

Religions   ☐ Religions and religious groups

    Judaism   the United Methodist Church

Deity   ☐ Names of the Deity and personal pronouns referring to the Deity

    God   Son of God
    Creator   His, Him, Thee, Thy, Thine

Bible   ☐ Bible, Scripture, and names of the books of the Bible These words are not italicized (underlined in handwriting).

    My favorite book in the Bible is Psalms.

Proper noun derivatives   **4.** Capitalize derivatives of proper nouns when used in their original sense.

    Chinese citizen   *but*   china pattern
    Salk vaccine   *but*   pasteurized milk

Pronoun I   **5.** Capitalize the pronoun *I*.

    In that moment of fear, I could not say a word.

Titles of publications   **6.** Capitalize titles of books, chapters, magazines, newspapers, articles, poems, plays, stories, musical compositions, paintings, motion pictures, and the like. Ordinarily, do not capitalize articles *(a, an, the)*, coordinate conjunctions *(and, or,* *but*), and prepositions (*or, by, in, with*), unless they are the first word of the title. It is acceptable to capitalize prepositions of five or more letters (*against, between*).

Newsweek (magazine)
"The Purloined Letter" (short story)
"Tips on Cutting Firewood" (article)
Godspell (musical composition)
Man Against Himself (book)
Madonna and Child (painting)

Titles with names

7. Capitalize titles immediately preceding or following proper names.

Juan Perez, Professor of Computer Science
Dr. Alicia Strumm
Dale Jaggers, Member of Congress

Substitute names

8. Capitalize words or titles used in place of the names of particular persons. However, names denoting kinship are not capitalized when immediately preceded by an article or a possessive.

Last week Mother and my grandmother gave a party for Sis.
Jill's dad and her uncle went hunting with Father and Uncle Bob.

Trade names

9. Capitalize trade names.

Dodge trucks   Hershey bars

Certain words with numerals

10. Capitalize the words *Figure, Number, Table*, and the like (whether written out or abbreviated) when used with a numeral.

See Figure 1. See the accompanying figure.
This is Invoice No. 6143. Check the number on the invoice.

School subjects

11. Capitalize school subjects only if derived from proper nouns (such as those naming a language or a nationality) or if followed by a numeral.

English   Spanish
history   History 1113
algebra   Algebra II

Numbers

The problem with numbers is knowing when to use numerals (figures) and when to use words.

NUMERALS (FIGURES)

Dates, houses, telephones, ZIPs, specific amounts, math, etc.

1. Use numerals for dates, house numbers, telephone numbers, ZIP codes, specific amounts, mathematical expressions, and the like.

July 30, 1987 *or* 30 July 1987   857-5969
600 Race Street   10:30 p.m.
61 percent   Chapter 12, p. 14

Decimals

2. Use numerals for numbers expressed in decimals. Include a zero before the decimal point in writing fractions with no whole number (integer).

12.0006
0.01
0.500 (The zeros following "5" show that accuracy exists to the third decimal place.)

10 and above; three or more words

3. Use numerals for number 10 and above or numbers that require three or more written words.

She sold 12 new cars in 2½ hours.

Several numbers close together

4. Use numerals for several numbers (including fractions) that occur within a sentence or within related sentences.

The recipe calls for 3 cups of sugar, ½ teaspoon of salt, 2 sticks of butter, and ¼ cup of cocoa.
The report for this week shows that our office received 127 telephone calls, 200 letters, 30 personal visits, and 3 telegrams.

Adjacent numbers

5. Use numerals for one of two numbers occurring next to each other.

12 fifty-gallon containers
two hundred 24 × 36 mats

WORDS

Approximate or indefinite numbers

1. Use words for numbers that are approximate or indefinite.

If I had a million dollars, I'd buy a castle in Ireland.
About five hundred machines were returned because of faulty assembling.

Fractions

2. Use words for fractions.

The veneer is one-eighth of an inch thick.
Our club receives three-fourths of the general appropriation.

Below 10

3. Use words for numbers below 10.

There are four quarts in a gallon.

Beginning of sentence

4. Use words for a number or related numbers that begin a sentence.

Fifty cents is a fair entrance fee.
Sixty percent of the freshmen and seventy percent of the sophomores come from this area.

Note: If using words for a number at the beginning of a sentence is awkward, recast the sentence.

UNACCEPTABLE	2175 freshmen are enrolled this semester.
AWKWARD	Two thousand one hundred and seventy-five freshman are enrolled this semester.
ACCEPTABLE	This semester 2175 freshmen are enrolled.

Adjacent numbers

5. Use words for one of two numbers occurring next to each other.

50 six-cylinder cars   four 3600-pound loads

Repeating a number

6 Except in special instances (such as in order letters or legal documents), it is not necessary to repeat a written-out number by giving the numerals in parentheses.

ACCEPTABLE	The drumpet was invented five years ago.
UNNECESSARY	The drumpet was invented five (5) years ago.

Plurals of Nouns

The English language has been greatly influenced not only by the original English but also by a number of other languages, such as Latin, Greek, and French. Some nouns in the English language, especially those frequently used, continue to retain plural forms from the original language.

Most nouns: *s* or *es*

1. Most nouns in the English language form their plural by adding *-s* or *-es*. Add *-s* unless the plural adds a syllable when the singular noun ends in *s*, *ch* (soft), *sh*, *x*, and *z*.

pencil	pencils	mass	masses
desk	desks	church	churches
flower	flowers	leash	leashes
boy	boys	fox	foxes
post	posts	buzz	buzzes

Nouns ending in *y*

2. If a noun ends in *y* preceded by a consonant sound, change the *y* to *i* and add *-es*.

history	histories
penny	pennies

For other nouns ending in *y*, add *-s*.

monkey	monkeys
valley	valleys

Nouns ending in *f* or *fe*

3. A few nouns ending in *f* or *fe* change the *f* to *v* and add *-es* to form the plural. These nouns are:

calf	calves	life	lives	shelf	shelves
elf	elves	loaf	loaves	thief	thieves
half	halves	leaf	leaves	wife	wives
knife	knives	sheaf	sheaves	wolf	wolves

In addition, several nouns may either add *-s* or change the *f* to *v* and add *-es*. These nouns are:

beef	beefs (slang for "complaints"), beeves
scarf	scarfs, scarves
staff	staffs (groups of officers), staffs or staves (poles or rods)
wharf	wharfs, wharves

Nouns ending in *o*

4. Most nouns ending in *o* add -*s* to form the plural. Among the exceptions, which add -*es*, are the following:

echo	echoes	potato	potatoes
hero	heroes	tomato	tomatoes
mosquito	mosquitoes	veto	vetoes

Compound nouns

5. Most compound nouns form the plural with a final -*s* or *es*. A few compounds pluralize by changing the operational part of the compound noun.

handful	handfuls	son-in-law	sons-in-law
go-between	go-betweens		(and other in-law compounds)
good-by	good-bys	passer-by	passers-by
court-martial	courts-martial	editor in chief	editors in chief

Internal vowel change

6. A few nouns form the plural by an internal vowel change. These nouns are:

foot	feet	mouse	mice
goose	geese	tooth	teeth
louse	lice	woman	women
man	men		

-*en* plurals

7. A few nouns form the plural by adding -*en* or -*ren*. These nouns are:

ox	oxen
child	children
brother	brothers, brethren

Foreign plurals

8. Several hundred English nouns, originally foreign, have two acceptable plural forms: the original form and the conventional American English -*s* or -*es* form.

memorandum	memoranda	memorandums
curriculum	curricula	curriculums
index	indices	indexes
criterion	criteria	criterions

Some foreign nouns always keep their original forms in the plural.

crisis	crises	die	dice
analysis	analyses	alumna	alumnae (feminine)
bacterium	bacteria	alumnus	alumni (masculine)
basis	bases	thesis	theses
ovum	ova		

Terms being discussed

9. Letters of the alphabet, signs, symbols, and words used as a topic of discussion form the plural by adding the apostrophe and -*s*.

The *i*'s and *e*'s are not clear.
The sentence has tóo many *and*'s and *but*'s.

Note: Abbreviations and numbers form the plural regularly, that is, by adding *s* or *es*.

The three *PhDs* were born in the 1950s.
The number has two *sixes*.

Same singular and plural forms

10. Some nouns have the same form in both the singular and the plural. In general, names of fish and of game birds are included in this group.

cod	swine	sheep
trout	cattle	Chinese
deer	species	Japanese
quail	corps	Portuguese

Two forms

11. Some nouns have two forms, the singular indicating oneness or a mass, and the plural indicating different individuals or varieties within a group.

a string of fish	four little fishes
a pocketful of money	moneys (or monies) appropriated by Congress
fresh fruit	fruits from Central America

Only plural forms

12. Some nouns have only plural forms. A noun is considered singular, however, if the meaning is singular.

measles (Measles is a contagious disease.)	mumps
economics	news
mathematics	physics
dynamics	molasses

A noun is considered plural if the meaning is plural.

scissors   (The scissors are sharp.)
pants

Punctuation

Punctuation is a necessary part of the written language. Readers and writers depend on marks of punctuation to help prevent vagueness by indicating pauses and stops, separating and setting off various sentence elements, indicating questions and exclamations, and emphasizing main points while subordinating less important sentence content.

Punctuation usage is presented here according to marks used primarily at the end of a sentence (period, question mark, and exclamation point), internal marks that set off and separate (comma, semicolon, colon, dash, and virgule), enclosing marks always used in pairs (quotation marks, parentheses, and brackets), and punctuation of individual words and of terms (apostrophe, ellipsis points, hyphen, and italics).

Period (.)

Statement, command, or request

1. Use a period at the end of a sentence (and of words understood as a sentence) that makes a statement, gives a command, or makes a request (except a short parenthetical sentence within another).

James Naismith invented the game of basketball. (Statement)
Choose a book for me. (Command)
No. (Understood as a sentence)
Naismith (he was a physical education instructor) wanted to provide indoor exercise and competition for students. (Parenthetical sentence with no capital and no period)

See also Sentence Fragments, pages 654–655.

Note: The polite request phrased as a question is usually followed by a period rather than a question mark.

Will you please send me a copy of your latest sale catalog.

Initials and abbreviations

2. Use a period after initials and most abbreviations.

Dr. H. H. Wright   p. 31
437 mi   no. 7

The abbreviations of organizations, of governmental divisions, and of educational degrees usually omit periods.

BA (Bachelor of Arts) degree
NFL (National Football League)
FBI (Federal Bureau of Investigation)

The U.S. Postal Service abbreviations for states (see pages 256–257 in Chapter 7, Memorandums and Letters, for a list of the abbreviations), contractions, parts of names used as a whole, roman numerals in sentences, and units of measure with the exception of in. (inch) and at. wt. (atomic weight) are written without periods. (Designations for units of measure are increasingly being regarded as symbols rather than as abbreviations. See item 9, page 718.)

Hal C. Johnson IV, who lives 50 mi away in Sacramento, CA, won't be present.

See also Abbreviations, item 6, pages 717–718.

Outline

3. Use a period after each number and letter symbol in an outline.

I.
  A.
  B.

For other examples see pages 128–129.

Decimals

4. Use a period to mark decimals.

$10.52
A reading of 1.260 indicates a full charge in a battery; 1.190, a half charge.

Question Mark (?)

Questions

1. Use a question mark at the end of every direct question, including a short parenthetical question within another sentence.

Have the blood tests been completed?
When you return (when will that be?), please bring the reports.

Note: An indirect question is followed by a period.

He asked if John were present.

Note: A polite request is usually followed by a period.

Will you please close the door.

Uncertainty

2. Use a question mark in parentheses to indicate there is some question as to certainty or accuracy.

Chaucer, 1343(?)–1400
The spindle should revolve at a slow (?) speed.

Exclamation Point (!)

Sudden or strong emotion or surprise

1. Use an exclamation point after words, phrases, or sentences (including parenthetical expressions) to show sudden or strong emotion or force, or to mark the writer's surprise.

What a day!
The computer (!) made a mistake.

Note: The exclamation point can be easily overused, thus causing it to lose its force. Avoid using the exclamation point in place of vivid, specific description.

INTERNAL MARKS THAT SET OFF AND SEPARATE

Comma (,)

Items in series

1. Use a comma to separate items in a series. The items may be words, phrases, or clauses.

How much do you spend each month for food, housing, clothing, and transportation?

Compound sentence

2. Use a comma to separate independent clauses joined by a coordinate conjunction (*and, but, or, either, neither, nor,* and sometimes *for, so, yet*). (An independent clause is a group of related words that have a subject and verb and that could stand alone as a sentence.)

There was a time when the homemaker had few interests outside the home, but today she is a leader in local and national affairs.

Note: Omission of the coordinate conjunction results in a comma splice (comma fault), that is, a comma incorrectly splicing together independent clauses. (See page 657).

If the clauses are short, the comma may be omitted.

I aimed and I fired.

See also Semicolon, item 1, page 730.

Equal adjectives

3. Use a comma to separate two adjectives of equal emphasis and with the same relationship to the noun modified.

The philanthropist made a generous, unexpected gift to our college.

Note: If *and* can be substituted for the comma or if the order of the adjectives can be reversed without violating the meaning, the adjectives are of equal rank and a comma is needed.

Misreading

4. Use a comma to prevent misreading.

Besides Sharon, Ann is the only available organist.
Ever since, he has gotten to work on time.

Number units

5. Use a comma to separate units in a number of four or more digits (except telephone numbers, ZIP numbers, house numbers, and the like).

2,560,781   7,868   *or*   7868

Note: The comma may be omitted from four-digit numbers.

Nonrestrictive modifiers

6. Use commas to set off nonrestrictive modifiers, that is, modifiers which do not limit or change the basic meaning of the sentence.

The Guggenheim Museum, *designed by Frank Lloyd Wright*, is in New York City. (Nonrestrictive modifier; commas needed)
A museum *designed by Frank Lloyd Wright* is in New York City. (Restrictive modifier; no comma needed)
Leontyne Price, *who is a world-renowned soprano*, was awarded the Presidential Medal of Freedom. (Nonrestrictive modifier; commas needed)
A world-renowned soprano *who was awarded the Presidential Medal of Freedom* is Leontyne Price. (Restrictive modifier; no commas needed)

Introductory adverb clause

7. Use a comma to set off an adverb clause at the beginning of a sentence. An adverb clause at the end of a sentence is not set off.

When you complete these requirements, you will be eligible for the award.
If given the proper care and training, a dog can be an affectionate and obedient pet.
A dog can be an affectionate and obedient pet *if given the proper care and training*.

Introductory
verbal modifier

8. Use a comma to set off a verbal modifier (participle or infinitive) at the beginning of a sentence.

Experimenting in the laboratory, Sir Alexander Fleming discovered penicillin. (Participle)
To understand the continents better, researchers must investigate the oceans. (Infinitive)

Appositives

9. Use commas to set off an appositive. (An appositive is a noun or pronoun that follows another noun or pronoun and renames or explains it.)

Joseph Priestly, *a theologian and scientist,* discovered oxygen.
Opium painkillers, *such as heroin and morphine,* are narcotics.

Note: The commas are usually omitted if the appositive is a proper noun or is closely connected with the word it explains.

My friend Mary lives in Phoenix.
The word *occurred* is often misspelled.

Parenthetical
expressions, conjunctive adverbs

10. Use commas to set off a parenthetical expression or a conjunctive adverb.

A doughnut, *for example,* has more calories than an apple.
I was late; *however,* I did not miss the plane.

See also Semicolon, items 2 and 5, pages 730–731.

Address, dates

11. Use commas to set off each item after the first in an address or a date.

My address will be 1045 Carpenter Street, Columbia, Missouri 65201, after today. (House number and street are considered one item; state and ZIP code are considered one item.)
Thomas Jefferson died on July 4, 1826, at Monticello. (Month and day are considered one item.)

Note: If the day precedes the month or if the day is not given, omit the commas.

Abraham Lincoln was assassinated on 14 April 1865.
In July 1969 humans first landed on the moon.

Quotations

12. Use commas to set off the *he said* (or similar matter) in a direct quotation.

"I am going," Mary responded.
"If you need help," he said, "a student assistant will be in the library."

Person or thing
addressed

13. Use commas to set off the name of the person or thing addressed.

If you can, Ms. Yater, I would prefer that you attend the meeting.
My dear car, we are going to have a good time this weekend.

Mild interjections	**14.** Use commas to set off mild interjections, such as *well, yes, no, oh*.

Oh, this is satisfactory.
Yes, I agree that farming is still the nation's single largest industry.

Inverted name	**15.** Use commas to set off, in an inverted name, a person's given name when the surname appears first.

Adams, Lucius C., is the first name on the list.

Title after a name	**16.** Use commas to set off a title following a name. (Setting off *Junior* or *Senior,* or their abbreviations, following a name is optional.)

Patton A. Houlihan, President of Irish Imports, is here.
Gregory McPhail, DDS, and Harvey D. Lott, DVM, were classmates.

Contrasting elements	**17.** Use commas to set off contrasting elements.

The harder we work, the sooner we will finish.
Leif Eriksson, not Columbus, discovered the North American continent.

Elliptical clause	**18.** Use a comma to indicate understood words in an elliptical clause.

Tom was elected president; Jill, vice-president.

Inc., Ltd.	**19.** Use commas to set off the abbreviation for *incorporated* or *limited* from a company name.

Drake Enterprises, Inc., is our major competitor.
I believe that Harrells, Ltd., will answer our request.

Note: Some companies omit the comma.

Introductory elements	**20.** Use a comma to introduce a word, phrase, or clause.

His destination was clearly indicated, New York City.
I told myself, you can do this if you really want to.

See also Colon, item 1, page 731.

Before conjunction "for"	**21.** Use a comma to precede *for* when used as a conjunction.

The plant is closed, for the employees are on strike.
The plant has been closed for a week. (No comma; *for* is a preposition, not a conjunction.)

Correspondence	**22.** Use a comma to follow the salutation and complimentary close in a social letter and usually the complimentary close in a business letter.

Dear Mother,	Sincerely,
Dear Lynne,	Yours truly,

Note: The salutation in a business letter is usually followed by a colon; a comma may be used if the writer and the recipient know each other well. Some newer business letter formats omit all punctuation following the salutation and the complimentary close. For a full discussion, see page 257, Chapter 7, Memorandums and Letters.

Tag question

23. Use commas to set off a tag question (such as *will you, won't you, can you*) from the remainder of the sentence.

You will write me, won't you?

Absolute phrase

24. Use commas to set off an absolute phrase. An absolute phrase (also called a nominative absolute) consists usually of a participle phrase plus a subject of the participle and has no grammatical connection with the clause to which it is attached.

I fear the worst, *his health being what it is.*

Semicolon (;)

Independent clauses, no coordinate conjunction

1. Use a semicolon to separate independent clauses not joined by a coordinate conjunction (*and, but, or*).

Germany has a number of well-known universities; several of them have been in existence since the Middle Ages.
There are four principal blood types; the most common are O and A.

Note 1: If a comma is used instead of the needed semicolon, the mispunctuation is called a comma splice, or comma fault (see page 657).
Note 2: If the semicolon is omitted, the result is a run-on, or fused, sentence (see pages 656–657).

Short, emphatic clauses may be separated by commas.

I came, I saw, I conquered.

Independent clauses, transitional connective

2. Use a semicolon to separate independent clauses joined by a transitional connective. Transitional connectives include conjunctive adverbs such as *also, however, moreover, nevertheless, then, thus,* and explanatory expressions such as *for example, in fact, on the other hand.*

We have considered the historical background of the period; thus we can consider its cultural achievements more intelligently.
During the Renaissance the most famous Humanists were from Italy; for example, Petrarch, Boccaccio, Ficino, and Pico della Mirandola were all of Italian birth.

See also Comma, item 10, page 728.

Certain independent clauses, coordinate conjunction

3. Use a semicolon to separate two independent clauses joined by a coordinate conjunction when the clauses contain internal punctuation or when the clauses are long.

The room needs a rug, new curtains, and a lamp; but my budget permits only the purchase of a lamp.

Students in an occupational program of study usually have little time for electives; but these students very often want to take courses in the humanities.

Items in series   **4.** Use a semicolon to separate items in a series containing internal punctuation.

The three major cities in our itinerary are London, Ontario, Canada; Washington, DC, USA; and Tegucigalpa, Honduras, Central America.

The new officers are Betty Harrison, president; Corkren Samuels, vice-president; and Tony McBride, secretary.

Examples   **5.** Use a semicolon to separate an independent clause containing a list of examples from the preceding independent clause when the list is introduced by *that is, for example, for instance,* or a similar expression.

Many great writers have had to overcome severe physical handicaps; for instance, John Milton, Alexander Pope, and James Joyce were all handicapped.

Colon (:)

List or series   **1.** Use a colon to introduce a list or series of items. An expression such as *the following, as follows,* or *these* often precedes the list.

A child learns responsibility in three ways: by example, by instruction, and by experience.

The principal natural fibers used in the production of textile fabrics include the following: cotton, wool, silk, and linen.

See also Comma, item 20, page 729.

Explanatory clause   **2.** Use a colon to introduce a clause that explains, reinforces, or gives an example of a preceding clause or expression.

Until recently, American industry used the English system of linear measure as standard: the common unit of length was the inch.

"Keep cool: it will be all one a hundred years hence." (Emerson)

Emphatic appositive   **3.** Use a colon to direct attention to an emphatic appositive.

We have overlooked the most obvious motive: love.

That leaves me with one question: When do we start?

Quotation   **4.** Use a colon to introduce a long or formal quotation.

My argument is based on George Meredith's words: "The attitudes, gestures, and movements of the human body are laughable in exact proportion as that body reminds us of a machine."

Formal greeting   **5.** Use a colon to follow a formal greeting (usually in a business letter).

Dear Ms. Boxeman: Dear Sir or Madam: Greenway, Inc.:

See also Comma, item 22, pages 729–730.

Relationships

6. Use a colon to indicate relationships such as volume and page, ratio, and time.

42:81–90 (volume 42, pages 81–90)   x:y
Genesis 4:8   3:1
2:50 a.m.

Dash (—)

The dash generally indicates emphasis or a sudden break in thought. Often the dash is interchangeable with a less strong punctuation mark: If emphasis is desired, use a dash; if not, use an alternate punctuation mark (usually a comma, colon, or parentheses). The dash is made with two hyphens when writing by hand or typing; there is no spacing before or after the dash within a sentence.

> I want one thing out of this agreement—my money.
> My mother—she is president of the company—will call this to the attention of the board of directors.
> If we should succeed—God help us!—all mankind will profit.

Note: In some sentences the writer may have a choice among dashes, parentheses, or commas. Dashes emphasize the words set off; parentheses subordinate them; and commas simply show that the words are not essential to the basic meaning of the sentence.

Sudden change

1. Use a dash to mark a sudden break or shift in thought.

> The murderer is—but perhaps I shouldn't spoil the book for you.
> And now to the next point, the causes of—did someone have a question?

Appositive series

2. Use a dash to set off a series of appositives.

> Four major factors—cost, color, fabric, and fit—influence the purchase of a suit of clothes.
> Because of these qualities—beauty, durability, portability, divisibility, and uniformity of value—gold and silver have gradually displaced all other substances as material for money.

Summarizing clause

3. Use a dash to separate a summarizing clause from a series.

> Tests, a term paper, and class participation—these factors determined the student's grade.

Note: The summarizing clause usually begins with *this, that, these, those,* or *such.*

Emphasis

4. Use a dash to set off material for emphasis.

> Flowery phrases—regardless of the intent—have no place in reports.

Sign of omission

5. Use a dash to indicate the omission of letters or words.

> The only letters we have in the mystery word are —a—lm.

Virgule (Sometimes Called "Slash") (/)

Alternative   1. Use a virgule to indicate appropriate alternatives.

Identify/define these persons, occurrences, and terms.

Poetry   2. Use a virgule to separate run-in lines of poetry. For readability, space before and after the virgule.

"Friends, Romans, countrymen, lend me your ears. / I come to bury Caesar, not to praise him." (Shakespeare, *Julius Caesar*)

Per   3. Use a virgule to represent *per* in abbreviations.

12 ft/sec   260 mi/hr

Time   4. Use a virgule to separate divisions of a period of time.

the fiscal year 1985/86

ENCLOSING MARKS ALWAYS USED IN PAIRS

Quotation Marks (" ")

Direct quotations   1. Use quotation marks to enclose every direct quotation.

The *American Heritage Dictionary* defines dulcimer as "a musical instrument with wire strings of graduated lengths stretched over a sound box, played with two padded hammers or by plucking."
"Marriage is popular," said George Bernard Shaw, "because it combines the maximum of temptation with the maximum of opportunity."

Note: Quotations of more than one paragraph have quotation marks at the beginning of each paragraph and at the end of the last paragraph. Long quotations, however, are usually set off by indentation, eliminating the need for quotation marks.

Titles   2. Use quotation marks to enclose titles of magazine articles, short poems, songs, television and radio shows, and speeches.

Included in this volume of Poe's works are the poem "Annabel Lee," the essay "The Philosophy of Literary Criticism," and the short story "The Black Cat."

Note: Titles of magazines, books, newspapers, long poems, plays, operas and musicals, motion pictures, ships, trains, and aircraft are italicized (underlined in handwriting).

Different usage level   3. Use quotation marks to distinguish words on a different level of usage.

The ambassador and his delegation enjoyed the "good country eating."

Note: Be sparing in placing quotation marks around words used in a special sense, for this practice is annoying to many readers. Rather than apologizing for a word with quotation marks, either choose another word or omit the quotation marks. Avoid using quotation marks for mere emphasis.

Nicknames

4. Use quotation marks to enclose nicknames.

William J. "Happy" Kiska is our club adviser.

Note: Quotation marks are usually omitted from a nickname that is well known (Babe Ruth, Teddy Roosevelt) or after the first use in a piece of writing.

Single quotation marks

5. Use single quotation marks (' ') for a quotation within a quotation.

"I was puzzled," confided Mary, "when Jim said that he agreed with the writer's words, 'The world wants to be deceived.' "
This writer states, "Of all the Sherlock Holmes stories, 'The Red-Headed League' is the best plotted."

Own title not quoted

6. At the beginning of a piece of writing, do not put your own title in quotation marks (unless the title is a quotation).

How to Sharpen a Drill Bit
The Enduring Popularity of the Song "White Christmas"

With other punctuation marks
Period or comma

7. Use quotation marks properly with other marks of punctuation.

☐ The closing quotation mark *always* follows the period or comma.

"File all applications before May," the director of personnel cautioned, "if you wish to be considered for summer work."

Colon or semicolon

☐ The closing quotation mark *always* precedes the colon or semicolon.

I have just finished reading Shirley Ann Grau's "The Black Prince"; the main character is a complex person.

Question mark, exclamation point, or dash

☐ The closing quotation mark precedes the question mark, exclamation point, or dash when these punctuation marks refer to the entire sentence. The closing quotation mark follows the question mark, exclamation point, or dash when these punctuation marks refer only to the quoted material.

Who said, "I cannot be here tomorrow"?
Was it you who yelled, "Fire!"?
Bill asked, "Did you receive the telegram?"

Parentheses ( )

Additional material

1. Use parentheses to enclose additional material remotely connected with the remainder of the sentence.

If I can find a job (I hope I won't need a driver's license), I will pay part of my college expenses.
Ernest Hemingway (1899–1961) won the Nobel Prize in literature.

Note: In some sentences the writer may have a choice among using parentheses, dashes, or commas. Dashes emphasize the words set off; parentheses subordinate them; and commas simply show that the words are not essential to the basic meaning of the sentence.

Itemizing

2. Use parentheses to enclose numbers or letters that mark items in a list.

Government surveys indicate that students drop out of school because they (1) dislike school, (2) think it would be more fun to work, and (3) need money for themselves and their families.

"See" references

3. Use parentheses to enclose material within a sentence directing the reader to see other pages, charts, figures, etc.

The average life expectancy in the United States is 70 years (see Figure 3).

Capitalization and punctuation with parentheses

4. Use capitalization and punctuation properly within parentheses.

Capitalization

☐ Do not capitalize the first word of a sentence enclosed in parentheses within a sentence.

The table shows the allowable loads on each beam in kips (a kip is 1000 pounds).

Period

☐ Omit the period end punctuation in a sentence enclosed in parentheses within a sentence.

The Old North Church (the name is now Christ Church) is the oldest church building in Boston.

Question mark, exclamation point

☐ Use a needed question mark or exclamation point with matter enclosed in parentheses.

Pour the footings below the frost line (what is the frost line?) for a stable foundation.

Separate sentence

☐ If the matter enclosed in parentheses is a separate sentence, place the end punctuation inside the closing parentheses.

The regular heating system will be sufficient. (Infrared heaters are available for spotheating.)

Brackets ([ ])

Insertion in quotation

1. Use brackets for parentheses within parentheses.

Susan M. Jones (a graduate of Teachers' College [now the University of Southern Mississippi]) was recognized as Alumna of the Year.

2. Use brackets to insert comments or explanations in quotations.

"Good design [of automobiles] involves efficient operation, sound construction, and pleasing form."
"It is this decision [*Miller* v. *Adams*] that parents of juvenile offenders will long remember," said the judge.

Sic in quotation

3. Use brackets to enclose the Latin word *sic* ("so," "thus") to indicate strange usage or an error such as misspelling or incorrect grammar in a quotation.

According to the report, "Sixty drivers had there [sic] licenses revoked."

Documentation

4. Use brackets to indicate missing or unverified information in documentation.

James D. Amo. *How to Hang Wallpaper.* [Boston: McGuire Institute of Technology.] 1985.

PUNCTUATION OF INDIVIDUAL WORDS AND OF TERMS

Apostrophe (')

Contractions

1. Use an apostrophe to take the place of a letter or letters in a contraction.

I'm (I am)
we've (we have)
Don't (Do not) come until one o'clock (of the clock).

Singular possessive

2. Use an apostrophe to show the possessive form of singular nouns and indefinite pronouns.

citizen's responsibility	Rom's car
someone's book	everybody's concern

Note: To form the possessive of a singular noun, add the apostrophe + s.

doctor + ' + s = doctor's, as in "doctor's advice"
Keats + ' + s = Keats's, as in Keats's poems

Note: The s may be omitted in a name ending in s, especially if the name has two or more syllables.

James's (or James') book
Ms. Tompkins's (or Ms. Thompkins') cat

Note: Personal pronouns (*his, hers, its, theirs, ours*) do not need the apostrophe because they are already possessive in form.

Plural possessive

3. Use an apostrophe to show the possessive form of plural nouns.

boys' coats   children's coats

Note: To show plural possessive, first form the plural; if the plural noun ends in s, add only an apostrophe. If the plural noun does not end in s, add an apostrophe + s.

Certain plurals

4. Use an apostrophe to form the plurals of letters and words used as words.

Don't use too many *and*'s, and eliminate *I*'s from your report.

Note: The apostrophe is usually omitted in the plurals of abbreviations and numbers.

Ellipsis Points (. . . or. . . .)

Omission sign in quotations

1. Use ellipsis points (plural form: ellipses) to indicate that words have been left out of quoted material. Three dots show that words have been omitted at the beginning of a quoted sentence or within a quoted sentence. Four dots show that words have been omitted at the end of a quoted sentence (the fourth dot being the period at the end of the sentence.)

"The average American family spent about $4000 on food . . . in 1984."
"The adoption of standard time in North America stems from the railroads' search for a solution to their chaotic schedules. . . . In November, 1883, rail companies agreed to set up zones for each 15 degrees of longitude, with uniform time throughout each zone."

Hesitation in dialogue

2. Use ellipsis points to indicate hesitation, halting speech, or an unfinished sentence in dialogue.

"If . . . if it's all right . . . I mean . . . I don't want to cause any trouble," the bewildered child stammered.

Hyphen (-)

Word division

1. Use a hyphen to separate parts of a word divided at the end of a line. (Divide words only between syllables.) Careful writers try to avoid dividing a word because divided words may impede readability.

On April 15, 1912, the *Titanic* sank after colliding with an iceberg.

Compound numbers; fractions

2. Use a hyphen to separate parts of compound numbers and fractions when they are written out.

seventy-four people   twenty-two cars
one-eighth of an inch   one-sixteenth-inch thickness

Compound adjectives

3. Use a hyphen to separate parts of compound adjectives when they precede the word modified.

an eighteenth-century novelist   40-hour week

Compound nouns

4. Use a hyphen to separate parts of compound nouns.

brother-in-law   U-turn
kilowatt-hour   vice-president

Note: Many compound nouns are written as a single word, such as *notebook* and *blueprint*. Others are written as two words without the hyphen, such as *card table* and *steam iron*. If you do not know how to write a word, look it up in a dictionary.

Compound verbs

5. Use a hyphen to separate parts of a compound verb.

brake-test   oven-temper

Numbers or dates   **6.** Use a hyphen to separate parts of inclusive numbers or dates.

pages 72-76   the years 1988-92

Prefixes   **7.** Use a hyphen to separate parts of some words whose prefix is separated from the main stem of the word.

ex-president   self-respect
semi-invalid   pre-Renaissance

Note: A good dictionary is the best guide for determining which words are hyphenated.

Italics (Underlining)

Italics *(such as these words)* are used in print; the equivalent in handwriting is underlining.

Titles   **1.** Italicize (underline) titles of books, magazines, newspapers, long poems, plays, operas and musicals, motion pictures, ships, trains, and aircraft.

At the library yesterday I checked out the book *Roots*, read this month's *Reader's Digest*, looked at the sports section in the *Daily Register*, and listened to parts of *Jesus Christ Superstar*.

Note: Use quotation marks to enclose titles of magazine articles, book chapters, short poems, songs, television and radio shows, and speeches.

Do not italicize or put in quotation marks titles of sacred writings, editions, series, and the like: Bible, Psalms, Anniversary Edition of the Works of Mark Twain.

Terms as such   **2.** Italicize words, letters, or figures when they are referred to as such.

People often confuse *to* and *too*.
I cannot distinguish between your *a*'s and *o*'s.

Foreign terms   **3.** Italicize words and phrases that are considered foreign.

His novel is concerned with the *nouveau riche*.
This item is included gratis. (The last word in this sentence is no longer considered foreign.)

Emphasis   **4.** Italicize a word or phrase for special emphasis. If the emphasis is to be effective, however, italics must be used sparingly.

My final word is *no*.

Spelling

Because of the strong influence of other languages, spelling in the English language is fairly irregular.

If you are currently having difficulty with spelling, here are several helpful suggestions:

1. *Keep a study list of words misspelled.* Review the list often, dropping words that you have learned to spell and adding any new spelling difficulties.

2. *Attempt to master the spelling of these words from the study list.* Use any method that is successful for you. Some students find that writing one or several words on a card and studying them while riding to school or waiting between classes is an effective technique. Some relate the word in some way, such as "There is 'a rat' in sep*arat*e."

3. *Use a dictionary.* It is the poor speller's best friend. If you have some idea of the correct spelling of a word but are not sure, consult a dictionary. If you have no idea about the correct spelling, get someone to help you find the word in a dictionary. Or look in a dictionary designed for poor spellers, which lists words by their common misspellings and then gives the correct spelling.

4. *Proofread everything you write.* Look carefully at every word within a piece of writing. If a word does not look right, check its spelling in a dictionary.

5. *Take care in pronouncing words.* Words sometimes are misspelled because of problems in pronouncing or hearing the words. Examples: *prompness* (misspelled) for *promptness, accidently* (misspelled) for *accidentally, sophmore* (misspelled) for *sophomore.* Pronunciation, of course, is not a guide for a small portion of the words in our language (typically words not of English origin). Examples: *pneumonia, potpourri, tsunami, xylophone.*

Although spelling may be difficult, it can be mastered—primarily because many spelling errors are a violation of conventional practices for use of *ie* and *ei* and of spelling changes when affixes are added.

USING *ie* AND *ei*

The following jingle sums up most of the guides for correct *ie* and *ei* usage:

Use *i* before *e*,
Except after *c*,
Or when sounded like *a*,
As in *neighbor* and *weigh*.

■ Generally use *ie* when the sound is a long *e* after any letter except *c*.

believe	chief
grief	niece
piece	relieve

■ Generally use *ei* after *c*.

deceive	receive	receipt

Note: An exception occurs when the combination of letters *cie* is sounded *sh*; in such instances *c* is followed by *ie*.

sufficient	efficient	conscience

■ Generally use *ei* when the sound is *a*.

neighbor	freight	sleigh
weigh	reign	vein

SPELLING CHANGES WHEN AFFIXES ARE ADDED

An affix is a letter or syllable added either at the beginning or at the end of a word to change its meaning. The addition of affixes, whether prefix or suffix, often involves spelling changes.

Prefixes. A prefix is a syllable added to the beginning of a word. One prefix may be spelled in several different ways, usually depending on the beginning letter of the base word. For example, *com, con, cor,* and *co* are all spellings of a prefix meaning "together, with." They are used to form such words as *commit, collect,* and *correspond.*

Following are some common prefixes and illustrations showing how they are added to base words. The meaning of the prefix is in parentheses.

ad (to, toward)   In adding the prefix *ad* to a base, the *d* often is changed to the same letter as the beginning letter of the base.

ad + breviate = abbreviate
ad + commodate = accommodate

com (together, with)   The spelling is *com* unless the base word begins with *l* or *r;* then the spelling is *col* and *cor,* respectively.

com + mit = commit
com + lect = collect
com + respond = correspond

de (down, off, away)   This prefix is often incorrectly written *di.* Note the correct spellings of words using this prefix.

describe	desire
despair	destroyed

dis (apart, from, not)   The prefix *dis* is usually added unchanged to the base word.

dis + trust = distrust
dis + satisfied = dissatisfied

in (not)   The consonant *n* often changes to agree with the beginning letter of the base word.

in + reverent = irreverent
in + legible = illegible

The *n* may change to *m.*

in + partial = impartial
in + mortal = immortal

sub (under)   The consonant *b* sometimes changes to agree with the beginning letter of the base word.

sub + marine = submarine
sub + let = sublet
sub + fix = suffix
sub + realistic = surrealistic

un (not)   This is added unchanged.

un + able = unable
un + fair = unfair

Suffixes. A suffix is a syllable added to the end of a word. One suffix may be spelled in several different ways, such as *ance* and *ence*. Also, the base word may require a change in form when a suffix is added. Because of these possibilities, adding suffixes often causes spelling difficulties. Learning the following suffixes and the spelling of the exemplary words will improve your vocabulary and spelling immeasurably. The meaning of the suffix is in parentheses.

able, ible (capable of being)   Adding this suffix to a base word, usually a verb or a noun, forms an adjective.

rely—reliable
consider—considerable
separate—separable
read—readable
laugh—laughable
advise—advisable
commend—commendable

sense—sensible
horror—horrible
terror—terrible
destruction—destructible
reduce—reducible
digestion—digestible
comprehension—comprehensible

ance, ence (act, quality, state of)   Adding this suffix to a base word, usually a verb, forms a noun.

appear—appearance
resist—resistance
assist—assistance
attend—attendance

exist—existence
prefer—preference
insist—insistence
correspond—correspondence

Other nouns using *ance, ence* include:

ignorance
brilliance
significance
importance
abundance
performance
guidance

experience
intelligence
audience
convenience
independence
competence
conscience

ary, ery (related to, connected with)   Adding this suffix to base words forms nouns and adjectives.

boundary
vocabulary
dictionary
library
customary

gallery
cemetery
millinery

efy, ify (to make, to become)   Adding this suffix forms verbs.

liquefy	ratify
stupefy	testify
rarefy	falsify
putrefy	justify
	classify

ize, ise, yze (to cause to be, to become, to make conform with)   These suffixes are verb endings all pronounced the same way.

recognize	revise	analyze
familiarize	advertise	paralyze
generalize	exercise	
emphasize	supervise	
realize		
criticize		
modernize		

Also, some nouns end in *ise*.

exercise	enterprise
merchandise	franchise

ly (in a specified manner, like, characteristic of)   Adding this suffix to a base noun forms an adjective; adding *ly* to a base adjective forms an adverb. Generally *ly* is added to the base word with no change in spelling.

monthly	surely
heavenly	softly
earthly	annually
randomly	clearly

If the base word ends in *ic*, usually you add *ally*.

critically	drastically
basically	automatically

An exception is *public—publicly*.

ous (full of)   Adding this suffix to a base noun forms an adjective.

courageous	outrageous	grievous
dangerous	humorous	mischievous
hazardous	advantageous	beauteous
marvelous	adventurous	bounteous

Other suffixes include:

ant (ent, er, or, ian)   meaning "one who" or "pertaining to"

ion (tion, ation, ment)   meaning "action," "state of," "result"

ish   meaning "like a"

less   meaning "without"

ship   meaning "skill," "state," "quality," "office"

Final Letters. Final letters of words often require change before certain suffixes can be added.

■ **Final *e*.** Generally keep a final silent *e* before a suffix beginning with a consonant, but drop before a suffix beginning with a vowel.

| use—useful | write—writing |
| love—lovely | hire—hiring |

Exceptions:

| true—truly | due—duly | argue—argument |

Note: In adding *ing* to some words ending in *e*, retain the *e* to avoid confusion with another word.

| dye—dyeing | die—dying |
| singe—singeing | sing—singing |

■ **Final *ce* and *ge*.** Retain the *e* when adding *able* to keep the *c* or *g* soft. If the *e* were dropped, the *c* would have a *k* sound in pronunciation and the *g* a hard *g* sound. For example, the word *change* retains the *e* when *able* is added: *changeable*.

■ **Final *ie*.** Before adding *ing*, drop the *e* and change the *i* to *y* to avoid doubling the *i*.

| tie—tying | lie—lying |

■ **Final *y*.** To add suffixes to words ending in a final *y* preceded by a consonant, change the *y* to *i* before adding the suffix. In words ending in *y* preceded by a vowel, the *y* remains unchanged before the suffix.

| survey—surveying | try—tries |

■ **Final consonants.** Double the final consonant before adding a suffix beginning with a vowel if the word is one syllable or if the word is stressed on the last syllable.

hop—hopping—hopped	occur—occurred—occurring
plan—planning—planned	refer—referred—referring
stop—stopping—stopped	forget—forgotten—forgetting

■ In adding suffixes to some words, the stress shifts from the last syllable of the base word to the first syllable. When the stress is on the first syllable, do *not* double the final consonant.

| prefer—preference | confer—conference |
| refer—reference | defer—deference |

Ceed, Sede, Cede Words. The base words *ceed*, *sede*, and *cede* sound the same when they are pronounced. However, they cannot correctly be interchanged in spelling.

■ *ceed:* Three words, all verbs, end in *ceed*.

| proceed | succeed | exceed |

■ *sede:* The only word ending in *sede* is *supersede*.

■ *cede:* All other words, excluding the four named above, ending in this sound are spelled *cede*.

recede   secede   accede

concede   intercede   precede

Symbols

Symbols are used mostly in tables, charts, figures, drawings, diagrams, and the like; they are not used generally within the text of pieces of writing for most audiences (intended readers).

Symbols cannot be discussed as definitely as abbreviations, capitalization, and numbers because no one group of symbols is common to all areas. Most organizations and subject groups—medicine and pharmacy, mathematics, commerce and finance, engineering technologies—have their own symbols and practices for the use of these symbols. A person who is a part of any such group is obligated to learn these symbols and the accepted usage practices. The following are examples of symbols common to these groups:

■ Medicine and Pharmacy

℞   take (Latin, *recipe*): used at the beginning of a prescription

℥   ounce

ʒ   dram

s   write (Latin, *signā*): on prescription indicates directions to be printed on medicine label)

■ Mathematics

+   plus   >   is greater than

−   minus   =   equals

×   times   ∫   integral

÷   divide   ∠   angle

■ Commerce and Finance

number, as in #7, or pounds, as in 50#

£   pound sterling, as in British currency

@   *at*: 10 @ 1¢ each

■ Engineering Technologies

More specialized fields, such as electronics, hydraulics, welding, and technical drawing, use standard symbols for different areas. These symbols are usually determined by the American Standards Association.

For example, one electricity text has nine pages of symbols used by electricians. There are resistor and capacitor symbols, contact and push-button symbols, motor and generator symbols, architectural plans symbols, transformer symbols, and switch and circuit breaker symbols. Some examples are illustrated below.

Contacts–N.O.   Ground   Conductor   Squirrel–Cage   Ceiling Outlet
(normally open)   Induction Motor

A technical drawing text has eight pages of symbols including topographic symbols; railway engineering symbols; American Standard piping symbols; heating, ventilating, and ductwork symbols; and plumbing symbols. Some examples are:

⊙ County seat   ─╫─ Flanged joint
---- National or state line   ⊙ Hot water tank
▨ Wood—with the grain   ⊗ Exit outlet

■ Nonspecialized Areas

Common among symbols that the average person would recognize and use are:

% percent   ' and "   feet and inches
° degree $ and ¢ dollars and cents
& and

See also units of measure as symbols, page 718, and flowchart symbols, page 137.

Most dictionaries have a section on common signs and symbols.

Application 1   Using Mechanics

Rewrite the following sentences, making corrections in the use of abbreviations, capitalization, and numbers.

1. Atmospheric pressure decreases with increase in Altitude, about 1 in of mercury for every 1,000 ft of altitude.
2. 9 out of 10 pounds of all the metal used today is steel.
3. In may 1927 Charles a lindberg flew the atlantic ocean nonstop from new york to paris.
4. The February issue of scientific monthly reported findings about the microbiology of the atmosphere.
5. The cashier counted two hundred one dollar bills in the register.
6. Hardness of tool materials is further explained in chapter 23, "heat treatment Of metals."
7. The correct point for a ruling pen is shown in fig. 40.
8. The phenomenon of radioactivity was discovered by a french scientist Henri becquerel.
9. Steel exists from eight hundredths percent carbon to two percent carbon.
10. On the drawing board were two twelve inch rulers.
11. Mister Curtis Berry owns Berry Appliance company.
12. Milling machines are provided with a large range of speeds, from fifteen to one thousand six hundred revolutions per minute.
13. In a desirable classroom for teaching drafting, window areas on the north side should be as large as practical and not less than twenty percent of the floor area.
14. The mechanic put a set of champion spark plugs in mister bronson's mustang.
15. Arthur Winchester manager of tri-d ranch in Cody Wyoming purchased a john deere tractor.
16. The measurement was within an in. of an acceptable length.
17. That small wheel may turn 100 RPM.

18. Many Engineers belong to ASTME.
19. Our company uses gmc trucks for delivering building supplies.
20. If I pass the state exams, I will become an rn.
21. Magnax Corporation was located at 501 south street, Chicago, Illinois, from may 15, 1965, to september 1, 1968.
22. Mr. Williams, instructor in refrigeration, said, "turn to page one hundred twenty-six and consult figure 14."
23. The lab team completed ½ the assigned experiments.
24. A survey revealed that a lab technician can see about 9 patients an hour.
25. We received your order for 5 50 h.p. boat motors.

Application 2   Using Mechanics

List at least ten terms (noun forms) used in your major field. Write the plural of each term.

Application 3   Using Mechanics

Using material in your major field, write sentences using the following punctuation marks.

POSSESSIVE FORMS

1. A sentence containing a singular possessive noun
2. A sentence containing a plural possessive pronoun
3. A sentence containing the possessive form of an indefinite pronoun

COLON

4. A sentence containing a colon to signal that a list or series of items is to follow an independent clause

COMMA

5. A sentence containing commas to separate items in a series
6. A sentence containing a comma before *and* joining two independent clauses
7. A sentence containing commas to set off a city from a state
8. A sentence containing commas to set off a month and a day from a year
9. A sentence containing commas to separate units in a number longer than four digits

DASH

10. A sentence containing a dash used to introduce a list of items following an independent clause
11. A sentence containing a dash (or dashes) to mark a sudden break in thought

12. A sentence (quoted from a source) containing ellipsis points to show that words have been left out of the quoted material

HYPHEN

13. A sentence containing a compound adjective preceding the word modified
14. A sentence containing a fraction expressed in words, not numerals

PARENTHESES

15. A sentence containing parentheses to enclose material not grammatically connected to the remainder of the sentence

PERIOD

16. A sentence containing a period to mark decimals

QUOTATION MARKS

17. A sentence containing quotation marks to enclose material quoted exactly from a printed source

SEMICOLON

18. A sentence containing a semicolon to separate two independent clauses not joined by a coordinate conjunction
19. A sentence containing a semicolon as a strong comma to separate items in a series that contains commas
20. A sentence containing a semicolon between two independent clauses containing commas that are joined by a coordinate conjunction

Application 4   Using Mechanics

Punctuate the following sentences.

1. When a large volume of liquid is to be lifted a short distance or pumped against relatively low pressure the centrifugal pump is generally used
2. The combustion chamber and general design of the jet are somewhat similar to the gas turbine but the thrust is delivered in a different manner
3. Modern Miss Shoe Company produces womens shoes to sell in a low price range
4. Since all assumptions cannot be verified there will always remain incomplete data errors in perception of facts and distortion in communication
5. Does the fact that annual reports show a high correlation of sales volume and advertising expense mean that more advertising will increase sales
6. American designers particularly the modern ones have learned to employ artificial light with excellent results

7. Persons engaged in industries dealing with paints batteries gasoline glazes for pottery and insecticides are exposed to unhealthy concentrations of lead unless suitable precautions are observed

8. The term sewage is applied to the pipes mains tanks etc that constitute a disposal system

9. Some decorators believe that the exposure of a room should influence its color scheme for example a north room should employ yellow to produce a feeling of sunshine

10. Finances are often referred to in terms of gain or loss time is referred to as future or past and temperature is denoted as above or below some preassigned zero

11. How many skilled men are now employed

12. The English system of linear measure measuring in a straight line is the standard adopted by American industry

13. A combination set measuring tool used in a machine shop consists of the following principal parts a steel rule a square stock incorporating a level a scriber and a 45 degree angle a protractor head and a center head

14. A taper or starting tap a plug tap and a bottoming tap are used in the order indicated when tapping a blind hole a hole that does not go entirely through the workpiece

15. Like charges repel unlike charges attract

16. Included in the broad classification of tool steels used for both cutting and noncutting purposes are water hardened air hardened shock resisting air hardened and hot worked types

17. Many of our chemical metallurgical and physical laws were discovered by the ancient Egyptians Greeks East Indians Chinese and Tibetans

18. In a drafting classroom an illumination of 50 footcandles at the drawing board level is generally specified as the minimum

19. Early electrical experimenters not fully understanding the nature of the flow of electricity believed that the electricity flowed from the positive terminal of the generator into the conducting wire this was given the name current flow

20. To thumbtack the paper in place on the drawing board the designer presses the T square head firmly against the working edge of the drawing board

Application 5   Using Mechanics

Review the papers that you have written this term. Make a list of the words that you misspelled, indicating words misspelled more than once. Try to determine why you misspelled each word.

Application 6   Using Mechanics

Find at least ten symbols used in your major field. Reproduce them as accurately as you can and explain the meaning of each.

Index

Checklist for Common Problems in Revising a Paper

(See also Revising, pages 681–682.)

TITLE   Clearly stated, in a phrase   *8*
Precise indication of paper emphasis   *8*
Correctly capitalized   *719–720*
No quotation marks around the title   *734*
See sample papers   20, 32, 49, 67, 68, 81, 105, 133, 155, 177, 199, etc.

ORGANIZATION
**(See "Procedure"
outline in each
chapter.)**
Introductory section *(See roman numeral I in each "Procedure" outline.)*
  Background information, overview of topic, or identification of subject
    353, 467, 682–683
  Obvious thesis statement or purpose statement (key sentence, controlling
    statement, central idea)   *353, 467, 669*
Body *(See roman numeral II in each "Procedure" outline.)*
  Adequate development of the thesis or purpose statement   *353*
  Topic statements of paragraphs or sections that point back to the thesis
    or purpose statement   *668–669*
  Carefully chosen details, specifics, and examples   *348–349, 670–671*
  Information arranged in logical sequence   *127–128, 159, 193–195,*
    357, 364, 370, 378, 387, 400, 404
Closing *(See roman numeral III in each "Procedure" outline.)*
  Compatible with purpose and emphasis of the paper   *467, 684–685*

CONTENT
**(See "Procedure"
outline in each
chapter.)**
Suitable for the intended audience and purpose   *4, 21–22, 41–50, 64,*
  96–98, 108–110, 173, 210–211
Accurate   *63–64, 348, 468*
Complete
  Adequate coverage of major and minor subdivisions   *679–680*
  Length of paper appropriate   *27, 51*
  No significant points omitted   *26, 127*
All information directly related to the topic   *349, 670–671*
Concise   *11–12, 349*

MECHANICS   Careful word choice
  Accurate, precise terminology   *8–12, 63–64, 348*
  Level of diction suited to the intended audience   *8–11, 96–98*
  Correct usage of words that are often confused   *705–712*
  No careless omission of needed words   *26–27, 348–349, 468, 703–704*
Correct punctuation
  Avoidance of comma splice, or comma fault   *657*
  No careless omission of end punctuation   *725–726*